World Sustainability Series

Series editor

Walter Leal Filho, Hamburg, Germany

More information about this series at http://www.springer.com/series/13384

Walter Leal Filho
Editor

Transformative Approaches to Sustainable Development at Universities

Working Across Disciplines

Springer

Editor
Walter Leal Filho
HAW Hamburg
Hamburg
Germany

and

Manchester Metropolitan University
Manchester
UK

ISSN 2199-7373 ISSN 2199-7381 (electronic)
ISBN 978-3-319-08836-5 ISBN 978-3-319-08837-2 (eBook)
DOI 10.1007/978-3-319-08837-2

Library of Congress Control Number: 2014948757

Springer Cham Heidelberg New York Dordrecht London

Printed on acid-free paper

Springer is part of Springer Science+Business Media (www.springer.com)

Preface

Sustainability in higher education is a fast growing field. From a slow start back in the late 1990s, where only a handful of people and institutions were engaged, this particular domain has evolved considerably over the past decades, and has now become mainstream. Moving on from a marginal area, with only a few researchers, sustainability as a whole and sustainability at university level in particular, has taken a central place in the scientific arena. As a result of this evolution, many people working with environmental education in the 1990s have chosen to embark on sustainability research, a trend which has been positive in the sense that a wide critical mass has now become available, across the continents, and many researchers are now following careers in this area.

But even though there are many actors now working in and around sustainability matters, and that much has been written about sustainability at universities, there is still a perceived need for forward looking, international publications, which go beyond the trivial and push the frontiers of this exciting field. Therefore, this book intends to fill in this gap and at the same time provide a timely contribution to the global debate on how to permeate sustainability across disciplines.

Consistent with this goal, this book presents a set of papers which show the effectiveness of transformative approaches towards sustainability in higher education, moving away from the theoretical discourse, and more into practice, illustrating how sustainability may be implemented within and between disciplines. With inputs from re-known experts and from emerging researchers from a wide range of higher education institutions across the world, the book demonstrates the value of transformative approaches to sustainable development, and the many benefits they may yield.

This book is structured along three parts:

- Part I: Deals with Integrative Processes and Concepts
- Part II: Handles Integrative Approaches in Teaching and Learning
- Part III: Discusses Problem-Solving and Integrative Practices

Papers in this book have derived from the "2nd World Symposium on Sustainable Development at Universities" (WSSD-U-2014), organised by Manchester Metropolitan University (UK) and the Research and Transfer Centre "Applications of Life Sciences" of the Hamburg University of Applied Sciences (Germany), in cooperation with the United Nations University initiative "Regional Centres of Expertise on Education for Sustainable Development" (RCE).

The Workshop, held in Manchester during 3–5 September 2014 and attended by over 120 delegates from 26 countries, was one of the last events to be held as part of the UN Decade on Education for Sustainable Development (UNDESD), with a focus on "transformative approaches to sustainable development across disciplines", contributing to the further development of this field.

I would like to thank the authors for their inputs and for the opportunity to access their experience and their wisdom. I hope this book will help to address the need for truly international works on integrative approaches to sustainability in higher education, and inspire further works in this still developing area.

Autumn 2014 Walter Leal Filho

Contents

Part I
Integrative Processes and Concepts

Education for Sustainable Development in Higher Education: Reviewing Needs

Walter Leal Filho

Abstract

This paper presents a review of the role of education for sustainable development and points out the benefits higher education institutions may gain, from implementing a holistic sustainability thinking as part of their work and institutional practice. It also outlines some of the needs seen in order to maximize the potential benefits of sustainable development to higher education institutions, with a special emphasis to the role of education for sustainable development in the process.

Keywords

Higher education · Education for sustainable development · Integration · Processes

1 Introduction: Education for Sustainable Development in Higher Education

The term education for sustainable development (ESD), coined in the mid-1990s as a replacement to the expression "environmental education" (i.e. education for, about and with the environment) widely used until then, is one of the results of the changes in the way of thinking catalyzed by the report "Our Common Future", also known as the "Brundtland Report". "Our Common Future" stated, in essence, that

W. Leal Filho (✉)
HAW Hamburg, Hamburg, Germany
e-mail: walter.leal@haw-hamburg.de

W. Leal Filho
Manchester Metropolitan University, Manchester, UK

W. Leal Filho (ed.), *Transformative Approaches to Sustainable Development at Universities*, World Sustainability Series, DOI 10.1007/978-3-319-08837-2_1

sustainable development is a process via which today's (environmental) resources need to be used with care, so that they are available to future generations. Consistent with this trend, ESD can be defined as an educational process characterised by approaches and methods aimed at fostering awareness about the issues pertaining sustainable development (e.g. social, political, economic and ecological matters). This differs from the previous approaches, where the emphasis was on environmental issues.

Evolving from environmental education, ESD widely enlarged its scope. Whereas environmental education had a formal concern with the environment and ways environmental resources are used, ESD's emphasis is on the means, tools and processes which may allow people to develop (or acquire) the values, competencies, skills and knowledge needed to contribute towards a more sustainable (i.e. more conscious) society.

The implications of this change of focus were significant. Instead of focusing on environmental protection or studies on matters of strict environmental concern, educators across the world were compelled to work towards the revision of teaching contents, so as to allow education systems to better respond to socio-economic (and not only environmental) challenges at the local, regional and global level. Furthermore, a new emphasis was given to the development of innovative teaching methods having sustainable development as focus, as advocated by UNESCO (2012a) on a report focusing on as education and learning in the context of sustainable development (SD) and involving a range of stakeholders (policy makers, practitioners, administrators, researchers, etc.) at different levels (local, regional and global) across all UN regions (Asia/Pacific, Africa, Europe, Arab Region, Latin and North America). This was complemented by a further document titled "Exploring Sustainable Development: A Multiple-Perspective Approach" (UNESCO 2012b), which states that the ability of educational institutions to respond to the complex expectations embedded in ESD can be enhanced, through a multiple-perspective approach to teaching and learning. The multiple-perspective approach, the document claims, promotes interdisciplinary and intercultural competencies, as it addresses challenges to local or planetary sustainability. Interdisciplinary thinking, in which concepts and knowledge from different academic traditions are used to analyze situations or solve problems, allows students to use knowledge in new and creative ways (UNESCO 2012b).

Over the past 15 years or so, educational institutions across the world have been encouraged to mobilise students and to take a more active participation in local, national and global processes towards sustainable development. Starting from a general confusion of what sustainability means (Jucker 2002), things have evolved greatly. As outlined by Brundiers et al. (2010), there are many opportunities to learn about sustainability, and more advantage may be taken from them.

Among the education sectors strongly influenced by the surge of ESD, mention can be made to higher education, which was all of a sudden expected to play a key role in the ESD debate. Due to their positioning and strategic nature, there is solid evidence that higher education institutions can and should make a strong contribution to sustainable development as a whole (Leal Filho 2010a), and to the implementation of education for sustainable development in particular. Publications such as "Sustainability at Universities: Opportunities, Challenges and Trends" (Leal Filho 2010b) or "Sustainability at Universities: New Horizons" (Leal Filho 2012) attest this. The latter book in particular, prepared in the context of the UN Conference on *Sustainable* Development (UNCSD) in Rio de Janeiro in June 2012, contains a comprehensive overview of the approaches, methods and tools currently being used, to integrate sustainable development in research programmes, curriculum and campus greening.

This is so, for two main reasons:

- firstly, it is partly thanks to the work of University researchers, that global threats such as high pollution levels, depletion of biodiversity, global warming or the damages to the ozone layer were identified;
- secondly, and closer to the point, university staff train millions of graduates each year—across all disciplines- and are hence in a strategic position to foster a broader awareness of what sustainable development is and of what it means means.

Not to be omitted is the fact that, due to the formidable body of knowledge and expertise they have, higher education institutions (especially their teaching and research staff) are uniquely placed to help society to identify and implement the social and technical solutions to the environmental challenges they have helped to identify. An interconnected approach to sustainable development may thus prove very useful.

As illustrated in Fig. 1, there are three main approaches currently being used by higher education institutions when implementing sustainability. They are, as Fig. 1 outlines, interrelated and characterized by a limited degree of overlapping, being at

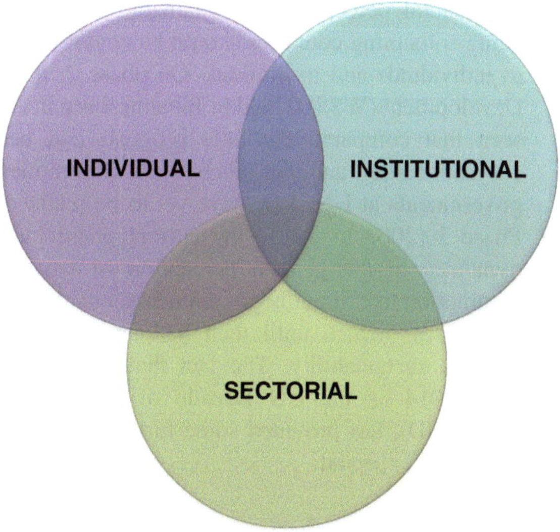

Fig. 1 Main approaches towards sustainability

the same time quite distinct. The first approach is the individual one, i.e. matters related to sustainable development are tackled by individual members of staff. The second is the sectorial one, whereby a Faculty (i.e. Natural Sciences) engage on sustainability, whereas others (e.g. Social Sciences) do not. Finally, the third approach is the institutional one, where there is a commitment from the whole university, towards sustainable development. Unfortunately only a small percentage of universities adopt the institutional approach, which shows that much still needs to be done.

The state of affairs here outlined suggests that, if ESD is to be properly integrated and embedded in higher education institutions, a change of perception is needed.

2 The Role of Education for Sustainable Development in Higher Education

The influence of sustainable development as a whole -and ESD in particular- in university systems, has evolved over time. Whereas sustainability was in the past a term hardly ever used, things have changed considerably since the Brundtland Report was issued in 1987. Leal Filho (2012) analysed the conceptual evolution of sustainability, and has identified the fact that it has gone through three main phases:

- Phase 1 (1987–1997)—on this initial phase, under the influence of the World Commission on Environment and Development (WCED), sustainable development was mostly regarded as a matter of concern to nations, as advocated by Agenda 21 (UN 1992) and as agreed by the Heads of States who attended the UNCED, held in Rio de Janeiro in June 1992.
- Phase 2 (1998–2002)—on this second phase, there was a noticeable change in the general perception of sustainable development, which has evolved from being something countries should be engaged with, towards a matter of concern to individuals and institutions. On phase 2, the World Summit on Sustainable Development (WSSD) held in Johannesburg in 2002 (also called Rio+10), it was seen that comparatively little progress had been made since UNCED held 10 years earlier, and that much of the commitments and pledges made by many governments at UNCED, were yet to be realized.
- Phase 3 (2003 to date)—the current phase has been characterized by a new dynamics in the general perception of what sustainability is, with a wide assumption that not only governments, but also individuals, institutions and even businesses—which until then had largely been kept aside- need to commit towards sustainability. The fact that the United Nations declared the period 2005–2014 as the UN Decade of Education for Sustainable Development (UNDESD), has provided some further impetus, albeit not to the extent it was originally expected.

The earlier mentioned World Conference on Sustainable Development, also known as Rio+20, held in Rio de Janeiro, Brazil in June 2012, offered an additional momentum to the current state of affairs. The publication of the report "The Future We Want" (United Nations (UN) 2012) may start a new phase, with a greater perception of what sustainability is, what it means and about what it can achieve. "The Future We Want" is the text of the Resolution by the General Assembly of the UN, inviting all relevant agencies of the United Nations system and other relevant international organizations to support developing countries and, in particular, the least developed countries in their efforts towards sustainable development—especially capacity-building for developing resource-efficient and inclusive economies.

The United Nations, which declared the years 2005–2014 as the "UN Decade on Education for Sustainable Development", also earlier referred to in this paper, hoped to foster the ideal of sustainable development in all areas of education, drawing attention to the need provide education opportunities to all people, enabling them to acquire knowledge and values and learn about behaviour and lifestyles which are needed to ensure a liveable future.

The higher education sector has not been indifferent to the many international developments seen in the field of sustainable development over past two decades. Rather, it has engaged to a considerable extent. Many universities and colleges across the world have critically analysed and substantially reduced the environmental impact of their operations. Many have engaged on new learning paradigms, using systems thinking (Habron et al. 2012).

In the United Kingdom, for example, the Higher Education Funding Council for England (HEFCE) document "Strategic statement and action plan on sustainable development" has reiterated hat individual higher education institutions should play a key role in their positions as centres of teaching and research (as well as managers of sometimes large campuses) both within the institution itself, but also in their local communities. The document outlines the elements which need to be considered by universities, in including sustainable development as part of their institutional plans.

The HEFCE is one of the few agencies across the world which specifically targets funding for such activities, working in partnership with its member universities and facilitating the sharing of good practice. However, this is not to say that current trends are all positive.

There are many reasons why many higher institutions have been lagging behind in the implementation of ESD. Some of them are:

- the lack of a critical mass of staff who work on the theme and may represent it in the relevant decision-making bodies at universities;
- the lack of strategic goals or action plans which may provide them with "a sense of direction" as to what they may want to achieve and why;
- the reduced willingness to engage on the structural changes required or pursue the investments needed so as to place sustainability more centrally in university programmes.

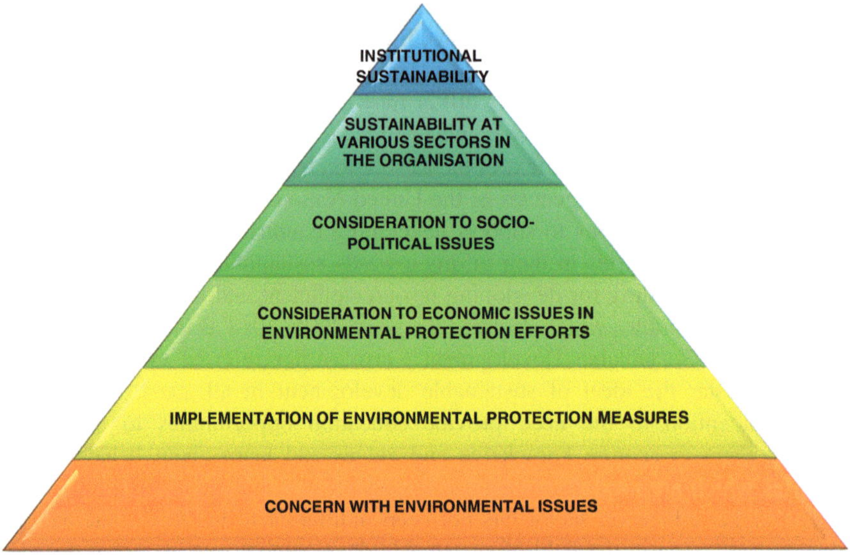

Fig. 2 Climbing the sustainability hierarchy

This list is by no means exhaustive, but offers an overview of the needs to be met. Moreover, the deep differences seen in the ways higher education institutions perceive ESD and the different levels of consideration to sustainability components as part of their activities, means that many need to "climb the hierarchy", i.e. they need to address the problems seen inthe past and intensify their efforts.

Figure 2 provides an overview of how this may take place. This paper defends the view that there should be an evolution from the basic concern with environmental issues (bottom level), up to the implementation of sustainability components as part of the operations of the whole institution (top level).

In doing so, higher education institutions can organize and undertake a wide range of measures, and take a set of concrete and effective actions, that can on the one hand help them to fulfill their institutional academic goals, but also create a new generation of students and help to promote a more sustainable and fair use of resources, on the other.

Drayson et al. (2012) on a study of students' attitudes towards and skills for sustainable development, report on an online survey of first and second-year students in 2011 across the UK. The overarching aims of the research were to understand any trends in new first-year student in 2011 as well as tracking the university careers of second-year students (surveyed as first-years in 2010) in terms of their demands and expectations. The study showed the value of giving the focus on the green economy as a solution to the economic situation currently facing the UK, the associated jobs creation linked with the green economy and the associated high levels of youth unemployment bringing a new focus to graduate skills in sustainability.

3 Education for Sustainable Development and Universities: Some Issues, Challenges and Needs to Be Met

The implementation of ESD at universities can take different shapes and formats. For instance, university campuses may be role models on how an institution can be both more sustainable and more resources efficient. If one picks an example, such as the reduction in energy use, this may lead to lower energy bills (an immediate economic benefit) but also to lower consumptions of fossil fuels. The same applies for areas such as waste management: reductions in the waste produced lead to less pressures in disposing it.

But the amount of progress reached to date should not overshadow the fact that there are many issues and challenges to be tackled. Some of the **needs to be met** are:

- the need for institutional guidelines to foster sustainable development within the institution
- the need for more engagement of senior staff, to broaden support for sustainability initiatives and foster capacity building
- the need for curriculum development—or in some cases curriculum modernisation- to cater for the inclusion of sustainability programmes at universities
- the need to make universities' footprint (i.e. environmental pressures caused by the operations) more sustainable, from energy use to CO_2 emissions from transport
- the need for local and regional partnerships on sustainable development, so as to yield results in the communities surrounding universities
- the need to mainstream sustainability research, possibly integrating it with organisational processes.

Moreover, it is a matter of fact that in order to yield the expected benefits, education for sustainable development at university level, as stated by Mehlmann et al. (2010) should be targeted towards:

(a) the development of interdisciplinary thinking,
(b) fostering skills in integrated planning,
(c) developing a broader understanding of complexity,
(d) understanding the role of decision-making processes.

A further need which exists is related to the shortage of initiatives to document and promote successful initiatives, so as to inspire others. Also, the fact that students' engagement is important, is often overlooked. Students have an important role to play in promoting sustainable development and encouraging behavioural change. Indeed, there are some research showing the potential of students in catalysing positive changes on environmental issues at higher education institutions–for instance at business schools. Jabbour (2010) in a study of greening of business

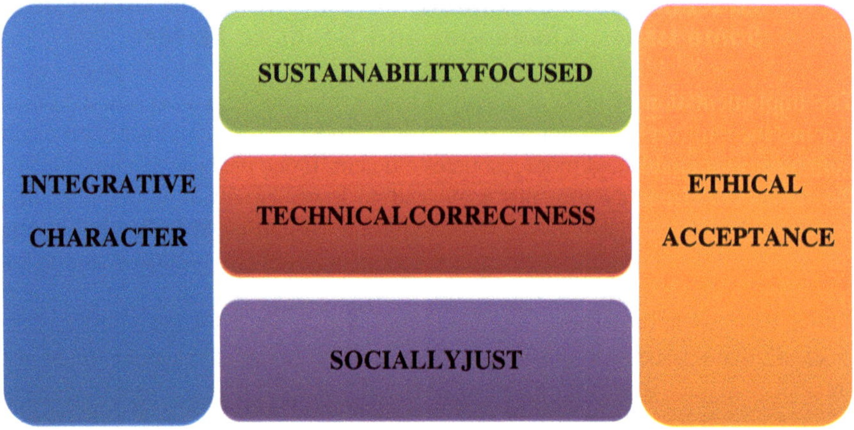

Fig. 3 Some elements catalysts of transformative sustainable development

schools specifies how the greening of business courses may be implemented, with a concrete analysis of the elements which influence the process.

Due to its scope, sustainable development is both a political goal and an institutional target. As such, it should have a greater degree of priority. There is no doubt that countries need innovative solutions which bring them economic wealth, social well-being and social equity, at the same time bearing in mind the need to afford due attention to the natural environment,

In order to achieve the goal of broadening the scope of sustainable development, and the effectiveness of ESD, there is a need to take into account a set of elements. Figure 3 shows that the conceptualisation and actual practice of sustainable development as a whole and ESD in particular, needs to be based on social justice, technical correctness and a sustainability focus, complemented by an integrative character and ethical acceptance. Their meaning is:

- **Integrative character**: meaning sustainability should be an integrated part societal efforts Ethical acceptance: meaning ethical values, need to be considered in the overall thinking
- **Socially just**: meaning social inequalities should be weighted and avoided
- **Technical correctness**: meaning that technical and technological components need to be adequate to specific situations and be methodologically sound
- **Sustainability focus**: meaning that a focus on the whole approach to sustainability is needed, instead of a sectoral one (e.g. environment or nature protection)

There are some good examples of universities, which have succeeded in integrating sustainable development as part of their programmes in a holistic way. In North America, mention can be made to Yale University and Columbia University (USA) as well as the University of British Columbia (Canada). In Europe, the University of Lüneburg, University Zittau-Görlitz or the Hamburg University of

Applied Sciences (in Germany), Polytechnic of Barcelona (Spain) and the Royal Institute of Technology (Sweden), offer good examples of what may be achieved. The University Saints Malaysia, University of Hong Kong or Tokyo University are exemplary of what is happening in Asia, whereas in Latin America and Africa the University of Sonora (Mexico) and University of Nairobi can be cited as other examples.

ESD as a modality of future-oriented education, needs to provide university students with the tools to know and be motivated towards engaging in global challenges such as biodiversity degradation and climate change on the one hand, but also use organisational learning towards sustainability on the other (Cebrian et al. 2013). Therefore, teaching sustainable thinking and promoting sustainability at universities—and other educational institutions- will catalyse the incremental changes which will lead to further progress towards fostering sustainability.

4 Conclusions

As this paper has tried to demonstrate, there are many problems which prevent universities from implementing education for sustainable development as part of their programmes. The issues and problems outlined in this paper, demonstrate that the elements related to the implementation of sustainable development at higher education institutions deserve more serious considerations.

The means to be used to implement ESD at universities vary and may include from better institutional and curricular integration, to improved institutional and decision-making mechanisms, which also cater for sustainable development thinking. Attempts to trigger the required transformative changes in a given institution need to be assessed based on a set of criteria, including—apart from and a sustainability focus- social justice, technical correctness and ethical acceptance. The absence of suitable arrangements for one or more of these elements may jeopardize the prospects for transformative change which is exactly what many higher education institutions need. Moves towards creating a sustainable future require people to fully understand the complexity of the work we live, and appreciate the need for more sustainable ways of living. ESD can help towards this understanding and facilitate this appreciation, and universities are in a good position to provide their contribution to this process.

The current emphasis to sustainable development as a whole and ESD at universities in particular, cannot however be regarded as sufficient, given the relevance, interdependence, complexity and impact of activities and targets associated with ESD. Therefore, further action is needed and the "sustainability hierarchy" (Fig. 2) may help to guide developments in this direction. The ground for this is very fertile, since the momentum for changes may continue and even increase, if higher education institutions intensify their efforts in helping society meet the challenge of sustainable development.

References

Brundiers K, Wiek A, Redman CL (2010) Real-world learning opportunities in sustainability: from classroom into the real world. Int J Sustain High Educ 11(4):308–324

Cebrian G, Grace M, Humphris D (2013) Organisational learning towards sustainability in higher education. Sustain Account Manag Policy J 4(3):285–306

Drayson R, Bone E, Agomba J (2012) Student attitudes towards and skills for sustainable development. Higher Education Academy, New York

Habron G, Goralnik L, Thorp L (2012) Embracing the learning paradigm to foster systems thinking. Int J Sustain Higher Edu 13(4):378–393

Higher Education Funding Council for England (HEFCE) (2012) Strategic statement and action plan on sustainable development. HEFCE, London

Jabbour C (2010) Greening of business schools: a systemic view. Int J Sustain High Educ 11(1):49–60

Jucker R (2002) Sustainability? Never heard of it! Some basics we shouldn't ignore when engaging in education for sustainability. Int J Sustain Higher Educ 3(1):8–18

Leal Filho W (2010a) Teaching sustainable development at university level: current trends and future needs. J Baltic Sea Educ 9(4):273–284

Leal Filho W (Ed) (2010b) Sustainability at Universities: Opportunities, Challenges and Trends. Peter Lang Scientific Publishers; 2010

Leal Filho W (Ed) (2012) Sustainability at Universities: New Horizons. Peter Lang Scientific Publishers

Mehlmann M, McLaren N, Pometun O (2010) Learning to live sustainably. In Global Environ Res 14:177–186

UNESCO (2012a) Shaping the education of tomorrow. Report on the UN Decade of Education for Sustainable Development. UNESCO, Paris

UNESCO (2012b) Exploring sustainable development: a multiple-perspective approach. UNESCO, Paris

United Nations (UN) (2012) The future we want. UN Publications Service, Geneva

The Matter of Geography in Education for Sustainable Development: The Case of Danish University Geography

Thomas Skou Grindsted

Abstract

Geographical imaginations are absolutely vital to make sense of sustainability challenges. Yet, a number of studies reveal that geography education has been slow in integrating issues of sustainability into curricula. Geography is particularly interesting in the context of ESD, due to its tradition of investigating human-environment interactions. In this paper we aim to contribute to this particular field of knowledge by providing an empirical analysis of ESD in Danish University Geography. In this paper it is examined how programs in Geography in higher education have taken different approaches to addressing issues of sustainability. Then, it is examined how geographers articulate their role and function as to addressing issues of sustainability. It is concluded that, though geographers generally are reluctant with using the concept of sustainability, and find it better serves as an implicit notion, geographers' find their discipline contributes considerably to ESD in three ways. First, geography's strong tradition in the human-environment theme provides a methodological basis for dealing with issues of sustainability. Second, the spatio-temporal dimensions of sustainability call for geographical approaches to be able to understand the dynamics, complexity and interactions in various scales. Third, geographers find their discipline provides an integrative knowledge platform between the natural and social sciences.

T.S. Grindsted (✉)
Department of Environmental, Social and Spatial Change, Roskilde University, University Road 1, Building 02, 4000, Roskilde, Denmark
e-mail: tskoug@ruc.dk

W. Leal Filho (ed.), *Transformative Approaches to Sustainable Development at Universities*, World Sustainability Series, DOI 10.1007/978-3-319-08837-2_2

Keywords
Geography · Higher education · Education for sustainable development · Curricula constructs · Sustainability discourses

1 Introduction: Greening Educational Policy and Geography in Higher Education

Since the Stockholm Conference on the Human Environment (1972) that first established a relation between education and sustainable development, the Rio Declaration (1992) and a number of subsequent declarations, policies and national strategies have promoted the idea of integrating ESD into all disciplines and academic traditions. Today, more than 31 declarations on sustainability in higher education have been made and during the past few years also declarations that address specific disciplines have developed (Grindsted and Holm 2012; Lozano et al. 2013). In 2007 the International Geographical Union Commission on Geographical Education (IGU CGE) officially announced their commitment to ESD with the "Lucerne Declaration on Geographical Education for Sustainable Development" in addition to UNDESD 2005–2014 (IGU-CGE 2007; Rienfrield 2009).

Declarations as well as the UN DESD, that will find new ways of continuation this year, are by nature designed to produce an impact on policy. Thus the development of declarations and the influence SHE declarations, have on educational policy and curricula constructs are closely interrelated (Adomßen 2013; Wals 2014). By way of example, the 2005 Graz Declaration on Committing Universities to Sustainable Development (Made under the umbrella of the European University Association and UNESCO) was developed to encourage the European Ministers of Education to integrate sustainability into the Bologna process. Thus the aim was to "Call on Ministers (…) to use sustainable development as a framework for the enhancement of the social dimension of European Higher Education as well as to contribute to the attractiveness of the European Higher Education Area" (Graz Declaration 2005). Sustainability issues have also gradually been incorporated in the Bologna process, and in Louvain-la-Neuve (2009) the European Ministers of Education decided to keep sustainability as a research topic for the next decade. Furthermore, the EU Commission has encouraged EU member states to use the UN Decade of Education for Sustainable Development (UNDESD) 2005–2014 as a point of reference in the development of national plans for ESD (EU Commission 2009). Correspondingly, The European University Association's annual rectors' conference (2012) carried the theme "Europe for Sustainable Universities". The EUA President, Maria Helena Nazaré, recognized that the challenges of sustainability is one of the greatest challenges for humanity: "Sustainability is the biggest issue for humanity on Earth; universities should be a role model to integrate sustainability into its activities, should contribute by informing (…) sustainable values and achievements should be part of education" (Personal Communication, Maria Helena Nazaré 2012).

In a Nordic context, national strategies have been prepared for implementing ESD. In 2007 the Nordic Council of Ministers (NCM) announced its commitment to education for sustainable development (ESD). The Nordic Strategy (NCM 2009) states that the Nordic countries should integrate knowledge on sustainable development into curricula from primary school, secondary school, adult education and higher education. In the NCM 2011 Strategy (NCM 2011) this was repeated, and national strategies, at least at a symbolic level, have been prepared for implementing ESD (Holm et al. 2012).

How the greening of educational policies influence various disciplines and academic traditions is particularly interesting in geography due to its strong tradition concerning the human-environment theme. Yet, a number of geographers (Yarnal and Neff 2004; Higgitt 2006; Chalkley 2006; Westawey 2009; Lui 2011; Kidmann and Papadimitriou 2012) reveal that in the US and UK the integration of sustainable themes into Geography curriculums is wanting: "Global warming presents an enormous threat to humanity, but the response from academia, including geography, has been relatively slow (…). I find this surprising, indeed astonishing, for there could hardly be a more important geographical topic" (Sayer 2009, p. 350). Sayer's statement is remarkable. Geographical representations are highly relevant to the agenda(s) of ESD. Disciplines like geography is expected to take a leading role on taking methodological approaches into consideration in dealing with the interface of society-environment interactions relevant to contemporary and future sustainability challenges (Yarnal and Neff 2004; Chalkley 2006; Clark and Button 2011).

Based on an empirical study (Grindsted 2013), we explore how sustainability challenges have been dealt with in the case of geography. The aim of this paper is to examine how Geography in Danish higher education contributes to ESD approaches. To do so, the following questions have been addressed: (1) Is the human-environment theme being reconfigured in geography? (2) How do Geography courses contribute to ESD? (3) How are issues of sustainability addressed in curricula? and (4) What is the influence of the Lucerne Declaration on Geographical Education for Sustainable Development? Answering these questions shows that the environmental theme is being reconfigured in geography paying greater attention to sustainability and how ESD has been introduced into curricula.

2 Is the Human-Environment Theme Being Reconfigured in Geography?

Three interesting perspectives substantiate the hypothesis that the human-environment theme is being reconfigured in geography, but also suggest a discrepancy between the "role(s)" of ESD in geography. However this does not imply that all geographers working on human-environment interactions conceive their research activities in terms of sustainability. Nothing could be more contradictory. Geography is much else than sustainability and most research geographers find their field of study have no relevance to the topic at hand. Nevertheless, the following three tendencies suggest changes in discourse coalitions toward being associated sustainability.

Table 1 The human-environment theme in the International Geographical Union Declarations

International Charter on Geographical Education (1992)	International Declaration on Geographical Education for Cultural Diversity (2000)	Lucerne Declaration on Geographical Education for Sustainable Development (2007)
Sustainable: 0	Sustainable: 1	Sustainable: 60
Pollution, Contamination, Hazards: 0	Pollution, Contamination, Hazards: 1	Pollution, Contamination, Hazards: 1
Climate change/global warming: 0	Climate change/global warming: 1	Climate change/global warming: 2
Ecology: 0	Ecology: 0	Ecology: 7
Environment: 2	Environment: 13	Environment: 13
Emission, greenhouse gas: 0	Emission, greenhouse gas: 0	Emission, greenhouse gas: 0
Nature: 0	Nature: 1	Nature: 8
Energy: 0	Energy: 0	Energy: 3
Biodiversity: 0	Biodiversity: 0	Biodiversity: 1
Human-nature interaction: 1	Human-nature interaction: 6	Human-nature interaction: 14

Firstly, the development of geographical declarations demonstrates a remarkable shift. Three declarations have been developed by the International Geographical Union, i.e. The International Charter on Geographical Education (1992), The International Declaration on Geographical Education for Cultural Diversity (2000), and The Lucerne Declaration (2007). The three declarations demonstrate a shift in the role of geography. From a shrinking and globalized world, with spatial transformations of economic, social and political significance, the Lucerne Declaration suggest the discipline performing a key role in solving sustainable challenges on Earth. By way of example, the International Charter of Geographical Education (1992) represents spatio-temporal tides and waves scarcely paying attention to the human-environment theme. Though human-environmental interactions are mentioned once, issues of globalization related to human rights remain the central focus (The Geographical Charter (1992) was developed the same year as the Rio (1992) conference and Agenda 21, Chap. "Experiences of 'Reflective Action': Forging Links Between StudentInformal Activity and Curriculum Learning for Sustainability"). The Lucerne Declaration by contrast states that the themes of the UNDESD 2005–2014 are very much in common with geography's objects of study; "the paradigm of sustainable development should be integrated into the teaching of Geography at all levels" (Lucerne Declaration 2007 p. 243). In Table 1 a word search condenses key aspects of the human-environment theme in the Geographic Education Declaration (1992), International Declaration on Geographical Education for Cultural Diversity (2000) and Lucerne Declaration (2007) sketching the role for geography.

As can be seen from Table 1 sustainability was not mentioned in 1992, once in 2000 and 60 times in 2007 which marks a noteworthy increase in the quantification of "sustainable related content". This illustrates that social-ecological and political-economic processes are not only intertwined, but also that core themes in geography are under reconfiguration. For whatever reasons they might be, the content analysis of declarations illustrates the down scaling of the human environment theme during the late 1980s and early 1990s, as Birkeland (1998), Fitzsimmons (1989) and Stoddart (1987) among others have argued.

Secondly, interviews demonstrate that sustainability issues are considered hugely important to geography. For example, one interviewee stated: "The concept of sustainability is of huge importance to geography at Copenhagen University, but also are related concepts such as resilience, vulnerability or ecology" (Interview 2012). Nearly all geographers interviewed found sustainability issues essential to geography, but remain critical of the concept itself and question of whether related concepts are better applicable for geographical analysis. The majority of the teachers claimed that sustainability is important to geography as an implicit notion, but when explicitly articulated many related concepts may better address particular phenomena (Interview 2012).

Compatibly, the international review suggests it is not hard to find geographers pushing the agenda for up scaling sustainability. By contrast, an interview with Danish geographers disclosed that sustainability best serves as an implicit or "hidden" curriculum. For example, Bednarz (2006, p. 239) states: "It seems that non-geographers also think that geography has an important role to play in environmental education (…) many geographers have defined geography as a discipline with a major, if not primary, interest in human—environmental interactions". Also in the context of ESD many researchers (e.g. Huckle 2002; Yarnal and Neff 2004; Chalkley 2006; Westaway 2009; Firth 2011; Morgan 2011; Cotton et al. 2013), offer an explanation of why geography plays an indispensable role in ESD. The recognition that geographical knowledge has importance for sustainable development makes Westaway (2009, p. 9) state that geography has a special role, maybe even above other disciplines: "Sustainable development is the extrinsic educational purpose that geography is best, indeed almost uniquely, equipped to serve (…). There is little doubt that geography is the best place to take the lead on sustainable development in schools." Such claims are indeed controversial, but authenticate the human environment theme gives geography its *raison d'étre* in the struggle for having a share in sustainability issues. Thus, the nature- society as well as the spatial dimension of sustainability becomes a major pillar that geography seeks to patentee.

Thirdly, Zimmerer (2010), Lui (2011) and Kidman and Papadimitriou (2012) demonstrate how geographers' research human-environment interactions particularly relating to sustainability issues has increased exponentially. Likewise Karatzoglou (2013) illustrates how leading ESD journals like International Journal of Sustainability in Higher Education and Journal of Cleaner Production reveal a similar growth. Despite the increasing numbers of articles, Lui (2011) shows how the number of articles contrast with efforts to integrate sustainability into curriculum in practice. "An examination of publications in sustainability education journals also reveals geography's lack of participation in sustainable education" (Lui 2011). This suggests a discrepancy between statements of the "role of ESD in geography" and geographers' research activity. Hence, there appears to be little evidence that ESD is recognized a central concern in geography within the US or UK (Higgitt et al. 2006; Morgan 2011). Also Turner (2002), Bednarz (2006), Yarnal and Neff (2004) among others argue that geography courses lag behind the growing environmental and sustainable research. But, what is the situation within Danish education in geography and might a similar tendency be identified in curricula?

3 How Does Geography Contribute to ESD?

Geography education shows its commitment to ESD in various ways. Examination of the study regulations, curriculum and interviews with chairs of the study boards and researchers reveal that the importance of geography to ESD demonstrates similarities to the Lucerne Declaration. "Geography has a major role in terms of sustainability. Many disciplines are experts on relatively narrowed subjects, whereas geography possesses the broadness which is an important dimension of sustainability. Geography is particularly potent because of its interdisciplinary approach as many other disciplines do not encompass. Moreover, geography merges the natural sciences and social sciences" (Interview 2012). Interdisciplinary approaches which integrate economic, social and physical aspects of sustainability are not only well suited for geography, but remain a pre-condition for understanding its multiple dimensions. Phenomena on global scales are caused by cumulative small scale activities in local places, and the impact of global processes exacerbates phenomena in specific localities (Morgan 2011). We may not appropriately understand sustainability issues if we ignore the climatological, hydrological or environmental processes that work in nature. Likewise, we misguide explanations of sustainability problems if we ignore social dynamics and economic activities. Geography knowledge is important to ESD and distinguished from other disciplines, because a narrow disciplinary focus may not unfold problems of sustainability that operates on multiple scales (Interview 2012). Another aspect found to be critical is that geography has a role in integrating perspectives from the natural and social sciences. "Geography can contribute in a unique way to sustainable development, especially regarding the integration of knowledge between social and natural sciences. In this way, geography plays a crucial role in dealing with sustainable challenges that you do not find in the tradition of many other disciplines, e.g. Sociology. Secondly, sustainability has an immanent spatial dimension" (Interview 2012). Thus it is argued that current environmental problems not only call for research and education that epistemologically transcend traditional disciplinary divides, but that geographers help to bridge the gap between natural and social sciences. To this may be added that geography has a distinct role being able to enrich related disciplinary discussions on ESD. "I find that geography has a responsibility to deal with issues of sustainability. We range competences and skills from the social and natural sciences—a holistic approach is imperative for dealing with sustainability" (Interview 2012). Additionally, complex interaction between nature and society and the spatio-temporal dimension of sustainability, requires methodological approaches to grasp such interactions that may even be impossible without geographical knowledge. "Before specialization, all students will acquire a holistic and broad basis of knowledge and approaches, about soil science, climate change, society and urban development (1.5 years of study red.). This broad foundation enables students to think critically and analyze side effects of a given phenomenon or human action. This body of knowledge is vital for sustainability, in order to understand side effects in very different areas and scales. Such questions I would say are only possible to deal with through geographical skills" (Interview 2012). Integrating the production of space and nature as a fundamental perspective of abstraction generates geographical knowledge and methodologies that make it possible to manage risks involved in the spatial distribution of problems. Non-geographical methodologies often fail to understand dynamics of spatial distribution. Though the interviews reveal recognition of the importance of sustainability issues to geography they also demonstrate that most geography teachers remain critical to the concept itself and/or find it better as an implicit basis for educating geographers (Interview 2012). "When I teach in accessibility for instance, then the aspect of sustainability is in the background. Whether or not sustainability is there [on the curriculum] depends how explicitly it should be mentioned. I rarely mention the term, but implicitly sustainability is the main objective for what we do and why we study it in this way. Sustainability is part of all geographers mindset I would say; sometimes so penetrated that one may not need to explicate it" (Interview 2012). This may be one of the reasons why the analysis of study regulation reveals that sustainability has a limited status in geography educations in practice.

4 How Are Issues of Sustainability Addressed in Curricula?

An examination of the preamble of the Aalborg, Copenhagen and Roskilde universities' curricula indicates a methodological foundation in which interconnectedness, processes and flows are given a primary status rather than fixed objects, direct causalities and permanencies. Ecological approaches focus on environmental problems from an interdisciplinary and normative angle (Rasmussen and Arler 2010). By way of example, the study regulation at Aalborg University requires that "students should acquire knowledge on human influences on ecosystems and the most important anthropogenic changes in history. They should be able to critically reflect on different philosophical views upon nature and its implications (...) understand concepts of sustainability and ecosystems in relation to elasticity and robustness to be able to analyze interactions between human activity (demands) and nature's capacity and limits" (Study regulation, Aalborg University 2010a, p. 29).

Second, there seems to be only little distinction between encouragements of the Lucerne Declaration and geographical educations as to interdisciplinary approaches. Thus study regulation requires problem based group work. "Students should be able to examine scientific problems and solutions using an interdisciplinary approach—not only from particular disciplinary premises, but also by including relevant theories, methods and philosophical interpretations from related disciplines"

(Study regulation, Roskilde University 2006, p. 23). Though study regulations do not go into detail in describing which methods are to be used, it gives a basis for enriching methodological reflection on real world problems (Clark and Button 2011). A holistic understanding of wicked problems like sustainability or climate change requires a reflective rather than an explanatory or commercial ambition (Interview 2012). In terms of sustainability, critical thinking is essential to understand different practices and agendas in play to be able to find possible solutions to sustainability challenges. However, though student driven projects on sustainability are few in practice (Interview 2012) similarities in learning objective and interdisciplinary approaches correspond to the Lucerne Declaration.

Third, the spatial-temporal dimensions of sustainability in curricula focus on processes in different time scales and spatial contexts. By way of example the graduate geographical qualification profile in *Ecological climatology and climate change, causes, effects, limitation and adaption* at Copenhagen University seeks to integrate the range from geological to economic time scales. The aim is to gain fundamental knowledge on climate change in history, relations between climate systems, ecosystems and land use, as well as relations between climate and the content of GHG gasses in the atmosphere. Thus the learning outcome is to be able to work with climate data and environmental observation in various scales to grasp complex relationship between physical and economic activities affecting the global climate (Study regulation, Copenhagen University 2009b, p. 8). The specialization illustrates how curricula seek to explain contemporary challenges to sustainability and climate change (e.g. desertification and deforestation) in its interconnectedness, complexity, aligned with the gradual perspective on climate change. If it is assumed that a dialectic approach comprehends the complexity of socio-spatial and economic-ecological processes, this, in turn, will make us recognize that environmental/social problems mutually interact, are spatially distributed, and produce different effects in different spatial scales. Relational time-space configurations vary considerably in different sciences, from geologists (million to billions of years) to economists (month, years and decades), even between physical and human geography (Rasmussen and Arler 2010). Geographical education programs emphasize how the matter of scale and the analysis of environmental problems are inseparable. Although the study regulations at Roskilde University, does not mention neither sustainability nor climate change, the Chair of the Study Board explains "students are given a body of knowledge on economic geography, natural processes and planning—also in an context of sustainability—so that they will be able to analyze the impact of a given plan or action in ecological, economic and social terms" (Interview 2012). However, it seems that Aalborg and Copenhagen Universities attach greater importance to the human environment theme as sustainability challenges than Roskilde University. The reason may be that urban geography is dominant at Roskilde University and has a strong spatial tradition. Aalborg University offers the most proactive geographical program dealing with issues of sustainability at BA level where 80 of 180 ECTS (and similarly at MA level) are allocated to various themes of the human-environment theme, mostly within issues of sustainability, climate change and anthropogenic environmental stress.

Geography at Copenhagen University offers a similar curriculum in terms of sustainability, both at BA and MA level. Thus it is only one of six qualification profiles (transformation of cities) that do not explicate sustainability, climate nor energy and resource themes (Study regulation, Copenhagen University 2006, p. 5). Here too, the reason may be that urban geography traditionally has identified itself as being closer to the spatial chorological theme.

5 What Is the Influence on Danish Universities of the Lucerne Declaration Concerning Geography?

The interviews revealed that the declaration was unknown for geographers at Copenhagen, Roskilde and Aalborg Universities and has neither been dealt with in the study boards nor on any other occasion (Interview 2012). "I have never heard of the declaration and it has neither been discussed in the study board nor at teachers meetings" (Interview 2012). It therefore goes without saying that the Lucerne Declaration has not produced an impact at the Danish universities. Instead, the declaration (and curricula) can be considered to be a product of specific history-geographical circumstances signalling geographical representations of human environment theme. As the Chair of the Study Board at Copenhagen University explains: "sustainability is not something we discuss, it is not necessary to discuss what you agree upon and which already is there" (Interview 2012). Though sustainability themes take various forms in geography at Copenhagen and Aalborg Universities, the chairs of the study boards do not find it is necessary to develop the study of sustainability further: "Well, I see no need for improving the sustainable content quantitatively. I would say it is saturated. Sustainability issues are filling out much of the education already, so I see no need of giving it more room, but sustainability will remain a core dimension of geography also in the long run" (Interview 2012). Also at Aalborg University geography seems to be saturated: "Sustainability is already integrated at many levels and has a strong critical mass. I do not see a point in enhancing it further. Environmental sustainability has the strongest position though, so in a way we could enhance aspects of social sustainability" (Interview 2012). Despite the fact that ESD at geography at Roskilde University is given less priority, the reconfiguration of the human-environment theme seems to be undergoing a process toward sustainability (certainly not definitely). Thus the gradualist perspective of climate change is easily identified. In contrast, the Lucerne Declaration resembles discourses of climate catastrophes and ESD as moral development (Lambert and Morgan 2009).

6 Conclusion

Analysis of study regulations, interviews and geography declarations suggest that the human-environment theme is under reconfiguration toward more notably being associated with sustainability. Thus, geographers find their discipline makes a particular contribution to ESD in three important ways. First, geography's strong tradition in the human-environment theme provides a methodological basis for dealing with issues of sustainability. Second, the spatio-temporal dimensions of sustainability call for geographical approaches to be able to understand the dynamics, complexity and interactions in various scales. Third, it is widely accepted that geography contributes by its interdisciplinary approaches to bridge the social and natural sciences, though it does not seem to be the case in practice. Dealing with geographical imaginations may not only better prepare students, teachers and practitioners in understanding sustainability challenges in various spatial contexts, but may help us better understand that, what appears to be a solution in one scale may produce sustainability challenges in another. To achieve such an understanding is not only relevant for geographers, but is relevant to ESD in various (inter) disciplinary contexts. Although issues of sustainability and climate change have been materialized in Danish geographical curricula, geographers remain critical to the concept itself and sometimes find it more suitable as an implicit notion. Still nearly all geographers interviewed find sustainability themes or related concepts important to geography.

References

Adomßent M (2013) Exploring universities' transformative potential for sustainability—bound learning in changing landscapes of knowledge communication. J Clean Prod 49:11–24

Bednarz R (2006) Environmental research and education in US Geography. J Geogr Higher Educ 30(2):237–250. doi:10.1080/03098260600717315

Birkeland IJ (1998) Nature and the 'cultural turn' in human Geography. Norw J Geogr 52:221–228. http://dx.doi.org/10.1080/00291959808552402

Chalkley B (2006) Education for sustainable development: continuation. J Geogr Higher Educ 30(2):235–236. doi:10.1080/03098260600717307

Clark B, Button C (2011) Sustainability transdisciplinary education model: Interface of arts, science, and community (STEM). Int J Sustain High Educ 12(1):41–54. doi:10.1108/14676371111098294

Cotton D, Winter J, Bailey J (2013) Researching the hidden curriculum: intentional and unintended messages. J Geogr Higher Educ 37(2):192–203. doi:10.1080/03098265.2012.733684

EU Commission (2009) The bologna process 2020—the european higher education area in the new decade communiqué of the conference of european ministers responsible for higher education. Leuven and Louvain-la- Neuve

Firth R (2011) The nature of ESD through geography. Teach Geogr 36(1):14–16

Fitzsimmons M (1989) The matter of nature. Antipode 21(2):106–120. doi:10.1111/j.1467-8330.1989.tb00183.x

Graz Declaration (2005) UNESCO

Grindsted TS, Holm T (2012) Thematic development of declarations on Sustainability in Higher Education. Environ Econ 3(1):32–40

Grindsted TS (2013) From the human-environment theme towards sustainability—danish geography and education for sustainable development. Eur J Geogr 4(3):6–20

Higgitt D (2006) Finding space for education for sustainable development in the enterprise economy. J Geogr Higher Educ 30(2):251–262. http://dx.doi.org/10.1080/03098260600717331

Holm T, Samalisto K, Virio T, Grindsted TS (2012) A model for enhancing Education for sustainabel development—examples from the nordic countries. In: Leal F (ed) Sustainable development at Universities: New horizons, Peter Lang

Huckle J (2002) Reconstructing nature: towards a geographical education for sustainable development. Geography 87(1):64–72

International Charter on Geographical Education (1992) The commission on geographical education of the international geographical union (IGU CGE)

International Declaration on Geographical Education for Cultural Diversity (2000) The commission on geographical education of the international geographical union (IGU CGE)

Karatzoglou B (2013) An in-depth literature review of the evolving roles and contributions of universities to Education for Sustainable Development. J CLEAN PROD, 49:44–53. http://dx.doi.org/10.1016/j.jclepro.2012.07.043

Kidman G, Papadimitriou F (2012) Content analysis of international research in geographical and environmental education: 18 years of academic publishing. Int Res Geogr Environ Educ 21(1):3–10. http://dx.doi.org/10.1080/10382046.2012.639152

Lambert D, Morgan J (2009) Corrupting the curriculum? the case of geography. London Rev Educ 7(2):147–157. doi:10.1080/14748460902990419

Lozano R, Lukman R, Lozano F, Huisingh D, Lambrechts W (2013) Declarations for sustainability in higher education: becoming better leaders, through addressing the university system. J Clean Prod 48:10–19. doi:10.1016/j.jclepro.2011.10.006

Lui L (2011) Where in the world of sustainability education is US geography. J Geogr Higher Educ, 35(2):245–263. http://dx.doi.org/10.1080/03098265.2010.548086

Lucerne Declaration (2007) International geographical union commission on geographical education (IGU CGE). Geographiedidaktische Forsch 42:243—250

Morgan A (2011) Sustaining ESD in geography. Teach Geogr 36(1):6–8

Nordic Council of Ministers (2009) Hållbar utveckling—En ny kurs för Norden, Köpenhamn

Nordic Council of Ministers (2011) Kunnskap for grønn vekst og velferd (MR-U) strategi for utdannings- og forskningsområdet i perioden 2011–2013, Köpenhamn

Rasmussen K, Arle F (2010) Interdisciplinarity at the Human-Environment Interface. Dan J Geogr 110(1):37–45. doi:10.1080/00167223.2010.10669495

Reinfried S (2009) Education for sustainable development and the lucerne declaration. Int Res Geogr Environ Educ 18(4):229–232

Sayer A (2009) Geography and global warming: can capitalism be greened? Roy Geogr Soc 41(3):350–353. doi:10.1111/j.1475-4762.2008.00867

Stoddard DR (1987) To claim the high ground: geography for the end of the century. Institute of British Geographers, NS 12:327–36

Turner B (2002) Contested identities: Human-environment geography and disciplinary implications in a restructuring academy. Ann Assoc Am Geogr 92(1):52–74. doi:10.1111/1467-8306.00279

Wals EJA (2014) Sustainability in higher education in the context of the UN DESD: a review of learning and institutionalization processes. J Cleaner Prod 62:8–15. http://dx.doi.org/10.1016/j.jclepro.2013.06.007

Westaway J (2009) A sustainable future for geography? Geography 94(1):4–12

Yarnal B, Neff R (2004) Whither parity? The need for a comprehensive curriculum in human-environment geography. Prof Geogr 56(1):28–36

Zimmerer KS (2010) Retrospective on nature–society geography: tracing trajectories (1911–2010) and reflecting on translations. Annal Assoc Am Geogr 100 (5):1076–1094. doi:10.1080/00045608.2010.523343

Additional Reference List

Interview with Maria Helena Nazaré (2012) the European University Associations Annual Conference

In depth interviews (2012) with the chair of the Study Boards in Geography at Aalborg University, Department of Development and Planning; Copenhagen University, Department of Geography and Geology; Roskilde University, Department of Environmental, Social and Spatial Change

Interviews (2012) with 31 of 42 full time associate professors and professors allocated education of geography at Aalborg, Copenhagen and Roskilde University

Study regulation for geography at Roskilde University (2005) Bachelor, Department of Environmental, Social and Spatial Change

Study regulation for geography at Roskilde University (2006) Bachelor and Master, Department of Environmental, Social and Spatial Change

Study regulation for geography at Copenhagen University (2009a) Bachelor, Department for Geography and Geology

Study regulation for geography at Copenhagen University (2009b) Master, Department for Geography and Geology

Study regulation for geography at Aalborg University (2010a) Bachelor, Department of Development and Planning

Study regulation for geography at Aalborg University (2010b) Master, Department of Development and Planning

Study regulation for geography at Aalborg University (2011) Bachelor, Department of Development and Planning

Study regulation for integrative geography at Aalborg University (2010) Master, Department of Development and Planning

Author Biography

Thomas Skou Grindsted is a Ph.D. student at the Department of Environmental, Social and Spatial Change, Roskilde University, Denmark. His area of expertise is the production of nature in educational policy, curricula and campus management in relation to climate change.

James Cook University's Holistic Response to the Sustainable Development Challenge

Colin J. Macgregor

Abstract

The Rio+20 Conference in 2012 affirmed that higher education should be one of the key forces central to the process of Sustainable Development (SD) during the 21st century. SD implies strategic change away from 'business as usual', where effort and investment is put into achieving a sustainable future. However, much of 'business as usual' for Higher Education Institutions (HEIs) involves generating income from learning/teaching and research programs so sustainability may not receive the support needed. James Cook University's (JCU) sustainability vision is to become an exemplar of a sustainable tropical university, i.e. "committed to the principles of sustainability, we will ensure that our actions today do not limit the range of social, cultural, environmental and economic options open to future generations." To support this vision many sustainability projects and initiatives are now underway at JCU. This paper provides an overview of the ways in which JCU is helping foster sustainability on its campuses, through its staff and students, through its operations and programs, and in the manner in which it engages with the northern Australia community. This paper may be of interest to other HEIs that are looking for practical ways to respond positively to the SD challenge.

Keywords

Education for sustainability pedagogies · Transdisciplinary learning · Action/participatory learning · Cross-school/faculty collaboration · Life index

C.J. Macgregor (✉)
College of Marine and Environmental Sciences, James Cook University,
PO Box 6811, Cairns, QLD 4870, Australia
e-mail: colin.macgregor@jcu.edu.au

W. Leal Filho (ed.), *Transformative Approaches to Sustainable Development at Universities*, World Sustainability Series, DOI 10.1007/978-3-319-08837-2_3

1 Introduction

Increasing public awareness of sustainability and the increasing urgency of global calls for the implementation of sustainable development (SD) has meant many universities and other higher educational institutions (HEIs) are responding to the challenge. At the Earth Summit in Rio in 1992, education was identified as one of the key forces central to the processes for SD and later the United Nations proclaimed the years 2005 to 2014 the *Decade of Education for Sustainable Development*. HEIs are being asked to do all they can to encourage education for SD (EfSD) within their curricula, as well as support sustainability-related initiatives and projects at their campuses.

Over the last 20+ years there have been some notable international declarations demonstrating broad-scale acceptance of sustainability across the higher education sector including:

- *The Talloires Declaration* (1990): a ten-point action plan for incorporating sustainability and environmental literacy in teaching, research, operations and outreach at colleges and universities. As of January 7, 2014 a total of 466 institutions from 54 countries around the world were signatories (ULSF 2014).
- *Kyoto Declaration on Sustainable Development* (1993): signed by members of the International Association of Universities (IAU), which has 1200 member institutions (Jones et al. 2010).
- *The University Charter for Sustainable Development* (1994): signed by the Association of European Universities (AEU), which has 305 European university members (Jones et al. 2010).
- *Association for the Advancement of Sustainability in Higher Education* (2006): mostly a US initiative, its main purpose is to make sustainable practices the norm within higher education. There are currently 821 HEI members (AASHE 2014).
- *UNESCO* (2004): Resolution (57/254)—*Decade of Education for Sustainable Development* (2005–2014).
- *Bonn Declaration* (2009): emerged from the *UNESCO World Conference on Ecological Sustainable Development* (150 countries). One of the actions urges stakeholders to 'Mobilise the core functions of universities: teaching, research and community engagement to strengthen global and local knowledge of E (Ecological) SD…' (German Commission for UNESCO, Action 15(o) 2009).

2 Traditional Teaching and Learning at Universities

Despite widespread acceptance of sustainability at many HEIs it is evident that adoption of sustainability within the curricula of universities appears to be slow. A number of writers have explored why this is the case (e.g. Cortese and McDonough 2001; Blewitt 2004; Dale and Newman 2005; Jones et al. 2010).

Blewitt (2004) agrees with Cortese and McDonough (2001) suggesting that higher education is one of the most conservative institutional sectors in society. Blewitt goes on to say that this conservatism supports disciplinarity which remains important in the intellectual organization of teaching, learning and, perhaps also, research funding (e.g. the Excellence in Research for Australia (ERA) initiative). Consequently, transdisciplinary activities tend to be seen as secondary to discipline-based approaches (Dale and Newman 2005). Cortese and McDonough (2001), p. 3 imply there is a deeper cultural issue, "traditional tertiary education emphasizes separation, compartmentalization, specialization and success, and there is an absence of explicit ethics and values, all of which reinforce an unsustainable worldview."

Despite the disciplinary tradition there is an increasing global demand for broader fields of study such as environmental science, informatics or media studies all of which lean towards transdisciplinarity and the social distribution of knowledge and knowledge production (Blewitt 2004). According to Pohl and Hirsch Hadorn (2007, p. 20) the transdisciplinary approach seeks to:

- "grasp the complexity of problems;
- take into account the diversity of life-world and scientific perceptions of problems;
- link abstract and case-specific knowledge; and,
- develop knowledge and practices that promote what is perceived to be the common good."

Sustainability has philosophical, economic, social, ecological, political, technological and scientific dimensions (Blewitt 2004) but being complex and complicated no single discipline can definitively address either the problems or the solutions, however it is normative in that it supports the common good. Nevertheless, disciplinarity is still the inviolable fact of university life. Universities administrative and financial arrangements are all organised around disciplines e.g. faculties, schools, departments etc. and transdisciplinary fields of study such as sustainability do not naturally fit with the compartmental arrangements of university institutions (Jones et al. 2010).

It may be argued that all those involved with HEIs have a responsibility, perhaps even a social mandate, to embrace and act on the sustainability challenge. The question is what are the most effective ways to bring about the cultural, bureaucratic and pedagogic shifts necessary? This paper cannot address all that is implied by this question but learning and teaching lies at the heart of university core business and EfSD is both an opportunity and a mechanism for HEIs to address the sustainability challenge. EfSD calls for a learning and teaching paradigm shift—from a teaching-centred, discipline-oriented approach, to a contextualized, learning-centred approach, based on implementation of problem/project-oriented curricula (Lehmann et al. 2008).

The purpose of this paper is to provide an overview of the ways in which JCU is utilizing problem-oriented and problem-based learning (POPBL) to enhance EfSD experiences and foster sustainability on its campuses, through its staff and students,

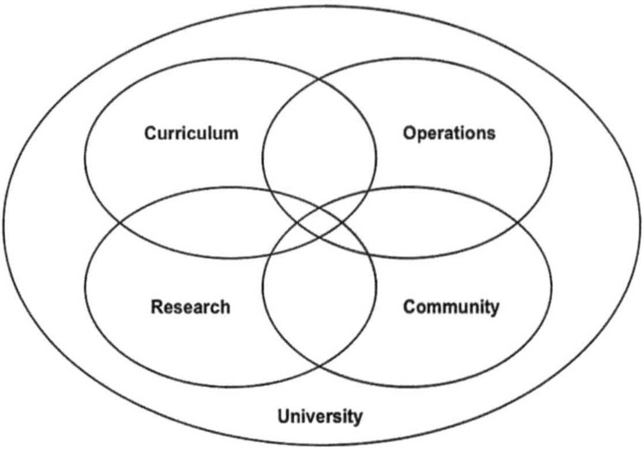

Fig. 1 Domains of campus sustainability (adapted from Cortese and McDonough 2001; Karlin et al. 2013)

through its operations and programs, and in the manner in which it engages with the northern Australia community. In general there are three components to the POPBL approach (de Graff and Kolmos 2003; Kolmos and de Graff 2007): *cognitive learning*, where learning is organized around problems and carried out in projects; *contents learning*, which emphasizes interdisciplinary learning and exemplary practice where theory and practice come together; and, *collaborative learning*, which emphasizes collective identification of problems and ownership of projects and the learning processes. In this context, capacity building for effective dialogue and communication is also encouraged. The POPBL approach is also innovative because it requires 'learning by doing' (Domask 2007, p. 55) i.e. students actively reflect on the process of both project development and planning, and on project implementation, thereby learning from the experience (Berkes 2009).

Before explaining JCU's approach in detail, it will be useful to the discussion to review the organizational structure of modern universities.

3 Organizational Frameworks for Considering Sustainability at Universities

Figure 1 was first presented by Cortese and McDonough (2001) but it was Karlin et al. (2013) that took the basic framework and developed a working model for testing the effectiveness of education for sustainability at the University of California, Irvine.

Figure 1 identifies four 'domains' for universities to consider sustainability. *Operations* typically refer to campus greening initiatives but these could include initiatives that seek to address social injustices. These may be formal policy shifts in how the institution operates, purchases, builds or even invests (Orr 1996) or it may refer to informal, perhaps student-led, initiatives and campaigns in which sustainability goals and objectives are sought (Karlin et al. 2013). *Community* refers to active participation by students and/or staff with the local, regional or even global

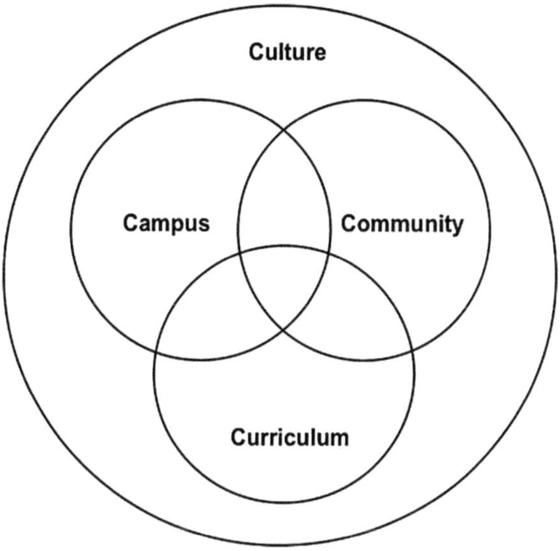

Fig. 2 The '4Cs' model (Jones et al. 2010)

communities (Karlin et al. 2013). *Research* in this context refers to "empirical analysis of sustainability issues, generally conducted by professors and graduate students, leading to new insights and solutions to sustainability problems" (Karlin et al. 2013, p. 2). Finally, *Curriculum* refers to the academic courses and subjects that support sustainability literacy or teach students about the theory and practice of sustainability and sustainable development. It will be apparent from Fig. 1 that while each of the four domains may be regarded as unique there are obvious synergies and overlaps, which supports the transdisciplinary nature of sustainability. In practice, many sustainability projects developed within the university have objectives that cut across more than one of the four domains; indeed, some may have elements in all four.

The '4 Cs' model (Fig. 2) also identifies four 'foci' for consideration (Jones et al. 2010). Developed by the Centre for Sustainable Futures (CSF) at the University of Plymouth, the intention of the model was to create a holistic approach to support the development of a *University Sustainability Policy* and *Action Plan*, which aimed at driving sustainability at the University of Plymouth. The four 'foci' for action were seen as 'mutually enfolded and complementary' (Jones et al. 2010, p. 7).

There are two notable differences between Figs. 1 and 2. First, the omission of the *Research* domain in the 4 Cs model. Jones et al. (2010) have not ignored this, rather they regard research is an integral part of *Curriculum*. This supports the notion that action-led research and teaching is critical in EfSD (see for example, Cortese and McDonough 2001; Sipos et al. 2008). The second notable difference is the addition of *Culture*. The concept of cultural sustainability is not defined by Jones et al. (2010) however a useful working definition was offered by Marens et al. (2010, p. 19); "a culture in which individuals are aware of major environmental,

Fig. 3 The LiFE Index (as adopted at James Cook University)

social and economic challenges, are behaving in sustainable ways, and are committed to a sustainable lifestyle for both present and future." Culture implies consideration of individuals' underlying knowledge/awareness, their values and attitudes, as well as their behavior. People within organizations resist or fail to respond to change for a variety of reasons (Beer and Nohria 2000; Kegan and Lahey 2001) but many prefer to continue with 'business as usual' simply because it is easier than adopting change. If sustainability is to become regarded as the dominant 'social norm' (Dressler and Carns 1969) within HEIs then cultural transformation is required. By displaying *campus*, *community* and *curriculum* within *culture* (Fig. 2) the significance of *culture* is emphasized; as Marens and Shriberg (2012, p. 557) explained, "nothing less than a paradigm shift in organizational thinking is needed for colleges and universities to promote cultural transformation." This transformation must find resonance with staff, students, and perhaps also external stakeholders of the university.

As indicated above, compartmentalization discourages sustainable development, however, compartmentalization is useful for strategic purposes because it helps ensure change is taking place across the various sectors of the university. To facilitate strategic planning and monitoring progress towards sustainability JCU is utilizing the LiFE (Learning in Future Environments) Index (Fig. 3).

While the title headings of the various 'foci' or 'domains' vary the similarities between LiFE's four primary themes and the two previous models (Figs. 1 and 2) will be apparent. According to the developers of LiFE (Environmental Association for Universities and Colleges), LiFE is a comprehensive performance improvement system to help tertiary education institutions manage, measure, improve and promote their social responsibility and sustainability performance (EAUC 2014). JCU

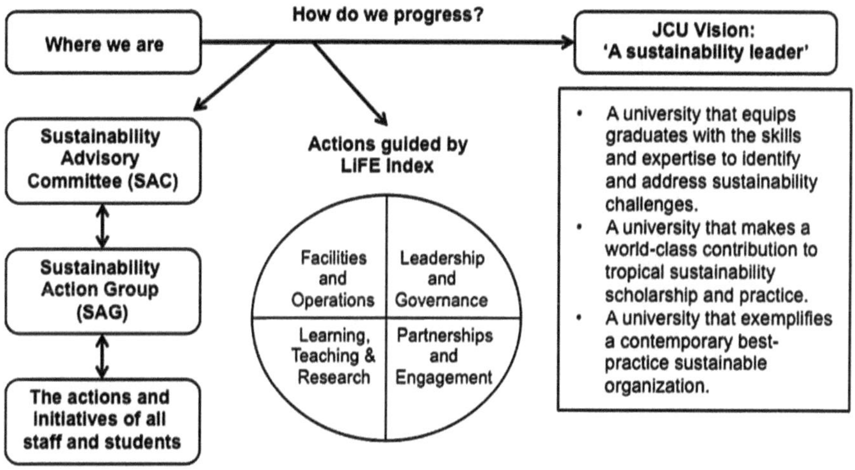

Fig. 4 James Cook University's sustainability vision and structure

adopted LiFE in 2012 and has been using it to identify and direct projects and other sustainability initiatives across LiFE's four thematic areas.

According to UNESCO (2013, paragraph 1) the goal of EfSD is to develop citizens who have the knowledge, skills, attitudes and values necessary to shape a sustainable future. Developing citizens is a notion that JCU has embraced. JCU is located in tropical northern Australia (Townsville and Cairns), with a third campus in Singapore. JCU's Strategic Intent emphasizes the need to improve the quality of life for people in the tropics, 'to create a brighter future for life in the tropics worldwide through graduates and discoveries that make a difference' (James Cook University 2013, p. 5). JCU has also developed a *Grand Challenges Framework* to support the Intent. Among the specific areas aimed at supporting sustainable futures are sustainability education, climate change, biodiversity, corporate sustainability, architecture, Indigenous language and culture, water, food, energy, and environmental policy.

Cortese and McDonough (2001) recognized that we tend to think of teaching, research, operations and relations with the community as separate activities but as indicated above JCU acknowledges that compartmentalization does not serve sustainable development; rather a holistic approach is required across all facets of the organization. Figure 4 demonstrates where the LiFE Index fits within the overall sustainability structure of JCU. If a shift from 'business as usual' is to occur then it must be directed top-down but in itself that would be insufficient. The shift also needs to emerge bottom-up from students and staff—almost out of the cultural transformation referred to above. Figure 4 demonstrates how top-down and bottom-up processes at JCU combine to support each other.

There are two sustainability leadership committees and both have representation from the faculties, schools and directorates of the University. Both also have student representation. The Sustainability Advisory Committee (SAC) reports directly to the Vice Chancellor (VC), making recommendations and providing sustainability leadership on policy and strategic matters. The primary functions of the SAC are to:

- Develop sustainability strategy and policy.
- Ensure University activities align with triple-bottom-line values and principles.
- Ensure the University addresses all relevant sustainability legislation, standards and agreements.
- Establish and implement sustainability objectives, targets and associated action plans and monitor and report against these.
- Establish key performance indicators and benchmark with other similar HEIs.
- Set and monitor annual planning and reporting for the Sustainability Action Group (SAG).

The Sustainability Action Group (SAG) reports to the SAC and with its guidance the Group creates and implements sustainability programs and projects, and provides guidance for the University's *Trop Eco* program (details of this program are provided below). The SAG's primary functions are to:

- Use the LiFE Index to embed sustainability across all aspects of the University.
- Develop cross-unit collaboration on sustainability initiatives.
- Fund projects, both internal and external.
- Implement sustainability programs and projects.
- Develop, monitor and report on key performance indicators.

In September 2013 JCU carried out a preliminary benchmarking assessment of the 16 sub-themes that make up LiFE. A detailed review of the methods and findings of this assessment is not possible within this paper but in summary, the assessment used 117 key performance indicators (KPIs) to represent the 16 sub-themes. It was found that 'good progress' has been made with three sub-themes, one of which is *learning and teaching*. Eight sub-themes may be described as 'upcoming/progressing' and another three are 'insufficient/lacking'. Two sub-themes have yet to be assessed. What follows now is a discussion about how JCU is responding to sustainability within each of the four primary themes of LiFE. Since learning and teaching for sustainability is one of the better performing sub-themes, and since this is a core area of business for universities, the discussion will demonstrate how student engagement in action-orientated projects is contributing and supporting EfSD pedagogies at JCU.

4 Leadership and Governance

(1) *Top–down, strategic efforts*: Leadership takes many forms at JCU and one of the most important is the sustainability vision, which in JCU's case has three components (see Fig. 4). The University's strategic intent also strongly supports sustainability, "We will produce graduates with the expertise and intellectual curiosity required for sustainable development of our communities, and we will conduct research to provide the knowledge and understanding needed to meet the challenges facing northern Australia and the tropics world-wide." The University also publically committed to supporting sustainability when the VC signed the *Talliores Declaration* at the University's annual *Sustainability Symposium and Fair* (see below) on March 8, 2013.

For sustainability to work in practice it must be sufficiently resourced. JCU has made a long-term financial commitment to support activities that fall within the University's *Sustainability* and *Trop Eco* programs. Sustainability funding is used to support salaried staff e.g. the Sustainability Project Officer and two Environmental Managers, as well as activities of the SAC and SAG.

(2) *Action for Sustainability Program*: Open to both staff and students this program provides grants for projects that result in a sustainable outcome for JCU and help raise the sustainability profile of the University. Grants of $5,000 to $30,000 may be awarded to a project. Examples from 2012/13 are provided in Table 1.

Most *Action for Sustainability* projects in 2012/13 were initiated and run by students. Following the POPBL approach, all addressed real-world problems, which provides a solid foundation for learning experiences (de Graaff and Kolmos 2003). And, like de Graaff and Kolmos, Cortese and McDonough (2001, p. 3) emphasize, "we must increase group work learning so students may be able to effectively collaborate as managers and leaders on complex problems." Nearly all *Action for Sustainability* projects involve group work.

5 Partnerships and Engagement

(1) *Annual Sustainability Symposium and Fair*: The symposium and fair was held at the Cairns campus for the first time on 8th March 2013. The topic of the Symposium was, *How we can live and eat sustainably in the tropics?* The Fair included arts performances, tropical food tasting, seed giveaways, and a number of expert-led sessions on food and sustainability. In 2014 (held 11th April) the Symposium and Fair considered *Consumerism and Sustainability* and it was extended to all three campuses (Cairns, Townsville, Singapore). Presentations and activities were delivered locally on each campus but some were video-linked across JCU's campuses. The *Annual Symposium and Fair* is a free public event providing an excellent opportunity to enhance linkages between the University and the local communities.

Table 1 Example projects from JCU's action for sustainability program (2012, 2013)

Project title	Summary of activities
GreenRide.com.au	This is an easy to use online carpooling network for staff and students travelling to and from the University. The scheme is expected to save 41 tonnes of CO_2 with an estimated fuel reduction of 28,000 L. Joining involves signing up for a free account using JCU's email. Users that want to offer seats in their cars have control over security; those that need a ride post a passenger request on the network
Permaculture edible food gardens	This project is reviving and redesigning gardens on the campuses to become productive food gardens. These also act as community learning hubs. Based on permaculture design principles the gardens minimize water use and demonstrate sustainable urban food production
JCU green bike fleet	This project is obtaining a stock of second hand bicycles, which are repaired and then sold cheaply to students ($50) for return at the end of their studies ($20 upon return). To date 30 bicycles have been obtained and repaired for sale. The goal is to provide 100 bicycles at the campuses over 2 years
Sustainable costuming (All's well that ends well)	This project supported the production of costumes for the 6th annual Shakespeare production of *All's well that ends well*, which took place at the Tanks Art Centre in Cairns (18–26 October 2013). All 52 costumes were hand made by the performers using recycled materials. The only new materials purchased were paint, thread and safety pins
Cairns campus arboretum	This project is highlighting local native flora at the Cairns campus. The arboretum consists of two walks, one an *Indigenous Use Walk* and the other a *Wet Tropics Walk*. The walks have been mapped and 45 trees have been given interpretive signs. Pamphlets and an interactive website provides cultural and botanical information about the trees

(2) *Trop Futures North Qld Schools Network*: Fang (2013) highlight the importance of partnerships with local communities. The aim of this project is to build a regional network of high school educators, and others e.g. Australian Association of Environmental Educators, CSIRO's *Scientist in Schools* who are passionate about sustainability. The project supports the *Queensland Sustainable Schools Initiative* and the Australian high school curriculum and encourages resource sharing between members. An online networking platform has been developed and a number of professional development, networking and capacity building workshops have been held. As part of the project the University provided six grants ($3,000) to members to assist them with sustainability efforts at their schools.

6 Learning, Teaching and Research

(1) *Curriculum Refresh*: This program began in 2011 with the aim to better align JCU's learning and teaching program with the University strategic intent. The refresh process targeted four pillars: tropical ecosystems, conservation and climate change; industries and economies in the tropics; peoples and societies in the tropics; and, tropical health, medicine and biosecurity. A number of themes were developed within the program and resources provided to support projects aimed at course development and alignment. Notable program themes included: embedding awareness of sustainability issues; embedding work integrated learning; and, embedding Indigenous perspectives. Work within *Curriculum Refresh* is on going but many undergraduate and postgraduate courses have now been 'refreshed'. At the last count there were 25 sustainability-related undergraduate or postgraduate courses or majors across the University. One of the completed projects within *Curriculum Refresh* with the sustainability theme was *Prospects for Establishing an Undergraduate Degree of Sustainability at JCU*. This ultimately led to the establishment of JCU's *Bachelor of Sustainability* in 2012.

(2) *Bachelor of Sustainability*: The B. Sustainability (Cairns campus) had an initial student intake of 36 students. The course has subsequently grown to nearly 100 students. A three-year transdisciplinary program the course is comprised of three majors: science (ecological/environmental); social science (equity, social justice); and, business (economic aspects). In Year 1 the underpinnings of the ecological, social and economic challenges confronting humanity are examined along with the history, philosophy, ethics, and political developments that have shaped current interpretations of sustainability. In Year 2 case studies from within North Queensland and around the World are used to explore the complex, transdisciplinary nature of sustainability and the inter-dependency of ecosystem functions and services (e.g. climate, soils, biodiversity), along with humanity's interactions with these. In Year 3 the efficacy of sustainability and SD is explored. The theory and practice of different approaches to SD are examined including market-based approaches, technocratic approaches and community approaches. The core and capstone subjects of the course provide students with opportunities to utilize on-campus sustainability projects (e.g. some of those described here) as case studies. These case studies help facilitate place-based and problem-based learning (Sipos et al. 2008) and wherever possible an adaptive management approach (Domask 2007) is encouraged.

(3) *Sustainability Internship Program*: The internship program provides students with unique opportunities to engage in meaningful collaboration with fellow students and University staff to bring about genuine sustainable outcomes (Fang 2013; Karlin et al. 2013). Students develop their projects with support from University staff (Fang 2013) and all are designed to support sustainability capacity building, professional development and knowledge sharing. In carrying out an Internship project students gain formal recognition (certification) of their work through JCU's Professional College and it is recorded on their academic record. Two examples are notable:

Green Switch: This project comprises a series of initiatives designed to improve the sustainability of the Student Association's operations. Activities revolve around awareness of consumerism, reducing waste and sourcing local food. *Green Switch: Cloths 'n Stuff Swap* provides students (and staff) with the opportunity to reuse and recycle cloths, books and just about anything that people would like to exchange. Through *Green Switch*: *Sustainable Student Kitchen* the Association hosts a free lunch for students with all food sourced locally. Finally, the Association's newspaper *Bull sheet* has gone on-line, partly to save paper but also to extend its coverage.

Atika Creek Restoration: This project aims to restore two riparian sites within the Cairns campus using species that belong to the Queensland endangered regional ecosystem register (7.3.12a and b). Cairns Regional Council supported the project by providing in-kind labour, seedlings and mulch. A very well attended community planting working bee completed the rehabilitation work in November 2013.

(4) *Extra-curricular activities*: There are a number of extra-curricula activities supporting sustainability at JCU including: sustainability master-class events/courses; sustainability-related movie nights and social events; and, the staff/student sustainability pledges. Two other activities are especially noteworthy because they are run by students for students.

The *JCU Sustainability Club* currently has 83 members. The management committee arranges weekly events during semester, recruits external speakers and arranges groups of students to volunteer in the community on a variety of sustainability related activities. Members use social media (*Facebook*) to share ideas and keep up to date with events and other sustainability information.

The Permaculture Society partners with a local food distribution organisation, the *Real Food Network* and at the last count there are 164 members. The Society hosts a weekly food market at the Cairns campus where students and staff may purchase locally produced fruits and vegetables. The Society also has links with local permaculture groups so the Society acts as a conduit linking students and staff with community groups.

(5) *Sustainability-related Research Centres*: The *Cairns Institute* pursues research and development activities around the human, social, economic and cultural dimensions of the tropics. It endeavors to have beneficial impacts on the livelihoods and communities of northern Australia and the global tropics. Staff are particularly concerned with Indigenous issues and needs, governance, sustainability in the tropics, tourism in the tropics, climate change adaptation, and social justice related issues. Currently there are 46 appointed staff and adjuncts associated with the Institute.

The *Centre for Tropical Environmental and Sustainability Sciences* (TESS) is concerned with providing a deeper understanding of the biological, physical and social processes that underpin natural and human systems in the tropics worldwide, and the rates and trajectories of change in natural and human systems in the tropics. There are five research themes within the Centre: ecology, biodiversity and conservation; environmental change; terrestrial biogeochemistry; sustainable landscapes and livelihoods; and, education, training and capacity building. The work of

TESS emerges from a variety of disciplines and currently the Centre has 50 staff and adjuncts and 40 research students.

7 Facilities and Operations

(1) *TropEco Program*: This program aims to involve staff and students in a range of sustainability-related activities, most of which address ecological/environmental aspects of sustainability. Notable target areas include: energy and water; sustainable transport; carbon management; recycling and waste; biodiversity and environment; *Green Colleges*; *Green Reps*; linking campus and curriculum; *TropEco News*; community gardens; and, the annual *TropEco Awards*. *TropEco* is an example of a crosscutting program i.e. many of its activities and projects address other sustainability themes within LiFE, so it is not concerned with facilities and operations only. For example, *TropEco News* and the annual *TropEco Awards* is shaping the attitudes, values, and perhaps even the lifestyles of students and staff, helping foster a sustainability culture at the University (Karlin et al. 2013).

(2) *Campus District Cooling (CDC) System*: The CDC at the Townsville campus is believed to be the largest system of its kind in the Southern hemisphere. A fully integrated cross-campus cooling system, this technology has reduced energy consumption at the University by and estimated 25 %, reducing carbon emissions by 10,600 tonnes per annum, and saving JCU over $2 million per annum in energy costs compared with comparable traditional systems.

(3) *Recycling and Waste*: Like all universities, JCU produces a considerable amount of waste, much of which is reused or recycled. E-waste is collected annually from each campus and sent to Brisbane (Queensland's capital city) for recycling by a reputable e-waste recycling facility. JCU also replaces thousands of fluorescent lights every year. These contain a variety of heavy metals that should not be disposed of in landfill and many of the components of these lights may be reused. Again, all used fluorescent lights are collected and sent to Brisbane for recycling. The *Second Life* program collects all second hand furniture on the campuses and this is then made available to staff and students for reuse within the University. *Bio-Regen* systems have been installed at University Halls and at the refectory areas. These units turn food waste from the kitchens and eating areas into a bio fertilizer. This fertilizer is used around the campuses (on gardens etc.) and it is made available for purchase by JCU staff, students and the wider community to help offset the costs of running the systems.

8 Conclusions

Universities may be thought of as microcosms of wider civil society but unlike wider society universities have a unique opportunity to progress sustainability. Karlin et al. (2013, p. 3) emphasized, "Universities sit in a unique position in their ability to integrate almost all operational areas, enabling a multi-faceted approach to sustainability that trains leaders, solves problems, and engages communities." Pedagogically, campuses are living laboratories providing universities with an ideal opportunity to 'practice what they preach'. Sustainability should be integrated into all parts of operations, purchasing and investments, and wherever possible efforts must be tied to the formal curriculum (Cortese and McDonough 2001).

JCU has allocated considerable resources to meet its sustainability ambitions and EfSD responsibilities. Results from a preliminary assessment of sustainability performance using sub-theme KPIs within LiFE confirm that JCU is still in the early stages of its SD program and evidently there is much to do. However, an area that is progressing well is teaching and learning for sustainability. Part of the reason is the problem-oriented and project-based learning approach where on-campus sustainability projects are utilized to support and enhance student learning. This approach means that many of JCU's sustainability projects have been developed and are run by students, especially those studying in the B. Sustainability. The B. Sustainability is a relatively new degree program but early student feedback on learning and teaching is demonstrating the pedagogical value of student engagement in active, real-world sustainability projects. On-campus sustainability projects and EfSD are mutually supportive, fulfilling learning and teaching, as well as broader sustainability objectives.

Other sustainability projects have been established under JCU's *Sustainability Internship Program* or through JCU's *TropEco Program*, providing the whole student body with work integrated learning opportunities. Sustainability projects may form part of formal course-work, or they may be extra-curricular; either way they have potential to deliver valuable sustainability outcomes for the University and its community partners. When projects utilize an adaptive management approach (learning by doing) they bring provide clear EfSD benefits. On-campus sustainability projects also offer considerable potential for internal cross-school/division collaboration.

Finally, extra-curricular sustainability-related activities are extremely important to transdisciplinary degree programs such as the B. Sustainability. Transdisciplinary courses require students to study subjects in different disciplines from a variety of schools and faculties, meaning students do not naturally build the sense of place/community that often goes with other more disciplinary-oriented programs such as engineering, law, medicine etc. The extra-curricular activities provide two important functions: they provide a mechanism to develop the social capital necessary to unite students in support of a common cause (studying and actively supporting sustainability), which means they have easy access to peer support while the study; and, they provide bottom-up inertia for sustainability initiatives, which encourages leaders of the university to meet them half way with top-down initiatives and support.

References

AASHE (2014) AASHE member directory. The association for the advancement of sustainability in higher education. http://www.aashe.org/membership/member-directory (Last Accessed 12 Mar 2014)

Beer M, Nohria N (2000) Cracking the code of change in HBR's 10 must reads on change. Harvard Business School Publishing. http://web.hbr.org/store/landing/must-reads/change.php (Last Accessed 20 Mar 2014)

Berks F (2009) Evolution of co-management: role of knowledge generation, bridging organizations and social learning. J Environ Manage 90(5):1692–1702

Blewitt J (2004) Introduction. In: Blewitt J, Cullingford C (eds) The sustainability curriculum: the challenge for higher education. Earthscan, London, pp 1–9

Cortese AD, McDonough W (2001) Accelerating the transition to sustainability through higher education. Second Nature, Inc. http://www.mma.gov.br/port/sdi/ea/deds/arqs/antcort_wilmcdon.pdf (Last Accessed 11 Mar 2014)

Dale A, Newman L (2005) Sustainable development, education and literacy. Int J Sustain High Educ 1(4):351–362

De Graaff E, Kolmos A (2003) Characteristics of problem-based learning. Int J Eng Educ 19(5):657–662

Domask JJ (2007) Achieving goals in higher education: an experiential approach to sustainability studies. Int J Sustain High Educ 8(1):53–68

Dressler D, Carns D (1969) Sociology, the study of human interaction. Alfred Knopf, New York

EAUC (2014) Learning in future environments. Environmental Association for Universities and Colleges. http://www.thelifeindex.com.au (Last Accessed 19 Mar 2014)

Fang C (2013) Ten Ways to integrate sustainability into the curriculum. The Association for the Advancement of Sustainability in Higher Education. http://www.aashe.org/blog/ten-ways-integrate-sustainability-curriculum (Last Accessed 12 Mar 2014)

German Commission for UNESCO (2009) Bonn declaration. United Nations Educational, Scientific and Cultural Organization. http://www.unesco.de/bonner_erklaerung.html?&L=1 (Last Accessed 12 Mar 2014)

Jones P, Selby D, Sterling S (2010) Sustainability education: perspectives and practice across higher education. Earthscan Publishing, London

James Cook University (2013) University plan 2013–2017. James Cook University, Cairns

Karlin B, Davis N, Matthew R (2013) GRASP: testing and integrated approach to sustainability education. Journal of Sustainability Education, Spring 2013: Experimential Education, Part I

Kegan R, Lahey LL (2001) The real reason people won't change in HBR's 10 must reads on change. Harvard Business School Publishing. http://web.hbr.org/store/landing/must-reads/change.php (Last Accessed 20 Mar 2014)

Kolmos A, de Graaff E (2007) Process of changing to PBL in: de Graaff E, Kolmos A (eds) Management of change: implementation of problem-based and project-based learning in engineering, SENSE Publishing, Rotterdam, p 31

Lehmann M, Christensen P, Du X, Thrane M (2008) Problem-oriented and project-based learning (POPBL) as an innovative learning strategy for sustainable development in engineering education. Eur J Eng Educ 33(3):283–295

Marens R, Levy B, Haven C, Bennett J, Bush K, Doman C, Holdstein B, Janiski J, Smith R (2010) Campus Sustainability Integrated Assessment: Culture Team Phase 1 Report. Graham Environmental Sustainability Institute and the Office of Campus Sustainability, University of Michigan. http://graham.umich.edu/media/files/CampusIA-InterimReport.pdf (Last Accessed 3/3/2014)

Marens R, Shriberg M (2012) Creating and assessing a campus culture of sustainability: The University of Michigan Experience. In Leal WF (ed) Sustainable development at universities: new horizons, Peter Lang Scientific Publishers, Vienna, p 994

Orr DW (1996) Educating for the environment. J Environ Educ 27(3):7–10

Pohl C, Hirsch Hadorn G (2007) Principles for designing transdisciplinary research: proposed by the swiss academics of arts and sciences. München, oekom Verlag
Sipos Y, Battisti B, Grimm K (2008) Achieving transformative sustainability learning: engaging head, hands and heart. Int J Sustain High Educ 9(1):68–86
ULSF (2014) Talloires Declaration institutional signatory list. Association of University Leaders for a Sustainable Future. http://www.ulsf.org/programs_talloires_signatories.html (Last Accessed 12 Mar 2014)
UNESCO (2013) Education for sustainable development. United Nations Educational, Scientific and Cultural Organization. http://www.unesco.org/new/en/education/themes/leading-the-international-agenda/education-for-sustainable-development/ (Last Accessed 12 Mar 2014)

Author Biography

Doctor Colin J. Macgregor originally joined James Cook University (JCU) in 1995 as a Lecturer, which is when he began studying for his Ph.D. in Sustainable Development. He took up a position with the Bureau of Rural Sciences (BRS) (Commonwealth Government) in 2000 where his research involved adoption of sustainable land management practices. He completed his Ph.D. in 2002 and moved to Scotland to take up a three-year Post-doctoral research position at the University of St Andrews. He then established St Andrews' undergraduate degree program in Sustainable Development before returning to Australia in 2006 as a Lecturer of Environmental Social Science at the University of Western Australia. Finally, Dr. Macgregor returned to JCU as a Senior Lecturer in 2012 to establish and coordinate JCU's Bachelor of Sustainability. Dr. Macgregor thanks staff and students at JCU for responding so positively to the sustainability challenge, and in particular he acknowledges the students of the B. Sustainability who put so much energy into supporting sustainability at JCU.

Triple Bottom Line: An Academic Perspective on Sustainability Practices and Accountability

Mohamed Saeudy

Abstract

This paper offers a critical analysis of sustainability in higher education institutions. This analysis will be based on the theoretical framework of triple bottom line (TBL) approach and academic accountability. It will explain how sustainability as an organizing and controlling framework may help these organizations to achieve institutional prosperity. This paper provides some guidelines on how to replicate sustainability teaching to an international audience. Semi-structured research interviews have been used as a qualitative research method to collect the empirical data from members of staff at Keele University. The research findings show that there are some sustainability tools that could be used to reinforce the academic practices in order to be more sustainable for the current and future generations. These tools include mixture between the sustainability themes and academic activities. Sustainability themes represent a wide range of institutional strides to assess the social, environmental and economic impact of the academic activities. The contribution of this paper focused on engaging the sustainability literature and the academic activities which represent the starting point for building up a generic framework of institutional prosperity. This paper explores not only an analysis of the TBL approach and academic accountability but also some critical lenses to develop sustainable practices. The innovative character of this research focuses on exploring the fundamental meaning of institutional prosperity and academic

M. Saeudy (✉)
Accounting Department, Keele Management School, Keele University,
Stoke on Trent ST5 5BG, UK
e-mail: m.h.saeudy@keele.ac.uk

W. Leal Filho (ed.), *Transformative Approaches to Sustainable Development at Universities*, World Sustainability Series, DOI 10.1007/978-3-319-08837-2_4

accountability in higher education institutions. In addition, it offers nuanced informed contribution by addressing sustainable academic practices, as an organisational tool, to help universities to be more sustainable.

Keywords
Sustainability • Triple bottom line approach • Sustainable teaching • Institutional prosperity • Sustainability tools

1 Introduction to Sustainable Development

There is no single explanation of the notion of sustainable development. This notion has been used in the Harlem Brundtland report to support the environmental policy agenda (World Commission on Environment and Development 1987, p. 8). This report has described the sustainable development as "the development that meets the needs of the present without compromising the ability of future generations to meet their own needs" (ibid). It seems to be that this definition does not provide clear explanation for the fundamental meaning of sustainability in terms of institutional practices or activities. Furthermore, it does not explain what sustainable development actually looks like to absorb the human activities and fulfil the needs of the future and current generation.

Organisational sustainability has been described in terms of organisational principles and values in order to explain the management process of organisational impact in the society e.g. product, services, process and other activities on managing the relationships with stakeholders (Epstin 2008, pp. 39–40). These principles focused mainly on explaining the quantified and monetized measurement of the important aspects of sustainability e.g. ethics, governance, transparency, business relations, financial returns, community involvement, employment practices and protection of the environment (ibid). But these values and principles overlooked the sustainability practices as organisational tools that could be used to achieve the institutional prosperity (Ferrer-Balas et al. 2010).

The ultimate purpose of organisational sustainability is the creation of a good quality of life for both current and future generations of human and nonhumans by achieving controlling balance between economic durability, ecological viability and social equitability (Carroll and Buchholtz 2012, p. 450). This kind of sustainability seems to be the base of restoring the institutional prosperity as one the favorable impact of innovative enterprise (Michie and Smith 2002, p. 203). It is notable that Brundtland definition (WCED 1987) holds some of this institutional prosperity and it gives more intellectual imagination to identify what is the sustainable and unsustainable organizational activity (Milne and Gray 2013).

However, sustainable development faces some fundamental challenges and contests (Bebbington 2009). These challenges embrace some political goals and measurement of moving set of practices (ibid). It entails some practical deficiencies of the systematic assessment of sustainable development e.g. the moving characteristics of sustainable performance, the role of sustainable development indicators (see DfEFRA 2005).

Accordingly, organizational sustainable practices refer to organisational activities, actions, day-to-day operations, decision-making process and circumstances that could be used to improve the living conditions of stakeholders throughout the society as well as intensify the natural environment, creating a society that can be sustained for longer period (Hopwood et al. 2010, p 250). This underlines the interaction between creating positive social, environmental and economic impact. Therefore, sustainability could be considered as one of the most promising framework within Higher Education Institutions (HEIs) to promote social, environmental and economic prosperity (Ferrer-Balas et al. 2010).

Based on the above mentioned challenges and characteristics of sustainable development in the academic literature, the main objective of this research is to clarify sustainable practices in academic work place. This research aims to offer such a rich empirical analysis through considering how academics understand and perceive sustainable development. It aims also to explain how to identify the desire to meet university's needs, particularly student and staff, without exceeding the environmental and social limits of institutional prosperity. The UK Higher education sector represents one of the main sources of future business developers and entrepreneurs as well as being institutional investors in sustainability (Lozano et al. 2013).

The current literature review shows a noteworthy demand on expanding the empirical researches to explore the unsustainable and sustainable academic practices (SAP) indifferent organisational settings (Spence and Gray 2007). These practices involve a contemporary promising discipline in UK Universities (Lozano et al. 2013).

2 Triple Bottom Line (TBL) as a Sustainable Development Approach

The TBL represents an organizing approach introduced by John Elkington (Elkington 2000). It has been used to refer to the management of corporate sustainability performance (ibid). This process involves the measurement of social, environmental and economic impacts. It presumes to manage the tension between the social and environmental activities with the economic and financial activities to create a new account of modern organisational capitalism (Gray and Milne 2007). This form of capitalism aims to engage, in open and free competition, the social responsibility of the business with the organisational activities to create profit (ibid). It examines how organizational activities translate the environmental and social issues into public relation issues (Marsden and Townley 2005, p. 406). This seems to be the consistent structure of the social contract theory as developmental process to empower the organisational legitimacy of business institutions (Mansell 2013, p. 32 and 57).

Gray and Milne (2007, p. 75) suggested that TBL reports should contain a full audited and substantial social and environmental disclosure to clarify zones of discretion, conflict with financial and proposed financial benefits or losses. This claim does not suggest a clear proposition for real accountability. In addition, it offers a deep debate about the nature of TBL activities and reporting practices to stakeholders. So, it seems that there is some rhetoric debate about how to create an institutional balance to apply the TBL approach. Furthermore, Milne and Gray (2013) argue that the TBL approach is insufficient conditions for organizations contributing to the sustaining of the Earth's ecology and it may create a greater level of unsustainability in terms of organisational practices and activities. This argument shows that TBL could represent an institutional barrier for sustainable development in some business organizations. The main reasons for this tension could be related to keen interests of business people to produce profit to grow their business and avoiding the demanding dilemmas of social and environmental issues (Young and Tilley 2006). This rational is not likely to be more warmly accepted in some institutions due to the international and national demands on creating more sustainable world (Savitz and Weber 2006, p. 250). In addition, some business institutions and societies trend to move toward more green economy and responding the climate changes (Bhimani and Soonawalla 2010, p. 184). These tendencies seem to offer new dimensions of the institutional accountability. These dimensions have been manifested by a notable increase in sustainability initiatives proposed by global international organizations e.g. United Nations (UN) and Global Reporting Initiative (GRI), Sustainability Context Group in USA and some governmental authorities e.g. UK Green Investment Bank.

More importantly, the current structure of TBL in the business institutions could be developed by exploring more sophisticated and pragmatic lenses to address sustainability practices in organizational setting (Unerman et al. 2007, p. 348). However, the successful mixture between the three main components of TBL could represent a new challenge of sustainable development within some industrial sectors e.g. health and public service; manufacturing; energy supply; retail and commercial enterprise; banking and financial services; and HEIs (see Lozano et al. 2013; Milne and Gray 2013; Unerman et al. 2007).

3 Sustainability Tools in HEIs

The fundamental meaning of TBL approach as a sustainability tool in HEIs is debated. The precise meaning of sustainable development, in these institutions, refers to creating TBL positive impact and achieving crucial prosperity for the current and future generations (Corcoran and Wals 2004, p. 92). So, it seems to be that there are so many different ways to achieve this broad ambition through developing the balance between social and economic development and ecological safety to maintain biological diversity and social well-being. There are many activities in HEIs and every aspect of these institutions operates in its own challenging environment, whether that be teaching, research or the other related activities e.g. services departments, campus activities, and operational facilities (Svanström et al. 2008).

Sustainable education represents four main activities (Huckle and Sterling 2001, p. 32). First, sustain people, caring of communities and ecosystems to ensure the viability of the intended learning outcomes. Second, work with integrity, justice, respect and inclusiveness to create human prosperity. Third, develop a healthy and viable system at workplace to embrace safe environment. Fourth, maintain durable teaching and learning practices to develop student learning experience. It is notable that these activities could be argued in terms of the stakeholder disillusion and lack of coherent performance indicators and competition.

The internationalisation of sustainable education could be seen as a sustainability dimension that could be used to reinforce the collaborative efforts and intensive sustainable development (Qiang 2003). HEIs seem to be affected by the globalisation of societies, economy and ecological concerns and thus they could be asked to consider these issues in the recruitment of foreign students and staff as well (ibid). Thus, HEIs become a real part of the globalisation process. Accordingly, there are some reasons for internationalisation of HEIs (Harris 2008). First, the interest of securing more financial and economic returns due to the global competitiveness and market mechanisms. Second, fostering the human prosperity across nations and developing the global diversity of local communities. It has been argued that business environment is viewed as a web of relationships and interrelated network, rather than traditional market with many independent stakeholder and beneficiaries (Johanson and Vahlne 2009). This implies more developmental emphasis on the role of HEIs to be more sustainable. Harris (2008) suggested a different interpretation of internationalisation of universities as a cultural structure rather than economic framework because internationalisation creates new forms of instrumentalism and diverts the national borders of these institutions. This dimension looks very plausible and could reflect more demands on the unique nature of HEIs as wider business communities. However, it is relatively worthy to realise that internationalisation of sustainable education could be perceived differently in the same country. So, this research paper take a closer look at the position of sustainability practices at one of the UK Universities to support the data analysis and interpretation.

Transdisciplinarity in research and teaching in HEIS has been argued as a sustainability tool (Lazano et al. 2013). This idea has imposed an intensive transition to new sustainable societal patterns to foster and implement new thoughts to embed sustainable development throughout the entire university activities and practices e.g. curriculum development, research, teaching and learning and engagement with stakeholders (ibid). But stakeholder identification represents an arguable contest for institutional leaders and sustainability setters in order to formulate the important bases of legitimacy in stakeholder-manager relationships (Mitchell et al. 1997). Furthermore, this process may impose some initial progress toward institutional social responsiveness, declaring the importance of clarifying the key stakeholders and exploring the source of their influence on the organization.

More importantly, sustainability in HEIs could be supported by looking at the efforts of corporate sustainability experience in business enterprises to develop these practices (Lozano 2011). This attempt seems to be a sustainability tool that could be used to enhance the societal impact of the university activities and practices to strengthen the societal business links with key stakeholders e.g. UK Higher Education Academy (HEA), current and prospective students, academic and administrative staff, financial institutions, alumni, student unions and business enterprises. Ultimately, this idea could be used to promote the reputational legitimacy of HEIs.

4 Academic Accountability

Traditionally, there are various forms of academic accountability articulated to maintain or develop the quality of teaching and learning activities in HEIs. Some of these forms focused on creating sustainable framework of knowledge for teaching and research activities as a learning organisation (Dill 1999). But this process does not offer a comprehensive framework for modifying the institutional activities and behaviour. It is not completely clear whether this is or could be a suitable way for the development of academic practices to support sustainable development in these institutions. However, there are some broad barriers to use accountability to describe sustainable development because sustainability involves some general elements rather than institution and its impacts (Bebbington 2001). These elements are focused mainly on how to understand institutional sustainability from environmental and societal point of view and explore the appropriate organisational activities and practices to achieve it (ibid). This suggests that accountability framework seems very narrow from sustainable development perspective. Some other literature has used accountability commitment as institutional approach to learn and understand the interactional relationships between profit and social responsibility as a new form of financial capitalism (Gray 2006). Furthermore, accountability has been used as a (un)sustainability tool to explain how to contribute to create a significant shareholder value (ibid). This approach does not describe the main organisational practices and characteristics of value creating institutions. It seems likely that accountability is a definitive framework of TBL and value creation.

Transparency has been used as an institutional tool in HEIs to disclose academic accountability (Neyland 2007). This effort explains the role of transparency to demonstrate the institutional accountability for environmental impact as a reformatted informational production processes (ibid). This process may be used to clarify the institutional internal dynamics of sustainable development.

Eventually, some practitioners clarify the distinction between learning about sustainable development (sustainability curriculum) and learning for sustainable development (sustainability practices and activities) form institutional point of view (The Quality Assurance Agency for Higher Education 2013). This attempt involves some clarification of the relevant factors to acquire the main attributes to add additional institutional value (e.g. student experience, employability skills, the impact of the research outputs and participation in community projects). It does not offer a clear institutional vision toward the academic accountability from societal and ecological point of view.

Ultimately, academic accountability and TBL approach could be used as institutional tools to develop and justify sustainable practices and activities to create more positive sustainable impact for current and future generations. This process intends to be a starting point to measure the progress of the institutional prosperity.

5 Research Method

The methodological approach of this study will be an interpretative structure and the methodology choice will be qualitative. The interpretative structure aims to identify motives, evidence, reasons, ideas and other subjective experience which motivate organizational practices (Duberley et al. 2012, p. 21).

The empirical research work of this paper has been divided into two main sections. The first section involves a preliminary study conducted by the researcher in order to explore the main themes of ASPs. This preliminary study has been published at Journal of Academic Development and Education in January 2014. The second section involves 10 semi structured in-depth interviews with nine academic staff from the School of Physical and Geographical Sciences (SPGS), Keele University as a case study. The interviews ranged from 40 min to one and half hours in duration. NVivo software has been used in sorting, coding and analyzing this data. Nvivo is qualitative data analysis software. This software has used for many reasons. It is a good time saving tool that has helped in the literature review to facilitate the recall of quotes from the main sources (Bazeley 2011). This has freed the researcher to explore numerous analytic questions with high capacity of richness to develop some theoretical accounts of ASPs (see Seale 2013, p. 269). Nvivo represents a dominant tool to ensure the anonymity and create coherent themes or Nodes. Therefore, this software has been used to analyse the interviews data as well as the collected sustainability documents. The analysis process was performed against three dimensions/sets of academic sustainable practices. These dimensions can be summarized into three main categories: educational, research and operational practices.

The in-depth semi-structured interviews have provided detailed insights into their university's operations and practices. These perceptions have helped to address the institutional accountability and sustainability practices. These interviews aimed to explore the fundamental meaning of sustainable practices and accountability in the HEIs. There were four main reasons for selecting academic members of staff from SPGS. First, some interviewees have been involved in sustainable education. Second, some academic staff at this school is responsible for managing sustainable education and green technology in the university

Table 1 The relationships between institutional prosperity, academic accountability and the institutional durability of ASPs

		Sustainability themes		
		Institutional prosperity	Academic accountability	Institutional durability of ASPs
Academic practices	Learning and teaching (L&T)	Developing sustainable oriented education	Stimulating sustainability norms, values and attitudes	Encouraging online materials and marking
	Scholarship activities	Assessing the research impact on the wider communities	Promoting human dignity and solidarity	Providing pragmatic solutions for real life business problems
	Operational facilities	Reducing the carbon footprint	Maintaining ecological viability	Introducing ecological facilities
	Service departments	Supporting value-oriented market structure	Developing common good activities	Identifying unsustainable practices

Source adopted from Saeudy (2014)

Sustainability Hub. Third, the majority of interviewees had broad teaching and research experience in the field of renewable energy projects, reducing institutional Carbon footprint, organic farms and biodiversity projects. Fourth, SPGS has participated in a 3 year research project funded by the Higher Education Academy's National Teaching Fellowship Scheme, entitled Hybrid problem-based learning—a scalable approach to sustainability education, a collaboration between Keele, Manchester, and Staffordshire Universities that aimed to explore, develop, and disseminate novel pedagogies and educational resources for sustainability education within universities based on problem-based learning (PBL).

The interview questions have been developed to explore the main academic sustainability and accountability practices from institutional point of view. The analysis of the interviews data has been supported by the sustainability documents and reports that have been received during interview process to reflect on emerging themes and provide generic analysis (Silverman 2010).

6 Research Findings

The findings of this research have offered an analysis of the research interviews through three key themes (Nodes); institutional prosperity, academic accountability and the institutional durability of ASPs (see Table 1). These findings have been articulated in the context of the relevant literature and research objectives.

6.1 Institutional Prosperity

The idea of institutional prosperity refers to the words of one of my interviewees (professor of environmental science).

> I 7: I have personally overseen the sustainability of any university or business institution to ensure we are in the right way for institutional prosperity for everybody in the world

This interviewee is apparently aware with the broad objectives of sustainability. Similarly, this framework seems to be consistent with the work of Carroll and Buchholtz (2012). But when questioned the measurement of this institutional prosperity, it may involves some personal bias in setting the assessment criteria for the institutional performance. However, the stakeholders could have greater power, and influence to set the strategic framework of institutional prosperity.

The mainstream of the organisational literature has also been keen to identify the key performance indicators to assess and measure this process.

It may be no surprise if academic members of staff, of an HEI, produce research publications, to pursue an individual or institutional recognition, without clear impact on the wider communities.

> I 2: Sustainability means for me creating self-activities to do my job without any harmful impact for the environment

Given the potential challenges of offering new thoughts of sustainability, some interviewees denied their responsibility for assessing the impact of their own practices and research outputs on the wider communities because of their limited experience in this field. In addition, it seems quite problematic to develop individual initiatives on measuring these issues e.g. Carbon footprint and supporting value-oriented market structure as well as meeting the university's sustainability targets in energy self-sufficiency.

6.2 Academic Accountability

As previously outlined in the accountability literature, academic staff is subject to align themselves to maintain the ethical and legitimate framework of their jobs. Some interviewees engaged in some institutional initiative to promote sustainable activities in the field of promoting student diversity & equality.

> I 2: Well, I am very keen to maintain actual activities to promote some sustainable values, attitudes and behaviours for our students as common good activities

This kind of engagement has been characterised as a form of institutional accountability (Dill 1999; Lozano 2011). It aims to introduce institutional accounts for maintaining human integrity and solidarity. The metaphor of common good could be extended to include the ecological considerations of the teaching and learning activities such as developing sustainable educational practices in the field of curriculum development and enhancing the current performance of the facilities departments.

6.3 Institutional Durability of ASPs

Some interviewees drew on the metaphor of durability to refer to the green practices and introducing real life solutions for the business problems. This includes some sustainable practices in the field of reducing Carbon emission, recycling, biofuel solutions and renewable energies across academic and professional services areas in relation to driving sustainability commitments.

More interestingly, some other interviewees have raised the idea of addressing the unsustainable academic practices to motivate sustainability.

> I 5: There are some unsustainable academic practices from my own experience such as misuses of the research funds, the lack of institutional assessment of the impact of this money. If we succeeded to spot these activities, we could be more sustainable, I think

More importantly, this repertoire has been supported by Gray (2006) and Gray and Milne (2007) to reflect the institutional attempt to promote accountability and sustainability to create more positive social and ecological impact from the organisational activities.

7 Conclusion

Based on the analysis of the in-depth semi-structured interviews with some academic members of staff in the SPGS, this research shows a notable emphasis on the crucial role of ASPs and academic accountability in promoting institutional prosperity. Through the analysis of the collected sustainability document, the research provides three main contributions. First, it explains how to explore the fundamental meaning of sustainability in order to craft the academic activities. Second, it gives a new institutional account of institutional durability to grasp the new challenges of HEIs. Third, it clarifies the unsustainability practices to allow more emphasis for institutional progress and improvement toward sustainable development (see Milne and Gray 2013; Lozano et al. 2013).

Finally, the main lesson that has been learned from this research is that ASPs will remain challengeable specifically in the field of managing the contextual balance between the pressures of political engagement and institutional involvement in sustainable development. In addition, the TBL approach is experiencing ongoing and arguable dimensions of legitimacy particularly in the UK HEIs because it seems to be challenged by some sort of institutional priorities and political agenda of the governmental directions toward sustainability. This rout may impose new articulations of the broad objectives and core values of ASPs. This includes deploying the institutional evidence of creating positive sustainable impact for current and future generations.

The future prospects of this research may focus on developing new institutional tools to construct official account or perspective of the unsustainable practices to help universities to be more sustainable. This effort has not been enthusiastically evidenced by some academic members of staff. These issues may require more statutory guidance form the supervisory bodies e.g. UK Higher Education Academy, student unions, government and local communities. Perhaps, developing further exemplars of this initiative and introducing a wider institutional accountability may entail a broader acceptance of the necessities of sustainable development. Future research might build on this attempt through explaining how the unsustainability account might be implemented. More specific examination may require a close analysis of the quality of the teaching and learning practices to capture the role and legitimacy of this account.

References

Bazeley P (2011) Qualitative data analysis with Nvivo. Sage Publications Ltd, London

Bebbington J (2001) Sustainable development: a review of the international development, business and accounting literature. Acc Forum 25(2):128–157

Bebbington J (2009) Measuring sustainable development performance: possibilities and issues. Acc Forum 33:189–193

Bhimani A, Soonawalla K (2010) Sustainability and organisational connectivity at HSBC. In: Hopwood A, Unerman J, Fries J (eds) Accounting for sustainability; practical insights, 1st edn. Earthscan, London

Carroll AB, Buchholtz AK (2012) Business and society: ethics, sustainability, and stakeholder management. South-Western Cengage Learning, Mason

Corcoran PB, Wals AE (2004) The promise of sustainability in higher education: an introduction. In: Corcoran PB, Wals AE (eds) Higher education and the challenge of sustainability: problematics, promise, and practice. Kluwer Academic Publishers, New York

Department for Environment, Food and Rural Affairs (2005) Securing the future: The UK government sustainable development strategy. Department for Environment, Food and Rural Affairs (DfEFRA), London

Dill D (1999) Academic accountability and university adaptation: the architecture of an academic learning organization. J Higher Educ 38(2):127–154

Duberley J, Johnson P, Cassell C (2012) Philosophies underpinning qualitative research. In: Organisational Qualitative (ed) Research: core methods and current challenges. Sage Publications Ltd, London

Elkington J (2000) Cannibals with forks—the triple bottom line of 21st century business. J Bus Ethics 23(2):229–231

Epstin MJ (2008) Making sustainable work: best practices in managing corporat social, environmental and economic impact. Greenleaf Publishing Limited, Sheffield

Ferrer-Balas D, Lozano R, Huisingh D, Buckland H, Ysern P, Zilahy, G (2010) Going beyond the rhetoric: system-wide changes in universities for sustainable societies. J Cleaner Prod 18:607–610

Gray R (2006) Social, environmental and sustainability reporting and organisational value creation? whose value? whose creation? Account Auditing Accountability J 19(6):793–819

Gray R, Milne M (2007) Towards reporting on the triple bottom line: mirage, methods and myths. In: Henriques A, Richardson J (eds) The triple bottom line: does it all add up?. London: Earthscan, p. 70–80

Harris S (2008) Internationalising the university. J Educ Philos Theory 40(2):346–357

Hopwood A, Unerman J, Fries J (2010) Accounting for sustainability: practical insights. Earthscan Publications Limited, London

Huckle J, Sterling S (2001) Education for sustaianbility. Earthscan Publications Limited, London

Johanson J, Vahlne J (2009) The Uppsala internationalization process model revisited: from liability of foreignness to liability of outsidership. J Int Bus Stud 40:1411–1431

Lozano R (2011) The State of sustainability reporting in universities. Int J Sustain 12(1):67–78

Lozano R, Lukman R, Lozano F, Huisingh D, Lambrechts W (2013) Declarations for sustainability in higher education: becoming better leaders, through addressing the university system. J Cleaner Prod 48:10–19

Mansell SF (2013) Capitalism, corporations and the social contract, 1st edn. Cambridge University Press, New York

Marsden R, Townley B (2005) The Owel of Minerva: reflections on theory in practice. In: Clegg SR, Hardy C (eds) Studying organisation; theory and method. Sage Publications Ltd., London, p. 406–407

Michie J, Smith JG (2002) Globalization, growth and governance: creating an innovative econmoy. Oxford University Press, New York

Milne MJ, Gray R (2013) W(h)ither Ecology? the triple bottom line, the global reporting initiative, and corporate sustainability reporting. J Bus Ethics 118(1):13–29

Mitchell RK, Agle BR, Wood DJ (1997) Toward a theory of stakeholder identification and salience: defining the principle of who and what really counts. Acad Manag Rev 22(4):853–886

Neyland D (2007) Achieving transparency: the visible, invisible and divisible in academic accountability networks. Organization 14(4):499–516

Qiang Z (2003) Internationalisation of sustainable education: toward a conceptual framework. J Policy Features Educ 1(2):248–270

Saeudy M (2014) Sustainable practices in the academic workplace. J Acad Dev 1:29–40

Savitz AW, Weber K (2006) The triple bottom line: how today's best-run companies are achieving economic, social and environmental success and how you can too. Jossey-Bass, San Francisco

Seale C (2013) Using computers to analyse qualitative data. In: Silverman D (ed) Doing qualitative research. Sage Publications Ltd, London, p. 268–269

Silverman D (2010) Doing qualitative research, 3rd edn. Sage Publications Ltd, London

Spence C, Gray R (2007) Social and environmental reporting and the business case. Certified Accountants Educational Trust, London

Svanström M, Lozano-García FJ, Rowe D (2008) Learning outcomes for sustainable development in higher education. Int J Sustain High Educ 9(3):339–351

The Quality Assurance Agency for Higher Education (2013) Education for sustainable development: draft guidance for UK higher education providers for consultation. The Quality Assurance Agency for Higher Education, Gloucester

Unerman J, Bebbington J, O'Dwyer B (2007) Sustainability accounting and accountability. Routledge, New York

World Commission on Environment and Development (1987) Our common future. Oxford University Press, Oxford

Young W, Tilley F (2006) Can business move beyond efficiency? the shift toward effectiveness and equity in the corporate sustainability debate. Bus Strategy Env 15:402–415

Author Biography

Dr. Mohamed Saeudy is a qualitative researcher in Social and Environmental Accounting. His research interest centres on how accounting practices may help business institutions to be more sustainable. His work has focused on developing organisational lenses to understand accounting for sustainable development practices in different organisational settings e.g. small and medium size businesses and higher education institutions. His current work explores the accounting for sustainable development practices in the banking sector. He has remained fascinated by the sustainability initiatives that have been developed by accountancy profession e.g. Global Reporting Initiative (GRI), Equator Principles, Global Alliance for Banking for Value and the UK Green Investment Bank. Dr. Saeudy is interested in exploring further exemplars of how the un-sustainability report might be implemented in organisational workplaces. Furthermore, his research interests focus around the main impediments and motivations of organisational sustainability. These efforts aim to explore the fundamental meaning of organisational prosperity and durability of business activities. Dr. Saeudy has spent many years working in the banking industry in Egypt and UK. He had also held a research and teaching commitments in accounting and finance at the University of Derby. He also leads and teaches the Green Accounting and Sustainability Modules at Keele University covering the topics ranging from accountability processes to sustainable business strategies and policies.

Engineering Education for Sustainable Industries: Approach, Implementation and Accreditation

Jurgis Kazimieras Staniškis and Eglė Katiliūtė

Abstract

Taking into account that a number of companies applying different preventive environmental measures grows up, there is an increasing demand for engineers possessing high competence in technologies for cleaner production and integrated environmental management. During the last decades the focus of environmental work has shifted from dealing entirely with the emissions and wastes of industrial production to including them into the total environmental responsibility and performance of all types of enterprises, where environmental properties of products become more and more important. The environmental issues today affect all types of enterprises. The Institute of Environmental Engineering (APINI) at Kaunas University of Technology has very strong links with industry and governmental institutions, has implemented a number of training programmes for industrial enterprises and has assisted more than 150 companies in developing and implementing cleaner production projects. M.Sc. and Ph.D. students have an opportunity to participate in the projects aimed at improvement of environmental performance in industry and to acquire valuable practical experience. Kaunas University of Technology was selected as a pilot institution for the QUESTE-SI project funded by the European Commission under the Lifelong Learning programme ERASMUS (2010–2012). QUESTE-SI stands for

J.K. Staniškis (✉)
Institute of Environmental Engineering, Kaunas University of Technology,
K. Donelaičio St. 73, 44029 Kaunas, Lithuania
e-mail: jurgis.staniskis@ktu.lt

E. Katiliūtė
School of Economics and Business, Kaunas University of Technology,
K. Donelaičio St. 73, 44029 Kaunas, Lithuania
e-mail: egle.katiliute@ktu.lt

W. Leal Filho (ed.), *Transformative Approaches to Sustainable Development at Universities*, World Sustainability Series, DOI 10.1007/978-3-319-08837-2_5

"Quality system of European Scientific and Technical Education for Sustainable Industry". The project was coordinated by EFMD, the Management Education Network, and ENQHEEI, the European Network for Quality of Higher Engineering Education for Industry. While the importance of evaluating the education of sustainable development programmes has been recognised, little information is available on the topic. The QUESTE-SI evaluation focuses on the institutional unit (department) that is responsible for one or more programmes. A key point is to assure that each graduate should have learned the sustainability aspects related to the concerned education domain. A fair evaluation of social responsibility and sustainability education was not limited to teaching and learning methods or curricular content—it depended upon parallel efforts in all dimensions. More than 10 European Science and Technical universities have been accredited in accordance with the QUESTE-SI requirements. Kaunas University of Technology is represented by the Institute of Environmental Engineering with the M.Sc. Programme "Environmental Management and Cleaner Production" and the Ph.D. Programme "Environmental Engineering and Landscape Management". The Institute of Environmental Engineering is awarded the highest ranking and has become a fertile basis for a larger pilot project. The programmes and staff experience facilitate the development of the university itself as a model for sustainable operations, research and studies. The paper presents an original education approach for sustainable industries, its implementation and experience in QUESTE-SI accreditation.

Keywords
Engineering education · Sustainable industries · Evaluation · Accreditation · Quality · QUESTE-SI

1 Introduction

An increasing number of corporations have come to the conclusion that for future development of the enterprise it is important to invest in developing the environmental work and profile of the company. The introduction and integration of environmental management systems into the management of companies are becoming a business necessity that requires a raise of the environmental competence at all levels within the company. This change is based on three key elements: (i) the recognition of the systematic thinking, (ii) the development and implementation of deregulation mechanisms and (iii) the growing concern for the long-term environmental cost benefit parameters and overall competitiveness in industrial companies. The recognition of the systems approach has been developed over a long period of time, representing a gradual change in thinking. Systematic thinking stands for taking into consideration different system relationships (product system, environmental system), e.g. the consequences of producing a given good under wider system borders (in space and time). The process of deregulation (the shift towards market-based instruments) started several years ago. Market forces are efficient in terms of combining responsibility and flexibility and they play an important role in parallel to the governmental regulation. Finally, cost-benefit criteria will always be of great importance to industrial companies.

The environmental work of a company must be an integral part of all its activities. Enterprises not producing goods are also strongly affected by environmental issues. Not least important are different types of consultancy companies. Additionally, such organisations as national, regional and local authorities are also strongly involved in the development towards an environmentally sustainable society.

To summarise, today we can foresee an increasing demand for engineers possessing high competence in technologies for cleaner production and integrated environmental management, because the number of companies applying different preventive environmental measures is growing up (Staniškis and Arbačiauskas 2002).

Moreover, sustainability must be seen by industrial engineering educators as an opportunity both to improve and to strengthen the field of humanising the engineering field by adopting the multidisciplinary learning and teaching to shift from unsustainable patterns of production to sustainable production and services through a better understanding of human health exposures and the environmental impacts of materials, processes and products. More sustainable industrial engineering tends to be a more humanistic term to refer to industrial engineering, potentially incorporating economic, environmental and social dimensions into the practises of industrial engineers. The new paradigm of Industrial Engineering considers social issues and the environment as fundamental rather than ancillary factors for industrial activity (Velazquez et al. 2011).

Initially, efforts to promote sustainability through universities were directed at greening the campus, demonstrating the commitment to sustainability in the way that an increased efficiency of the resource and waste management on campus contributed to the culture of trained engineers (de Ciurana and Leal Filho 2006; Sammalisto and Lindhqvist 2008; Krizek et al. 2012). Alternatively, Anne Sibbel tested a teaching and learning intervention to develop students' capabilities required for sustainability. The tasks included in the intervention had been carefully adapted for the classroom within the existing curriculum framework. The results showed that students participating in this intervention would finish their studies much better prepared for sustainability (Sibbel 2012).

Education for sustainability must therefore be transformational in character, aimed primarily at changing behaviour (both individual and collective), instead of just increasing knowledge about sustainability. Given the need for global sustainability, it must also support a similar transformation in behaviour across all existing social and cultural boundaries. What is needed is a simple, clear and flexible approach that enables individuals to make better choices today than they did yesterday (Denton 2011).

Education for sustainability must be at the forefront of the minds of all educators, especially in this challenging situation. This action at universities will require curriculum modification in some cases or new curricula development in other cases. Barrett (2012) has elaborated a 45 h (three credits) course that consists of six main topics. The objectives of the course are to: (i) outline and justify a perspective in environmental education (EE), (ii) recognise the potential of EE within all curricula, (iii) use information relative to the nature of EE to make decisions, (iv) draw on a range of appropriate EE concepts, materials, teaching strategies, learning activities and evaluation strategies for EE, (v) display competence in using a range of skills related to teaching EE and (vi) exhibit a mastery of the knowledge of the causes and consequences of selected national, regional and global environmental problems. The target group of that course were educators delivering a Bachelor's degree programmes. This course has certainly caused transformation in the attitudes of the participants. Some of them have already started to teach the EE issues in their schools. (Barrett 2012).

Programmes on sustainable education are subject to the increasing attention around the world. For example, monitoring and evaluation are identified as a key strategy for implementation of the UN Decade of Education for Sustainable Development (UNESCO 2005). However, little information is available or accessible on evaluating the education for sustainable development, despite the continued recognition of its importance. In fact, the ESD programmes should distinguish between input, activity, output, outcome and impact objectives, and should articulate the assumed causal connections among these objectives. This will provide clarity about what should be evaluated both in terms of objectives and the processes believed to result in these objectives. Programme evaluation has much to offer for the ESD programmes and the ESD field as a whole. The growing interest in the ESD programme evaluation is encouraging, as well as the quality of the identified ESD programme evaluations and their findings in regard to the benefits of the ESD programmes are (Zint 2011).

2 Approach and Guiding Principles for Engineering Education

Sustainability education programmes in Lithuania could be considered as one of the outcomes of cleaner production programmes implemented in the country. The Institute of Environmental Engineering (APINI), Kaunas University of Technology, initiated implementation of cleaner production (CP) in Lithuania two decades ago. It is the main institution in the country to work in the area and these activities have been operatively supported by the Lithuanian Confederation of Industrialists. In Lithuania most of CP programmes have been developed and implemented in support of donors and international organisations. Several countries, particularly the USA, Norway, Denmark, Sweden, the Netherlands, were supporting activities in this field. All programmes pursued similar goals but from different approaches. Lithuanian experience, where different donors/organisations have been active, shows the complementarity of these approaches for introducing preventive environmental practices into the industrial sector.

In the period of 1994–2012, in support of different donors and in co-operation with different foreign and local partners, APINI supported more than 200 Lithuanian companies in implementing the resource efficient and CP projects. Most active in this respect were textile and food industries. In long-term training programmes 245 CP experts were trained.

CP assessment is based on several logical stems embedded in an organised procedure. This procedure is instrumental in organising the CP efforts, informing necessary stakeholders within the company and bringing together the persons that can develop, evaluate and implement CP innovations. CP assessment is most often divided into five phases:

(1) *Planning and organisation.* CP planning is a systematic method for identifying options to reduce or avoid the generation of waste.
(2) *Pre-assessment.* The prime objective of the phase is the selection of one or a few assessment focuses. It requires a preliminary identification and evaluation of the CP potential at the plant level. While doing so, first inventories of obvious options as well as a preliminary estimate of the waste generation costs are made.
(3) *Assessment.* The phase focuses on the production processes that involve waste streams and is based on examination and re-evaluation of the production processes. The re-evaluation consists of "source identification" followed by "cause diagnosis" and "option generation".
(4) *Feasibility studies.* These studies have to prove whether each of the options is technically and economically feasible and whether each option contributes to the environmental improvement. The level of detail in the feasibility studies needs to be tailored to the nature of the option, since options may vary from simple operational improvements, the use of alternative materials to installation of advanced equipment.
(5) *Implementation.* In the framework of the UNEP and UNIDO projects, Lithuanian experience was transferred to several countries in Africa, Central America, South-East Asia and to China and Russia.

Experience from the implemented CP programmes has clearly shown that to sustain preventive environmental activities in industry, there is an urgent need to improve the education process of future specialists. As a first step, a new M.Sc. programme in Environmental Management and Cleaner Production has been introduced at Kaunas University of Technology.

Multidisciplinary research and environmental education may be integrated by focusing on the issues of global concern, but it needs to be done in an integrated way, not from their separate industrial or merely pedagogical angles, as has largely been the case so far.

The researchers who conduct such studies may not necessarily relate their research to the goals of sustainable development, because many scholars outside the mainstream environmental field do not yet recognise the significance of sustainable development for their research. The higher education sustainability movement is relatively new and has not been able to reach out to all scholars and university managers yet. It may also happen that many, if not most, advocates of sustainability in higher education have tended to come from the fields of environmental studies, education and facilities' management and, thus, have tended to concentrate on the economic and ecological pillars of sustainability, and have not often seen the relevance of sociological, political science and cultural studies research to their goals.

Consequently, much research on sustainability in higher education does not address the pillars of sustainability in a holistic, interdependent and systemic way. This is a key problem that the attention to alternative paradigms of research may help redress. It may also help redress several related problems that characterise much of current research in this area.

Research funds are always limited and research policies increasingly tend to fund single purpose, short-term studies often with a strictly problem-solving approach and targeted objectives. The supreme role of research, namely, the advancement of knowledge, and, thus, the transfer of this knowledge advancement, is often overlooked or forgotten. Therefore, the Institute of Environmental Engineering Kaunas University of Technology is emphasising the dictum "*research and education always go together*" and searches for the promotion and improvement of the transfer efforts and mechanisms.

Science is aimed at generating true knowledge, while engineering is at changing the world. Therefore, the world of technology is far more directed towards changing our society than the sciences and the arts are. It is therefore especially important that engineers are taught to deal with the problems of society. Often, engineering deals with its responsibilities by defining three separate stages (society is responsible for the demand of technology, the engineers create it, and society is again responsible for its application). This self-proclaimed docility is empirically untenable and morally doubtful. However, sustainable development is not a technological problem and the engineers must learn to be susceptible to non-technical issues and must be able to communicate with both the citizens involved and other experts. The world needs the engaged engineers that are willing and able to contribute to sustainable development.

Therefore, the educational programme at Kaunas University of Technology is structured in accordance with the following main guiding principles:

(1) *Interdisciplinary approach*. The environmental field is in its nature multidisciplinary, which is mirrored in the course structure of the Master's course. Today engineers need to know how to work along with economists, biologists, ecologists and many other professionals in order to avoid the negative externalities and/or side effects generated by their economic activities. More sustainable industrial engineering requires the professional formation of industrial engineers concerned who shall have the competence to develop and implement the strategies of incorporating economic, environmental and social dimensions into the practice of their profession (Velazquez et al. 2011). The compulsory

part of the course includes the areas such as environmental technologies, management, policy, law and economics and environmental strategies such as cleaner production and eco-design. The optional course program widens interdisciplinarity of the Programme with the areas such as sustainable development, systems analysis, monitoring and modelling as well as advanced courses in technical areas. To summarise, the Programme has a strong emphasis on engineering, management and policy to make sure that graduates have a broad understanding and capability to work with complex issues of sustainable industrial development.

(2) *Strong research connection.* Most of the optional courses treat advanced topics with close research connections. The subject matter as well as the topics of the course assignments and projects are closely linked to the research of the Institute giving the course, or to its R&D cooperation with industry, with the projects defined by the need for solutions to current environmental or environmental management problems in the industrial company. The connection with the research is in particular ensured in thesis projects for which active researchers are preferentially engaged as supervisors.

(3) *Practical education.* Kaunas University of Technology has very strong links to industry and governmental institutions. The Institute of Environmental Engineering has implemented a number of training programmes for industrial enterprises and assisted more than 150 companies in development and implementation of cleaner production projects. M.Sc. students of the Programme in environmental management and cleaner production have an opportunity to participate in the projects aimed at improvement of environmental performance in industry and to acquire valuable practical experience.

(4) *Integrated programme.* The aim is to provide M.Sc. students with a solid theoretical knowledge and hands-on experience in the real world. The goal of integration can be achieved in various ways, for example, by integrating the principles of environmental communication and informatics into the normal business practice in industry. Comparison of the perspectives provided by the multidisciplinary research and systems theory with some of the perspectives of environmental education shows a number of interesting similarities. The Programme achieves it by a balanced and integrated theoretical and practical education. M.Sc. students use theoretical knowledge in their practical assignments accomplished in industrial companies that result in practically applicable sustainability performance improvement projects. By integrating the systems theory (research) with environmental education and other programmes, students can learn more about the relationships between natural resources, environmental sustainability and human well-being. This knowledge accompanied by appropriate action is prerequisite for maintaining not only our natural resources but also our way of life. Research and environmental education thus require closer cooperation and more involvement in expanding their combined effect. With the support provided by these two areas, industrialists, on the one hand, and educators, on the other, can do more than ever to help people understand and act on environmental concerns.

2.1 Training in Decision Making Process

The increased inter-linkage between physical-ecological and socio-economic sectors of the society has placed a variety of pressures on decision makers in government, society and industry. The decision making system is a resource allocation mechanism, which attempts to rationalise and optimise the system performance to achieve a given set of goals. In this context, the decision maker performs one essential link between the control targets, the control costs and the individuals, groups and components which make up the system being controlled. Additionally, in many cases it is essential to construct a model and out of the existing information system to generate the forecasts of future systematic behaviour. Without a modelling system to allow forecasts, the decision making process can act only as a feedback control system, which is capable of counteracting deviations from a target only after they have occurred.

The education provides a systematic approach as a powerful vehicle for the statement of environmental situations, for reducing the areas of uncertainty in our increasingly complex decision making situations and different possible control systems, for instance, feedback, feedforward, cascade, adaptive etc. In advanced engineering education a key issue is "contextual awareness", i.e. the ability of engineers to view actions, problems and possible solutions in a broad context in which the consequences may at once be scientific, technical, economic, or social in nature.

To some extent, engineering education is already dealing with the environment, professional ethics, safety and health, and discipline-specific problems related to engineering process and practice. This content has been often integrated with other subjects, not least the engineering design. All too often, engineering students are taught how to react to situations, rather than how to anticipate the consequences of decisions and to proceed with them responsibly. This should be regarded as a key point, when defining the learning objectives and outcomes.

Several areas have attracted a considerable attention at the university level, for example, management of the environment in a broad sense. All engineering processes, projects and products, starting with the extraction of raw materials to the final production of goods and generation of wastes, tend to have an impact on the environment, in most cases negative. Energy production and environmental management are fair subjects for sustainability as well. Various actions could be taken for the limitation of greenhouse gas emissions in the production of energy. Related choices pertain to the use of alternative energy sources, the approach to building design and construction, the production of low-emission vehicles, the rational energy use and effective use of material resources.

Sustainable consumption and consumer's protection in a broad sense, including health and safety issues, and the provision of timely and useful public information are other important areas, which have attracted the attention at the university level. Sustainable consumption is defined by the United Nations Environmental Programme as "the use of services and related products, which respond to basic needs

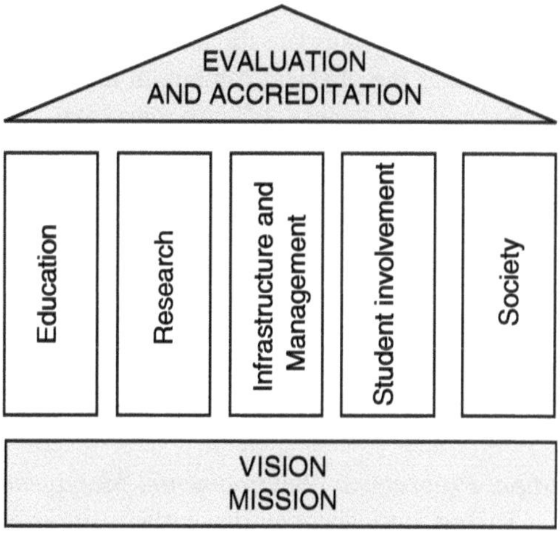

Fig. 1 The five roles of the university

and bring a better quality of life while minimising the use of natural resources and toxic materials as well as emissions of waste and pollutants over the life-cycle so as not to jeopardise the needs of future generations".

Corporate social responsibility and ethics in relation to positive or negative consequences and social impacts are fundamental to the current debate over sustainable development. It should be reflected in the choices that must be made by authorities in technology, government, business and industry. It may appear that choices are made in a scientific and technical context, even when they have an impact on the economy, the environment, public health and societal conditions.

Kaunas University of Technology has undertaken some successful actions at the levels mentioned above and in all five roles (see Fig. 1). The education was constructed and related to the following contributing areas for sustainability activity: the university as a whole, education and curriculum, student involvement, research and innovation in cooperation with industry and society. At the university level, the following strategic principles have been applied:

- University authorities view social responsibility and sustainability as integral to the institutional identity and values;
- University management treats sustainability as part of the operational strategy;
- Evidence of good practises, for instance, in energy efficiency, conservation, environmental protection, waste management, sustainable campus etc.;

At faculty/institute level professional development activities include sustainability as a university issue and as an area with distinct approaches for teaching, learning and research:

- Research activities include subjects related sustainability;
- Students involvement and using the campus for learning about sustainability;

- Sharing with students what sustainability issues the university is addressing;
- Adopting appropriate didactics approaches;
- Utilising the physical campus in teaching and training for sustainability;
- Raising awareness of environmental, social and economic dimensions of sustainability,
- Inviting students to consider their own values and behaviour;
- Facilitating students to contribute ideas for improving sustainability on campus;
- Measuring, evaluating sustainability achievements.

3 Education for Sustainable Industries

M.Sc. Programme "Environmental Management and Cleaner Production"

Taking into account the earlier discussed issues and associated challenges, technical universities of the Baltic Sea region, in the framework of the BALTECH consortium, on an initiative of the KTU Institute of Environmental Engineering (APINI), decided to develop and implement a new M.Sc. Programme in environmental management and cleaner production based on an integrated approach of industrial ecology towards current and long term/strategic environmental issues, focusing on technologies and concepts in environmental planning and management for a sustainable industrial development. This is a 2 year (120 ETCS credits) programme suitable for graduates with qualifications in many engineering fields, such as chemical engineering, energy and electrical engineering, mechanical engineering, civil engineering, environmental engineering and the others. The Programme started at Kaunas University of Technology (KTU) in September 2002.

The programme development was driven by the fact that the number of companies applying different preventive environmental measures is growing, there is an increasing demand for engineers with high competence in technologies for cleaner production and integrated environmental management. The presented M.Sc. Programme makes it possible to satisfy this demand.

The contents of the Programme are based on an industrial ecology approach i.e. on industry—environment interactions to aid industry in evaluating and minimising the impacts on the environment. The Programme courses reflect one of the most important concepts of industrial ecology which, like in the biological system, is rejecting the concept of waste. The Programme covers technologies in coping with industrial residues, particularly those aimed at reuse and recycling (Environmental Technologies course); identifying, evaluating and implementing technical and managerial options for improving environmental and economic performance (Environmental Assessment, Cleaner Production and Environmental Management courses); design of industrial processes and products from dual perspectives of product competitiveness and environmental impact (Eco-design course), and development of policy framework which provides appropriate incentives for enterprises to adopt preventive environmental management practices and to increase their efficiency (Environmental Policy, Law and Economics courses).

The compulsory courses of the Programme cover all basic aspects of an industrial ecology approach. Optional courses are used to discuss these issues in more detail and to provide additional knowledge to ensure that graduates of the Programme will be able to apply an industrial ecology approach, i.e. they will be capable both to conduct systematic analysis of industrial activities and to find optimal solutions for many problems related to sustainable industrial development (Staniškis 2012).

Ph.D. Programme "Environmental Engineering and Landscape Management"

The first Ph.D. programme in Environmental Engineering and Landscape Management at Kaunas University of Technology started in 1994. The main research and education topics of the programme were initially related to environmental technologies. Fast development of preventive concepts and the idea of sustainability gradually led to the change of priorities, contents and structure of the Ph.D. education. Therefore, a new interdisciplinary Ph.D. programme in Environmental Engineering and Landscape Management was launched in 2012. The programme is designed to educate researchers and university teachers in engineering, social, economic and natural science disciplines that underpin sustainable development.

The research results presented in the programme's doctoral dissertations are based on integrative, interdisciplinary research that is needed to explore science and policy issues in the area of sustainable development. Integrated assessment methods and concepts (e.g. transitions, modelling and scenario analysis) are instrumental to provide answers to the central questions of sustainable development.

The objectives of the courses are:

- To familiarise students with the main contemporary social, economic and environmental theories and concepts related to sustainability science, policy and practice;
- To familiarise students with the challenges, constraints and interactions of the academic definitions, decision making tools and practices of the principles of governance, sustainable development and sustainability science;
- To provide students with skills to perform a quantitative/qualitative analysis of 'sustainable development issues' by using a set of 'Integrated Assessment' methods and tools;
- To enable students to conduct original research in the field of sustainability science, policy and practice by solving existing problems, modifying existing approaches, identifying ways to obtain more relevant empirical evidence, developing new concepts/models and deriving new policy implications.

The Ph.D. candidates have opportunity to choose from a selection of courses offered by KTU and other Universities involved in the Programme. For each Ph.D. student a tailor-made programme is designed.

4 Evaluation and Accreditation "Engineering Education for Sustainable Industries" by QUESTE-SI

This is not the standard accreditation program, where institutions or programs either pass or fail. QUESTE-SI is concerned with development in an area that is large, diverse, evolving and marked by efforts at many levels from the student and a new graduate to departments, institutions and, not least, industries. This prime objective can be met in many ways and to varying degrees. The starting point here is the setting of credible teaching and learning objectives for sustainability.

Overall objective: Sustainability efforts proceed from a visible sense of social responsibility and are treated as a strategic concern for an engineering institution, in terms of education, research and engagement with the community.

Criteria are indicators that something has been or should be done to meet strategic objectives. An important indicator is the formulation of strategic objectives, plans and their implementation. The existence of a coherent program is a more significant indicator, especially when viewed in an institutional context.

A centrally-important criterion or indicator is that "the institution visibly recognises the concept of social responsibility as reflected by sustainability activity at the program level." Such commitments may appear in the institutional mission, policy statements and support for program development.

A sustainability strategy for engineering education should first identify the areas in which it will have some effect and rationale on such choices. The depth and breadth of efforts may be defined by the dimensions affected.

There are four dimensions for sustainability efforts. Activity in each Dimension contributes to the overall expression of social responsibility and commitment to sustainability.

- The Institution and Department or Faculty
- Education and Curriculum
- Student Involvement
- Research and Innovation in Cooperation with Industry

Topical target areas: These are topics to which sustainability objectives, actions and outcomes are directed. Topical goals may be translated into coherent programmes concentrations, elective options, or individual courses and projects. However defined, the learning experience should be related to well-defined topical objectives.

5 Indicators

Programmes should be an expression of strategic objectives, with progress measured in terms of strategic fulfilment. A strategic plan should define what will serve as evidence of progress toward the achievement of goals. The indicators shown below pertain to a full programme, but may be adapted or scaled to fit a concentration, option, a cluster of elective courses or a module within an interdisciplinary program.me. Other indicators may be added by planners (QUESTE-SI 2012).

Indicators for the educational dimension:

- There is a rationale for selecting topics, methodologies, tools and techniques. For example, the use of impact studies, multidisciplinary approaches to problem-solving, life cycle analysis, new or "best" approaches to design and new approaches to project-based or problem-led learning.
- Scientific or technical knowledge and practical skills related to sustainability are addressed in curricular planning and student assessment plans.
- New transverse skills are addressed in the teaching/learning plan; e.g. crisis communication, risk management, policy analysis-formulation-implementation, breadth of vision etc. Attention is given to skills in finding and using sustainability research materials. There is evidence of input from industrial experts.
- Learning objectives comprehend new professional profiles. There is consideration of what a graduate should know (with an eye to the future), and again, the breadth of vision of a 'sustainable engineer."
- Context awareness. For example, acquired familiarity with pertinent international standards (e.g., the ISO series). The use of exercises and problems that require a broader understanding of problems, solutions and consequences. Projects done in cooperation with other disciplines, e.g., business and management, informatics and the "pure" sciences.
- The department has taken steps to address sustainability learning objectives, outcomes and competencies in the quality assurance and program review systems.

Indicators for the Institution and Faculty:

- Institutional leaders view social responsibility and sustainability as integral to the institutional identity and values.
- Institutional management treats sustainability as part of the operational strategy.
- The evidence for such views may include minutes of meetings, white papers, mission and policy statements, reports.
- Faculty professional development activities include sustainability as an institutional issue and as an area with distinct approaches for teaching, learning and research.
- Evidence of good practices; e.g., in energy efficiency, conservation, environmental protection, waste management, safety and health of staff and students etc. Normally, such practices are accompanied or followed by documentation.
- Research activities include subjects related to sustainability

The Student Dimension:

- Social responsibility and sustainability are presented to students in attitudinal and cultural terms. Student awareness of these issues is cultivated through a student advising and information process that reaches students individually or in groups.
- There is a productive dialogue between the engineering school, its departments, employers and practitioners concerning sustainability and the aptitudes, attitudes and capacities sought in candidates for employment. This produces information that is communicated to students and applied through the teaching and learning plan.
- There is encouragement and support for student activities inside and outside the institution; e.g., participation in cleaning, remediation, restoration and conservation campaigns; outside involvement in relevant NGOs, networks, community organisations and humanitarian efforts; student field work and projects. Such activities may complement, but are not part of the curriculum.
- The institution or engineering school tracks the job placement of graduates and actively seeks their feedback on work-related experiences that bear on sustainability.

Research and Innovation in Cooperation with Industry:

- The engineering school and department(s) have strong and active linkages with industry, the profession, technological business.
- The engineering school, department(s) and relevant faculty are significantly involved in associations, organisations and professional or technical bodies that are actively committed to sustainability.
- There is an operational strategy for sustainability-related research, development and innovation projects in partnership with one or more industrial or business firms and, as appropriate, peer institutions, research institutes and suitable organisations.
- The results of sustainability research and cooperative development efforts are used to enrich the teaching and learning process.
- The plan for sustainability research, innovation and cooperative development is sustainable in itself. There is adequate research and project management support and the capacity needed to secure grants and subsidies.

QUESTE-SI is not about accreditation, nor is it a pass/fail proposition. The issue is development with respect to social responsibility and sustainability education. The self-assessment calls for ratings of progress or the application of certain principles, response to specific questions and succinct descriptions of achievements, strengths and areas under development. This will help us focus on significant choices, actions and results, as we prepare for the site visit.

In a developing and evolving field, it is difficult to define what amounts to a pass or fail, which misses the point when the real issues are strategic, attitudinal and developmental. Sustainability has many derivatives, more than will fit into a conventional accreditation scheme. Thus, QUESTE-SI is more interested in evidence of strategic progress marked by well-considered choices, careful planning, action and measurable results. It is assumed that university has identified a unit/faculty that is its best or most developed model of sustainability education.

6 Results and Discussion

For the pilot evaluation and accreditation, one technical university from each EU country was invited to participate. Eleven of them have participated in the final stage of the process:

- KTH Royal Institute of Technology, Department of Industrial Ecology, Sweden,
- Brno University of Technology, Faculty of Architecture, Faculty of Business and Management, Faculty of Electrical Engineering and Communication, Faculty of Information Technology, Faculty of Mechanical Engineering, Czech Republic,
- Wroclaw University of Technology, Faculty of Computer Science and Management, Poland,
- Kaunas University of Technology, Institute of Environmental Engineering, Lithuania,
- Vienna University of Technology, Continuing Education Center, Austria,
- Instituto Superior Tecnico, Mechanical Engineering Department, Portugal,
- Czech Technical University in Prague, Faculty of Mechanical Engineering, Czech Republic,
- Riga Technical University, Institute of Applied Computer Science, Latvia,
- Telecom Bretagne, General Engineering Programme, France,
- Cyprus University of Technology, Department of Civil Engineering and Geomatics, Cyprus and
- European University—FYR of Macedonia, Faculty of Informatics, Republic of Macedonia.

Three best evaluated institutions were KTH Royal Institute of Technology (Department of industrial Ecology), Kaunas University of Technology (Institute of Environmental Engineering) and Brno University of Technology (Faculty of Architecture).

In traditional engineering education much is said about response to needs or demands—those of employers, industry and the marketplace. Such needs or demands, when focused on the acquisition of specific Scientific and technological knowledge, practical skills and competencies may result in a concise description of a limited set of learning objectives for assessment. Today there is a growing

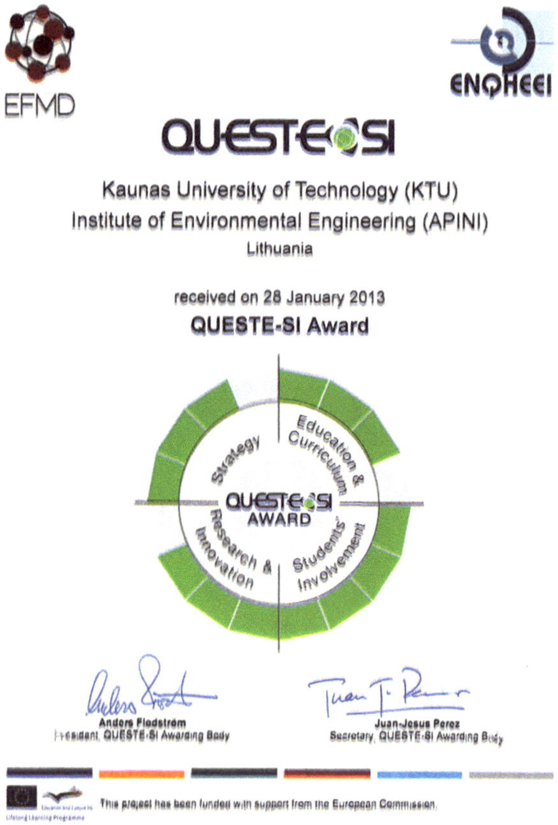

Fig. 2 QUESTE-SI certificate

concern about responsibility for decisions that may have a negative impact on society, the environment and resources.

Kaunas University of Technology has undertaken some successful actions at the levels in all four roles: education, research, infrastructure and management and society. The QUESTE-SI project has developed along three axes:

- Share best practice between Business and Scientific and Technical education using the examples of EQUIS (the European Quality Improvement System for business school), exchanges with ENQA (the European Association for Quality Assurance in Higher Education) and ABET (the Accreditation Board for Engineering and Technology, USA);
- Set up referentials in Scientific and Technical higher education for sustainable industry;
- Develop, implement and evaluate a quality improvement system.

The QUESTE-SI evaluation focused on the institutional unit (department) that is responsible for one or more programmes. All submitted programmes should at least be at formal or implementation stage. A key point is to assure that each graduate should have learned the sustainability aspects related to the concerned education domain. A fair evaluation of social responsibility and sustainability education is not limited to teaching and learning methods or curricular content—it depends upon parallel efforts in all dimensions.

The project has surveyed various networks of academic institutions, rankings, quality assurance systems, models to enhance the development of the triangle of knowledge to identify the awareness of and commitment to sustainability like strategic approach to sustainability, including sustainability in the study programmes, involvement in KICs of the European institute of Innovation and Technology, membership in certain networks etc. (QUESTE-SI 2012).

Kaunas University of Technology was presented by the Institute of Environmental Engineering with the M.Sc. programme "Environmental Management and Cleaner Production" and Ph.D. programme "Environmental Engineering and Landscape Management". The Institute of Environmental Engineering is awarded by the highest ranking and has become a fertile basis for a larger pilot project. Programmes and staff experience facilitated the development of the university itself as a model for sustainable operations, research and study (see Fig. 2).

7 Conclusions

In many countries around the world, education for sustainable development (ESD) is understood as integration of sustainable development into education, research and operations as a starting point in the strategic policy of the university. In other words, university has five roles in society. First, university acts as an educational institution and its impact on society is to guide and assist students with their learning processes and thus deliver sustainably educated professionals. Second, university is a research institution and from this perspective it delivers the results of fundamental or applied scientific research to society. Third, university is an organisation which as all organisations has all kinds of operational interactions with the outside world, like procurement, the employment of staff members, the use of materials, energy and water, the transport of people and goods, the production of waste, air pollution and other ways of environmental impacts. The fourth role is its direct interaction with society, for instance, participation with the local community in the implementation of joint local Agenda 21 projects, cultural events etc. The fifth role is the efforts of the institution to engage students and graduates in the culture of social responsibility and sustainability.

One of the main barriers to ESD is the university organisation with its sharp division between faculties and departments where the various courses are "owned" by the separate units. To avoid the risk of internal conflict and extra costs of changes, which could happen during the integration of sustainable development principles into a course, departments are generally very reluctant to change curricula. Secondly, university lecturers are supposed to be active scientists, working on the scientific frontline and this implies that disciplinary approach is perceived as essential. Interdisciplinary work is only perceived as important for applied projects, not for scientific progress, as it does not contribute to the conceptual core of the discipline. This creates a strong barrier for the introduction of sustainable development in the education. Scientists and lecturers often interpret sustainable development too narrow, often too specifically focusing on only one of three dimensions of sustainable development, mostly the environmental aspect in the case of technical disciplines.

ESD is an increasingly important feature of higher education which poses new challenges for teaching and learning. There are two options: ESD might focus entirely or mainly on developing new education, if this type of education is completely absent or ESD has to be to change the existing education. On the one hand, sustainable development itself is a field of study that is complex by its very nature, while, on the other hand, the systems of education are also highly complex. So, it is no surprise that education for sustainable development appears to cause many barriers and resistance. The process of such education evaluation and accreditation at the Institute of Environmental Engineering has identified the following success factors:

- Sustainability has a strong presence both in the Institute's teaching and learning and research and innovation strategies. Teaching and learning activities explore themes as environmental impact, social and economic wellbeing, volunteering, community involvement and corporate social responsibility.
- Education for sustainable development is achieved through adoption of appropriate pedagogic approaches which promote collaborative learning, discovery and problem based individual work with real problems etc.

Strengthening links between education and campus management promote greater understanding of the real challenges for sustainability implementation. Student involvement in campus projects and management provides the potential in using performance data for teaching statistics, well-being and inclusivity in the social sciences and procurement in the business disciplines.

The Research has shown that to perceive sustainability uni-dimensionally in relation to the environment at the expense of social and economic dimensions is a wrong approach. Encouraging students to learn about sustainability in its broader sense by involving them in appropriate research projects enables them to develop their awareness of different dimensions as well as interdimensional connections.

The measurement of resources and activities in the area of sustainability by using internal and external indicators is increasingly important. Students' involvement in monitoring increases their awareness and possibilities to engage them in campus sustainability management.

Finally, effective implementation of education for sustainable development needs adoption of sustainability as a university-wide agenda which links curriculum, research, campus, community and culture.

References

Barrett CL (2012) Preparing teachers to educate for sustainable development. In: Leal Filho W (ed) World trends in education for sustainable development, Vol 32. Peter Lang GmbH, New York p 425–440

De Ciurana A, Leal Filho W (2006) Education for sustainability in university studies. Int J Sustain High Educ 8(1):53–68

Denton HP (2011) Making better choices today: ethics, technology and sustainability. In: Leal Filho W (ed) World trends in education for sustainable development, Vol 32. Peter Lang Gmbh, New York, p 219–234

Krizek K, Newport D, White JA, Townsend A (2012) Higher education‘s sustainability imperative; how to practically respond? Int J Sustain High Educ 13(1):19–32

QUESTE–SI (2012) Kit for advisers, quality referents and auditors (Prepared by Tom Philips and Stanislaw Butkowski). EFMD. p 33

Sammalisto K, Lindhqvist T (2008) Integration of sustainability in higher education: a study with international perspectives. Innov High Educ 32(4):221–233

Sibbel A (2012) Bridging the gaps to develop graduate capabilities for sustainability. In: Leal Filho W (ed) Sustainable development at universities: New Horizons, Vol 34. Peter Lang GmbH, New York, p 70–84

Staniškis J, Arbačiauskas V (2002) Sustainable industrial development. In: Review of progress in the process of sustainable development in Lithuania, Vilnius

Staniškis JK (2012) International M.Sc. Programme environmental management and cleaner production: Success Story//Chapter 21 in sustainable development at Universities: New Horizons. Peter Lang GmbH, New York, Frankfurt am Main. ISBN-978-3-631-62560-6, p 247–261

UNESCO (2005) United Nations decade of education for sustainable development (2005–2014) International Implementation Scheme. UNESCO, Paris

Velazquez L, Munguia N, Esquer J (2011) Engineers for more sustainable industrial engineering. In: Leal Filho W (ed) World trends in education for sustainable development, Vol 32. Peter Lang GmbH, New York. p 173–184

Zint M (2011). Evaluating education for sustainable development programs. In: Leal Filho W (ed) World trends in education for sustainable development, Vol 32, Peter Lang GmbH, New York, p 328–347

Author Biographies

Prof. Dr. hab. Jurgis Kazimieras Staniškis is a director of the Institute of Environmental Engineering, Kaunas University of Technology, full member of Lithuanian Academy of Sciences. Active for over two decades on sustainable development strategies and their application in industry, universities and governmental institutions in Lithuania, Central and Eastern Europe, Africa, Latin America and Asia. He is an author of 12 books, more than 250 publications and 140 inventions.

Prof. Dr. Eglė Katiliūtė is a professor at the Department of Management, School of Economics and Business, Kaunas University of Technology. Her research interests are sustainable university, sustainable development, quality management systems, higher education quality research, education policy research methodology and interdisciplinary research. She is an author of more than 40 publications.

Sustainable University for Regional Development: Quality Management Model that Integrates Employer and Social Partner Attitudes

Eglė Katiliūtė and Jurgis Kazimieras Staniškis

Abstract

Sustainable development, promoting responsible, environmentally friendly conduct and formation of ecological culture is one of the universities' priorities in the 21th century. The position of the universities is to ensure that new knowledge and technologies serve to improvement of environmental well-being. This paper presents a model of how university quality management system integrates the key elements of sustainable university to support regional development. A special emphasis is given to employer and social partner's attitudes and how they recognize the impact of university activities on regional and national development. The model is based on the experience of Kaunas University of Technology in improving its performance in the area of sustainability. The University has understood the importance to reorganize University's activities and strengthen the cooperation with partners for the unity of economic, environmental, social and cultural objectives and values, to identify the most important development issues of the region and country. The University emphasizes the necessity to constantly develop a systematic education and consulting for companies, organizations and business, and to support activities within the University that are responsive to the problems of sustainable development of the region and country including quality of life.

E. Katiliūtė (✉)
School of Economics and Business, Kaunas University of Technology,
K. Donelaičio St. 73, 44029 Kaunas, Lithuania
e-mail: egle.katiliute@ktu.lt

J.K. Staniškis
Institute of Environmental Engineering, Kaunas University of Technology,
K. Donelaičio St. 73, 44029 Kaunas, Lithuania
e-mail: jurgis.staniskis@ktu.lt

W. Leal Filho (ed.), *Transformative Approaches to Sustainable Development at Universities*, World Sustainability Series, DOI 10.1007/978-3-319-08837-2_6

Finally, some specific recommendations on how universities can increase the effect of regional development via involvement of their faculties, students and staff in sustainable development programmes and processes are outlined.

Keywords
Sustainable university · Sustainable development · Quality · Management · Regional development

1 Introduction

University faculties and administrators are important actors for regional development processes, because they provide their expertise and contribute to economic and human well-being. Universities working for sustainable development have three significant components: (1) universities must promote greater interaction between the political and private sectors by creating and developing knowledge through research and having a multidisciplinary approach; (2) universities must introduce sustainability to all of the education that they provide. In the future, students from all disciplines should be educated in sustainability to make sure that the society benefits from the knowledge they bring; (3) universities must transfer knowledge and collaborate fully with society to make sure that society takes into account the social, economic and ecological dimensions of sustainable development (Goldstein 2009; Lehmann et al. 2009; Sedlacek 2013).

The lack of scientific data in regards to the university quality management systems that integrate the key elements of sustainable university effects to regional development leads to the following question: what are the links between sustainable university quality management system and university effects on regional development?

The aim of the paper is to identify the theoretical links between sustainable university quality management system and university effects on regional development, and to create the conceptual model on how university quality management system need to integrate the key elements of sustainable university.

The paper is organized in 4 sections. Section 1 provides the theoretical background and is structured according to the different roles of universities. In Sect. 2 the applied methods and proposed analytical framework are discussed. Section 3 presents the case of Kaunas University of Technology (KTU) and research data from the social partners as well as entrepreneurs' survey. The final section outlines a conclusion based on the main findings.

2 Theoretical Framework

While the role of education in sustainability processes is clear, the ways to approach this issue are evolutionary. Some scholars (Warburton 2003; De Haan 2006; Shephard 2008; Svanström et al. 2008) recognize education for sustainable development as an interdisciplinary and holistic approach, while others discuss the values-driven approach (Kopnina 2012). Sterling and Scott (2008) recognize that higher education contribution to sustainable development process is achieved through its major functions of research and teaching. However, sustainable development process touches higher education as an institution with existing management and academic structures. Accordingly, university role in education for sustainable development processes becomes more than teaching and research alone. Tilbury's (2011) research based on 200 articles shows that education for sustainable development practices covers "processes of collaboration and dialogue (including multi-stakeholder and intercultural dialogue); processes which engage the 'whole system'; processes which innovate curriculum as well as teaching and learning experiences; and processes of active and participatory learning" (Tilbury 2011, p. 7).

Suggestions on the role of universities and sustainable development to tackling these issues include: "a change in the universities' own management practices, for instance their involvement in recycling schemes, energy efficiency initiatives, or the implementation of an environmental management systems; promotion of integration, synthesis, critical reasoning, and system-thinking skills, supporting students and researchers beyond skill development to cope with the future multidisciplinary complex challenges of sustainability; the assumption of a leading role in coordinating, promoting, and enhancing the engagement of local authorities and other societal stakeholders to design and implement regional sustainability plans by acting as sources of technical expertise; and a new research and teaching agenda for universities as centres of development of the sustainability science as an innovative scientific field defined by the problems it addresses" (Karatzoglou 2013, p. 45).

University role in sustainable development typically includes management practices such as recycling or energy efficiency initiatives, support to students to develop multidisciplinary knowledge of sustainability, act as technical experts in dealing with regional sustainability issues, etc. It grounds university responsibility to come up visibly accountable to stakeholders and wider society. Jongbloed et al. (2008) state, that legitimacy urgency and power of higher education to society could be analysed as a level of the higher education institutions' commitment to its stakeholders.

3 Methods

The research methods applied in this study include the analysis of research literature with the purpose to reveal the characteristics of sustainable university for regional development and a case study. A case study is defined as a qualitative strategy of inquiry "which involves an empirical investigation of a particular contemporary phenomenon within its real life context using multiple sources of evidence" (Robson 2011:163; Seadlacek 2013:78). For the case study, the analysis of KTU documents and the questionnaire based survey of employers and social partners have been used. The case study is based on the following criteria: university with the vision and mission to sustainable development.

The empirical survey aimed to reveal the key impact features of sustainable university for regional development in the viewpoint of employers and social partners.

The design of the questionnaire was based on separate questions disclosing what factors are important for collaboration among the university, employers and social partners.

To evaluate the factors, which disclose the university sustainability manifestation, 4 points Likert scale was used (from "1"—as do not agree at all, to "4"—totally agree). More than that, the respondents could point out that they do not have opinion on a given question. Actually, just several respondents pointed out that they do not know whether they agree or not.

The initial step of the survey analysis was grouping of answers. In analysis of the most manifested factors, the answers with "3" (agree) and "4" (totally agree) have been joined. The factors with very low manifestation ("2") or ("1") were added together in the findings of the survey.

Methods of descriptive statistics have been used in data analysis (using SPSS 16.0). The survey data was collected in the end of 2012 (November–December). The questionnaires were prepared in an electronic form. The total sample consisted of 308 employers from companies of Lithuania and 100 social partners of KTU.

Since the focus of this research has been the university quality management system, we shortly introduce to its context. As KTU has been conducting a very large number of operations under contracts with companies/organisations, there was a necessity to implement quality management system for scientific and technological research activities. In 2000 quality management system in accordance to the international standard ISO 9001 has been developed and implemented in the university. The first certificate was given by the State of Nuclear Power Safety Inspectorate. For the years 2000–2006, four testing laboratories have been certified. In 2007 this system has been licensed by the KTU and carried out for the metrological monitoring and inspection activities. In 2004 the System of Internal Study Quality Assurance has been implemented and approved by the Senate of KTU. This system ensures the internal study quality. Quality management system at KTU is adapted to the learning process (Guidelines for the application of ISO 9001:2000 in education). Some elements are adopted from scientific and technological research quality assurance system (for example, contract management, infrastructure management, procurement, etc.). Quality management system at KTU covers all university's operations. The normative acts have been developed and adopted for each activity in the system (the Senate resolutions, the Rector's orders, regulations, rules, procedures, lists of forms, etc.).

The new strategy of KTU approved in 2012 was aimed at the enhancement of the University's responsibility to society and the country, by consolidating its activities to the improvement of human life quality and strengthening the development of statehood. The main actions in this area are enhancement, formation and delivery of the University's contribution to the country's vitality and sustainable economic, social and cultural, knowledge-based development. The University's strategic activities reflect the mission of the University with a focus on research-based studies at international level to create and transfer knowledge and innovative technologies for sustainable development of the state and development of innovations.

Strategic development priorities of the impact on the region and the country are realized through the University's strategic programmes and objectives:

1. Student competence development, insurance of their self-expression and a successful career. Goals of the University: to ensure readiness to study and availability of studies for the Lithuanian citizens from all country regions and across different social strata, to provide conditions for foreign students to study; to provide informal education services widely and continuously for individuals of all ages and experience seeking to improve or retrain.
2. Development and transfer of internationally recognized expertise and technology. Goals of the University: to develop the trans-domain and interdisciplinary research in physical, biomedical, technology, social international level sciences and humanities and create knowledge significantly affecting the society and economy; to focus the research and experimental development on the demands of industry, business, and social and cultural development; to promote the development of high international level research centres.
3. Concentration of University's activities on human well-being and sustainable development of the country. Goals of the University: to develop activities of the University and co-operation with partners aiming at the unity of economic, environmental, social and cultural goals, to solve relevant problems of the city, the region and the country in economic, social and cultural development issues on the basis of sustainable development principles, to educate and advise the public; to organize lifelong studies, which promote socially and morally correct, ethically acceptable sustainable development and economic growth preserving the country's natural resources.

The University's Statute lists the University aims and objectives emphasizing the impact on the region and the whole country in educating the public and promoting its humanitarian, technological and engineering culture, moral values, civic consciousness and citizenship, equal opportunities, creating a diverse learning environment in which individual characteristics—sex, age, nationality, social status, disability or ethnic origin, religious beliefs, political views—are recognized and valued, fostering the identity of the nation and the state, the Lithuanian scientific and technical terminology, the European humanistic traditions, linking fundamental research and applied research of the University with the country's economic needs; joining the common European Research Area; promoting international integration of science, developing relationships with the global academic community, disseminating Lithuanian cultural identity, protecting and nurturing cultural heritage and raising society's awareness, disseminating ideas for development of science, engineering, technology and innovation, popularizing advanced technology, developing patent and licensing activities, educating society to accept science and knowledge, ability to compete in the market with a high level technologies, products and services; contributing creative and intellectual input to the country's economic, social, and economic development, while maintaining close ties with industry, business representatives, national and local governmental authorities.

The University as an institution of higher education organizes studies and scientific activities in a constructive and open dialogue with the University's community. Popularization of science is not included in the structure of formalized work relations but is closely associated with this activity and its results.

Sustainable development, promoting responsible, environmentally friendly conduct and the formation of ecological culture is one of the University's priorities. The position of the University is to ensure that new knowledge and technology must serve to society and improvement of the environmental well-being. The University's Sustainable Development Strategy and activities in 2012 were positively recognized internationally during assessment of the European Union—QUESTE-SI accreditation. The University has become a member of the Global Compact agreement. In 2012, the University launched a coordinated systematic waste sorting project, led by the Institute of Environmental Engineering (APINI) of KTU (Staniškis 2012).

Unconventional education promotional events have a major impact on society. Therefore, in 2012 the University celebrated the traditional scientific festival "Spaceship Earth", and traditional popular science events "Night of the Nights" and the "Researchers Night". During the latter event, KTU lecturers delivered 30 different types of lectures for society, conducted more than 70 workshops, experiments and demonstrations aimed at promoting research and studies on environmental and green energy topics.

The University lecturers and researchers lectured in many Lithuanian schools, gymnasiums and the KTU Children University and the KTU University of Pupils. During the summer semester of 2012, 5 lectures have been delivered at the KTU Children University operating in the progress centre (attended by 591 children), 19 classes have been delivered in the autumn semester (attended by 716 children). Even more listeners attended the KTU University of Pupils. In spring semester 48 lectures have been delivered (attended by 975 pupils) and in the autumn semester 28 classes involved 569 pupils.

Despite of experience ant best practices of the University, it is important to ensure that these good initiatives become common routines and systematic practices in the university quality management system. This model will be based not only on theoretical framework, but also using research data, employers and social partners' survey results.

4 Results

Knowledge transfer has become a strategic issue and covers various direct (i.e. collaborative research projects, intellectual property rights and spin offs, labour and student mobility, consultancy, etc.) as well as indirect or 'soft' activities (i.e. attendance at conferences, creation of electronic networks, etc.). To establish the cooperation areas between the University and employers, to ensure closer

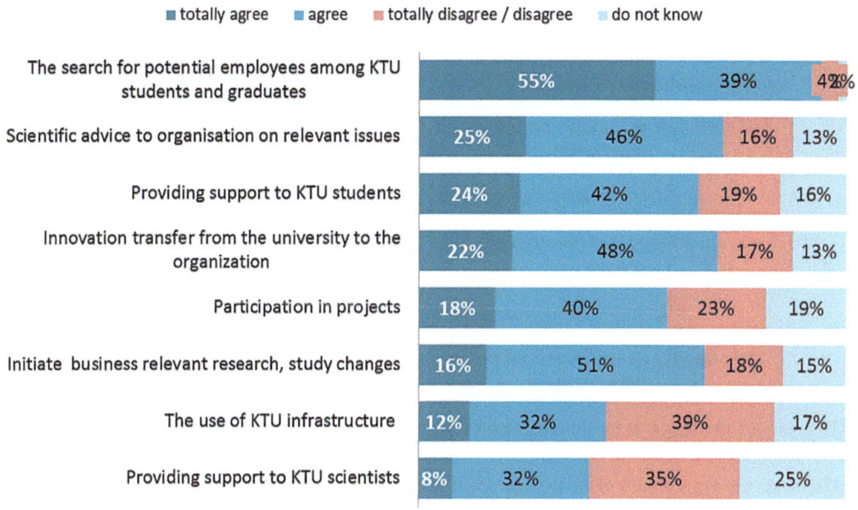

Fig. 1 Incentives and reasons for cooperation with the university (% from all respondents, n = 308)

interaction between the two partners in strengthening the constituents of the knowledge triangle—research and development, innovation and studies—the University carried out the surveys of employers in 2012 (Fig. 1).

The summarised employers' survey results: (1) The main determinant of a partnership with the University is the search for potential employees among KTU students and graduates (55 % of employers completely agreed); (2) Among the most rarely mentioned reason for partnership is the use of infrastructure (12 %) and providing support to KTU scientists (12 %); (3) The most important areas of cooperation for employers—employment (34 % of employers completely agreed), while the least important—cooperation in the sphere of support (5 %); (4) The major obstacles for closer cooperation with the KTU are the lack of a clear mechanism for co-operation (67 % of employers completely agreed) and a lack of communication between KTU and organizations (54 %); (5) The main tools that enable organizations to improve cooperation with KTU: KTU students visits (17 % of employers completely agreed) and regular KTU teachers and researchers visits to the organization (13 %), KTU activities, that are relevant to the organization, publicity (15 %) and informal KTU and organizations representatives meetings (13 %); (6) The biggest benefit of cooperation with KTU: increasing the organization's staff expertise (19 % of employers completely agreed), increasing the awareness of the organization in Lithuania (11 %), facilitating the transfer of knowledge to the organization of KTU and vice versa (10 %); (7) According to the employers, KTU students have sufficient theoretical and professional knowledge of the general university education (16 % of employers completely agreed), but they lack the practical skills (5 % of employers completely agreed); (8) Essential skills of a potential employee: specialized skills (58 % of employers completely agreed),

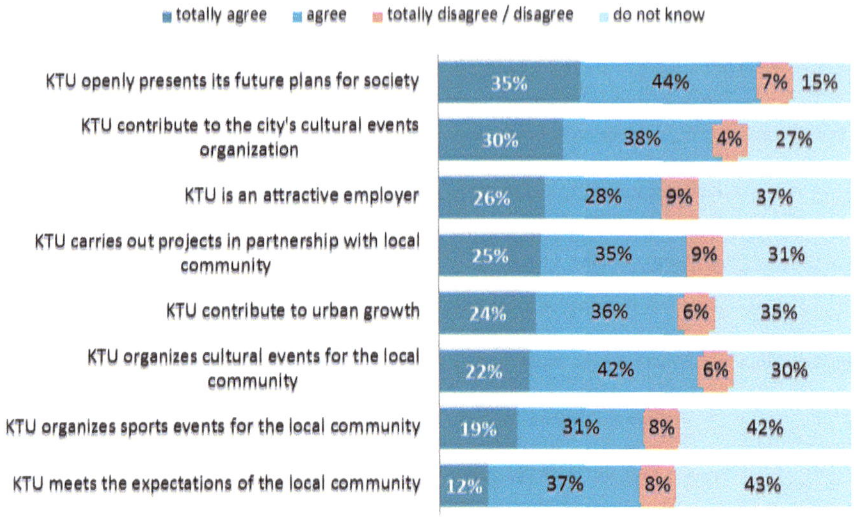

Fig. 2 KTU current role in the local community

teamwork (41 %) and communication skills (40 %). The least often employers take care about career management skills (14 % of employers completely agreed) and sustainable development competencies (12 %).

The results of social partners' survey demonstrated that 35 % of the University's social partners fully agree with the statement that the University openly presents its plans for the future to the public, 30 % agree with the statement that the University contributes to the city's cultural events. Somewhat smaller share of social partners fully agrees with the statement that the University is an attractive employer (26 %) and is engaged in the projects together with the local community (25 %), contributes to the city's economic growth (24 %) and organises cultural events for the local community (22 %) (see Fig. 2).

The least often social partners agree that KTU organizes sports events for the local community (19 %) and that KTU meet local communities' expectations. In most cases the social partners cooperate with KTU in the project activities ("often" and "average"—25 %) and the enrolment of students in practice (17 %). Three of the four social partners have never participated in the final thesis defence committees (76 %) and the evaluation of KTU study programmes in terms of the needs of the public (75 %). In general, cooperation between the social partners and KTU is limited to certain areas.

The summarised social partners survey results reveal that: (1) Social partners cooperate with KTU mostly in the project activities ("often" and "average"—25 %) and during the enrolment of students in practice (17 %), almost never—take part in the final thesis defence committees ("never": 76 %) and evaluate KTU degree programmes in terms of the needs of the public (75 %); (2) From the point of view of the social partners, the following partnerships are the most important: participation in project work ("completely agree" and "agree": 67 %) and KTU students practice (63 %). The least important form of partnership—participation in the final thesis defence committees ("completely agree" and "agree": 41 %); (3) Participation in project work and KTU student's admission to practice are the strengths of KTU and the social partner cooperation. The greatest potential lies in combining participation in the system of student internship programmes and developing new study programmes in KTU; (4) In terms of the directions in which the social partners are foreseeing the targeted cooperation, most respondents focused on education, knowledge exchange and collaboration.

5 Model and Recommendations

The quality management system of the University is based on the Bologna process legislation. This process is designed to introduce a system of academic degrees that are easily recognisable and comparable, to promote the mobility of students, teachers and researchers, to ensure high quality teaching and incorporate the European dimension into higher education. The Prague Communiqué (2001), Berlin Communiqué (2003), Bergen Communiqué (2005), London Communiqué (2007), Leuven-Louvain de Nueve (2009), Budapest-Vienna declaration (2010) are the part of quality assurance system of the University.

The standards and guidelines of European higher education quality assurance (2009) are based on a number of basic quality assurance principles, both internal in and external to higher education. These include: (1) providers of higher education have the primary responsibility for the quality of their provision and its assurance; (2) the interests of society in the quality and standards of higher education need to be safeguarded; (3) the quality of academic programmes need to be developed and improved for students and other beneficiaries of higher education; (4) there need to be efficient and effective organisational structures within which those academic programmes can be provided and supported; (5) transparency and the use of external expertise in quality assurance processes are important; (6) there should be encouragement of a culture of quality within higher education institutions; (7) processes should be developed through which higher education institutions can demonstrate their accountability, including accountability for the investment of public and private money; (8) quality assurance for accountability purposes is fully compatible with quality assurance for enhancement purposes; (9) institutions should be able to demonstrate their quality at home and internationally; (10) processes used should not stifle diversity and innovation.

European standards and guidelines for internal quality assurance within higher education institutions: (1) policy and procedures for quality assurance; (2) approval, monitoring and periodic review of programmes and awards; (3) assessment of students; (4) quality assurance of teaching staff; (5) learning resources and student support; (6) information systems; (7) public information.

Quality management model is based on the European higher education quality assurance guidelines and regulations, and the criteria of the Excellence Model of the European Foundation for Quality Management (Fig. 3). Based on the guidelines

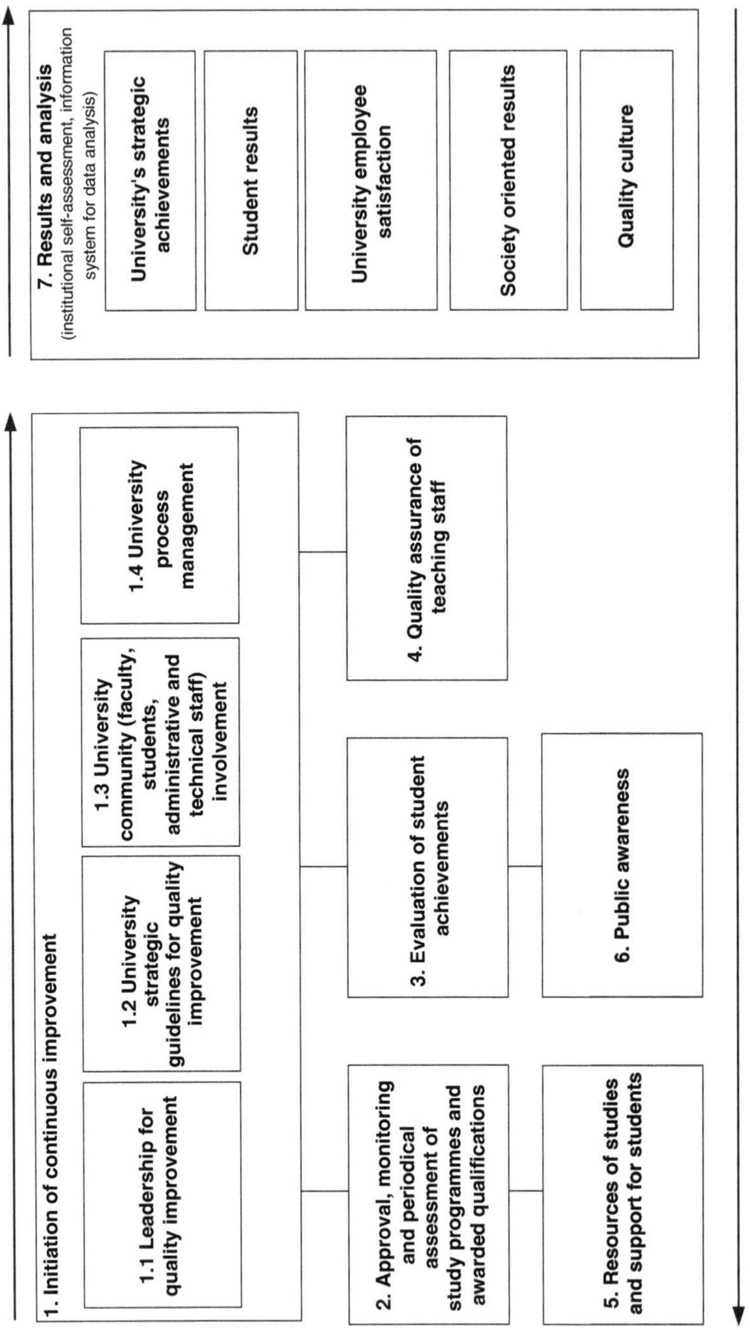

Fig. 3 Sustainable university quality management system model

and regulations, areas for long-term improvement have been identified, i.e. quality assurance; compliance of avoided qualifications with the national and European qualification framework; assessment of curricula quality and student achievement; advancement of teaching excellence; assurance of support to students; data collection and public awareness organization. The Excellence Model helps to explain the procedure for initiating continuous improvement process, i.e. leadership of quality improvement, knowing university strategic guidelines; inclusion of the university's community, and management of the university's processes. The quality policy, which needs to be reviewed every year, identifies relevant areas at university, where development efforts are concentrated on for the respective year (such as the education based on the learning outcomes, academic integrity, and partnership with employers and social partners). In this way, quality management model ensures both the consistency of improvement effort and the necessary flexibility to identify the relevant areas of improvement at the university every year.

Quality policies and procedures. University strategy, Quality system model and quality policy are the main tools to target improvement efforts. University policy needs to be written without duplicating the strategy and quality system model statements. The approved quality policy summarizes the main quality system principles: quality system model and the defined concept of quality in the context of higher education. Current topical areas, that target improvement efforts, need to be identified in partnership with all stakeholders, especially with employers and social partners.

Approval, monitoring and periodical assessment of study programmes and awarded qualifications. Study programme committees or other study management boards must ensure the content and structure quality of learning programmes provided. Annual monitoring and review of curricula have to be conducted. Study modules must be reviewed once a year, while every 2 years the study programme committee or other study management board assesses the description of learning modules. Periodic assessment of study programmes needs to be performed through self-analysis and external assessment. According to the research data, employer and social partner participation in this process is required.

Evaluation of student achievements. Learning outcomes of each study programme, each module and the applied assessment methods need to be published on the university's website. University documents must describe the main procedures associated with the learning processes and have to be publicly available on website. Research data shows that only 12 % of employers are interested in the new employees' sustainability competences, but university needs to integrate these competences in each study programme.

Quality assurance of lecturers' performance. University needs to introduce three-dimensional tuition quality assessment, covering the lecturer's self-assessment, student assessment, opinion of co-workers and managers to ensure and to improve the quality of teaching. Assessment allows lecturers to identify strengths and improvement areas of professional activity individually, and to formulate personal competence development goals. Once a year, all University lecturers can assess their professional performance in accordance to the lecturer self-assessment methodology, covering the analysis of lecturer's planning skills, application of teaching methods, student learning outcome assessment, leadership and other competences. Additional information about the quality of teaching is provided by colleagues lecturers, heads of departments, and deans who may assess: (a) the organization of the subject taught (whether the course content meets the requirements and relevance), (b) effectiveness of communication (whether tasks and goals are clearly defined for students), (c) subject conversance and transfer of knowledge quality, (d) impartiality of examination and assessment; (e) didactic system used by lecturers (whether the applied methods are flexible and promote creativity), (f) student performance and progress (whether the student's work and examination results are consistent with the module/ course objectives). Sustainable university needs to evaluate sustainable education competences of the teachers and to assess the ways they integrate sustainability issues in their modules and evaluate the reached learning outcomes.

Resources of studies and support for students. University's learning environment, which consists of the academic and scientific facilities, residential environment (dormitory premises), information technology infrastructure and recreational resources, needs to be continuously monitored, evaluated and improved, taking into account university student surveys on the satisfaction with the supply of learning resources, based on the university-approved feedback compilation method. University needs to provide opportunities for students to realize their interests in various student organizations, art groups, physical education and sports clubs, etc.

Information systems. Academic information system needs to be applied in the learning process and allow students, lecturers and administrative staff to use the feedback-related information collected in the databases and associated with the studies, students, lecturers; e-learning software and hardware that move the learning process into cyber space. The Library Information System provides access to online education and scientific information resources. Information related to the university's performance and management, education, research and innovation, international relations, etc., is published on university website. The university needs to analyse qualitative and quantitative performance indicators covering the following areas: student progress and learning outcomes, employment of the university's graduates, student feedback on the studies and study programmes, lecturer performance quality, and other data related to students, study infrastructure and teaching materials as well as other university performance results.

Public awareness. Information on the university's web site needs to be continuously updated and adjusted. On its internet website and elsewhere, the university continuously needs to publicise quantitative and qualitative information about study programmes, higher education degrees awarded, scientific (art) activities, self-assessment results, opinion of students, graduates and other stakeholders on the quality of higher education, university performance appraisals conducted by the recognized institutions, graduate career indicators, and other study-related information for public. Public information about areas and programmes of studies includes: degree/professional qualification conferred, programme objectives and expected results, professional career opportunities, enrolment requirements, list of study modules, knowledge and skill assessment procedures, grading requirements and information related to software updates. Along with the information on study programmes, the university needs to publicise statistics about all students and graduates, student and alumni feedback, graduate careers, achievements and innovations, sustainability initiatives and projects.

6 Conclusions

This paper presents a new insight to the expression of sustainability in university quality management system by integrating point of view of employers and social partners. While summarising the theoretical points of the interaction between sustainable university practices and the university quality management system, it might be stressed that development and the use of sustainable university practices by university top management in everyday activities forms new values and attitudes, which are reflected in behaviour of university top management and culture of quality. It changes university sustainability ideas and practices delivered to the community and influences regional and national development.

Quality management model could be based on the European higher education quality assurance guidelines and regulations, and the criteria of the Excellence Model of the European Foundation for Quality Management. Based on the guidelines and regulations, areas for long-term improvement have been identified, i.e. quality assurance; compliance of avoided qualifications with the national and European qualification framework; assessment of curricula quality and student achievement; advancement of teaching excellence; assurance of support to students; data collection and public awareness organization. The Excellence Model helps explaining the procedure for initiating continuous improvement process, i.e. leadership of quality improvement, knowing university strategic guidelines; inclusion of the university's community, and management of the university's processes. In order to emphasize University impact in community grow, university needs to invest more in co-operation with the local community (joint projects, the organization of various events in the community) and its ongoing activities (cultural, sports events, etc.), and to strengthen the contribution of university to urban economic growth.

The quality policy, which needs to be reviewed every year, identifies relevant areas at university, where development efforts are concentrated on for the respective year (such as the education based on the learning outcomes, academic integrity, and partnership with employers and social partners). In this way, quality management model ensures both the consistency of improvement effort and the necessary flexibility to identify the relevant areas of improvement at the university every year.

The scientific insights in regards to the interaction between university sustainability and the quality management system of university prescribe the necessity of further research. Additional quantitative and, especially, qualitative studies based on the theoretical model of the link between university sustainability and the quality management system of the university will provide new knowledge on the interaction of concepts delivered within the model based on employers and social partners' actual collaboration with university that conditions new quality attitudes and values.

References

Bergen Communiqué (2005) The European Higher Education Area–Realising the goals. Bologna Secretariat, European Commission. http://www.enqa.eu/wpcontent/uploads/2013/03/050520_Bergen_Communique.pdf. Accessed 25 Jan 2014

Berlin Communiqué (2003) Realising the European Higher Education Area. Bologna Secretariat, European Commission. http://www.ehea.info/Uploads/about/Berlin_Communique1.pdf. Accessed 15 Jan 2014

Budapest-Vienna declaration on the European Higher Education Area (2010) Bologna Secretariat, European Commission. http://www.ehea.info/Uploads/news/Budapest-Vienna_Declaration.pdf. Accessed 15 Jan 2014

De Haan G (2006) The BLK '21' programme in Germany: a 'Gestaltungskompetenz'-based model for education for sustainable development. Environ Educ Res 12(1):19–32

Goldstein HA (2009) What we know and what we don't know about the regional economic impacts of universities. In: Varga A (ed) Universities, knowledge transfer and regional development. Geography, Entrepreneurship and Policy. Edward Elgar, Cheltenham, pp 11–35

Jongbloed B, Enders J, Salerno C (2008) Higher education and its communities: interconnections, interdependencies and a research agenda. High Educ 56(3):303–324

Karatzoglou B (2013) An in-depth literature review of the evolving roles and contributions of universities to education for sustainable development. J Clean Prod 49:44–53

Kopnina H (2012) Education for sustainable development (ESD): the turn away from 'environment' in environmental education? Environ Educ Res 18(5):699–717

Lehmann M, Christensen P, Thrane M, Jorgensen TH (2009) University engagement and regional sustainability initiatives: some Danish experiences. J Clean Prod 17:1067–1074

Leuven-Louvain de Nueve (2009) The Bologna Process 2020 - The European Higher Education Area in the new decade. Bologna Secretariat, European Commission. http://www.ehea.info/Uploads/Declarations/Leuven_Louvain-la-Neuve_Communiqu%C3%A9_April_2009.pdf. Accessed 15 Jan 2014

London Communiqué (2007) Towards the European Higher Education Area: responding to challenges in a globalised world. Bologna Secretariat, European Commission. https://www.eqar.eu/fileadmin/documents/bologna/London-Communique-18May2007.pdf. Accessed 15 Jan 2014

Robson C (2011) Real world research, 3rd edn. Wiley, Chichester

Sedlacek S (2013) The role of universities in fostering sustainable development at the regional level. J Clean Prod 48:74–84

Shephard K (2008) Higher education for sustainability: seeking affective learning outcomes. Int J Sustain High Educ 9(1):87–98

Staniškis JK (2012) International MSc programme "Environmental Management and Cleaner Production": Success story. In: Filho WL (ed) Sustainable development at universities: new horizons. vol 34, pp 247−260

Sterling S, Scott W (2008) Higher education and ESD in England: a critical commentary on recent initiatives. Environ Educ Res 14(4):386–398

Svanström M, Lozano-García FJ, Rowe D (2008) Learning outcomes for sustainable development in higher education. Int J Sustain High Educ 9(3):339–351

The Prague Communiqué (2001) Towards the European Higher Education Area. Bologna Secretariat, European Commission. http://www.enqa.eu/wpcontent/uploads/2013/03/010519PRAGUE_COMMUNIQUE.pdf. Accessed 15 Jan 2014

Tilbury D (2011) Education for sustainable development: an expert review of processes and learning. UNESCO, Paris

Warburton K (2003) Deep learning and education for sustainability. Int J Sustain High Educ 4(1): 44–56

Author Biographies

Dr. E. Katiliūtė is a professor at the Department of Management, School of Economics and Business, Kaunas University of Technology. Her research interests are sustainable university, sustainable development, quality management systems, higher education quality research, education policy research methodology and interdisciplinary research. She is an author of more than 40 publications.

Prof. Dr. hab. Jurgis Kazimieras Staniškis is a director of the Institute of Environmental Engineering, Kaunas University of Technology, full member of Lithuanian Academy of Sciences. Active for over two decades on sustainable development strategies and their application in industry, universities and governmental institutions in Lithuania, Central and Eastern Europe, Africa, Latin America and Asia. He is an author of 12 books, more than 250 publications and 140 inventions.

Investigating the Relationship of Wellbeing with Sustainable Values, Attitudes and Behaviour

Mark Giltrow

Abstract
The current study investigated the relationship of wellbeing with personal values, attitudes and behaviour. It was found that individuals' with sustainable values and attitudes had higher levels of wellbeing. These relationships were found to be especially strong when wellbeing was disaggregated into its four related but distinct elements, namely, personal feeling, personal functioning, interpersonal feeling and interpersonal functioning.

Keywords
Wellbeing · Sustainable values · Sustainable attitudes · Sustainable behaviour

1 Introduction Wellbeing and Sustainability

Seligman (2002) states that we should "learn how to build the qualities that help individuals and communities not just endure and survive, but also to *flourish*". We flourish when our lives are going well, when we both *feel* good and are functioning *effectively* (Huppert 2009). The *feeling portion* of the concept incorporates not only happiness and contentment, but also such emotions as interest, engagement, confidence, and affection. The *effectiveness portion* of the concept includes developing one's potential, having control over our lives, having a sense of purpose (e.g. working towards valued goals), and experiencing positive relationships (Huppert 2009). As this research area evolves and broadens what started with the roots of

M. Giltrow (✉)
British Columbia Institute of Technology, 3700 Willingdon Avenue,
Burnaby, BC V5G 3H2, Canada
e-mail: mark_giltrow@bcit.ca

W. Leal Filho (ed.), *Transformative Approaches to Sustainable Development at Universities*, World Sustainability Series, DOI 10.1007/978-3-319-08837-2_7

happiness and satisfaction has grown to include positive emotion, engagement, relationships, meaning, purpose and accomplishment. This broader concept has been termed *wellbeing* or *flourishing*.

In 1987 the Brundtland Commission published the report *Our Common Future* (World Commission on Environment and Development 1987), which clearly defined the harm human development was having on our planet, including life that was yet to be born. In 1992, at the first Earth Summit in Rio de Janeiro world leaders co-created *Agenda 21* (United Nations 1993), outlining their plans to address the impacts of development on biodiversity, trade, debt, deforestation, poverty, education, agriculture, desertification, human settlements, and consumption. While little tangible success has come from these meetings, one step forward has been a clearer definition of sustainable development (O'Brien 2008). More recently and in a similar vein, those researching wellbeing have also begun to see the relationship of the human pursuit of happiness on both society and the environment (O'Brien 2008).

There have been efforts to measure the two concepts and move away from our traditional (and overly materialistic) conception of progress—namely the Gross Domestic Product (GDP). Two alternate streams to GDP are the Genuine Progress Indicators and the Well-Being Indicators. Most of the organizations working on genuine progress indicators focus upon environmental and sustainable factors. In addition, they sometimes include the factors of human or social capital. Those developing well-being indicators, place greater emphasis on the human and social factors yet often measure environmental and sustainable factors as well (Thompson 2005). Regardless of the emphasis, of import is the close relationship of these two subject matter domains. To mitigate our impact on the planet (environment/sustainability) we will have to redefine our definition of wellbeing and happiness (human/social). We will have to evolve our definition from a consuming, materialistic self-focus to a giving, humanitarian, and pluralistic focus. To do this perhaps the best place to begin is with our values.

2 Sustainable Values, Attitudes and Behaviour

Values are consistent knowledge and belief about the worth or importance of an object, person or concept. Values associated with sustainable development were first formulated by the Earth Charter (2000), the World Summit on Sustainable Development (WSSD 2002) and the Global Scenario Group (Raskin et al. 2002). Leiserowitz et al. (2006), states that the foundation of sustainable development must rest upon the values of environment, economy and equity. Shepherd has pointed out that there has been scant research on how these values impact sustainable development. The belief that wealth leads to happiness while perhaps subjectively valid, runs counter to the research literature. At best there has been demonstrated a small relationship between wealth and happiness (Diener et al. 1993). What has been found is that valuing wealth over other more flourishing values such as autonomy, relatedness, and competence can lead to negative outcomes such as mental illness, physical illness, alienation, and interpersonal problems (Kasser 2002). Certainly in our society there has been too great an emphasis placed on economic output and productivity; and not enough placed on the desired outcome of those activities i.e. human wellbeing. Yet it is clear that individuals can fulfill their values through positive wellbeing (Sagiv et al. 2004). The overconsumption by our consumer societies is neither the ideal path to happiness nor the best path to sustainability (O'Brien 2008). In fact individuals who have more materialistic values score lower on scales that measure subjective well-being than those who hold less materialistic values (Kasser 2006). As such it is clear that we will not be able to consume our way to sustainability. Development even if combined with institutional and technological progress, will not have much effect on improving the environment without significant changes to our consumer lifestyles (Trainer 2010) and how we define our own personal wellbeing and success. When we look at the basic human values (values that include self-direction, stimulation, hedonism, achievement, power, security, conformity, tradition, benevolence, and universalism) universalism stands out as one likely associated with sustainable attitudes and behaviour. Universalism is defined as understanding, appreciation, tolerance, and protection for the welfare of all people and for nature. It is associated with being broadminded, concerned about social justice, equality, world peace, world of beauty, unity with nature, wisdom, protecting the environment, inner harmony, and a spiritual life. Conversely we would anticipate that Power as defined as valuing social status and prestige, control or dominance over people and resources would be negatively associated with sustainably attitudes and behaviour. Finally we would expect Benevolence as defined as voluntary concern for others' welfare, such as being helpful, honest, forgiving, responsible, loyal, and a true friend would be associated with the sustainable element of pro-social behaviour.

Empirically, sustainable values are expressed through our attitudes, actions, and behaviours (Leiserowitz et al. 2006). Attitudes have been defined as a state of readiness for mental or physical activity, or how individuals evaluate objects or others in terms of being favorable or unfavorable (Dawes 1972). Attitudes and wellbeing have been found to be mutually dependent. Wellbeing has been found to have a positive impact on attitudes, behaviour and performance (Luthans and Yousef et al. 2007) and attitudes have also been found to positively affect wellbeing. Positive attitudes such as happiness appear to "buffer" the impact of negative life events, and for those less fortunate, chronic dissatisfaction increases our vulnerability to disease and premature death (Veenhoven 1991). The importance of attitudes on wellbeing and environmental sustainability is clear when we recognize that attitudes towards life, once established, may be largely insensitive to living conditions, and that positive attitudes enable individuals to cope more effectively with the problems of life. Ellen (1994) has found that the attitude termed "environmental concern" has been a significant predictor of recycling, source reduction, and political action. Environmental concern has been defined as the extent to which the consumer is worried about threats to the environment, the consequences of such threats for the harmony of nature and future generations, and the lack of human action to protect the environment for future generations (Bohlen et al. 1993).

The ultimate measure of individual sustainability is our behaviour, and consumption is perhaps one of the best indicators of how environmental values and attitudes get translated into behaviour. Research has been found that behaviour is evolving toward more environmental ends on larger investments, but less so on day-to -day costs. For example GlobeScan (2002) found that 44 % of respondents in high-income countries were very willing to pay 10 % more for an environmentally friendly car, yet only 28 % of respondents were willing to pay 10 % more for gasoline. As a share of gross domestic product, worldwide household consumption has hovered close to 60 % from 1965 through 2002 (World Bank 2004). While intensities of use are declining (e.g., energy and water use per unit gross domestic product), these trends have been outpaced by increases in population and affluence. Individuals often believe that environmental problems are a function of our industrial processes, yet much of our emission come from individual household behaviour (Svensson 2012). For example 46 % of greenhouse gas emissions in Canada can be attributed to households (Stats Canada 2004). Even though nearly 1/2 of emissions have been the responsibility of homeowners, this group has been hesitant to change their behaviour. This is despite the fact that the majority of households express a concern for the environment and have stated a willingness to change (Naturva°rdsverket 2007). It seems clear then that sustainable change will require both a change in our governments and market sand in our own personal household behaviour (Sustainable Development Commission 2006).

3 Sustainability Though Wellbeing

There are four important points required to tie wellbeing to sustainable development (Coulthard 2011). First, we need to recognize that there is more to people's lives than their livelihoods. While people are certainly concerned about their livelihoods (especially those in need) there is much more to their lives that matters to them than just the material aspects of their lives. Second, wellbeing is more than material achievement; it also involves individuals' differential capabilities and position in society. Third, the pursuit of wellbeing by one can affect the wellbeing of another. And fourth, people's conception of wellbeing is a primary driver of their decisions and behaviours. Tying sustainability to wellbeing helps us understand sustainable behaviour beyond the narrow conceptualization of cost savings and efficiency. It demonstrates how sustainability impacts our values, social identity and our relationship with others. When we consider sustainable development and those that become active in this field, it becomes apparent that the passion is more than a means to save (or earn) money, it is rather a 'way of life', invoking a strong sense of who we are and our place in this world.

Research has supported the strong relationship between well-being and sustainable behaviour. Kasser and Sheldon have found that individuals report more satisfaction and less stress during the holidays when engaged in more environmentally friendly behaviors (e.g. consuming organic or locally grown foods, giving environmentally friendly presents etc.). Teenagers who are engaged in environmentally friendly behaviour (e.g. turning off electric lights, recycling, and reusing paper, aluminum foil, and plastic baggies) report higher level of happiness. And people engaged in ecologically sustainable behaviours and who have significantly lower ecological footprints experience higher life satisfaction and higher levels of happiness (Brown and Kasser 2005).

While improving individual or aggregate levels of wellbeing may not in itself be enough to achieve sustainability, it would be interesting to see if there is a strong relationship in the reverse, that is, can individuals' wellbeing be enhanced by helping individuals become more sustainable in their behaviour, attitudes and values. Kasser (2006) in a review of the literature found that individuals who are more careful in their consumer choices and attempt to reduce their material dependency show higher levels of wellbeing. Enhancing wellbeing through simplicity has been termed 'voluntary simplicity' or 'down-shifting' in the literature. It is rare for individuals to change purely for altruistic reasons; however if mutual benefit can be demonstrated, people are more inclined to change their behaviour (Ferdjani 2010). Interventions that help save the individual money are social, and that make people feel good will have more traction in society. If there are immediate, measureable, and tangible benefits, behavioural change will be enhanced (Jones et al. 2002). Tying individual goals to climate goals is referred to as the health dividend of climate change—improving the biosphere also improves individuals health, as such, indicators that measure the improvement in personal wellbeing are as needed for global change as the macro indicators used currently to report changes in the earth's health (Costello et al. 2009).

To facilitate change we will have to transform the definitions that people have of happiness, from an external goods and consumption focus to an internal wellbeing focus. Lenzen and Cummins (2011) state that there has been no published research that links subjective well-being and the negative environmental impact of household consumption. Such research would help guide decision making toward fostering lifestyle aspects that offer the double dividends of increased well-being and reduced environmental impact (Lenzen and Cummins 2011). It is the intent of the current study to begin to address this void.

This study will explore the relationships between sustainable values, sustainable attitudes, reported sustainable behaviour and actual sustainable behaviour. The study will also determine the relationship of wellbeing with sustainable values, attitudes and behaviour.

Hypothesis 1: Valuing Universalism is positively associated with wellbeing, while valuing Power is negatively associated with wellbeing. Valuing Benevolence is also positively associated with interpersonal wellbeing.

Hypothesis 2: Individuals with stronger environmental attitudes have higher level of wellbeing.

Hypothesis 3: Individuals with stronger environmental behaviour have higher level of wellbeing.

Hypothesis 4: Individuals with stronger pro-social behaviour have higher level of interpersonal wellbeing.

4 Methods

To test these hypotheses, in the Fall of 2013 a website was created for presenting various surveys to participants. These surveys included the Environmental Attitude Measure, the Environmental Attitude and Ecological Behaviour Measure, the Human Values Scale, the ESS Well-being Module (Hubbart et al. 2009), demographic variables from the 2007 Wellbeing Research Synthesis report collated by Newton. The Environmental Attitude measure, was used to measure participants attitudes toward the environment. The Environmental Attitude and Ecological Behaviour measure was used to measure environmental attitudes and reported behaviour on two dimensions—environmental behaviour and pro-social behaviour. The Human Values Scale was used to provide a measure of participants' values. The ESS Well-being Module (Hubbart et al. 2009) provided a measure of participants' level of wellbeing. Many of the demographic variables of interest were obtained from the 2007 Wellbeing Research Synthesis report collated by Newton.

Participants were recruited in lectures of various programs across the British Columbia Institute of Technology campus. The variety of recruitment venues was utilized in an attempt to increase the heterogeneity of the sample. Students were encouraged to participate through the raffling of a $50 campus store gift certificate (along with other runner up prizes). Students were asked to complete the 5 measures online in their spare-time. Participants had the opportunity to "opt out" of the study at any time and were able to skip over any questions that they preferred not to complete. The consent form was posted on the "splash page" of the project and participants were required to indicate their willingness to participate before they could begin the questionnaires. The surveys were set up so that participants could return to the project after taking (a) break(s) and continue at the place they had stopped. All data was recorded into a data file on the server. Once all the data was collected it was downloaded into SPSS for analysis.

Of the six hundred and five individuals who completed the questionnaires 54 % identified themselves as female and 46 % identified themselves as male. Six percent were under 19, 30 % were between the ages of 18 and 22, 28 % were between the ages of 23 and 28, 16 % were between the ages of 27 and 30 and 19 % over the age of 30. Sixty five percent were born in Canada, while 35 % were born outside of Canada. Most were white (49 %) with the next largest group being Chinese (22 %) and then South Asian (9 %). Most were none religious (38 %) followed by Christian (19 %) and Catholic (14 %). Most (76 %) were earning less than $25000 per year, which is to be expected as they vast majority surveyed were students. With regard to highest earned degree, 39 % of the participants reported they had a high school diploma, 35 % had a bachelor's degree, 19 % had a diploma and 8 % had a university certificate. Students came from all of the schools across the campus including the School of Business (40 %), School of Health Sciences (18 %), School of Computing and Academic Studies (16 %), School of Construction and the Environment (12 %), School of Energy (8 %), School of Transportation (3 %), and the remainder were from outside of BCIT (3 %).

5 Results

From the work undertaken, a set of results were achieved that allowed us to investigate the relationship between wellbeing and the values, attitudes and behaviour associated with sustainability. These findings are as follows:

5.1 Values and Wellbeing

It was predicted that valuing Universalism is positively associated with wellbeing, while valuing Power is negatively associated with wellbeing and valuing Benevolence is also positively associated with interpersonal wellbeing. The correlation matric in Table 1 shows the relationship of the values (and the other fundamental human values) with wellbeing.

As anticipated the greatest relationships between our human values and wellbeing appear to be with Benevolence, Power and Universalism, interestingly there was also a strong relationship between Stimulation and elements of wellbeing.

Table 1 Correlation of basic human values and elements of wellbeing

	Personal feeling	Personal functioning	Interpersonal feeling	Interpersonal functioning
Achievement			−0.04*	0.06*
Benevolence	0.17**	0.14**	0.21**	0.16**
Conformity	0.08*			0.06*
Hedonism	0.07*			
Power	−0.08*	−0.12**	−0.15**	0.15**
Security	0.08*		0.04*	0.06*
Self-direction	0.06*	0.08*	−0.07*	0.15*
Stimulation	0.20**	0.16**	0.04*	0.18**
Tradition	0.12*		0.04*	0.10*
Universalism	0.11*	0.15**	0.14**	0.16**

*p > 0.05, **p > 0.001

Table 2 Correlation of sustainable attitudes and behaviour with wellbeing

	Personal feeling	Personal functioning	Interpersonal feeling	Interpersonal functioning
TantawiEnv. Attitude scale		0.20**		
Kaiser ecological behaviour	0.18**	0.21**		0.22**
Kaiser pro-social behaviour	0.12*	0.27**	0.19**	0.12*

*p > 0.05, **p > 0.001

When regression analysis was used to determine what values were the best predictors of the four elements of wellbeing it was found that for Personal Feeling—Stimulation, Power and then Tradition were the best predictors (R = 28, R^2 = 08), for Personal Functioning—Stimulation and Power were the best predictors (R = 23, R^2 = 06), for Interpersonal Feeling—Benevolence and Power were the best predictors (R = 26, R^2 = 07),and for Interpersonal Functioning—Stimulation and Benevolence were the best predictors (R = 22, R^2 = 05).

5.2 Environmental Attitudes and Behaviour and Wellbeing

Hypotheses two through four predicts that (2) *individuals with stronger environmental attitudes have higher level of wellbeing*, (3) *individuals with stronger environmental behaviour have higher level of wellbeing*, and (4) *individuals with stronger pro social behaviour have higher level of wellbeing.* Table 2 portraits the relationship between these variables and the four elements of wellbeing.

The best predictor of the Personal Feeling aspect of wellbeing was Kaiser's Ecological Behaviour (R = 0.18, R^2 = 0.03). The best predictors of the Personal Functioning aspect of wellbeing was Kaiser's Pro-social Behaviour and Kaiser's Ecological Behaviour (R = 0.30, R^2 = 0.09). The best predictor of the Interpersonal Feeling aspect of wellbeing was Kaiser's Pro-social Behaviour (R = 0.19, R^2 = 0.04). The best predictor of the Interpersonal Functioning aspect of wellbeing was Kaiser's Ecological Behaviour (R = 0.22, R^2 = 0.05).

6 Conclusions

Many have called for research that addresses the relationship between environmental sustainability, quality of life, and wellbeing. Arguably determining the relationship between wellbeing and sustainable behaviour will become the basis for more sustainable lifestyles. It was the thesis of this paper that to affect a change in our personal and social wellbeing will require a change in our values, attitudes and behaviour.

This thesis was confirmed, the current study found that wellbeing is influenced by our sustainable values, attitudes and self-reported behaviour. Maintaining strong sustainable values, attitudes and behaviour improves our sense of wellbeing. It was predicted and found that Universalism and Power—values associated with sustainability are closely associated with individuals' wellbeing. Individuals who value Universalism (understanding, appreciating, and protecting the welfare of all people and nature) the value most associated with sustainability did have a higher level of wellbeing. Those who valued Power (i.e. seeking social status and prestige, and wanting control or dominance over people and resources). A value incongruent with sustainable behaviour reported lower levels of wellbeing. These findings are illuminating in determining the mechanism(s) that drive those toward sustainable activities. This research indicates that those involved in the field gain a sense of satisfaction and purpose from their sustainable values, attitudes and behaviors.

From a social perspective it was anticipated and found that individuals who were more Benevolent (preserving and enhancing the welfare of those with whom they are in frequent personal contact) also had higher levels of wellbeing. Again this is encouraging to know that the good Samaritan members of our community are being rewarded intrinsically for their social efforts.

One unanticipated result was the strong relationship between Stimulation (as defined by excitement, novelty, and challenge in life) and wellbeing. While at first overlooked as a value associated with sustainability, Stimulation is perhaps one of the most important sustainable values of all, as it is the value associated with acting and making change. The fact it is associated with wellbeing is not surprising as those who are motivated and effective at making change are rewarded through their engagement with others and the associated sense of purpose that comes from these efforts.

Given the different levels of complexity and the need to include a diversity of stakeholders, sustainable development is often conceptualized as a wicked problem, and wicked problems required the motivations and skills inherent in the value of Stimulation. As Batie (2008) has stated the pursuit of sustainability is a complex sociobiological technological problems that eludes easy definition. Sustainability problems are in fact a host of interrelated issues that are not readily solved, but at best are managed or mitigated. Wicked problems addressing sustainable development require individuals who value Universalism, Benevolence and Stimulation, and are not driven by Power. Fortunately these values bring the rewards of personal and social wellbeing to those who enter into this field.

As our understanding of wellbeing becomes more sophisticated it has been determined that the construct is comprised of four discrete elements namely *feeling, functioning, the personal and the interpersonal* (Huppert 2009). Feeling (having, being) at the personal level is related to satisfaction, positive and negative affect, optimism, and self-esteem. Feeling at the interpersonal level reflects belonging, support, respect, fair treatment, and social progress. Functioning (doing) at the personal level is defined by ones' autonomy, competence, interest in learning, goal orientation, sense of purpose, and resilience. Functioning at the interpersonal level reflects ones caring, social engagement and altruism. When wellbeing is disaggregated into these elements we find that individuals' personal feelings and their functioning are driven by Stimulation and a low emphasis on Power. At the social level individuals' interpersonal feeling is driven by Benevolence and Power and their interpersonal functioning is driven by Stimulation and Benevolence.

These findings are important as we determine what elements of our wellbeing are driven by our own personal satisfactions and what element are driven by the concern for others. It is logical that our personal feelings, satisfaction, self-esteem, competence, interest in learning, goal orientation and sense of purpose are strongly influenced by Stimulation a value that captures our need for excitement, novelty, and challenge in life. It is also reassuring that Power or the drive toward status, prestige, and the control over people and resources diminishes our wellbeing. Such findings suggest that people who are truly satisfied and grounded are well position to address sustainable issues and help the world mitigate the wicked problems facing humanity. Further as the interpersonal elements of wellbeing were found to be determined by individuals' Benevolence. Those individuals focused on helping and enhancing the welfare of others are rewarded through higher levels of interpersonal wellbeing or the sense of belonging, respect, support, social engagement and altruism.

These results concerning values were to a degree reflected in the findings of the attitude and behavioural measures. Namely individuals' personal ecological behaviour was most associated with the personal feeling element of wellbeing and the individuals' pro-social behaviour affected individual interpersonal feelings of wellbeing. The current findings supplement Ellen's (1994) research in that not only will individuals with greater environmental and perhaps social concern be more inclined to recycle, reduce and be socially active but they will also gain the intrinsic and social rewards from these pro-environmental and pro-social attitudes and behavior through an enhanced sense of personal and interpersonal wellbeing.

It is known that wellbeing is a complex multifaceted construct comprising personal, interpersonal, feeling and functioning components. Further these elements of wellbeing are independently influenced by specific sustainable values, attitudes and behaviours. Tying sustainability to wellbeing through individuals' values, attitudes and behaviour helps us understand sustainable behaviour beyond the narrow conceptualization of cost savings and efficiency. This research path demonstrates how we can enhance the quality of individuals' lives by getting them involved and showing them the importance of sustainability. Such education could help refocus our western culture from a focus on the material to one more aligned with self and social enhancement. When we step back and see the emerging field of sustainable development and those that are becoming active in this field, we are now better positioned to understand why these individuals are so passionate, it is not because they are saving money or creating new efficiencies, it is because they are wrestling with wicked problems and reaping the rewards of enhancing their competence, sense of purpose, social engagement and altruism.

References

Batie SS (2008) Wicked problems and applied economics. Am J Agric Econ (Impact Factor: 0.99). 90(5):1176–1191
Bohlen G, Schlegelmilch BB, Diamantopoulos A (1993) Measuring ecological concern: a multi-construct perspective. J. Mark Manage 9:415–430
Brown KW, Kasser T (2005) Are psychological and ecological well-being compatible? The role of values, mindfulness and lifestyle. Soc Indic Res 74:349–368
Costello A, Abbas M, Allen A, Ball S, Bell S, Bellamy R, Friel S, Groce N, Johnson A, Kett M, Lee M, Levy C, Maslin M, McCoy D, McGuire B, Montgomery H, Napier D, Pagel C, Patel J, de Oliveira JA, Redclift N, Rees H, Rogger D, Scott J, Stephenson J, Twigg J, Wolff J, Patterson C (2009) Managing the health effects of climate change: lancet and UCL Institute for Global Health Commission. Lancet, May 16; 373(9676):1693–1733
Coulthard S, Johnson D, McGregor JA (2011) Poverty, sustainability and human wellbeing: a social wellbeing approach to the global fisheries crisis. Glob Environ Change 21(2):453–463
Dawes RM (1972) Fundamentals of attitude measurement. Wiley, New York
Diener E, Sandvik E, Seidlitz L, Diener M (1993) The absolute relationship between income and subjective well-being: relative or absolute? Soc Ind Res 28:195–223
Earthcharter (2000) Values and principles for a sustainable future. The earth charter initiative. http://www.earthcharterinaction.org/content/. Accessed 30 Oct 2013
Ellen OS (1994) Do we know what we need to know? objective and subjective knowledge effects on pro-ecological behaviours. J Bus Res 30:43–52
Ferdjani J (2010) Improving global wellbeing, improving personal wellbeing. 7(1):33–35
GlobeScan (2002) International environmental monitor. GlobeScan, Toronto
Huppert FA, Marks N, Clark A, Siegrist J, Stutzer A, Vittersø J, Wahrendorf M (2009) Measuring well-being across Europe: description of the ESS well-being module and preliminary findings. Soc Ind Res 91(3):301–315
Huppert FA (2009) Psychological well-being: evidence regarding its causes and consequences. Appl Psychol Health Well-Being 1(2):137–164
Jones L, Sidell M, Douglas J (eds) (2002) The challenge of promoting health: exploration and action, 2nd edn. Macmillan Press Ltd, Basingstoke
Kasser T (2002) The high price of materialism. MIT Press, Cambridge
Kasser T (2006) Psychological need satisfaction, personal well-being and ecological sustainability. A report for the New Economics Foundation. London
Leiserowitz AA, Kates R, Parris T (2006) Sustainability values, attitudes and behaviours: a review of multinational and global trends. In: Annual review of environment and resources, vol 31, pp 413–444
Lenzen M, Cummins RA (2011) Lifestyles and well-being versus the environment. J Ind Ecol 15(5):650–652
Luthans F, Youssef CM, Avolio BJ (2007) Psychological capital. Oxford University Press, New York
Naturva˚rdsverket (2007) Allma¨nheten och klimatfo¨ra¨ndringen 2007. Allma¨nhetens kunskap om och attityder till klimatfo¨ra¨ndringen med fokus pa˚ egna a˚tga¨rder och fo¨retagens ansvar. Stockholm: Naturva˚rdsverket, rapport 5760
O'Brien C (2008) Sustainable happiness: how happiness studies can contribute to a more sustainable future. Can Psychol 49(4):289–295
Raskin P, Banuri T, Gallopin G, Gutman P, Hammond A, Kates R, Swart R (2002) Great transition; the promise and lure of the times ahead. Stockholm Environment Institute, Global Scenario Group, Boston
Sagiv L, Roccas S, Hazan O (2004) Value pathways to well-being: healthy values, valued goal attainment, and environmental congruence. In: Linley PA, Joseph S (eds) Positive psychology in practice. Wiley, New York, pp 68–87

Seligman MEP (2002) Positive psychology, positive prevention and positive therapy. In: Snyder CR, Lopez SJ (Eds) Handbook of positive psychology. Oxford University Press, New York. pp 3–9
StatsCanada (2004) Greenhouse gas emissions—a focus on Canadian households. www.statscan.gc.ca/pub/16-002-x/2008004/article/10749-eng.html. Accessed 27 Oct 2013
Sustainable Development Commission 2006 I will if you will: towards sustainable consumption. Sustainable consumption roundtable. http://www.sd-commission.org.uk/publications/downloads/I_Will_If_You_Will.pdf. Accessed 27 Oct 2013
Svensson E (2012) Achieving sustainable lifestyles? Socio-cultural dispositions, collective action and material culture as problems and possibilities. Local Environ 17(3):369–386
Thompson FG (2005) Around the world, genuine progress and well-being indicators. Futures Res Q, Winter 2005, 21(4)
Trainer T (2010) Can renewables etc. solve the greenhouse problem? The negative case. Energy Policy 38(8):4107–4114
United Nations (1993) Agenda 21: earth summit-the united nations programme of action from Rio. unp.un.org/details.aspx?entry = E93020
Veenhoven R (1991) Questions on happiness. Classical topics, modern answers, blind spots. In: Strack F, Argyle M, Schwarz N (eds) Subjective wellbeing, an interdisciplinary perspective. Pergamon Press, London, pp 7–26
World Commission on Environment and Development (WCED) (1987) Our common future. Oxford University Press, Toronto
World Bank (2004) World development indicators http://go.worldbank.org/VXW7J2NON0
WSSD (2002) Report of the world summit on sustainable development. CONF. 199/20 Johannesburg

Author Biography

Mark Giltrow is the Chair and a Full Professor for the Sustainable Business Leadership Program in the School of Business at the British Columbia Institute of Technology (BCIT). His research covers a number of areas including, the relationship of sustainable values, attitudes and behaviour with wellbeing, sustainability in the curriculum, creativity and innovation within organization, competencies leading to the advancement of mid-level managers into executive leadership positions, and memory and social cognition in the elderly. Prior to BCIT, Dr. Giltrow was a corporate psychologist at BellSouth, United Parcel Service and Florida Power Corporation and co-founded a number of companies including Organizational Diagnostics Online and the Canadian Center for Local Government Research. He serves on the Board of Directors of the Sunshine Coast Credit Union in Gibson's British Columbia and the Sunshine Coast Natural Resources Advisory (NRAC) Committee. He also serves on BCITs' Sustainability Steering Committee, the Pacific Spirit Project Steering Committee (building sustainability into the curriculum) and the BCIT School and Institute Research Committee. He received his PhD in Organizational Psychology from Georgia Institute of Technology, Masters from the University of Akron and Bachelor-Honours from Simon Fraser University.

Green Fever EcoCycle and Sustainability in Action

Kirsten Davies and Elisabeth Blik

Abstract

Green fever is infecting Macquarie University as it champions an innovative, people-driven, campus-wide approach to sustainability that will inspire universities across the world. A coupled human-nature framework, The EcoCycle is introduced in this paper, which will assist Universities to develop and evaluate future programs for behavioural change. The EcoCycle framework is applied to evaluate one of the University's many sustainability activities, the Annual Department Sustainability Challenge. This program commenced in 2009 and operates at a minimal cost. In 2013, 37 Departments and 1,815 or (approx.) 81 % staff members voluntarily participated, competing for 'sustainability stars' awarded for their activities, including greening their offices. They received prizes, such as trees planted in critical koala habitat. The findings from this study demonstrate how the EcoCycle can be employed to understand and evaluate variables that drive or inhibit human actions towards a sustainable planet. Analyses of The Challenge highlight the importance of a whole-of-university approach that is peer and incentive driven and, most importantly, playful. Universities, as community leaders in research, technology and education are strategically placed to address global sustainability challenges. In the ominous context of the impacts of Global Climate Change, positive, constructive sustainability activities, such as the Department Sustainability Challenge provide an exemplar for University communities investing in a sustainable planet.

K. Davies (✉) · E. Blik
Department of Environment and Geography, Macquarie University,
Balaclava Rd, North Ryde, NSW 2109, Australia
e-mail: kirsty.davies@mq.edu.au

E. Blik
e-mail: elisabeth.blik@students.mq.edu

W. Leal Filho (ed.), *Transformative Approaches to Sustainable Development at Universities*, World Sustainability Series, DOI 10.1007/978-3-319-08837-2_8

This paper argues that potential program participants should be empowered to design their activities. This strategy ensures their engagement, emphasising that human-centric solutions are critical to addressing human-caused problems.

Keywords
Sustainable development · Behavioural change · Decisions · Environment · University · Coupled human and nature systems · CHANS · Socio-ecological · Gender

1 Introduction

Australia's Macquarie University (MQ), is suffering from a severe case of 'green fever' as it champions an innovative, people-driven, campus-wide, approach to sustainability which will inspire universities across the world. This human-centric approach can be understood in terms of a Coupled Human and Nature System (CHANS) (Liu et al. 2007). A new, CHANS framework, the EcoCycle, is introduced in this paper. It is applied to evaluate one aspect of the University's sustainability activities, the annual Department Sustainability Challenge (DSC) as a case study. The EcoCycle, designed by one of the papers authors, will assist Universities to design and evaluate programs for behavioural change towards sustainability outcomes.

Universities across the world, as leaders in communities, research, technology and education are strategically placed to address sustainability challenges especially those presented by the impacts of Global Climate Change (GCC). According to the United Nations Intergovernmental Panel on Climate Change (IPCC), human activity is almost certainly responsible (95 % confidence rate), for the increase in ocean, land and atmospheric temperatures. The effect will be cataclysmic, as the disruption of weather and climate heavily impact agriculture, economies and human society (Cubasch et al. 2013). This paper argues that human-centric solutions are critical to addressing human-caused problems. The DSC demonstrates the power of community driven activities based on positive action.

2 Case Study-Department Sustainability Challenge

Macquarie has significant challenges. The ecological footprint of the University was measured in 2011 as occupying 1.3 planets (Macquarie University 2012a). Many steps have been taken to reduce this footprint. For example, a cogeneration system has been installed, with MQ becoming one of the first universities in Australia to tackle problems related to Green House Gas emissions and the rising costs of electricity. The system provides abatement of 5,400 tonnes of CO_2 emissions annually. Macquarie measures progress, reporting annually applying the Sustainability Indicator Framework, developed utilising the Campus Sustainability

Assessment Framework (Cole 2003), in conjunction with Global Reporting Initiative (GRI) indicators. Significant progress has been achieved. MQ was a recipient of a Green Globe Public Sector Sustainability Award (2010) and the Green Lifestyles' Large Business Award (2012).

The landmark ***Our University: A Framing of Futures*** provides a roadmap for the University's future. One of its seven strategic **priorities is "Developing a vibrant and sustainable campus..."** (Macquarie University 2013b p. 22). The University has three strategic domains, 'people, planet and participation' that ensure a campus-wide approach to implementing the sustainability vision. The DSC assists the delivery of the 'participation' domain. Part of the whole-of-university sustainability approach includes the establishment of The Sustainability Representative Network (SRN) which identifies staff and student champions and enables a platform of activities, knowledge exchange and support to achieve MQs goals. Vice-Chancellor, Professor Dowton noted, "Sustainability initiatives in large, complex organisations will only ever be as successful as to the degree in which there is alignment and personal commitment of all staff members from the leadership of the University, across academic and professional staff, to make the necessary goals become a reality on a daily basis" (Macquarie University 2012b).

Macquarie is a large organisation, with many staff members only interacting within their Departments, and not engaging with the broader campus community. The Sustainability Department has been integral in breaking down silos and building a more connected community, with the DSC being one avenue in which outcomes have been achieved.

The program aims to "inspire and engage staff at all levels to undertake activities that foster a sense of community on campus as well as benefit the environment " (Macquarie University 2013a). The genesis of the DSC dates back to 2009 when the MQ Director of Sustainability and Sustainability Officer facilitated a workshop with senior managers and sustainability experts from across the University. A proposal from the workshop was a program where incentives and challenges promoted engagement and action. From this idea the DSC emerged. Program objectives include: cultural change, up-skilling of champions, improving sustainability literacy, action in specific areas and staff engagement benefits (e.g. satisfaction, retention, increased workplace engagement). The Department of Sustainability's budget and team are small, however its impact is high. The total estimated budget for the DSC in 2013 was $AUD39,970,[1] including prizes, the awards night and staff wages and excluding overheads.

Over a two–three month period, during second semester (July–September), staff within Departments voluntarily form teams and participate in various challenges competing for prizes. Activities change from year to year. Examples include 'nude

---

[1] $AUD 6,800 = prizes, awards, and awards night catering. Three staff members worked on the DSC. The Sustainability Officer (3 months @ 5 days per week = $21,500), Sustainability Engagement Officer (3 months @ 2 days per week = $7,700), Multimedia Coordinator (1 month @ 3 days per week = $3,970).

Fig. 1 Office greening DSC activity

food' team lunches; promotion of 'Ride2UniDay'; picking up MOOP (Matter Out Of Place), office greening initiative (Fig. 1), community volunteering efforts; scavenger hunting for campus sustainability initiatives; mobile phone recycling collections; or swapping items at a Swap Party. Engagement in activities earns 'sustainability stars' (or points).

There is an additional 'bonus' category that must demonstrate the social, environmental and economic benefits of the initiative. Creativity and quality of the activity (number of people involved and impacted) are rewarded with bonus stars. Charity fundraising, arboretum walks with colleagues, 'Bushcare' volunteering, and Department collections (e.g. for Second Chance Toys) are suggested activities. Most Coordinators are members of the SRN. They usually self-nominate then motivate their colleagues to participate. The Coordinator is responsible for organising their team's involvement in the activities. When teams have completed activities, they enter their involvement on a 'live' online scorecard (accumulated stars) entered into a 'leaderboard'. All teams have access to the leaderboard on the web site where they can monitor their team's progress.

Progress is monitored and recorded by University student interns, known as Challenge Liaisons, who are supported by sustainability staff. Liaisons assist the Coordinators to implement the DSC. They support Departments via the online challenge scorecard, meet regularly with their Departments and with the Sustainability Team, blog about their Department's achievements, document scores, audit Sustainability Challenge actions as a process of validating Department scores and issue prizes at certain achievement levels. The internship provides students with the opportunity to gain experience and enhance their skills in organisational change management and staff engagement for sustainability. 'Spot' prizes during the Challenge, and 'final' prizes are awarded to the winning team as well as the winning 'faculty' during a celebratory awards night (Macquarie University 2013a).

In 2012, Macquarie's total number of non-casual (full-time and fractional full-time) staff was 2,350 (Macquarie University 2012b). A total of 1,436 staff members volunteered to participate in the DSC from 41 Departments (Macquarie University 2012a) or 61 % of all staff.

3 The EcoCycle and the DSC

Sustainability issues and activities, such as the DSC, are increasingly being examined through coupled Human and Nature Systems (CHANS) (Liu et al. 2007; McConnell et al. 2009; Wiek et al. 2012). The EcoCycle (Fig. 2) is a theoretical framework that rethinks human behaviour and interactions with all components (living and non-living) in the earth's biosphere. The framework has been developed by Davies (2012) based on a literature review pertaining to human environmental behaviour (Lucas et al. 2008; Darnton 2008; Kollmuss and Agyeman 2002). The EcoCycle has been adapted from the Social Indicator Planning and Evaluation System (SIPES) (Genskow and Prokopy 2008). The title 'EcoCycle' was adopted as a substitute for 'EcoSystem', stressing "the importance of cyclical regenerative processes while recognising that natural systems and webs exist" (Davies 2012, p. 198).

The framework includes five phases that represent the progressive steps towards human behaviour, commencing with 'Humans & Nature' which provide the context where human decisions and behaviour are based. This first phase is where issues or problems are identified (e.g. GCC). 'Awareness & Knowledge' is the second phase, highlighting the level of comprehension required to understand the problem. 'Capacity & Skills' is the third EcoCycle phase that recognises the ability of people to perform an action (e.g. they may require tools, training, financial support). 'Conduits & Barriers' is the next phase which includes components that facilitate and/or prevent the 'Decision & Behaviour', the final phase. McKenzie-Mohr (2000) recognised many 'Conduits & Barriers' are contextual and specific to certain activities. What may be considered a barrier to some could be a conduit for another (e.g. legislation, incentives). This completes the EcoCycle which is a continuous loop as the behaviour then impacts on the first phase of 'Humans & Nature' (e.g. addressing the impacts of GCC).

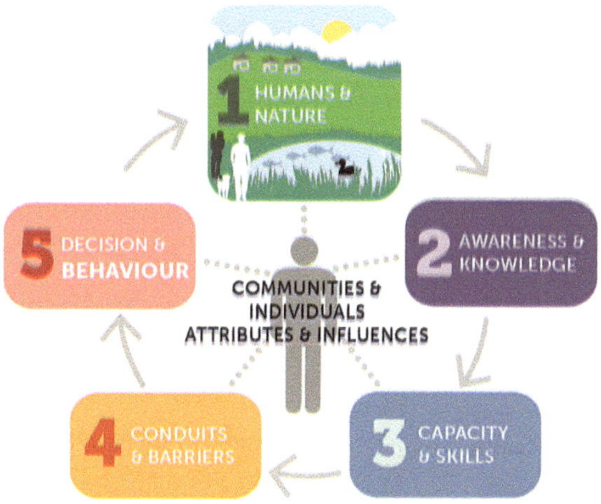

Fig. 2 The EcoCycle (Davies 2014 unpublished)

Central to the EcoCycle are community and individual attributes and influences that influence each EcoCycle phase. ‘Attributes’ are those intrinsic to each person e.g. personal values, attitudes, gender, age, education level and personality. ‘Influences’ are those external to the individual but directly impact on their identity e.g. community values, social norms. Davies (2012) argues that the most influential of these are values which are often linked to social norms (Nye and Hargreaves 2010; Davies 2012). Psychological ownership (self-efficacy, accountability, sense of belonging and identity) and engaging ‘actors’ in the process (e.g. participation in the DSC), support positively-oriented behaviour (Avey et al. 2009; Stern 2000). Having sufficient time to engage is recognised by Nye and Hargreaves (2010) who state the importance of sustainability champions being relieved from their daily activities to engage with such tasks. Secondary variables of significant influence include: age, gender, personality, time, education and socio-economic status. There appears to be a trend indicating that educated, well paid, politically liberal females are more likely to engage in environmentally responsible behaviour (Derksen and Gartrell 1993; Gilg et al. 2005; Martin et al. 2006). However, Olli et al. (2001) argue that correlations between environmentally responsible actions and demographic factors remains unconvincing.

Human-Nature relationships and human behaviour is complex involving multiple influential variables in any given situation and community. The EcoCycle was developed by Davies (2012) to assist understanding of the broad groupings of components that lead human behaviour. The framework can be applied to examine coupled human-nature relationships and influences and to design and evaluate human behavioural interventions. When designing an intervention or program, after the issue has been identified (Phase 1), the key phases to consider are: Awareness & Knowledge, Capacity & Skills and Conduits and Barriers.

4 The EcoCycle and the DSC

This paper focuses on EcoCycle phase four ‘Conduits & Barriers’ analysing how they influence ‘Decisions & Behaviour’ in the context of the DSC case study. Conduits pertaining to the DSC are numerous on personal, institutional and national and international levels. For example a participant might be motivated by personal relationships and/or career positioning. The institution may embed some conduits in the program such as providing funding and relief from usual duties to enable participation. Nationally and internationally conduits may exist, for example, ethical and/or legal requirements surrounding sustainability performance and reporting. A key conduit is the development of team spirit that may have a catalytic effect on local action, thus being essential in facilitating change (Chavis and Wandersman 1990). Aligned with this, Nye and Hargreaves (2010) describe how momentum can build by drawing people together with common goals. Barriers to behaviour can be categorised into the same groupings as conduits. Institutional factors relating to the DSC could include a lack of: valuing, training, education and resources (Cowell et al. 1998 cited in Hayles and Holdsworth 2008).

5 Methods

This study involved a desktop review and in-depth qualitative interviews, centred on testing the EcoCycle framework's capacity to analyse and evaluate the DSC. In particular it examined the EcoCycle's Phase 4, 'Conduits & Barriers' that facilitate 'Decisions & Behaviour' (Phase 5) towards sustainability outcomes. A key objective was to identify aspects of the program that have been successful and those that require further consideration, with the aim of continuous improvement. A subset of five service Departments was selected for this study as a focus sub-group. Interviews were conducted with Coordinators and Directors representing these Departments. Interview questions were designed to capture data pertaining to the study's objectives. Human Ethics approval was granted for this project. Data were de-identified and analysed manually and thematically.

6 Limitations

Comparable University programs have not been included in the analysis of the DSC, nor a detailed description of the range of activities. This study involved the initial testing of one phase of a new theoretical framework and its application to an existing program. As this framework includes five integral phases, testing one limits the scope of the investigation. Detailed demographic information was not available for the whole cohort of DSC participants (e.g. gender, age) limiting quantification.

7 Results

As a precursor to the 2013 DSC, the SRN facilitated the 'Get green with your team program' which initiated 'green fever' across the campus as offices were filled with plants. This program promoted that "Plants can reduce potentially harmful volatile organic compounds emitted from plastics and synthetics by up to 80 % and can also reduce negativity and stress in the workplace by 60 %" (Macquarie University 2013e). In 2013, 37 Departments and 1,815 or (approx.) 81 % of staff members voluntarily participated in the DSC (Fig. 3).

Challenges were organised across five categories: M-Power, Campus, Transport, Food and Waste. Departments represented Professional Services and the four teaching faculties: Arts; Business and Economics; Human Sciences; and Science. Department teams varied from four to 500 people. Of the 37 participating Departments, seven Coordinators were male and 30 female.

The Faculty of Human Science won the Most Sustainable Faculty award. Executive Dean Professor Janet Greeley, noted "All those stair climbs and extra steps walked along with the day to day mindfulness about not wasting paper and electricity, the planting of gardens (Fig. 4), the support for refugees, and the many other activities taken up by Departments contributed to the Faculty's award" (Macquarie University 2013d).

Fig. 3 Logo for 2013 sustainability challenge

Fig. 4 Creating an herb garden at work

Most Sustainable Department awards went to Human Resources and Ancient History.

Overall 1,135 sustainability stars were earned. Thirty two Departments achieved over 15 stars each, for which Macquarie Sustainability organised 15 trees planted in critical Koala habitat. Sixteen Departments achieved over 30 stars, for which they received Oxfam sponsorship of sixteen piglets for families in Vietnam. Other prizes included cases of organic wine, indoor office plants and recognition plaques.

7.1 Sub-Group Results

The EcoCycle model's 'Conduits & Barriers' (Phase 4) was applied to evaluate the DSC through a sub-group of five services Departments which were: Learning and Teaching Centre, Human Services, Child Services, Open Learning Centre and Property. In 2013 a total of 236 staff members representing these five Departments participated in the DSC of which a notable 80 were males and 156 were females (Table 1).

From the sub-group, the Human Resources Department attained the highest scores (Table 2).

Table 1 Sub-group departments, number and gender of participants

Department	Male	Female	Total
Learning and teaching centre	40	42	82
Human resources	15	37	52
Centre for open education	7	21	28
Property	18	4	22
Children services	0	52	52
Total	80	156	236

Table 2 Final scores for sub-group 'professional service departments'

Department	Scorecard	Bonus	Total
Human resources	39	20	59
Property	21	20	42
Children's services	26	0	26
Centre for open education	11	0	11
Learning and teaching centre	8	0	8

At the completion of The Challenge, the Sustainability Officer, Coordinators and Directors from the subgroup participated in interviews. Data collected from these interviews examined the EcoCycles (Phase 4) 'Conduits & Barriers' influencing participation in the DSC.

7.2 Conduits

The Sustainability Officer found a 'foot in door' engagement technique (small, easy to achieve prizes and progress recognition mechanisms), motivated Departments to continue achieving and added to the fun factor (Figs. 5 and 6).

Prizes were commensurate with levels of achievement and facilitated further engagement, providing Departments with the opportunity to extend The Challenge beyond two months such as through the sponsorship of an animal. The most significant factor to successful participation was identified as 'enjoyment'. Stars and the leaderboard (friendly competition) contributed to this. A Coordinator noted that even vicarious enjoyment was sufficient: "…folks who don't garden get a kick out of seeing it, or us working in it."

Alignment of activities and prizes with the interests of the group was identified as very important and having lasting impacts. An interviewee said: "It has been a fun thing to participate in…. I have recently attended a more childcare specific training on embedding sustainable practice within our service. The activities we competed in during the sustainability challenge gave me a more holistic experience

Fig. 5 Dressing up for the sustainable scavenger hunt

Fig. 6 'MOOP' litter collection day

with sustainability and I'm now in the process of working with my manager in getting a really visible, embedded program running…"

Within the Departments and the University, it was noted that communication was a critical conduit. Clear communication and practical strategies were required to convey the message that sustainability is not merely the rhetoric, but foundational to all practices.

Experiencing a cleaner environment as a result of clean-up days inspired further action by implementing a regular clean-up. This activity demonstrated that immediate tangible results can provide stimulus for longer term behavioural change. Interviewees expressed concern that sustainable practices can often seem pointless in terms of global impacts, "as though it were just a drop in the ocean". They described how immediate results can motivate behaviour such as, "experiencing a cleaner area has been very beneficial".

Directors considered it important that staff should feel a sense of 'ownership', as it provided them with a greater sense of worth and pride. Interviewees stated that support from upper management was crucial. High level vision and ownership modelled to staff the importance of sustainability, and gave credence to their efforts. Having additional support from Liaisons was considered to be helpful. "Encouraging and supporting is very important… discussing obstacles and sharing ideas on how to overcome them… being focused on positive results …" said one Coordinator.

Coordinators stated the reason for involvement in The Challenge was to participate in a University-wide activity that was not directly work related. The opportunity to network with other enthusiasts across campus gave them confidence that they were not alone in their sustainable practices. Team Coordinators generally fulfilled the role of champions. Their level of enthusiasm for The Challenge encouraged participation and was found to have a strong correlation with their team's performance.

Unsupported by upper management at a time of very low morale due to high staff turnover, one Coordinator overcame initial team reluctance and feelings of despair. She was able to see The Challenge as an opportunity to unite the Department and have fun. Whilst their team did not score particularly highly, they finished the DSC satisfied and unified.

In 2014 conduits for the DSC will be strengthened. There will be more opportunities to earn stars through the Bonus mechanism—the one content category that has grown year by year. It allows Departments to develop their own initiative, an approach which cultivates ownership. This evolution of the program will build relevancy and context, empowering teams to be rewarded for creative and meaningful action. Sustainability Staff have observed that many people participate for the sake of competing, not for the prizes. In response they will be offering more meaningful, altruistic prizes such as increasing animal sponsorship, donations and tree plantings. This will have the added bonus of providing The Challenge with more 'feel good stories' which will assist the program's profile through media opportunities and general communications channels.

7.3 Barriers

Respondents recognised lack of time was one of the most limiting factors pertaining to their participation and selection of activities. A Coordinator said, "any activity that required leisure time went down like a lead balloon", and another, "With so many issues contending for attention there needs to be a very good justification for the inclusion of sustainability added onto people's already busy work schedules". Involvement in the DSC activities was designed not to cause additional workload for staff. However, the role of the Coordinator did incur a significant amount of additional work. Though they enjoyed the challenge, many believed it created performance pressure as they balanced responsibilities of the DSC and their usual workload. The short time frame of the DSC was both a conduit and a barrier. On one hand it facilitated an active competition. From another perspective, it detracted from the continuous engagement of staff.

Performance anxiety was another barrier. A Coordinator said "I initially felt lousy when I saw our team had not participated in many of the activities and had a low score. I felt guilty until I was affirmed by the Liaison that this challenge was not about doing all the activities, but having fun and getting the issues of sustainability out there". Some interviewees described how the points system was an effective motivator, while others disagreed stating the need for more flexibility. Another believed the scorecard was irrelevant and that the focus should be on the outcome. When activities and prizes were not aligned with the interests of the groups there was little participation. The 'swap party' activity was not of interest to one Department while another still had last year's prize in the cupboard—unused. Lack of knowledge of sustainability impacted heavily on engagement. A Coordinator felt "like they are sometimes flying in the dark". This Department requested access to a consultant with specialist knowledge in sustainability. The process of registration (after which details of the specific challenge activities are revealed) was identified as a barrier because some Departments wanted to know what the challenges were prior to making a commitment.

In 2014 the DSC will be further developed to address the identified barriers and strengthen the conduits. Sustainability Staff are planning to remove registration barriers as participants have requested more specific details of The Challenge activities before committing to participation. They are looking to space the DSC over a longer time frame to enable continuous engagement and provide more time to plan events. This will require a careful balancing act so as not to lose the competitive element.

8 Discussion

Specific 'Conduits & Barriers' impacting on behaviour were identified in this evaluation of the DSC, applying the EcoCycle framework. Interview responses from Directors indicated they were not really aware of the specific activities of the 2013 Challenge however they did describe their passion for sustainability and personally participated in many of the activities. This high level engagement and leadership was the conduit for broader participation across the University as described by Cowell et al. (1998 cited in Hayles and Holdsworth 2008).

The advocacy of Directors was found to directly impact on the buy-in of Coordinators. Chavis and Wandersman (1990) described the increased likelihood of building momentum when an individual participant is part of a greater whole focused on the same goal. This collective approach was found to be effective as collaboration engendered a spirit of collegiality within and across Departments. The system of awarding stars, offering incentives and playful activities culminating in the awards night were found to provide strong conduits for participation and the success of the DSC as described by Chavis and Wandersman (1990).

Fostering of shared commitment and ethical responsibility for the DSC was evident, reinforcing the powerful role of values as 'Conduits & Barriers' supported by the literature (Darnton 2008; Genskow and Prokopy 2008). DSC participant variables such as gender, knowledge level, interest in activity and prizes for participants were all found to be influential. It was notable that of the 236, 156 participants were female and 80 male and of the 37 Coordinators 30 were female and 7 male. This female dominance in participation in the DSC subgroup is supported by the literature (Derksen & Gartrell 1993; Gilg et al. 2005; Martin et al. 2006). A gender split for the entire cohort of participants would have been useful to validate this finding however this data were not collected.

Directors and Coordinators described the lack of dedicated resources as a barrier to the program. This concern was aligned to the time pressures participants experienced trying to balance the usual jobs and their DSC activities. They identified the need for additional resources to relieve a staff member to work on sustainability activities. This finding was supported by Nye and Hargreaves (2010) who highlighted the need for dedicated resources and time to participate. Knowledge of sustainability impacted heavily on engagement. One Director commented, "If the practices make sense, they were readily accepted". What 'makes sense' however is ambiguous especially if there is a "lack of confidence in the efficacy of the sustainability practice" as expressed by this Director. For example staff members' level of understanding of Climate Change could be a significant conduit/barrier to participate in change towards sustainable practices. This observation is aligned to phase two of the EcoCycle 'awareness & knowledge', supporting the importance of this previous phase in the whole picture of understanding the components of behaviour.

9 Conclusion

The EcoCycle framework introduced in this paper, will assist universities as they design, implement and evaluate sustainability activities that engage campus-wide activities. It is important to note that designing behavioural interventions requires the inclusion of the five consecutive EcoCycle phases of: 'Humans & Nature', 'Awareness & Knowledge', 'Capacity & Skills', 'Conduits & Barriers' and 'Decision & Behaviour'. The application of the EcoCycle to the DSC, examined the impacts of the 'Conduits & Barriers' on subsequent 'Decisions & Behaviour'. In future studies, the collection of empirical data alongside each phase of this model will further validate its theoretical structure and capacity to accommodate specific issues, communities and a targeted behaviour. Future investigations of the model should also analyse the relationships between interplaying influential variables to observe any correlations. A future phase for the development of the EcoCycle will include designing and testing indicators for each phase that will enable robust measurements through the collection of qualitative and quantitative data.

The DSC case study has provided a shining beacon for positive, inclusive action across a large institution. This program highlights the importance of activities that are designed by the people they are targeting as a method to ensure their engagement. Human-centric solutions are critical to addressing human-caused problems especially those presented by Global Climate Change. The Challenge demonstrated the capacity of an activity to lift the institutions spirit through playfulness and collaboration. Universities across the world, as leaders in communities, research, technology and education are strategically placed to address sustainability challenges. Universities 'grow' global leaders of the future, a role that requires nurturing them through positive actions in the present. Contagious 'green fever' is spreading across universities through the power of fun nurturing human connections with nature.

Acknowledgments The authors would like to thank Belinda Bean, Sustainability Officer and Leanne Denby, Director of Sustainability from Macquarie University for their assistance in preparing this paper.

References

Avey J, Avolio B, Crossley C, Luthans F (2009) Psychological ownership: theoretical extensions, measurement and relation to work outcomes. J Organ Behav 30(2):173–191

Chavis D, Wandersman A (1990) Sense of community in the urban environment: a catalyst for participation and community development. Am J Community Psychol 18(1):55–81

Cole L (2003) Assessing sustainability on Canadian university campuses: development of a campus sustainability assessment framework. s.n., Quebec

Cubasch UD, Wuebbles D, Chen MC, Facchini D, Frame N, Mahowald Winther JG (2013) Introduction. In: Stocker TF, Qin D, Plattner G-K, Tignor M, Allen SK, Boschung J, Nauels A, Xia Y, Bex V, Midgley PM (eds) Climate change 2013: the physical science basis. Contribution of Working Group I to the Fifth Assessment Report of the Intergovernmental Panel on Climate Change. Cambridge University Press, Cambridge, United Kingdom and New York, USA

Darnton A (2008) GSR behaviour change knowledge review practical guide: an overview of behaviour change models and their uses. Government Social Research Unit, UK. http://www.civilservice.gov.uk/wp-content/uploads/2011/09/Behaviour-change_practical_guide_tcm6-9696.pdf. Accessed 2 Feb 2014

Davies K (2012) Intergenerational democracy: rethinking sustainable development. Galla A (ed) Common Ground Publishing, Illinois, USA

Derksen L, Gartrell J (1993) The social context of recycling. Am Sociol Rev 58:434–442

Genskow K, Prokopy L (eds) (2008) The social indicator planning and evaluation system (SIPES) for nonpoint source management: a handbook for projects in USEPA region 5. Great Lakes Regional Water Program. Publication Number: GLRWP-08-SI01, p 169

Gilg A, Barr S, Ford N (2005) Green consumption or sustainable lifestyles? Identifying the sustainable consumer. Futures 37(6):481–504

Hayles C, Holdsworth S (2008) Curriculum change for sustainability. J Educ Built Environ 3(1): 25–48

Kollmuss A, Agyeman J (2002) Mind the gap: why do people act environmentally and what are the barriers to pro-environmental behavior? Environ Educ Res 8(3):239–260
Liu J, Dietz T et al (2007) Complexity of coupled human and natural systems. Science 317(5844):1513–1516
Lucas K, Brooks M, Darnton A, Jones J (2008) Promoting pro-environmental behaviour: existing evidence and policy implications. Environ Sci Policy 11(5):456–466. http://dx.doi.org/10.1016/j.envsci.2008.03.001
Macquarie University (2012a) Annual sustainability report 2012. http://www.mq.edu.au/search/search-results.html?cx=018228073491741698199%3Abz06o_gto80&cof=FORID%3A9&ie=UTF-8&q=annual+suatianability+report+2012&sa=Search&siteurl=www.mq.edu.au%2Fabout_us%2Fstrategy_and_initiatives%2Fsustainability%2F&ref=&ss=12528j15919830j35. Accessed 28 March 2014
Macquarie University (2012b) Macquarie University annual report 2012. www.mq.edu/annual-reports/MU-AR2012.pdf. Accessed 10 March 2014
Macquarie University (2013a) Department sustainability challenge. http://www.mq.edu.au/about_us/strategy_and_initiatives/sustainability/get_involved/department_challenge/. Accessed 24 July 2013
Macquarie University (2013b) Our university: a framing of futures. http://www.mq.edu.au/about/strategy/. Accessed 4 March 2014
Macquarie University (2013c) Sustainability page. http://www.mq.edu.au/about_us/strategy_and_initiatives/sustainability/. Accessed 8 March 2014
Macquarie University (2013d) Human sciences web page. http://www.humansciences.mq.edu.au/news_and_events/news/2013_news/weve_won_the_2013_sustainability_challenge. Accessed 8 March 2014
Macquarie University (2013e) Get green with the team. https://www.facebook.com/media/set/?set=a.565091573529875.1073741829.176729995699370&type=3. Accessed 21 March 2014
(MQ see Macquarie University)
Martin M, Williams I, Clark M (2006) Social, cultural and structural influences on household waste recycling: a case study. Resour Conserv Recycl 48(4):357–395
McConnell W, Millington A et al (2009) Research on coupled human and natural systems (CHANS): approach, challenges, and strategies. Bull Ecol Soc Am 2011:218–228
McKenzie-Mohr D (2000) Fostering sustainable behavior through community-based social marketing. Am Psychol 55(5):531
Nye M, Hargreaves T (2010) Exploring the social dynamics of proenvironmental behavior change. J Ind Ecol 14(1):137–149
Olli E, Grendstad G, Wollebaek D (2001) Correlates of environmental behaviors bringing back social context. Environ Behav 33(2):181–208
Stern P (2000) New environmental theories: toward a coherent theory of environmentally significant behavior. J Soc Issues 56(3):407–424
Wiek A, Farioli F et al (2012) The gap between science and society. Sustain Sci 7(1):1–4

Author Biographies

Dr Kirsten Davies is an ethno-ecologist who holds a Master's Degree and PhD in Sustainable Management from the University of Sydney. In 2002, she was awarded a Winston Churchill Fellowship to conduct research in USA, UK and Japan. In 2009, Kirsten was awarded an (Australian Government) Australia Research Fellowship to undertake sustainability research in Vanuatu. She is the author of the (2012) book, Intergenerational Democracy, rethinking sustainable development. In 2013 Dr Davies was appointed by the Australian Government as Chair of the Small Pelagic Fishery Resource Assessment Group. She is employed as an academic in The

Department of the Environment and Geography, (Faculty of Science) at Macquarie University, Australia where she is a member of the Human Ethics Sub-Committee. Kirsten is currently undertaking a second PhD in Law (MQU).

Ms Elisabeth Blik is a Masters student at Macquarie University, Australia studying Sustainable Development.

The Processes of Integrating Sustainability in Higher Education Curricula: A Theoretical-Practical Experience Regarding Key Competences and Their Cross-Curricular Incorporation into Degree Courses

M. Novo and MªÁ. Murga-Menoyo

Abstract

This paper presents the experience of the UNESCO Chair in Environmental Education and Sustainable Development in the process of integrating sustainability in university curricula. Our points of reference are both theoretical and practical. With regard to the former, we concentrate our work on the principles and values of sustainable development. At the practical level, the Chair constitutes part of an Interdisciplinary and Inter-University Working Group brought into being by the Conference of Rectors of Spanish Universities ('CRUE') which works on defining key competences in sustainability, and on strategies to incorporate these competences in a cross-curricular manner into all degree qualifications. This paper presents the theoretical approaches and goals achieved by the above-mentioned Working Group, which has succeeded in reaching a consensus on the definition of a number of key competences in sustainability. Our basic model of competences is based on systemic thinking, critical thinking, anticipative thinking, social commitment and ethical commitment. At the same time, as an example of the practical effects of these competences, the Chair presents an educational scheme to train our students along these sustainability lines. The scheme consists of a stimulus activity, in the framework of a teaching strategy oriented towards meaningful learning. Rubrics

M. Novo (✉) · MªÁ. Murga-Menoyo
UNESCO Chair in Environmental Education and Sustainable Development,
National University of Distance Education, Juan del Rosal 14, 28040 Madrid, Spain
e-mail: mnovo@edu.uned.es

MªÁ. Murga-Menoyo
e-mail: mmurga@edu.uned.es

W. Leal Filho (ed.), *Transformative Approaches to Sustainable Development at Universities*, World Sustainability Series, DOI 10.1007/978-3-319-08837-2_9

(performance indicators) are used as an instrument to facilitate the training process in the context of distance-learning university teaching. The proposal can be implemented in other educational contexts of similar characteristics.

Keywords
Key competences • Scoring rubric • Development • Sustainability • Higher education • Curriculum • Learning for sustainable development • Formative assessment • Stimulus activities • Distance-learning university teaching • Degree studies • Systemic thinking • Anticipatory thinking • Critical thinking • Ethical commitment • Social commitment

1 Introduction

The importance and seriousness of the global change our planet is experiencing, and the way in which the resulting effects are already affecting our lifestyles, puts forward the proposition that *as a priority, higher education needs to be open and creative so that new challenges can be viewed from a fresh perspective* (Novo 2012, p. 243). This evidence is taking form, amongst other schemes and avenues of action, via the integration of sustainability into the curricula of higher education degree studies. This entails a new focus on the content of the different degree courses offered from the perspective of a new paradigm, designed to foment in new generations of students the interest and the necessary competences to evaluate, make decisions and act within contexts affected by the problems of development, resource exploitation and the impacts of our way of life on the Biosphere.

It is fair to say that the introduction of sustainability concepts at university level has received a fair amount of attention over the past years, especially in Europe and in North America, but not only there (Leal-Filho 2002). In Spain, research into integrating sustainability into higher education curricula gained an impetus since the appeal at the General Assembly of the Conference of Rectors of Spanish Universities (CRUE 2005), which urged universities to take a three-fold initiative. Firstly, "the comprehensive review of curricula from the perspective of Sustainable Human Development, to ensure the inclusion of sustainability in basic cross-curricular content in all degree courses, aimed at the acquisition of necessary professional, academic and disciplinary skills". Secondly, "the inclusion of sustainability criteria in the systems to evaluate university quality". And thirdly, "the inclusion of sustainability criteria in the process of evaluating lecturers, with the aim of ensuring that teaching and research be conducted in accordance with the principles of Sustainable Human Development". In the three aspects cited above, educational research must be responsible for offering technical solutions to the pedagogical problems and questions raised, and this should be done with due regard to the political dimension involved in such decision-making, since this dimension has to be acknowledged as a necessary pre-requisite for the implementation of such solutions.

The above petition gave additional support to those research teams at the National University of Distance Education who already had wide experience in the area of environmental teaching and education for sustainable development. One such team was the EDUCAMDES group, within the UNESCO Chair of Environmental Education and Sustainable Development. The National University of Distance Education, following the establishment of the European Higher Education Area, had already reinforced the orientation of its research and teaching teams with the aim of improving training strategies to inculcate students' acquisition of the principles and values of sustainable development (Novo et al. 2010; Murga-Menoyo 2007, 2011; Murga-Menoyo et al. 2011; Melendro et al. 2008).

In addition, to meet the demands of CRUE, the EDUCAMDES group undertook to promote the process of integrating *sustainability into the curricula* of the various disciplines taught in higher education. Other research groups from other Spanish' universities have followed the same path (Aznar et al. 2011; Barrón et al. 2010; Gomera Martínez et al. 2012; Segalàs et al. 2009; Aznar Minguet et al. 2014; Azcárate et al. 2012; Sáenz-Rico de Santiago, 2009; Bonil et al. 2012; Vilches Peña and Gil Pérez 2012, 2013; Junyent, 2007; Junyent and Geli de Ciurana 2008; Aznar Minguet et al. 2014, among others).

In this work area, a relevant issue is *the operative definition of competence in sustainability*, and, in line with it, the construction of *a system of performance indicators* to enable the evaluation of students' results, attained via the training processes. The predicament of the evaluation of quality in education constitutes a focal point in the context of current pedagogical research.

This paper details a proposal of a theoretical-practical nature, which has been realised via the integration of both approaches. Building on a clarification of what we understand to be competence in sustainability, we present the competence model which we take as the basis for the design and planning of the training processes pursued in the courses of the Chair. The model also contains the corresponding indicators system and the attainment levels (scoring rubric), illustrated via an actual case study of one of the activities (included in the teaching project) carried out by students. The activity in question can be considered as belonging to the pedagogical strategy based on stimulus activities. It gathers the necessary characteristics to be considered a significant example of the kind of activity that facilitates educational processes for the acquisition of competences in sustainability. And, additionally, it offers the advantage of being able to be implemented, with minor adaptations, in different educational contexts of higher education.

2 What Do We Mean by Competences in Sustainability?

"The strategic move towards a sustainable university begins with efforts of the institution to accomplish the mission of the university in a sustainable manner. Thus, in their mission and vision statements and policy documents universities should define what a sustainable university means for them" (Ciegis and Gineitiene

2006). This follows on from the Lisbon European Council (European Parliament 2000) announcement about the necessity to orientate training processes towards students' acquisition of competences, hence leading to numerous attempts in the academic sector to clarify and define the competence in sustainability operatively and to identify its components.

The OECD (2005, p. 4) states that "sustainable development and social cohesion depend critically on the competences of all of our population—with competences understood to cover knowledge, skills, attitudes and values". To that effect, the *Final Report* of the *Tuning* project (funded to articulate the bases for the European Higher Education Area) adopts the same stance when it links competences with the theme of "what its graduates should be able to *know, understand and do*" (European Commission 2002, p. 20); or, according to the *Tuning* project co-ordinators, it is accepted that "skills represent a dynamic combination of knowledge, understanding, capabilities and capacities" (González and Wagenaar 2003, p. 8).

Subtle nuances should be added to these basic definitions to clarify what we mean by skills, in more detail. Hill et al. (2010) justly underline that competency (which they denominate *environmental competency*) means that an individual acts effectively (an adaptive response) in three scenarios: to resolve problems, to improve and optimise a situation, or to avoid or prevent a negative development. In the last case we must refer to the precautionary principle, which is so significant in the sustainability approach (Cózar Escalante 2005). However, motives, attitudes and environmental beliefs are also psychological factors that function as personal pro-environmental preconditions (Corral-Verdugo 2002), although there is also a correlation with social status (values, norms and access to resources).

Another important aspect to consider in the concept of competence is the versatile nature of its character: these capacities permit individuals to adapt their actions when faced with changes in environmental requirements and problems, as emphasised by Fraijo-Sing et al. (2010, p. 231). Contrary to skills, which almost always do not vary, competences are adaptable. A skill manifests itself in the same way whatever the circumstance faced, whereas competences tailor (i.e., change) their behaviour in response to environmental changes and problems; this inherent property means that competences are extremely pertinent when dealing with one of the most significant characteristics of the ecological crisis: its changeable nature.

In conclusion, it is evident that training in competences in sustainability exhibits multiple, inter-related aspects and, as a consequence, requires a multifaceted teaching programme with innovative pedagogical strategies. If we examine sustainability competences, it seems obvious that it is necessary to encourage their adoption at moments when changing our thinking models and reorienting our professional and human practices requires many readjustments in the ways that humanity today seems to understand the world and to act on it (Novo 2002).

The question of changing our educational models towards incorporating sustainability competences was first suggested some decades ago, but everything would seem to suggest that the next 30 years will be as crucial to the future of the human species as the last 30 years have been in contributing to the current systemic crisis, which is already evident (Millennium Ecosystem Assessment 2005; UNEP

Table 1 Basic competences in sustainability

Basic competence areas	Competence in sustainability (CADEP 2011)	
Systemic thinking	S1—critical contextualisation of knowledge, establishing inter-relationships between social, economic and environmental problems, local and/or global	
Critical thinking		
Anticipative thinking	S2—sustainable use of resources and prevention of negative impacts on the natural and social environment	Activated by performance
Social commitment	S3—participation in community processes which promote sustainability	
Ethical commitment	S4—application of ethical principles related to sustainability values in personal and professional behaviour	

2011; The World Future Council 2012; Worldwatch Institute 2013). If current tendencies are not reversed, then not only will previous problems intensify, but other new problems will emerge as a form of synergetic effect (UNEP 2012).

A call has been made to higher education to train people in the aforementioned competences (UNESCO 2005; UNECE 2011), facilitating people's acquisition of a combination of basic pro-environment skills. In addition, as already pointed out by Corral-Verdugo et al. (2004), people should be trained in how to apply these skills when faced with changing demands and requirements, and how to generate new skills as a response to variations in environmental problems.

From these premises, the recent work of the EDUCAMDES group is allowing to strengthen the work of the "sustainable-ization" of the curriculum in our university (Murga-Menoyo et al. 2013; Olalla-Tárraga et al. 2013; Bautista-Cerro and Murga-Menoyo 2013). To do this, we are advancing in the integration of basic competences in sustainability as main goals of our training model.

In this process, we have taken as our own the thesis that defends the approach of complexity applied to education (Morin 1992, 1994, 1999a, b, 2007, 2008a, b). This approach has global, intra-systemic and inter-systemic interdependence as a powerful cognitive organizer of the dense network of links between living beings, whatever their species, between them and their ecosystems; between inert elements themselves. Additionally, it is understood that competences in sustainability include a values dimension, which is reflected in the *Earth Charter* (ECI 2000), a code of ethics recognized by Unesco (2003) as ethical model for the education for sustainable development that provides a solid axiological basis for building a sustainable world ruled by the respect for nature, universal human rights, economic justice and a culture of peace.

In this framework, we propose the training on competences in sustainability, at the university level, to be focused on five main areas of competence. Three of them-systemic thinking, anticipatory thinking and critical thinking-match the results of an extensive empirical research that has allowed to compare European and Latin American contexts (European and Latin-American perception) (Rieckmann 2012). The other two specify two important areas related to competences: ethical commitment and social commitment. The performance of competences in sustainability occurs on the different dimensions of sustainable development: environment, economy and society, with culture as a transversal dimension (ONU 1992, 2002, 2012; UNESCO 2005, 2011, 2012).

Taking the five previously-mentioned areas of competences, together with the proposal of CADEP (2012)—one of the sectorial committees of the Conference of the Rectors of Spanish Universities (CRUE)—we considered the basic competences in sustainability described in Table 1 to be a strong point of reference (admittedly subject to constant revision) of the specific training processes in sustainability encouraged by the Chair.

In relation to the performance levels of the competences referred to, a revision of the recognised taxonomy of Bloom (Anderson and Krathwohl 2001, pp. 67–68) sets six levels of competence in the cognitive domain: remember, understand, apply, analyse, evaluate, and create. We shall use this revised taxonomy as the basis for setting the scoring rubrics, which are described subsequently in the case study.

The previously described theoretical framework reflects basic elements (epistemological and value principles) that form the basis of the teaching programme of the Chair, and on which the courses and disciplines taught are subsequently outlined. As an example, we present an actual case that helps us exemplify the innovative teaching direction that we are developing.

3 Case Study: Implementation of the Model in Educational Practice

This section presents a case study that consists of an educational proposal ora training process based on a stimulus activity, which responds to a pedagogical and enabling strategy of significant learning centred on the student. This technique has been considered as relevant to the achievement of competences in sustainability by the group of experts tasked by Unesco to monitor The United Nations Decade of Education for Sustainable Development (2005–2014) (UNESCO 2011, p. 26).

3.1 Stimulus Activity

The documentary *Darwin's nightmare* (Sauper 2004), a resource available on the Internet, was used as a stimulus, as it offers a great potential to conduct an active learning practice in the context of training processes which, in higher education, contribute to students acquiring the necessary competences for sustainable development in society. This documentary was awarded a prize at the 2004 Venice Festival and subsequently, in 2006, it was an Oscar nominee for the Academy

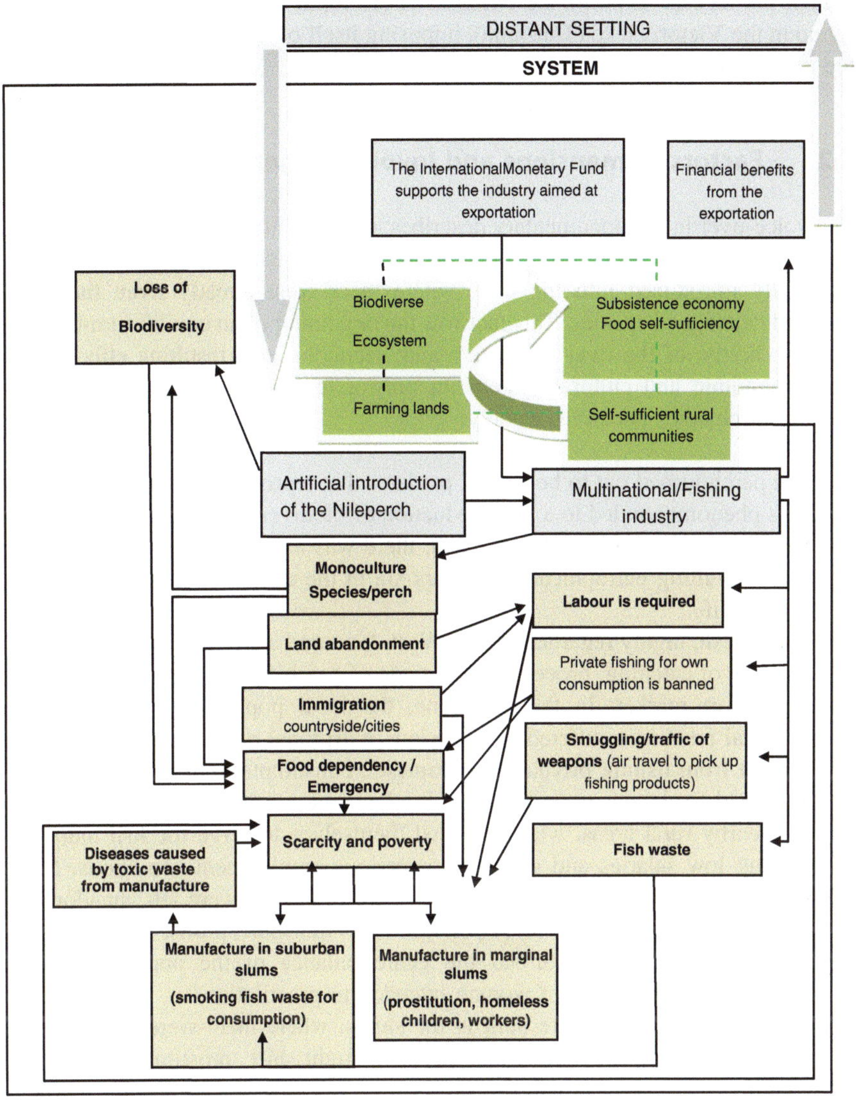

Fig. 1 Interrelations and flows among variables. Case study. *Source* Authors' own—created from *Darwin's nightmare*

Award for Best Documentary Feature. It reflects superbly, as shown in Fig. 1, the extraordinary complexity of the problems affecting sustainable development, the numerous interrelated factors and, consequently, the necessity to adopt a systemic approach towards development, capable of embracing the dense mesh that interrelates the economic, socio-cultural and ecological dimensions of every human reality; in this case in point, the effect of an economic activity, the fishing industry set up in the Victoria lake (Tanzania), imposing itself over people, communities and the natural ecosystem.

3.2 Factors, Dimensions and Involved Agents

The voice-over in the documentary describes a real event, whose origins go back to the middle of last century, when the Nile perch, a non-indigenous fish species, was artificially introduced into the lake, where there were already more than three hundred catalogued species. This fact was the beginning of an equilibrium-breaking process: firstly, of the river ecosystem and, secondly, as a resulting effect, of the socio-economic and cultural system. In only a few decades, a high number of significant changes occurred, namely:

(1) The perch turned out to be a great predator that reproduced itself at a high rate. This phenomenon led to a great reduction of biodiversity in fish species, and as a result of this process of extinction, there was an increase of algae, bringing about resulting eutrophication and anoxia in the depths of the river, thus preventing life.
(2) Perch flesh, highly regarded in The West, favoured a thriving multinational fish industry of catching, processing and trade of perch fillets geared exclusively to the export market. In the meantime, the local population that had scarce financial resources suffered from serious nutritional deficiencies, having been banned from fishing privately for domestic consumption.
(3) The need for labour provoked a 'pull effect' on poorly qualified farmers from the nearby rural areas, who displaced themselves to serve the fish industries, earning low salaries and suffering precarious employment conditions. Meanwhile, the farmlands, basis of a subsistence economy, were left abandoned.
(4) The increase in the male population (air transport pilots, fishermen and immigrant workers) disturbed the genre balance of the population, hence encouraging the arrival of women intended to prostitution.
(5) Marginal communes were built in the slums, where there were three types of inhabitants: day-shift unskilled labour, night-shift prostitutes and street children.
(6) The unskilled population consumed the waste from the fish industry (fish heads and bones piled up in the suburbs), previously prepared via smoking processes which, as they used plastic, produced health-damaging acids.
(7) The plastic remnants themselves were used by homeless children to obtain a liquid glue that they inhaled, hence adding drug addiction to the other afflictions that they suffered, such as sexual abuse and destitution.
(8) The constant air traffic was soon associated with the introduction and trade of arms destined for the bloody wars of the African continent.

(9) Meanwhile, the country was periodically hit by severe famines, and received food via international aid donations. The International Monetary Fund supported private initiatives, and export-oriented production.

The documentary reveals that the aforesaid factors operated in a systemic dynamic, with flows and interrelations, with global effects (synergic) and the emergence of new factors and problems (system emergencies). Thanks to its versatility to exemplify situations and interdependent phenomena, the documentary was selected as a resource to be used within the training process planned for the acquisition of competence in sustainability. It acted as the basis (or supporting framework) to the activity that students had to complete, and that was included in the teaching programme in the degree subjects of the Chair.

3.3 Context and Characteristics of the Training Activity

This scheme of teaching innovation has been designed with certain students in mind, that is, the students that the Chair deals with in the degree qualifications offered by the UNED, a distance-learning university that uses ICT (Information and Communication Technologies) intensively in its training processes. Most students have no previous systematic training with regard to the characteristic difficulties of sustainable development.

The activity constitutes part of an ample set of activities included in the teaching programme. It is designed so that all the tasks can be completed on the virtual platform, supported by the Internet.

3.4 Activity Goals

The activity has seven objectives, which are defined as:

1. Understand the case presented by the documentary from a sustainable development perspective.
2. Identify the active agents involved, and their motivations and responsibilities.
3. Recognise the interrelations between different human groups and significant factors.
4. Understand side-effects, local effects and global effects.
5. Reinforce relational and systemic thinking processes, a critical judgement and debating skills.
6. Practise collaborative work, using technology via forums.
7. Consolidate language abilities and the correct use of written language.

Table 3 Rubric 2 (report)

Indicators	Performance levels		
	Level 1	Level 2	Level 3
Critical thinking	Exhibits the strengths and weaknesses of a situation	Explain arguments of strengths and weaknesses in a reasoned manner	Issues reasoned judgements from individual argumentations
Anticipatory thinking	*Establishes linear cause-effect relationships*	*Anticipates possible consequences and effects of actions*	*Suggests alternative ways of proceeding*
Ethical and social commitment	Identifies the value of the human behaviour involved	From a local perspective, logically explains the social and ethical implications of the identified values	From a 'glocal' perspective, logically explains the social and ethical implications of the identified values

4 Conclusion

The described activity revolves around a case study that reflects key issues of sustainable development. It enables to put the spotlight on the various nested factors in the ecological, economic and social dimensions of development. And above all, it allows to appreciate the effects of reciprocal relationships. It is a prime example of complexity, very suitable in promoting relational and critical thinking and social consciousness, which serve as basis for competences in sustainability. Moreover, it is a simple activity of implementation at the level of higher education, even in diverse contexts and with different student profiles.

Embracing different paths to progress towards sustainable development is an urgent task, indispensable for our societies. Some of the environmental problems we are creating present irreversible signs (for example, global warming) and they simply increase the uncertainty with which today's world population lives, with regard to the effects that could emerge in the future and that we have suffered in recent times (for example, sea level rise and flooding of lower areas, or the loss of territory in some islands).

In this scenario, evidence emerges that our actions, whatever their nature, are not neutral (Novo 2006, p. 233) and with it, the necessity that designers, managers and educators accept the importance of working on sustainability.

In the last decade there has been an educational research and innovation trend that has worked in the direction referred to above in the field of higher education, and that aims to address the design and development of training processes based on the competence in sustainability. Its objective and implementation are complex, and together they present varied and multiple facets, as well as numerous factors to bear in mind.

Surely, the theoretical corpus of an educational model coherent with sustainable development, which serves as a fundamental reference in a work of this nature, has been built in the last decades and it is fair to state that today it holds a solid and well-argued epistemological and axiological framework. Similarly, the conviction that training in competences is the educational goal to aim for is practically unanimous; nobody debates the new terminology, which is now generally accepted. However, as we descend towards more concrete and practical levels, difficulties seem to increase.

Thus, when the moment comes to deal with the theoretical-practical area that requires the operative definition of competences, difficulties start to arise, so far preventing a consensus. The term is applied to a complex concept, with numerous components and qualitative aspects that are difficult to pin down. Above all, it refers to behaviours that are 'observable' only whilst the pertinent competence is performed.

For that reason, current competence models have a temporary nature, are subject both to permanent examination and to an active process of research-action, in an attempt (apparently permanently unfinished) to describe behaviours that on many occasions imply a high degree of complexity. The issue affects directly, though not only, the evaluation process and, specifically, the creation of a system of performance indicators, the conception of rubrics or evaluation protocols.

It is quite likely that more means and dedication will be required in the near future. The work presented here aims to contribute to the already open debate, and to offer a proposal that we understand could lead to good practice in the field of training in competence in sustainability.

References

Anderson L, Krathwohl DA (eds) (2001) Taxonomy for learning, teaching and assessing: a revision of Bloom's taxonomy of educational objectives. Longman, New York, p 352

Azcárate P, Navarrete A, García E (2012) Aproximación al nivel de inclusión de la sostenibilidad en los curricula universitarios", Revista de curriculum y formación del profesorado 16(2):105–119. http://www.ugr.es/~recfpro/rev162ART7.pdf. Accessed 3 Feb 2014

Aznar Minguet P, Ull A, Piñero A, Martinez-Agut MP (2014) La sostenibilidad en la formación universitaria: desafíos y oportunidades, Educación XX1 17(1):131–158. doi:10.5944/educxx1.17.1.10708

Aznar P, Martinez-Agut MP, Palacios B, Piñero A, Ull A (2011) Introducing sustainability into university curricula: an indicator and baseline survey of the views of university teachers at the University of Valencia. Environ Educ Res 17(2):145–166

Barrón Á, Navarrete A, Ferrer-Balas D (2010) Sostenibilización curricular en las universidades españolas. ¿Ha llegado la hora de actuar?, Revista Eureka sobre Enseñanza y Divulgación de las Ciencias 7:388–399. doi:10498/9877

Bautista-Cerro MªJ, Murga-Menoyo MªÁ (2013) Evaluation as a key element of the learning process: contributions from environmental education. European Conference on Educational Research. European Educational Research Association, Istambul. 9–13 Sept 2013

Bonil J, Calafelll G, Granados J, Junyent M, Tarín RM (2012) Un modelo formativo para avanzar en la ambientalización curricular, Revista de curriculum y formación del profesorado 16 (2):145–163. http://www.ugr.es/~recfpro/rev162ART9.pdf. Accessed 3 Feb 2014

CADEP (2012) Resolución aprobada en plenario. Jornadas del Seminario Permanente de la Comisión Sectorial de la CRUE para la Calidad Ambiental, Desarrollo Sostenible y la Prevención de Riesgos (CADEP), Valencia, 8–9 March 2012

Ciegis R, Gineitiene D (2006) The University in the Promotion of Sustainability. In: Leal Filho W (ed) Innovation, Education and Communication for Sustainable Development. Peter Lang GmbH, Frankfurt am Main, p 763

Corral-Vedugo V (2002) A structural model of pro-environmental competency. Environ Behav 34:531–549

Corral-Verdugo V, Varela C, Gonzalez D (2004) O papel da Psicologia Ambiental na promoçao de competenica pró-ambiental. In: Tassara ETO, Rabinovich EP, Guedes MC (eds). *Psicologia e ambiente*. Editorial da PUC-SP Sao Paulo, Brazil, p 412

Cózar Escalante JM (2005) Principio precautorio y medio ambiente. Revista Española de Salud Pública 79(2):133–144

CRUE (2005) Directrices para la Sostenibilización Curricular. Document approved by the Executive Committee of the Working Group on Environmental Quality and Sustainable Development of the Conference of Rectors of Spanish Universities, held at Valladolid on April 18th 2005. http://www.uah.es/universidad/ecocampus/documentos/C3Sostenibilizacion Curricular_CRUE.pdf. Accessed 3 Feb 2014

ECI/Earth Charter Iniciative (2000) The earth charter. http://www.earthcharterinaction.org/invent/images/uploads/echarter_english.pdf. Accessed 3 Feb 2014

European Commission (2002) Tuning educational Estructures in Europa. Closing Conference, Brussels, 31 May. http://www.aic.lv/bolona/Bologna/Reports/projects/Tuning/Tun_Book.pdf. Accessed 3 Feb 2014

European Parliament (2000) "Presidency Conclusions" Lisbon European Council 23 and 24 march. http://www.europarl.europa.eu/summits/lis1_en.htm. Accessed 3 Feb 2014

Fraijo-Sing B, Corral-Verdugo V, Tapia-Fonllem C, González-Lomelí D (2010) Promoting pro-environmental competency. In: Corral-Verdugo V; García-Cadena CH, Frías-Armenta M (eds) Psychological Approaches to Sustainability: Current Trends in Theory, Research and Applications. Nova Science Publishers, Inc., New York, p 473

Gomera Martínez A, Villamandos de la Torrre F, Aquero Abellán M (2012) Medición y categorización de la conciencia ambiental del alumnado universitario: contribución de la Universidad a su fortalecimiento, Revista de curriculum y formación del profesorado 16 (2):213–228. http://www.ugr.es/~recfpro/rev162ART11.pdf. Accessed 3 Feb 2014

González J, Wagenaar R (2003) Tuning educational structures in Europe. Universities' contribution to the Bologna Process, p 160. http://www.unideusto.org/tuningeu/images/stories/Publications/ENGLISH_BROCHURE_FOR_WEBSITE.pdf. Accessed 3 Feb 2014

Hill D, Figueredo AJ, Jacobs WJ (2010) Contextual influences on sustainable behaviour. In: Corral-Verdugo V, García-Cadena CH, Frías-Armenta M (eds) Psychological approaches to sustainability: current trends in theory, research and applications. Nova Science Publishers, Inc., New York, p 473

Junyent M (2007) The ACES network: greening the curriculum of higher education. In: Hopkins C, McKeown R (eds) Good practices in teacher education institutions. UNESCO, Paris, France, p 54. http://unesdoc.unesco.org/images/0015/001524/152452eo.pdf. Accessed 3 Feb 2014

Junyent M, Geli de Ciurana AMª (2008) Education for sustainability in university studies: a model for reorienting the curriculum. Br Educ Res J 34(6):763–782. doi:10.1080/01411920802041343

Leal-Filho W (2002) Teaching sustainability: some current and future perspectives. In: Leal Filho W (ed) Teaching sustainability at Universities. Peter Lang GmbH, Frankfurt am Main, p 576

Melendro M, Murga MªÁ, Novo M, Bautista-Cerrro MªJ (2008) Estrategias formativas innovadoras en Educación Ambiental y para el desarrollo sostenible. Revista Iberoamericana de Educación a Distancia (RIED) 11(2):15–40

Millennium Ecosystem Assessment (2005) Ecosystems and human well-being: synthesis. Island Press, World Resources Institute, Washington. http://www.millenniumassessment.org/documents/document.356.aspx.pdf. Accessed 3 Feb 2014

Morin E (1992) From the concept of system to the paradigm of complexity. J Soc Evol Syst 15 (4):371–385

Morin E (1994) La complexité humaine. [Human complexity]. Flammarion Publishing, Paris, p 380

Morin E (1999a) Seven complex lessons in education for the future. UNESCO, Paris, France, p 67. http://unesdoc.unesco.org/images/0011/001177/117740eo.pdf. Accesed 2 Jan 2014

Morin E (1999b) Les Sept savoirs nécessaires à l'éducation du future. Unesco, Paris, p 67. http://unesdoc.unesco.org/images/0011/001177/117740fo.pdf. Accessed 3 Feb 2014

Morin E (2007) Restricted complexity, general complexity, In: Gershenson C, Aerts D, Edmonds B (eds) World-views science, and us: philosophy and complexity, World Scientific Publishing, New York, p 351

Morin E (2008a) The reform of thought, transdisciplinarity, and the reform of the university. In: Nicolescu B (ed) Transdisciplinarity: theory and practice. Hampton Press, Cresskill, United States, p 320

Morin E (2008b) On complexity. Hampton Press, Cresskill, United States, p 127

Murga-Menoyo MªÁ (2007) Potentiality of collaborative learning to train behaviours oriented towards sustainability. In: Leal Filho W, Manolas EI, Sotirakou MN, Boutakis GA (eds) Higher education and the challenge of sustainability: problems, promises and good practice. Environmental Education Center of Soufli, Orestiada, Grecia, p 320

Murga-Menoyo MªÁ (2011) Infotecnología e innovación docente. Los editores de mapas conceptuales: posibilidades y límites. Revista Española de Pedagogía 249:273–288

Murga-Menoyo MªÁ, Bautista-Cerro MªJ, Novo M (2011) Mapas conceptuales con Cmap tools en la enseñanza universitaria de la educación ambiental. Estudio de caso en la UNED. Revista Enseñanza de las Ciencias 29(1):47–60. http://ensciencias.uab.es/index.php?option=com_content&task=view&id=105&Itemid=37. Accessed 3 Feb 2014

Murga-Menoyo MªÁ, Bautista-Cerro MªJ, Novo M (2013) Técnicas docentes orientadas al logro de competencias en sostenibilidad. I Encuentro Universitario de Sostenibilización Curricular, UEM/CADEP-CRUE. Madrid, 24–25 Jan 2013

Murga-Menoyo MªÁ, Bautista-Cerro MªJ, Olalla MA, Novo M (2013) Competencias en sostenibilidad: modelización, indicadores de logro y criterios de evaluación. I International Congress of Educational Sciences and Development. Santander, 8–11 de octubre de 2013. Fundación General UGR-Empresa/ Universidad de Granada/ Sociedad Española de Pedagogía

Novo M (2002) Higher environmental education in the XXI century: toward a new interpretative paradigm. In: Leal Filho W (ed). Teaching Sustainability at Universities. Peter Lang GmbH, Frankfurt am Main, p 576

Novo M (2006) El desarrollo sostenible. Su dimensión ambiental y educativa, Pearson/UNESCO, Madrid 431 p

Novo M, Murga-Menoyo MªÁ, Bautista-Cerro MªJ (2010) Educational advances and trends for sustainable development: a research project on educational innovation. J Baltic Sci Educ 9(4): 302–314. http://journals.indexcopernicus.com/abstracted.php?level=5&icid=925829. Accessed 3 Feb 2014

Novo M (2012) Higher education, modernity and a new paradigm. In: GUNI: higher education in the world 4 (higher education's Commitment to sustainability: from understanding to action. Palgrave/Macmillan, London, p 341

OECD (2005) The definition and selection of key competences. Executive Summary, p 20 http://www.oecd.org/pisa/35070367.pdf. Accessed 3 Feb 2014

Olalla-Tárraga MA, Bautista-Cerro MªJ, Díaz MªJ, Salmerón P, Murga-Menoyo MªÁ, Novo M (2013) Ambientalización de la gestión y sostenibilización curricular en la UNED. Análisis de situación. In: 7th WEEC, World Environmental Education Congress. WEEC International Network, p 9–14 junio. Marrakech

ONU (1992) Report of the united nations conference on environment and development, vol 1, Rio de Janeiro. A/CONF.151/26 Annex I. 3–14 June 1992. http://www.un.org/documents/ga/conf151/aconf15126-1annex1.htm. Accessed 13 Feb 2014

ONU (2002) Johannesburg declaration on sustainable development. In: World Summit on Sustainable Development (A/CONF.199/20) http://www.un-documents.net/jburgdec.htm. Accessed 13 Feb 2014

ONU (2012) United nations conference on sustainable development (Río+20) Agenda item 10, Outcome of the Conference, 20–22 de junio, A/CONF.216/L.1 https://rio20.un.org/sites/rio20.un.org/files/a-conf.216l-1_english.pdf.pdf. Accessed 13 Feb 2014

Rieckmann M (2012) Future-oriented higher education: which key competences should be fostered through university teaching and learning? Futures 44:127–135

Sáenz-Rico de Santiago B (2009) Ejemplificación del diseño curricular por competencias en el contexto del Espacio Europeo de Educación Superior. Grupo Editorial Universitario, Granada

Sauper H (2004) Darwin's nightmare. Documentary 107'. http://www.youtube.com/watch?v=fir29x8EkBc. Accessed 3 Feb 2014

Segalàs J, Ferrer-Balas D, Svansrom M, Lundqvist U, Mulder K (2009) What has to be learnt for sustainability? a comparison of bachelor engineering education competences at three European universities. Sustain Sci 4:17–27. doi:10.1007/s11625-009-0068-2

The World Future Council (2012) Ombudspersons for future generations: tackling sustainability implementation gaps. http://www.uncsd2012.org/content/documents/World_Future_Council_Zero_Draft.pdf. Accessed 3 Feb 2014

UNECE (2011) Learning for the future: competences in education for sustainable development. p 11. United Nations Economic Commission for Europe/ Steering Committee on Education for Sustainable Development. http://www.unece.org/fileadmin/DAM/env/esd/ESD_Publications/Competences_Publication.pdf. Accessed 3 Feb 2014

UNEP (2011) Keeping track of our changing environment: form Rio to Rio+20 (1992–2012). (Nairobi: Division of Early Warning and Assessment (DEWA), United Nations Environment Programme, p 99. http://www.unep.org/geo/pdfs/Keeping_Track.pdf. Accessed 3 Feb 2014

UNEP (2012). GEO 5. global environment outlook. Environment for the future we want, p 22. United Nations Environment Programme. http://www.unep.org/geo/pdfs/geo5/GEO5_report_full_en.pdf. Accessed 3 Feb 2014

UNESCO (2003) Records of the 32nd general conference, vol 1. http://unesdoc.unesco.org/images/0013/001331/133171e.pdf. Accessed 13 Feb 2014

UNESCO (2005) United Nations decade of education for sustainable development 2005–2014. Draft international implementation scheme. Unesco, Paris, p 31. http://usnesdoc.unesco.org/images/0014/001486/148654e.pdf. Accessed 3 Feb 2014

UNESCO (2011) Education for sustainable development. An expert review of processes and learning, p 132. Unesco, Paris. http://unesdoc.unesco.org/images/0019/001914/191442e.pdf (Last. Accessed 3 Feb 2014

UNESCO (2012) Education for sustainable development sourcebook, p 51. http://unesdoc.unesco.org/images/0021/002163/216383e.pdf. Accessed 13 Feb 2014

Vilches A, Gil Pérez D (2013) Creating a sustainable future: some philosophical and educational considerations for chemistry teaching. Sci Educ 22(7):1857–1872. doi:10.1007/s11191-011-9404-x

Vilches Peña A, Gil Pérez D (2012) La educación para la sostenibildiad en la universidad. El reto de la formación del profesorado. Revista de curriculum y formación del profesorado 16(2):25–43. http://www.ugr.es/~recfpro/rev162ART3.pdf. Accessed 3 Feb 2014

Worldwatch Institute (2013) State of the world 2013: is sustainability. still possible? p 464

Author Biographies

María Novo Ph.D. in Philosophy and Educational Sciences, is a Senior Professor and the Chairholder of the Unesco Chair in Environmental Education and Sustainable Development at the National University of Distance Education (Spain) [www.uned.es/catedraunesco-educam]. She has worked as a Unesco Consultant in relation to environmental issues, environmental education and sustainable development. Member of the Association of Environmental Scientists (CIMA). Analyst Member of the Institute of Transnational Studies (INET). Member of the International Network of Writers for The Earth (RIET). Special Advisor to the Earth Charter International Initiative. Lecturer in more than 180 national and international conferences and meetings. As well as many scientific articles in specialised magazines, she has published 16 books on environmental matters, environmental education and sustainable development, six of which were sponsored by Unesco. She is Director of the Research Group EDUCAMDES and of the ECOARTE Project, which aims at integrating Science, Art and Environment [www.ecoarte.org].

MªÁ. Murga-Menoyo Ph.D. in Philosophy and Educational Sciences, is Professor and Head of the Department of Theory of Education and Social Pedagogy at the National Distance Education University, UNED, Madrid (Spain). She is a member of Unesco Chair in Environmental Education and Sustainable Development and of the EDUCAMDES research group. She carries out her research mainly in two areas: environmental education for sustainable development, and key competences for sustainable development in higher education. She represents the UNED in the working group *Curricular sustainability* of the Conference of Rectors of Spanish Universities (CRUE). The book entitled *Desarrollo sostenible, problemáticas, agentes y estrategias* (2013, Madrid: Mac Graw Hill) is among her recent publications.

Integrating a Triple Bottom-Line Approach into the Management System: A Framework for Institutions and Businesses Alike

Claire Bennett and Geoffrey Lewis

Abstract

Organizations are increasingly transitioning to a more sustainable management approach as a result of drivers related to corporate reputation, operational efficiency, and regulatory compliance. Accordingly, organizations that are proving resilient to this trend are gaining a competitive edge. This paper draws from the field of industrial ecology in terms of material flow analysis to set a foundation for organizations that wish to integrate sustainability principles into the management system. Though research has linked industrial ecology to sustainable business operations, there is a void on the practical level in terms of implementation and maintenance strategies, leaving businesses unequipped to adapt to a more sustainable approach, now commonly known as triple bottom line. In response, the framework of this paper uses the concept of material flow analysis, to guide a detailed process for achieving more sustainable business operations.

Keywords

Sustainable management · Triple bottom-line · Material flow analysis · Indicators

C. Bennett
Sustainability Office, Physical Resources, Wilfrid Laurier University,
75 University Ave W, Waterloo ON N2L 3C6, Canada
e-mail: cbennett@wlu.ca

G. Lewis (✉)
Faculty of Environment, School of Planning, University of Waterloo,
200 University Ave W, Waterloo ON N2L 3G1, Canada

W. Leal Filho (ed.), *Transformative Approaches to Sustainable Development at Universities*, World Sustainability Series, DOI 10.1007/978-3-319-08837-2_10

1 Introduction

Organizations are facing new challenges when it comes to profit. Tighter legislation and standards, stakeholder pressure, and increasing commodity prices are driving more sustainable operations, often called triple bottom-line management (Blizzard and Klotz 2012; Kaiser 2011; Carroll and Buchholtz 2011; Rasmussen 2011; Hoffman and Woody 2008; Walton and Galea 2005; Azapagic 2003; Karapetrovic and Jonker 2003; Azapagic and Perdan 2000). Still, triple bottom-line management systems remain fragmented and ineffective (Esquer-Peralta et al. 2008; Hoffman and Woody 2008; Azapagic 2003; Karapetrovic and Jonker 2003; Kirkland and Thompson 1999), largely due to the fact that this type of management has not been clearly defined in a common operations framework (Esquer-Peralta et al. 2008; Azapagic 2003). Additionally, existing research in this area focuses on assessments and reporting rather than on the process of action planning and implementation (Rasmussen 2011; Bertels 2010; Hoffman and Woody 2008; Kirkland and Thompson 1999; Azapagic 2003). Frameworks that do exist, such as ISO's environmental (14001) and energy management standard (50001), do not embrace triple bottom line management in a holistic manner and tend to be impractical for many organizations because they are resource intensive and focus largely on the manufacturing sector.

Accordingly, this paper offers a structured management framework for organizations that seek to achieve a comprehensive triple bottom-line approach to business, both strategically and operationally. Specifically, the paper outlines a process for assessment, indicator and target setting, and reporting that is necessary to achieve an embedded result, yet largely absent in both academic and professional contributions. The argument is that a triple bottom-line approach to management improves an organization's ability to maintain efficiency, compliance, and reputability to be resilient in terms of external requirements and emerging trends (Holdsworth 2003).

1.1 Background: Industrial Ecology, Sustainability, and Triple Bottom-Line

The notion of sustainability is strongly connected to ecology, a field that focuses on the interdependence of entities through material and energy flows in nature. Initially, the field studied individual organisms, but later broadened to include physical processes such as food chains and material cycling, from which the contemporary field of industrial ecology developed (Odum 1953). The focus on physical processes, specifically material analysis, results in a reduction in harmful emissions and wastes.

As the field of industrial ecology progressed, the notion of sustainability in business became a common phenomenon (Blizzard and Klotz 2012; Kaiser 2011). In business, sustainability implies that management practices are based on environmental knowledge and social concerns, in addition to economics (Carroll and Buchholtz 2011; Rasmussen 2011). With the triple bottom-line approach, the economy is reliant on society, and both the economy and society are contained within the environment (Rasmussen 2011; Willard 2009).

In the triple bottom line approach to business there are several levels of capital. Human capital exists in the form of labour, culture, and organization; economic capital includes financial elements such as cash, investments, and monetary instruments as well as manufactured elements fundamental to infrastructure, machines, tools and factories necessary for existing levels of production; and natural capital are ecological resources systems and services (Carroll and Buchholtz 2011; Rasmussen 2011).

1.2 Organizational Management Systems

Using industrial ecology's approach as inspiration, an organization's business plan is approached as a system, which is made up of the common structures, processes, hardware, software, facilities, and people that work together to produce the product, service, and culture of an organization. Accordingly, the management system is understood in the context of the relationships it contains rather than in isolation (Blizzard and Klotz 2012; Esquer-Peralta et al. 2008; Azapagic 2003; Clayton and Radcliffe 1996).

The systems approach uses key inputs and outputs to guide the design, management, and communications to integrate sustainability within the business plan (Azapagic 2003; Esquer-Peralta et al. 2008; Karapetrovic and Jonker 2003). Using material flow analysis (MFA) to quantify inputs and outputs, this paper defines three main stages in creating a management system: (1) assessing the organization; (2) the inventory process; and (3) identifying goals, actions and indicators. With the MFA approach, the organization is viewed as a system made up of many parts, all of which function because of materials, energy, and knowledge that flow through the organization based on desired outputs.

2 Method for Creating a Triple Bottom-Line Management System

2.1 Stage 1: Assessing the Organization: Define the System Boundaries

2.1.1 Scope Analysis

Before an assessment of inputs and outputs can take place, the organization must define the measurement boundaries, including the overall organizational mission and core administrative functions under consideration. The first step is to determine the scope of business operations in terms of its spatial, temporal, and administrative parameters. These parameters act as a guide for defining and organizing the capabilities of an organization when considering the day-to-day inputs and outputs under investigation (Shriberg 2002; Chee Tahir and Darton 2010; Bell and Morse 2008).

Temporal

The temporal scale is the periodover which the impacts of operation are considered. This period must accommodate emerging as well as existing trends (Shriberg 2002; Chee Tahir and Darton 2010; Bell and Morse 2008).

Span of a management plan—the life-span intended for a major guiding management document, such as master plan, academic plan, transportation management plan, waste management plan, climate plan, etc.

Time management—time designated for in and out of office work hours as well as breaks, lunches, social events, etc.

Period of review—time given to the life of a plan, policy, job, committee, etc. based on such things as years of service, experience, predestined service length, votes, etc.

Spatial

The spatial scale is the physical extent of the system and its components, and can include factors that impact the business in terms of geography (Chee Tahir and Darton 2010).

Capacity of space and people—the expected or planned growth of an organization and what that means in terms of physical size (e.g., square footage or employees).

Function of space and people—the planned use of physical space and employee roles.

Organizational structure and scope—the planned size, hierarchy, and responsibility of departments, roles, and relationships.

Demography

Demography includes trends that drive business in terms of stakeholder base and changes to the economy driven by the future environment. More specifically, it includes information on employees such as employment type, qualifications, and other profile information.

Internal stakeholders—the members of the organization that directly contribute to its operational function (e.g., employees, managers, board and committee members).

External stakeholders—the members who affect the state of an organization (e.g., auditors, customers, and media).

Trends—past, current, and future drivers affecting the market (e.g., age, gender, economic standing, location, and marital status).

2.2 Stage 2: Inventory Process

2.2.1 Gap Analysis

A gap analysis brings the organization's management structure into perspective by identifying strengths and weaknesses in the flow of information (Chee Tahir and Darton 2010; Holdsworth 2003). The concept is to make sure an organization is making the best use of resources in relation to its business plan (Chee Tahir and Darton 2010). Once gaps are identified, those deemed to be significant can be categorized and associated with measurable indicators, supporting the development of a set of metrics aligned with a triple bottom-line management plan (Bell and Morse 2008).

The gap analysis is performed on both the strategic and operational level. The strategic level analysis identifies where existing management and outcomes fall short in terms of potential structural performance (Holdsworth 2003). The operational level gap analysis follows and confronts the inputs and outputs of the organization. Throughout this process, the organization must continually reference its overall mission and guiding principles to fully capture where current administration falls short and to guide the improvement process, which in this case emphasizes triple bottom line management.

Strategic Gap Analysis

Strategic business practices help respond to driving forces in industry by directing how inputs and outputs are governed (Pojasek 2009; Holdsworth 2003; Kirkland and Thompson 1999). Therefore, the business must identify the central areas and supporting functions necessary to achieve the overall mission. The strategic scope analysis includes the necessary features of business, but is not drawn so widely that inessential activities confuse subsequent analysis (Chee Tahir and Darton 2010; Bell and Morse 2008; Shriberg 2002). The organization must first identify its mission and guiding principles as well as the administrative parameters by which it operates via policies, plans, procedures, regulations, and measurement tools.

The mission of a business is an overarching directive that is marketed widely. Triple bottom-line principles fit very easily into the mission, vision, and guiding principles of an organization because they help bring resources back to deliverables (Chee Tahir and Darton 2010). Missions are generally supplemented with more specific values and guiding principles that allow a business to highlight important organizational objectives. While some activities are paramount to the central goals of the organization, others will be considered supporting activities. Most organizations have similar support activities, such as purchasing, facilities management and customer service (Walton and Galea 2005).

2.2.2 Administration

The core mission of an organization drives the strategic systems, which can be categorized as:

- Tangible products manufactured, distributed, resold, etc.
- Services, such as consulting, education, labour
- Education
- Administration

Strategic considerations of an organization are very much consistent across all organizational sectors, as the mission statement succinctly captures the overall direction and objective of the organization. Each organization differentiates itself though the guiding principles and goals that supplement the organizational mission. Guiding principles are the ideals that an organization seeks to embody in all its operations, for example:

- Leadership
- Socially responsible
- Environmentally responsible
- Quality control and continuous improvement
- Customer Service

The mission of the organization is to provide excellence in the product, service, or experience being offered and the guiding principles identify the chief components of the organization that will help achieve the mission. The triple bottom line approach is included at the highest-level of the strategic system because it will guide all that follows. To best achieve the mission, the organizational policies, procedures and programs must translate the message into clear processes, which assume forms such as:

- Master plans
- Standards
- Work procedures
- Internal and external programs

Finally, the administration of a business must have an integrated way of tracking operations. Measurement tools monitor the inputs and outputs of the management system and include such things as utility metering, building automation systems, human resources information systems, energy management systems, and financial and capital planning systems. Reporting methods use information from the measurement systems to communicate with stakeholders. Board level stakeholders may need only high-level information, and middle management will want detailed reporting on day-to-day operations.

Operational Gap Analysis

An operational gap analysis captures key business inputs and outputs with an eye to the strategies driving operations. Inputs and outputs include materials, energy and knowledge that flow through the organization (Chee Tahir and Darton 2010). Material flows include physical resources that enter and exit an organization; energy flows are the means by which materials are transformed. Knowledge is closely related to the strategic elements of the organization, such as policies and procedures, standards, meetings, and the outreach initiatives that frame the triple bottom line message in a manner that reflects the organization's values (Chee Tahir and Darton 2010; Willard 2009). All three resources are related: materials require energy to transform, which takes place via the knowledge driving the transformation.

2.2.3 Identify Transformations

Once the gap analysis of the inputs and outputs is complete, the next step is to establish the relationship between all flows. Transformations are how inputs turn into outputs. When analyzing and identifying this process, they note how elements function individually, interdependently and as a whole. These considerations provide insight into the value of the functions in terms of the ability to operate within the system, which ultimately provides insight into strategic level changes that may be necessary (Clayton and Radcliffe 1996).

2.2.4 Identify Subsystems

Each subsystem will provide functions of the overall product or service provided by the organization. Each function of the product should have at least one system, though often a given subsystem will contribute many different functions, for example, the subsystem of business operations includes finance and administration, facilities management, procurement, legal, and human resources. To understand the subsystems involved, connect the flows of materials, energy, and information and define the relationships.

2.3 Stage 3: Identify Organizational Goals, Actions, and Indicators

Like other performance metrics, sustainable development indicators provide means to target, measure, and report on progress towards a goal. They help assess conditions and trends, enabling businesses to formulate and communicate strategic mandates through measurable goals and associated actions (Azapagic and Perdan 2000). This information is used to establish baselines, a necessary step in monitoring, evaluating, and reporting on programs and overall operations (Singh et al. 2009).

All organizations share the desire to improve efficiency, compliance levels, and reputational standing, so indicators for triple bottom line management will be similar across organizations and sectors, facilitating comparison and benchmarking (Chee Tahir and Darton 2010). As with MFA, indicators are designed to be traceable to a particular process, which is especially useful in determining impact and association with an organization's inputs and outputs (Chee Tahir and Darton 2010; Azapagic and Perdan 2000). Therefore, there must be a clear association between what the indicators are measuring and the outputs that are desired (Azapagic and Perdan 2000). Considering the assessment process in Stage 1 and 2, the next step is to summarize the areas of the organization that have significant consequences to business processes in terms of continual improvement (Azapagic and Perdan 2000).

Indicators can be qualitative or quantitative; they result from attributed values (Meadows 1998). Quantitative indicators are the easiest to calculate since they measure a physical quantity, such as the size of a building. Qualitative indicators also have a numeric value attached to them, but rather than measuring something concrete, like size, they measure, for example, the change of judgment or perception and evaluated through surveys and questionnaires (Church and Rogers 2006).

There are several criteria to consider when developing indicators, and a comprehensive understanding of the organization is beneficial. The first step is to identify what should be measured based on what is valued in terms of maintenance or alteration. The second step defines the scope of the indicators so they provide a comprehensive picture of organizational goals. The scope informs the elements and structure that make up a standard of change, which includes items such as target demographics, time frames, and scale (Church and Rogers 2006; Holdsworth 2003). The next step is to allocate a metric so all parties understand the scope of the intended change.

Figure depicts the steps to create a standard for an organization. With standards created, indicators then measure, influence, and report on progress.

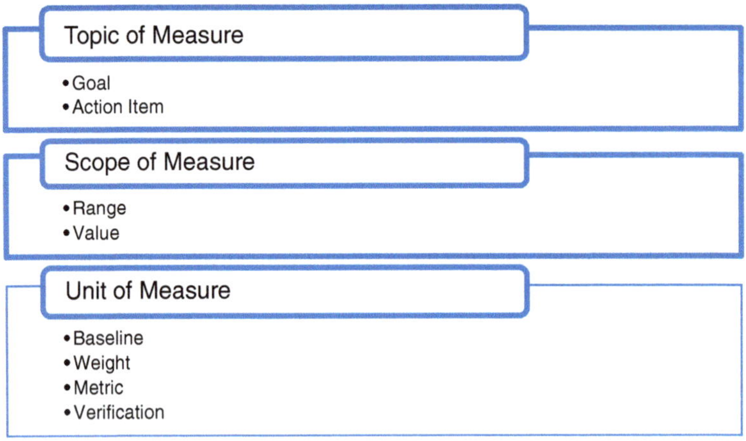

2.3.1 Topic of Measure

To identify key information used for indicators, the organization must focus on areas that have a high degree of influence on business performance (Holdsworth 2003). In most cases, the indicator themes will relate closely to the strategic level items discovered in the assessment stage.

Goal

Transitioning business operations to a more formal approach requires organizations to set clear goals and objectives (Holdsworth 2003). Goals can be at the organization, department or individual level and arise from particular values (Hoffman and Woody 2008; Meadows 1998). With a mind to triple bottom-line management, goals will be more successful if they fit the nature of the company (Hoffman and Woody 2008).

Using the assessment as a guide, the goals frame the specific areas or items being measured. The goals include such categories as business operations, facility operations, and customer service. Subsystem goals are more specific and include areas such as production, quality assurance, and marketing. Each subsystem will have associated items that are geared to specific actions and indicators.

Action Item

Action items provide operational information on how the organization must work to achieve strategic goals, and include actions taken in day-to-day business operations that have an effect on achieving the overall mandate of the organization. Once the goals and associated action items are developed, the organization is able to define indicators that will help manage triple bottom-line principles within the management plan.

Each action item does not need its own indicator; the indicators should be broad enough to deal with several actions within a goal. The indicator provides an overall assessment on the success of on area, which the action items all help achieve. Indicators must be easily understandable and limited in number so not to become overly complicated or unrealistic considering already extensive business reporting requirements (Spangenber et al. 2002). Additionally, attaching milestones to the process of implementing change provides both a plan of action as well as a reporting structure (Bertels 2010).

2.3.2 Scope of Measure

The scope of measure includes information that will allow the indicator to guide to the desired output. The scope provides the specifications that make up the indicator and inform the next step, which is attaching a unit of measure. To attach a meaningful measure, it is necessary to know as much information about the indicator as possible, which is made possible by the inventory process.

Range

Indicators should be based on existing data or data that is relatively simple to gather and analyze (Spangenber et al. 2002). With this data, businesses can organize indicators by common characteristics or activities; though, each indicator must have unique characteristics; otherwise there will be redundancies that are inefficient to a business's operations. Additionally, in order to be widely applicable, the indicators must be generic enough to not become overly dependent on specific context and they should also have the ability to withstand minor changes in business operations, changes to available data, or the approach for using the indicators (Spangenber et al. 2002).

Value

The next step studies the object of assessment and its impact in order to assign a value. Value is based on a business's mission and provides a means to understand operations by laying out the value hierarchy. With triple bottom-line concepts and the organizational mission as drivers, the value scheme captures the energy and environmental impacts and social costs that are implicit in material choices, which are not often fully reflected in the economic analysis decision-making process (Singh et al. 2009).

The three-pillared value scheme of triple bottom-line management considers how resource flows (materials, energy, and knowledge) are understood within economic, environmental, and social capital.

2.4 Economic Capital

2.4.1 Inputs

Economic capital inputs involve the management of supply-chains, risk and other governance, and resource productivity. More specifically, this includes administrative and production materials, human resources, utilities and services as well as money flows from capital providers, customers, ancillaries, and agencies (Chee Tahir and Darton 2010). With these in mind, a business will consider such things as ethical investing, integrated reporting, health and safety, and continuous improvement (Chee Tahir and Darton 2010; Azapagic and Perdan 2000).

2.4.2 Outputs

Economic outputs include the undesired by-products of operations. With economic capital, by-products include waste resulting from inefficiency. Major outputs include flows of money to employees for salaries, to suppliers and contractors for products and services, to capital providers for loan repayment or interest, and to governmental agencies for taxes and fees (Chee Tahir and Darton 2010; Azapagic and Perdan 2000).

2.5 Environmental Capital

2.5.1 Inputs

With environmental (or natural) capital inputs, consideration goes to the impact that operations have on ecosystems. Inputs include land access and management, raw material extraction, biodiversity and water management, and environmental performance of suppliers and other stakeholders (Chee Tahir and Darton 2010; Azapagic and Perdan 2000). Specific considerations are product durability, recyclability and other life-cycle considerations, performance of suppliers, and workplace procedures.

2.5.2 Outputs

Environmental capital outputs include solid and liquid waste and atmospheric emissions as well as damage to land and biodiversity (Chee Tahir and Darton 2010; Azapagic and Perdan 2000). Particular concerns include carbon intensity, conservation in terms of material and energy intensity, and pollution and damage to ecosystems. Environmental capital outputs have the can elicit fines for infractions as well as negative attention from stakeholders.

2.6 Social Capital

2.6.1 Inputs

Inputs from social capital contribute to the collective well-being of the constituents (Chee Tahir and Darton 2010). Specifically, inputs include policies that promote diversity and equality, socially responsible business operations, human resources management, and community partnerships. Considering these flows, a business will consider such things as ethical investing, procurement and other standards of practice, stakeholder inclusion, diversity programming, and hiring practices (Chee Tahir and Darton 2010; Azapagic and Perdan 2000).

2.6.2 Outputs

Social capital outputs have the ability to strengthen workplace morale, leading to better retention and attraction of stakeholders. In organizations, outputs stem from stakeholder satisfaction related to fair pay, training programs, volunteering, and professional development opportunities (Chee Tahir and Darton 2010; Azapagic and Perdan 2000). Social capital values society by investing in human stakeholders to get a return on investment.

2.6.3 Unit of Measure

Indicators and metrics go hand in hand; metrics are necessary to quantify actions and provide progress information (Singh et al. 2009). Though metrics are a daily part of business operations, industry has been very slow to track and measure progress relating to sustainable development (Holdsworth 2003; Azapagic and Perdan 2000). For this reason, indicators must not only be objective, they must also be based on easily accessible data and easy to calculate (Azapagic and Perdan 2000).

Baseline

Indicators inform an organization on progress in a particular area. They require a baseline against which progress will continually be measured and reported on. An organization must choose baseline information, which should include the output of materials, energy, and knowledge over a defined period of time. For a more comprehensive baseline, several years may be averaged in order to accommodate variability in, for example, weather patterns or commodity prices.

Weight

A weight shows the importance of the indicator and is directly related to its value to the organization. The process of choosing a weight, in terms of selection, scope, and significance, must be transparent in order to be widely understood by its audiences (Spangenber et al. 2002). At this stage, the organization determines the indicators that have a positive impact in increasing efficiency, compliance, and reputation (Spangenber et al. 2002). Weights can take many forms, including numeric ranking, visual ranking (colours or images), or they can simply be labeled from low to high priority. The organization can make comparisons by asking which of the two indicators are more important to overall goals (Spangenber et al. 2002).

Metric

Once the organization determines its baseline information and weighting process, the next step is to attach a metric. Without a value attached, indicators are unable to provide determine progress. Though indicators should be quantitative whenever possible, qualitative descriptions are more appropriate for some aspects of triple bottom-line, particularly concerning social capital (Azapagic 2003). With any indicator, the metric will take into account baseline information, trends, and goals. Baseline information provides the reference point while trending information allows the organization to see progress; the goals influence how aggressive or conservative an organization is with targets. For example, if an organization chooses a moderate baseline year for electricity consumption, say 2010, and finds that the average progress of consumption has been a 1 % a year reduction, then a conservative target would be to reduce electricity consumption by a total of 7.5 % over the next 5 years, at 1.5 % annual reduction.

At this point, indicators are equipped with a weight, as well as a unit of measurement. The final stage of attaching a unit is made possible with the topic and scope information from the first two stages, and is to provide a means of continual verification of the indicators to ensure relevancy.

Verification

In order to remain viable, organizations must continually scan for opportunities and threats. Scanning includes both internal and external influencers and is focused on keeping up to date on industry best-practices. This scan reveals how organizations may be leading or lagging the efforts and achievements of other organizations and provides a means of assessment (Bertels 2010). Therefore, rather than continually performing major assessments, the verification stage can ensure that the major elements of the organization, as reflected in the indicators, are contextually relevant to business. In addition to using available data to validate indicators, organization can perform surveys, and stakeholder meetings to collect feedback (Bertels 2010; Azapagic and Perdan 2000).

3 Conclusion

The development of a management framework reveals the systematic and cyclical nature of the process, which follows a natural progression of goal setting, data gathering and review. Once the boundaries of the management system are defined, an extensive assessment process takes place, resulting in key areas to guide the direction and measurement of the management system. This cyclical process is quite natural. An organization begins and directs the process guided by its strategic mandates in the form of goals and mission, and then defines policies and procedures, and the strategic elements then guide the day-to-day operations. The clearer the process, the more adept the management approach will be in confronting existing challenges based on issues of efficiency, compliance, and reputation that arise from the flow of material, energy, and knowledge throughout the lifecycle of operations.

In summary, this paper provides foundational information for organizations to create a triple bottom-line management approach with the intention that further research will expand on key operational elements. Specifically, a detailed guide to identify material, energy, and knowledge flows, including subsystems, will direct the inventory process in a clear and consistent manner. All organizations experience these categories of flows so this process can be generic in order to make a wider-scale impact. Similarly, there are no set of established indicators in businesses or institutions, which leads to inconsistencies and often an unmanageable amount of information to report on. By developing a concise standard of practice for each sector (or even across sectors), particularly in the key areas of inventory gathering and indicators, organizations will be better prepared to create and apply the triple bottom-line approach.

References

Azapagic A (2003) Systems approach to corporate sustainability: a general management framework. Process Saf Environ Prot 81(5):303–316

Azapagic A, Perdan S (2000) Indicators of sustainable development for industry: a general framework. Trans IChemE (Process Saf Environ Prot) Part B 78(B4):243–261

Bell S, Morse S (2008) Sustainability indicators: measuring the immeasurable. Earthscan, London

Bertels S (2010) Embedding sustainability in organizational culture. Network for Business and Sustainability, London. www.nbs.net/knowledge/culture

Blizzard JL, Klotz LE (2012) A framework for sustainable whole systems design, United States. J Hazard Mater 104(2003):193–205

Carroll A, Buchholtz A (2011) Business and society: ethics, sustainability, and stakeholder management. Cengage Learning, Stamford, CT, USA

Chee Tahir A, Darton RC (2010) The process analysis method of selecting indicators to quantify the sustainability performance of a business operation. J Clean Prod 18:1598–1607

Church C, Rogers M (2006) Designing for results: Integrating monitoring and evaluation into conflict transformation programs. Search for Common Ground, Washington, DC

Clayton AM, Radcliffe NJ (1996) Sustainability: a systems approach. Earthscan, London

Esquer-Peralta J, Velazquez L, Munguia N (2008) Perceptions of core elements for sustainability management systems (SMS). Manag Decis 46(7):1027–1038
Hoffman AJ, Woody JG (2008) Climate change: what's your business strategy? (memo to the CEO). Harvard Business School Press, Boston
Holdsworth R (2003) Practical applications approach to design, development and implementation of an integrated management system. J Hazard Mater 104:193–205
Kaiser J (2011) 10 billion plus: why world population projections were too low. ScienceInsider. http://news.sciencemag.org/scienceinsider/2011/05/10-billion-plus-why-world-population.html. Accessed 5 Dec 2013
Karapetrovic S, Jonker J (2003) Integration of standardized management systems: searching for a recipe and ingredients. Total Qual Manag Bus Excell 14(4):451
Kirkland LH, Thompson D (1999) Challenges in designing, implementing and operating an environmental management system. Bus Strategy Environ 8:128–143
Meadows D (1998) Indicators and information systems for sustainable development. Sustainability Institute, Hartland Four Corners
Odum EP (1953) Fundamentals of ecology. Saunders, Philadelphia
Pojasek RB (2009) Using leading indicators to drive sustainability performance. Environ Qual Manag 18(4):87–93
Rasmussen JE (2011) Transitioning to green: implementing a comprehensive environmental sustainability initiative on a university campus. California State University, Long Beach (Copyright 2011 by ProQuest LLC)
Shriberg M (2002) Institutional assessment tools for sustainability in higher education: strengths, weaknesses, and implications for practice and theory. Int J Sustain High Educ 3(3):254–270
Singh RK, Murty HR, Gupta SK, Dikshit AK (2009) An overview of sustainability assessment methodologies. Ecol Ind 15(1):281–299
Spangenberg JH, Pfahl S, Deller K (2002) Towards indicators for institutional sustainability: lessons from an analysis of agenda 21. Ecol Ind 2(1):61–7
Walton SV, Galea CE (2005) Some considerations for applying business sustainability practices to campus environmental challenges. Int J Sustain High Educ 6(2):147–160
Willard R (2009) The Sustainability champion's guidebook: how to transform your company. New Society Publishers, Gabriola Island

Author Biographies

Claire Bennett is a sustainability professional who is currently the manager of the Sustainability Office at Wilfrid Laurier University in Ontario Canada as well as a sustainability consultant in the area of sustainable operations and programming.

Dr. Lewis's research focuses on environment, policy and sustainable energy, including siting and integration of renewable energy systems energy use in buildings, green building, and the use of life-cycle assessment in decision-making. Dr. Lewis is a member of the Waterloo Institute for Sustainable Energy.

Sustainable Development at Universities in the Sultanate of Oman: The Interesting Case of Sultan Qaboos University (SQU)

Sadiq Hussain and Thuwayba Al Barwani

Abstract
While the implementation of sustainable development in higher education is a global trend, the case of Sultan Qaboos University (SQU) and its College of Education in particular is significant given the implementation speed, depth, and intensity of SQU's sustainable development practices; and that SQU experiences may be useful elsewhere. This paper adds to Sustainable Development at Universities (SDU) literature by analyzing the methodological approaches and concrete steps undertaken, allowing SQU to integrate sustainable development in its curriculum, research, innovation, practical projects, and in Omani regional and national development endeavours. This paper will be useful to anyone interested in the different levels of emphasis given to SDU in the Arabian Gulf region.

Keywords
Oman · Universities · Sustainable development · Innovation

S. Hussain (✉)
Sultan Qaboos University, PO Box 32123, Al Khodh, Muscat, Sultanate of Oman
e-mail: profsadiq@yahoo.com

T. Al Barwani
College of Education, Sultan Qaboos University, Muscat, Oman

W. Leal Filho (ed.), *Transformative Approaches to Sustainable Development at Universities*, World Sustainability Series, DOI 10.1007/978-3-319-08837-2_11

1 Introduction: Aspects of Sustainability Values in SQU and Oman

According to the "Rio+20" National Report, sustainable development aspects are 'key elements' of the overall Omani development plan; that Oman has achieved all aspects of the eight MDG goals; committed (p. 79) to all relevant UN conventions; implementing Agenda 21 (Environment-Ministry 2012, p. 7); overseen by the Cabinet and the Scientific Research Council (SRC). According to its royal charter, SQU as in the present study, has to relate its research to Omani national development through the identification of the kind of research to be financed by His Majesty (HM) funds. In a royal line stretching back four centuries, the contemporary Omani renaissance with its emphasis on sustainable development dawned with the accession of HM Sultan Qaboos Bin Said in 1970 making Oman stable (in the middle of a conflict-prone area) with rapid development. Three years ago HM had envisioned that "Collaboration and cooperation between all responsible bodies and exchange of opinions and consultation among those in charge is the way that leads to the success of national plans and programs in playing their desired role in comprehensive development and serving the present and future generations", HM Sultan Qaboos, (Sultan 2011). Since then HM's call for collaboration towards sustainable development implementation in Oman has made speedy progress.

According to (Albemani 2012) SQU owes its progress to HM funds and support through royal visits. Royal Camp visits have also benefited sustainable development advances in provincial colleges such as Ibri where HM was accompanied by Her Excellency Dr Rawya Al Busaidi, Minister of Higher Education and Chair of SQU Council who recognised IT students graduation projects with excellent Oman-based sustainable development technology applied in-practice (Albusaidi 2010). In many ways SQU with more than 17,000 students, is the crown jewel of the Omani higher education system and the "House of expertise and consultancy for the whole nation" (His Excellency Yahya Al Manthari personal communication October 2013).

This study arose from first the new communication channels opened for one of the authors (SH) as SQU Office of External Cooperation (OEC) research visitor to strengthen the SQU internationalization profile. Second the passion for SDU at SQU across its disciplines which attracted us as authors to take on this task of analysing the SQU strong sustainability values; and third from Leal's WSSDU-2014 call for papers.

2 The Context of the SDU-SQU Case Study and Previous Work

In Oman assumptions are that rapid sustainable development and operationalising education, research, and innovation for sustainable development are necessary for the speedy fulfillment of successful modern statehood; and that older greening forms as inscribed in UNESCO World Heritage List of the Bahla Fort oasis

(Culture-Ministry 2003) and traditional Omani Falaj (canals) water distribution systems are emblems of Oman's prestigious past environmental values and knowledge (Limbert 2001). The way the discipline of education forms and studies its systems stems from nation-state history, national policy documents and culture (Lawn 2012; Hussain 2013c). In Oman SQU is the only world-ranked university, in a league of billion dollar enterprises (Hazelkorn 2011); one of the few in the GCC region; was established with government funds in 1986. In 2012 as part of SQU Modus Vivendi, Key Performance Indicators have been internally measured according to international standards. All of this is what in the first place attracted us as authors to take on this task of strongly analysing aspects of SQU sustainability values.

Examples of the vast international literature on sustainable development in the UN-defined educational regions of the world include UK, Europe, and North America, (Lozano 2011; Leal Filho 2009; Martin 2012; Thomas 2013; Eweg 2010).

Asia Pacific (Naeem 2012; Aberdeen 2013, p. 418); South America and the Caribbean (Wals 2009); Sub-Saharan Africa (Leal Filho 2014); Arab States (Alhumoud 2005; Al-Ghamdi and Bakry 2008; Donn 2013). In Oman sustainable development aspects cover socioeconomic development, environment, culture and education (Environment-Ministry 2012, p. 7). In this study we concentrate on first sustainable development-related curriculum content, second, application-oriented sustainable development research and innovation; and third the speed of sustainable development progress. Like many universities historically before it in other places (Luzano 2006), SQU has played many roles in transforming Omani society and has supported sustainable development scientific research (Environment-Ministry 2012, p. 11). Within internationalization (the fourth clause in the SQU Charter) and the research visitor framework; the Office of External Cooperation OEC opened up a space for one of the authors (SH) to participate in this study aiming to strengthens the SQU internationalization profile. These SQU experiences may be usefully replicated elsewhere, considering local needs while paying attention to global perspectives (Shiel 2012; Eweg 2010); and to fill the international literature gap, noted by Leal quoted in Lee (2014) of empirical works trying to ascertain firsthand what is expected of university sustainability policies. So far as is known, there are no previous published international sustainable development works with the SQU institution itself as the topic of research; but there has been interesting studies on particular aspects of SQU academic programs, see for example (Alharthy 2012) on engineering (Alsalmy 2013) teacher education (Alkhanjari 2013) e-learning; (Lane-Kelso 2012) digital citizens. Also our field-collected data and content analysis of the e-recorded oral answers to the open-ended research questions revealed that (T4), SQU College of Engineering CENG, College of Medicine CMED programs have been, and College of Education CEDU is on its way to being internationally professionally-accredited as being sustainable development-compliant—a necessary pre-condition for accreditation. This is in line with (Mcdevitt 2008) assertion that for universities to remain competitive they must first meet the demands of international accrediting bodies and second be more innovative. It has also been argued that accreditation is a first step to accomplish higher performance levels

(Sahney 2014), and to being a role model for replication in institutions elsewhere (Alrawas 2012).

Recent examples of works in the realms of sustainable development and innovation include north of the Gulf studies on system approach for sustainable development (Hussain 1981, 1982); and Oman-based projects (Hussain 2006, 2007a, 2008a, 2009a, b, 2012a, c, 2011, 2013a, b, d, 2014a, b, c, d; Donn 2010, 2013; Albarwani 2009, 2010, 2012a, b). Also recently (Hussain 2012a) studied the role of SQU in Omani sustainable development; and according to the editors (Law, private communication August 2013) the work describes an appropriate topic for scholarly analysis; and the article would be useful in informing SQU faculty, students, staffs, and the wider public to develop interest in the SQU-sustainable development subject. Another recent study (Hussain 2012b) investigated SQU's World-Ranking specific Omani development planning issues; again in the views of the editors (Cuthbert, private communication August 2012) the ranking position of SQU, and the disadvantaging of SQU by global ranking systems are of interest to higher education scholars and researchers. Our study benefited from (Sarkis et al. 2013) sustainability-based research overview; that the ordinary management of people (human) problems is a top challenge in sustainable development advances; in agreement with (Lee 2014), that UN International Decade of Education for Sustainable Development (IDESD) was designed to encourage changes in human behaviour; literature and practice shows wide success for sustainable development policies, and that sustainable development has become "all things to all mankind".

3 Methodology

The objective of this study is to analyse implementation speed, depth, and intensity of SQU's sustainable development practices. To achieve this we designed interview protocols with an institution-wide sample of selected respondents based in the SQU institution itself lead by the Vice Chancellor (VC), three Deputy VC (DVC), OEC Assistant VC OEC (AVC), a VC's Advisor, and all college deans; SQU has nine colleges each lead by a Dean and three Assistant Deans. This study gives a first hand sense of how successful SQU-sustainable development has been so far and what SQU needs to do to progress further in this important SDU field. The interview schedule for top SQU management across disciplines consisted of two open-ended questions. Respondents answered questions RQ1 and RQ2 in interviews with the vice chancellor VC, two deputy VC (DVC), one Assistant VC AVC, one VC's Advisor (VCA); and to answer RQ3-RQ7 nine college deans, one dean for research and another for PG studies. The interview schedule for college deans consisted of a set of open-ended questions that asked respondents how they interpret sustainable development in curricula, research, and innovation whether sustainable development is currently understood globally as a driver for university innovation, whether SQU is integrating into all aspects of education and learning the sustainable development principles, values, and practices. This study assesses, compares, and

Table 1 Research questions RQ for category T (Top management) and category D (Deans)

Respondent category	Research questions RQ
Top SQU management category T	RQ1 What are the concrete steps which have been undertaken in order to allow SQU to integrate sustainable development at universities SDU in its (1) curricula (2) research (3) practical projects (4) regional and national development endeavors RQ2 What are the concrete steps which have been undertaken in order to allow SQU to integrate innovation in its (1) curricula (2) research (3) practical projects (4) regional and national development endeavors
SQU College and university deanship category D	RQ3 How does your deanship interpret sustainable development at universities SDU in its undergraduate and postgraduate courses RQ4 How does your deanship interpret SDU in its research projects RQ5 Do you think that SDU is currently understood globally as a major driver at universities with promising implications for university innovation? RQ6 Is SQU integrating the principles, values, and practices of SDU into all aspects of education and learning RQ7 What are the concrete steps needed in order to allow SQU to integrate SDU in its (1) research (2) innovation (3) practical projects (4) regional and national development endeavors?

gives a sense of perceived level SQU-sustainable development awareness and what SQU needs to do to progress further by answering the research questions RQ shown in Table 1.

To obtain relevant answers, to establish the state of current knowledge, to identify areas of consensus between respondents the research instrument in this study draws on primary data from multiple sources including first semi-structured protocol voice recorded interviews; second follow-up on interviews with respondents, third computer data base searches and SQU official reports. This enabled us to determine the extent and types of sustainable development engagement at SQU, its determinants, and its impact on students, academics, SQU management and other stakeholders. We focused our analysis on five respondents from the top SQU management, the nine college deans, and others because the decision to engage in sustainable development is a decision that, in the SQU context, is primarily at these levels. We applied the following procedure. To galvanize respondents support we sent with the RQs, a covering letter that identified SQU-sustainable development being the 1st such research to be published internationally, with an executive summary as background material and assuring anonymity for respondents. We use a mixed qualitative and quantitative methodology to analyze through the lens of SQU-SDU online literature subscription databases exhaustive searches as well as official SQU primary sources. The framework for our SQU-sustainable development curricula investigation used the Registration Deanship computer database and outputted lists shown in the results section below of SQU courses which have in the title or description of the course the following keywords that the authors identified: "Sustainable" and / or "development" in the definition (course description).

4 Analysis and Results

Through our interviews and analysis, we have found that respondents agreed that education at all levels is necessary for making sense of sustainable development in Oman, for communicating important information about sustainable development, and for creating a collective cross-disciplinary institution-wide space where SQU academics communicate sharing together something of the experience of sustainable development. All these uses of sustainable development by SQU academics help to form and define sustainable development in Oman.

Table 2 shows the codes used for interview data of the 16 semi-structured protocol voice recorded interviews, while Table 3 gives respondents details of the selected sample of SQU academic management.

We also identified the fact that SQU is sustainable development-active in teaching, learning, research, and innovation; in contrast with universities in other less-knowledge intensive countries like Poland where official policy documents criticized Polish universities for their low level sustainable development performance in teaching and research (Kwiek 2012). Six out of the nine SQU colleges (CASS, CAMS, CENG, CEPS, CLAW, CSCI) had between them thirteen active courses out of a total of 25 hits for courses that showed in our results; while 12

Table 2 Codes for interview data October–November 2013

	Interview respondent category	Code given for interview respondent
1	Top SQU management	T1
2		T2
3		T3
4		T4
5		T5
6	College dean	D1
7		D2
8		D3
9		D4
10		D5
11		D6
12		D7
13		D8
14		D9
15		D10
16		D11

Table 3 Respondents SQU SDU interview

	Interviewee position abbreviation	Name	Date Day, Time
1	Deputy VC DVC (academic)	Said Al Yahyaee	21 October 2013 Monday 9 am
2	VC's advisor VCA	Taher Ba Omar	22 October 2013 Tuesday 9 am
3	Vice chancellor VC	Ali Al Bemani	22 October 2013 Tuesday 10 am
4	DVC post graduate DVC PG	Amer Al Rawas	22 October 2013 Tuesday 11 am
5	Assistant VC AVC	HH Dr Muna Al Said	24 October 2013 Thursday 9 am
6	Dean, college of engineering, CENG	Ali Al Harthy	24 October 2013 Thursday 9 am
7	Head of state council EX VC SQU EX minister higher education	HE Yahya Al Manthari	30 October 2013 10 am
8	Dean college of science CSCI	Salma Al Kindy	31 October 2013 Thursday 9 am
9	Dean college of medicine CMED	Omar Al Rawas	3 November 2013 Sunday 10 am
10	Dean college of law CLAW	Ashraf Wafa	3 November 2013 Sunday 1 pm
11	Dean college of agriculture and marine science CAMS	Anvar Kacimov	6 November 2013 Wednesday 2 pm
12	Dean college of arts and social science CASS	Abdulla Al Kindi	10 November 2013 Sunday 9 am
13	Dean college of economics political science CEPS	Fahim Al Murhooby	11 November 2013 Monday 9 am
14	Dean college of nursing CNUR	Esra Al Khasawna	29 November 2013

coursed turned out to be non-active hits, that showed for the title or description of the course having the following keywords that the authors identified: "Sustainable" and/or "development" in the course description. This compare very well with a California university (Lee 2014) at CSUEB where only ten individual courses use the words sustainable or sustainability in their title or course description. Unsurprisingly, at SQU, because their titles do not contain the searched-for key words, some technology courses apparently relating to sustainable development did not appear in our search results for instance the 'Desalination' CENG CHPE4302 course. At first, it may seem surprising perhaps that such a course description 'missed' these keywords; On the other hand further investigation may be necessary to determine the centrality of sustainable development to this course.

In our computer database search, we also found two innovation & entrepreneurship courses MNGT3559 run by CEPS; and a university-wide general education elective across colleges run at SQU since 2012 so that students have been able to satisfy part of their undergraduate UG education by passing a university cross-disciplinary elective on innovation co-taught by academics from three colleges CENG, CSCI, CEPS, and a professional innovation expert from outside SQU. Important elements of this course include developing students life skills in communication, team-work and entrepreneurship.

Content analysis of the SQU top management respondents e-recorded oral answers to the open-ended research questions RQ1 revealed the insights that follow on the concrete steps taken at SQU to integrate sustainable development in its research portfolios, innovation, practical projects, regional and national development endeavours.

Insight 1 (Respondent T3): Responding to Omani national needs, sustainable development is built-in the SQU system of receiving fresh raw material from schools (pupils) as intake and producing excellent graduates in all disciplines ready to join the workforce and to serve the Omani community—this is sustainability. For instance, SQU takes pupils from schools and few years later send some of them back to the schools as graduate teachers to eventually illuminate every school in Oman and change the pathways of education systems for the better (T3). This is in line with (Pinheiro 2012) finding that societal demands have influenced the way in which universities have evolved in different places. Insight 2 (Respondent T5) sustainable development is there throughout SQU but we are not consciously working on it. Now comes the time we really should work to embed sustainable development in SQU strategy (T5). Insight 3 sustainable development as a learning outcome is a pre-requisite for international professional accreditation of SQU academic programs (T4). Insight 4 sustainable development is a continuous holistic initiative(not ad hoc and then it stops!) spread through SQU programs leading to fruitful results fuelling further Omani progress in human resource development and generating wealth. (T1) This is in line with (Wangenge 2012) that universities are seeking better engagement with their communities while also moving ahead with their traditional functions of training the human capital base. Also, Content analysis of the SQU college deans respondents e-recorded oral answers to the open-ended research questions RQ3 and RQ4 revealed that SDU may be interpreted as being widely practiced "between the lines".

Prospective SQU international research visitors may consider benefiting from the framework of the new OEC communication channels opened for one of the authors (SH) as research visitor to strengthen the SQU internationalization profile; with the insight (Al-Said 2014) that we are not consciously aware that sustainable development is already in SQU strategy—now the time has come for both sustainable development and internationalization to come out loud and clear. According to (Lawn 2012) the process of education internationalization involves Anglicization, and that political borders are suspended in away similar to borders being dissolved for international economic processes and e-commerce.

This study provided a fresh space for top management and deans to articulate their perspectives on SQU-sustainable development leading to critical reflections on the experience of their own units which in turn may lead to deriving insights to further advance the topic of SQU-sustainable development. While the research opportunities offered by the Omani sustainable development realm have been well explored in the literature the use of an sustainable development lens to analyze the SQU management of resources across colleges has been less well developed. Significantly, in the second half of the last century there was growing interest to advocate a "Systems Approach" tool for developing countries like Oman to be used for solving sustainable development problems including concurrently with the birth of the Gulf Cooperation Council (GCC) a systems approach for a north Gulf water resources management development project (Hussain 1981). Recent trends focus more on innovations and advancing science and technology and on the integration of the needs of science, technology, and socio economic practice.

5 Discussion and Conclusion

The above Omani sustainable development achievements are all the more important in the light of UNESCO world-wide review report of contexts and structures for ESD citing manifestations of 'unsustainability' as deeply rooted in local histories, political and cultural traditions (Wals 2009, p. 16) in the UN-defined region of Arab States where there is lack of relevant, culturally appropriate educational materials, scientific research, and studies, p. 20.

Recently, SRC has funded university students in a national green environment standards house design, construction and evaluation competition, see for example (Albadi 2014). The Ministry of Education is providing environment education tools to assist school curriculum designers (p. 93). Also, since 1980's, Oman has been funding first the International UNESCO Sultan Qaboos Prize for the preservation of the environment (p. 30); and second, international research to combat desertification through two Sultan Qaboos Academic Chairs (SQAC) chairs one in Europe, @uu.nl, and the other at Bahrain's Arabian Gulf University, p. 41. Furthermore, in academic and technological aspects of sustainable development, the report cites the following, p. 86, MOHE's external scholarships funds, university graduate surveys, Oman Academic Accreditation Authority (OAAA) regulating the quality of Oman-based higher education, and the new Oman University for Science and Technology projected to open in 2016 (Environment-Ministry 2012), 9th five-year development plan. These five-yearly national development plans started in 1975, whence the first five plans covered the last quarter of the twentieth century focusing on diversification of the economy to achieve faster and greater sustainable development (Algafry 2002, p. 236). In the new millennium the eighth five-year plan which ends in 2015 includes economic diversification as well as transitioning the Omani economy to knowledge-based for greater presence in the global economy (p. 10 of the report). All in all, citations of SQU sustainable development contribution items add up to a total of 12 pages out of the 98 page report–one would be hard pressed to refute that SQU had a striking effect on advancing sustainable development in Oman.

The flowing resource of water is one of the substantive (thematic) sustainable development concepts running across SQU disciplines along with oil, innovation, indigenous knowledge, and technology. Diversity, interaction, interdependence, and change as second-order concepts helping to organize the SDU-SQU curricula content and shape the line of research with ideas and processes. Linking together substantive and second-order concepts in context, results in 'generalisations' that can be transferred across disciplines (Sasson 2012); with greater background knowledge making for more sophisticated generalizations. Third order concepts are simultaneously foundational to the fabric of SDU-SQU including the fundamentals of SDU education policy subspace, place, and time.

Based on the analysis, results, and discussion above our conclusion is that sustainability implementation at Sultan Qaboos University (SQU), its College of Education, and the Sultanate of Oman has been rapid, deep, and intense.

Acknowledgment Sincere thanks are due to Walter Leal, Ali Al Bemani, and Dan Hammett for their helpful comments on earlier versions of this manuscript; to Muna Al Said for the research visitor 'acceptance letter' of one of the authors (SH), and to one and all SQU colleagues for their generous support that made this work possible.

References

Aberdeen L (2013) Rocking the foundations: the struggle for effective indigenous studies in Australian higher education. High Educ Rev 45(3):36

Al-Ghamdi A, Bakry SH (2008) King Saud University drive toward the knowledge society: a stope view. In: Lytras MD, Carroll JM, Damiani E et al (eds) Open knowledge society: a computer science and information systems manifesto. CCIS, vol 19. Springer, Berlin, pp 297–307

Al-Said M (2014) International Cooperation Office at SQU. In: Al Rawas A (ed) SQU Research Strategy, 2nd edn. SQU

Albadi AH (2014) Eco house hybrid energy system. Int J Sustain Eng (Submitted for publication)

Albarwani T (2009) Sustainable capacity building in a CCM-oriented EFL methods course at SQU. In: Proceedings 54th world assembly of the international council on education for teaching, Muscat, Oman, SQU

Albarwani T (2010) Cross-border collaboration for quality assurance in Oman: contested terrain. In: Sadiqi F (ed) Women in MENA agents of change. Routledge, New York, pp 215–231

Albarwani T (2012a) Sustainable innovation: where is the missing link. In: Townsend T (ed) International handbook on leadership for learning. Springer, New york, Part 2, pp 845–865

Albarwani T (2012b) Innovation wheel: towards systemic sustainable innovations in the Gulf. In: 3rd Gulf comparative educational society (GCES) symposium, Bahrain Teachers college, Bahrain

Albusaidi A (2010) Formal letter of thanks from her excellency Dr Rawya Saud Al Busaidi minister of higher education archived at The Minister's Office. MHE MO 2010 2/4/267

Algafry HH (2002) The Impact of national development plans on the education system in Oman. PhD thesis, Exeter University UK

Alharthy A (2012) Sustainable outcome-based assessment SQU engineering programs. In: Proceedings 7th international conference on technology, education and development, Valencia Spain

Alhumoud JM (2005) Municipal solid waste recycling in the Gulf Co-operation Council states. J Resour Conserv Recycl 45(2):142–158

Alkhanjari A (2013) SQU future from e-learning to m-learning application development. In: Ali S (ed) Information systems for sustainable business development. Universal e-book, Brown Walker, New York
Alrawas O (2012) Dean's forward SQU CMED annual report. SQU, Muscat
Alsalmy A (2013) Degree of professionalism in teacher preparation programs. SQU J Educ 7 (1):214
Bemani A (2012) Forward: SQU key performance indicators. SQU, Muscat
Donn G (2010) Globalisation and higher education in the Arab Gulf states. Symposium Books, Oxford
Donn G (ed) (2013) Education in the broader middle east. Symposium Books, Oxford
Environment M (2012) The Omani National Rio-2012 Sustainable Development Report Muscat, Ministry of the Environment
Eweg M (ed) (2010) Meeting the sustainability challenge in research and education, Utrecht, HU University of Applied Sciences, Utrecht @hu.nl
Hazelkorn E (2011) Rankings and reshaping of higher education battle for world-class excellence. Palgrave MacMillan, Basingstoke
Hussain S (1981) Case study for automatic station for resources development in Iraq. In: Proceedings of IFAC system approach for development: third IFAC IFIP IFORS international conference. Pergamon, Oxford
Hussain S (1982) Design of electronic optimizer for solar electric drive system. Energy Res (John Wiley) 6:367–376
Hussain S (2006) Information technology and education of the Mellenium generation. In: Almusawi A (ed) CEDU SQU ILT international symposium role of IT in education. SQU College of Education, Muscat, SQU
Hussain S (2007) Invited paper Arabian gulf development internet and innovation dynamics. In: IEEE first Asia international conference on modelling & simulation (AMS'07), Phuket, Thailand
Hussain S (2008) Arabian gulf development internet and innovation dynamics. SSST Simul Syst Sci Technol, Special Issue—Simul Ind Bus Serv 8(4):1
Hussain S (2009a) ICCCP09-025 e-commerce case study: Omani electricity e-billing and e-payment paradigm shift based on smart kilowatt hour (kwh) meters. In: IEEE international conference on SQU CENG ICCCP09, SQU
Hussain S (2009b) ICCCP'09 energy management case study: smart electrical energy management paradigm shift for Oman. In: SQU CENG ICCCP'09, SQU
Hussain S (2011) Omani e-government innovative trends. In: Proceedings of CEPS international information systems conference IISC, SQU Muscat, SQU College of commerce and economics
Hussain S (2012a) The role of Sultan Qaboos University SQU in Omani sustainable development. Perspectives: Practice and Policy in Higher Education (Taylor and Francis, London) (Submitted for publication)
Hussain S (2012b) Sultan Qaboos University SQU: a world-ranking development planning case study. HEP Higher Education Policy
Hussain S (2012c) CC Arabian universities world ranking sustainability insights and implications. SHE Sustainability in higher education IJSHE-Aug-2012-0071 submision ID
Hussain S (2013a) The connected city: how networks are shaping the modern metropolis. Int Dev Plann Rev (Liverpool University Press) 35(3):306
Hussain S (2013b) The politics of education reform in the Middle East: self and other in textbooks and curricula. Comp Educ Rev CER (Chicago University Press) 57(4):737
Hussain S (2013c) Education conflict and development. CIES Comparative and International Education Society CIES Perspectives Newsletter
Hussain S (2013d) Omani community-based emerging intelligent power and machinery systems applications in agricultural engineering. In: SQU Proceedings international agriculture engineering conference CAMS, Feb 2013

Hussain S (2014a) Globalisation and higher education in the Arab gulf states. Perspect Pract Policy High Educ doi:10.1080/13603108.2013.863237
Hussain S (2014b) Sustainable development at universities in Oman: The case of Ibri CAS (College of Applied Sciences). WSSD-U-2014. In: Leal F (ed) Transformative approaches to sustainable development at universities: working across disciplines. Springer International Publishing Switzerland Scientific, Frankfurt
Hussain S (2014c) Sustainable development at universities: recent joint pilot WSSD-U-2014 Oman/Netherlands water resources innovation initiatives. WSSD-U-2014
Hussain S (2014d) Arabic/English language and higher education in the third millenium. In: Proceedings of SQU CEDU symposium
Kwiek M (2012) Universities regional development and economic competitiveness the Polish case. In: Pinheiro R (ed) Universities and regional development. Routledge, London
Lane-Kelso M (2012) Global digital citizenship becoming a world university in Oman. World universities forum (accepted for publication)
Lawn M (ed) (2012) Europeanizing education governing a new policy space. Symposium Books, Oxford
Leal Filho W (2009) Towards the promotion of education for sustainability. Revista De Educacion: 263–277
Leal Filho W (2014) Special issue on climate change in Africa. Int J Clim Change Strat Manage 5 (2):129–139
Lee R (2014) Manifestos for sustainable development: Student visioning of change through a geography course. In: Leal F (ed) Transformative approaches to sustainable development across disciplines. Peter Lang Scientific, Frankfurt
Limbert ME (2001) The senses of water in an Omani town. Social Text 19(3):35–55
Lozano R (2011) Assessing Cardiff University's curricula contribution to sustainable development. Educ Sustain Dev 5:1
Luzano R (2006) Incorporation and institutionalization of SD into universities. Cleaner Prod 14 (9):787–796
Martin L (2012) Innovation great expectations or bleak house. Czech Republic Parliament Document
Mcdevitt R (2008) Strategy revitalization in academe: a balanced scorecard approach. Educ Manage 22(1):32
Ministry OH (2003) Bahla, Bat and Frankincense Trade Route in the World heritage list
Naeem M (2012) Sustainability in Asia Pacific education. Sustain High Educ 13(1):60
Pinheiro R (ed) (2012) Universities and regional development. Routledge, London
Sahney V (2014) Developing a high performing teaching hospital. In: SQUH Workshop, SQUH Muscat, SQU
Sarkis J, De Bruijn T et al (2013) Guest editorial: sustainability in engineering management-setting the foundation for the path forward. IEEE Trans Eng Manage 60(2):301–314
Sasson H (2012) Teaching the geography of development from the 'big picture'. Teach Geogr 37:113–115
Shiel C (2012) Global vision, local action: education for sustainable development and global citizenship. In: Proceedings of the fourth international conference. Bournemouth University, Bournemouth
Sultan HT (2011) Royal Speeches by His Majesty Sultan Qaboos Bin Said at Oman Council. Muscat, Media Department of State Council
Thomas C (2013) Air transport in an environmentally constrained world. Airport Manage 5:4–6
Wals A (2009) Review of contexts and cultures for esd, UNESCO
Wangenge G (2012) National universities and the regional dimension. In: Pinheiro R (ed) Universities and regional development. Routledge, London

Author Biographies

Sadiq Hussain Professor Sadiq Hussain has spent decades in development technology, teaching, researching, and consulting; spanning UK, Cardiff University (BS), Warwick University, (PhD) and Resident-Tutor; Brunel University (AERE, Harwellpost-doctoral), Durham University (Lecturer); Oman, SQU Visiting Professor, Ibri College, and Majan College; Iraq, Baghdad University, University of Technology, and Alnehrain University, research center director, board-member, Iraq Science Research Council; visiting fellowships USA, California, (NASA Ames Research Contract); Budapest Technical University (Hungary), Montpellier University(France), Knowledge Village, Dubai, UAE. Awarded Iraq Republic President's Medal (*Wisam Al-Ulema*).

Thuwayba Al Barwani Professor Thuwayba Al Barwani is State Council Member; Ex Deputy Minister (Social Development); BA, English, Kuwait University; SUNY, NY, USA, Curricula MA and Doctorate; member Higher Education Council; Trustees Board, AlSharqiya University. Honoured by His Majesty Second Class Order Culture, Sciences and Arts; 1998; Gulf Award for Science and Research; Arab Women's Renaissance Award, Arab Parliamentary Union, Women's Committee in Kuwait; Honorary Doctorate Sports Culture National Academy Physical Culture, Newcastle, UK; ICET Leadership Award; published internationally in areas such as higher education, teacher education, gender issues, family issues, and quality and innovation in education.

Enhancing and Monitoring Sustainability Culture at the University of Michigan

Robert W. Marans, John Callewaert and Michael Shriberg

Abstract

As the need to address major environmental challenges increases, the focus is turning from technological to social solutions. While technologies largely exist to address climate change, the question is whether there is a cultural consensus to use those technologies and make other behavioral adjustments. Sustainability culture is defined as a culture in which individuals are aware of major environmental (and social/economic) challenges, are behaving in sustainable ways, and are committed to a sustainable lifestyle for both the present and future. Colleges and universities are at the forefront of this cultural transition. In this paper, we review the University of Michigan's (U-M) Campus Sustainability Integrated Assessment as background to our current work. We then describe new initiatives designed to change the culture of sustainability on campus. Next, the Sustainability Cultural Indicators Program (SCIP) is reviewed including the design and use of sample surveys of U-M students, faculty, and staff. Selected

R.W. Marans (✉)
Institute for Social Research, University of Michigan,
426 Thompson Street, Ann Arbor, MI 48104, USA
e-mail: marans@umich.edu

J. Callewaert
Graham Sustainability Institute and College of Literature, Science,
and the Arts, University of Michigan, 625 E. Liberty Street, Suite 300,
Ann Arbor, MI 48104, USA
e-mail: jcallew@umich.edu

M. Shriberg
Graham Sustainability Institute and Program in the Environment,
University of Michigan, 625 E. Liberty Street, Suite 300, Ann Arbor
MI 48104, USA
e-mail: mshriberg@umich.edu

W. Leal Filho (ed.), *Transformative Approaches to Sustainable Development at Universities*, World Sustainability Series, DOI 10.1007/978-3-319-08837-2_12

findings including indicators covering the initial year (2012) are then presented as well as preliminary results from SCIP's second year (2013). Finally, the uses of these initial findings are discussed along with the potential for replicating SCIP at other universities and in other settings.

Keywords
Sustainability · Cultural indicators · Community awareness · Surveys

1 Introduction

As issues of climate change and resource depletion become more pressing, major societal changes are needed. In academia and public policy, discussions about these issues are now considering social/cultural as well as technological approaches. While technologies are available to address climate change and other pressing environmental concerns, the question is whether there is a societal willingness or cultural consensus to use those technologies and make necessary lifestyle adjustments. That is, there is a need to make a shift to a sustainability culture from one that has been based largely technological solutions and consumption. The need for this change has been documented by several writers during the past decade (see Kunstler 2005; Friedman 2009) and more recently by De Young (2013) and Erik Assadourian, senior fellow at the Worldwatch Institute (2013). De Young for instance argues that there are inevitable societal changes and the challenges society faces—including consuming less—in accepting a "resource-limited future".

In his chapter in *State of the World 2013: Is Sustainability Still Possible?* Assadourian posits that a culture of consumerism has been nurtured by business and government leaders for centuries and has become the dominant paradigm among most societies. Moreover, he states that:

> Consumerism is not a viable cultural paradigm on a planet whose systems are deeply stressed and that is currently home to 7 billion people, let alone on a planet of 8 to 10.6 billion people, a population that the UN projects for 2050. Ultimately, to create a sustainable human civilization…consumer cultures will have to be re-engineered into cultures of sustainability, so that living sustainably feels as natural as living as a consumer today (2013).

Assadarian goes on to provide examples of efforts to makes these cultural changes within the corporate world and at colleges and universities. Indeed, colleges and universities including the University of Michigan have been at the forefront of this cultural transition.

For more than a decade, the University of Michigan (U-M) has been making focused efforts at sustainability cultural change through its educational programs and in the operations of its Ann Arbor campus. Furthermore, the University has been measuring and monitoring that change since 2012. In an earlier book, we described two outreach activities aimed at changing sustainability culture. The paper also presented a conceptual framework for measuring and monitoring changes (Marans and Shriberg 2012).

This paper elaborates on U-M's transformational efforts and reports progress on the University's Sustainability Cultural Indicators Program (SCIP)—an innovative approach to measuring sustainability cultural change on campus. SCIP is an outgrowth of the 2011 Campus Sustainability Integrated Assessment (CSIA) and is intended to inform campus operations and educational programs. SCIP can also serve as a model demonstrating how changes in sustainability culture can be measured and monitored at other universities and in other settings. Following a review of CSIA and current outreach activities, SCIP is described including the design and use of sample surveys of U-M students, faculty, and staff. Selected findings including indicators covering the initial year (2012) are then presented as well as preliminary results for SCIP's second year (2013). Finally, the paper discusses uses of these initial findings along with the potential for replicating SCIP at other universities and in the corporate world and in urban settings.

2 Campus Sustainability Integrated Assessment

In 2009, U-M President Mary Sue Coleman elevated the University's commitment to sustainability in teaching, research, operations, and engagement by creating the U-M Environmental Sustainability Executive Council. One of the first actions of the Council was endorsing a Campus Sustainability Integrated Assessment (CSIA) to analyze U-M's sustainability efforts to date, benchmark against other institutions, and chart a course for the future through identifying long term goals for sustainable operations on the U-M Ann Arbor campus, including the Athletic Department and the Health System.[1] The CSIA builds on a long history of sustainability commitments in U-M campus operations, such as implementing cogeneration technology at the Central Power Plant in the 1960s, adopting the EPA Green Lights and Energy Star programs in the 1990s, and more recently establishing LEED Silver certification as the standard for new non-clinical construction projects where the construction value exceeds $10 M.

During the first phase of the CSIA, faculty-led and student-staffed analysis teams focused on: *Buildings, Energy, Land & Water, Food, Transportation, Purchasing & Recycling, and Culture.* A unique aspect of the Culture Team was that it placed members within each of the other analysis teams to support coordination across teams. At the conclusion of Phase 1, the teams submitted comprehensive reports and suggested ideas for further study in Phase 2. An integration team reviewed the reports and identified areas of intersection across these ideas. This review resulted in a priority list of proposed sustainability ideas that required further analysis during Phase 2.

[1] The Council is comprised the University President, the Provost and Executive Vice President for Student Affairs, the Vice Presidents for Research, Student Affairs, Development, and Global Communications & Strategic Initiatives, the Executive Vice President for Medical Affairs, and the Executive Vice President and Chief Financial Officer.

During Phase 2, the analysis teams conducted detailed analyses that included costs, benefits, technical guidance, uncertainties, and reasonable implementation timeframes for potential actions. The final CSIA report contains the recommendations developed by the integration team, and informed by the Phase 2 Analysis Team reports. The report outlines four high level themes—*Climate Action*, *Waste Prevention*, *Healthy Environments*, and *Community Awareness*. Accompanying the themes are Guiding Principles to direct U-M's long-range strategy and 2025 Goals that are time-bound and quantifiable.[2] Table 1 provides an overview of the U-M's 2025 Sustainability Goals.

Table 1 CSIA themes, guiding principles, and 2025 goals

Theme	Guiding principle	2025 goals
Climate action	We will pursue energy efficiency and fiscally-responsible energy sourcing strategies to reduce greenhouse gas emissions toward long-term carbon neutrality	Reduce greenhouse gas emissions (*scopes 1&2*) by 25 % below 2006 levels. Decrease carbon intensity of passenger trips on U-M transportation options by 30 % below 2006 levels
Waste prevention	We will pursue purchasing, reuse, recycling, and composting strategies toward long-term waste eradication	Reduce waste tonnage diverted to disposal facilities by 40 % below 2006 levels
Healthy environments	We will pursue land and water management, built environment, and product sourcing strategies toward improving the health of ecosystems and communities	Purchase 20 % of U-M food from sustainable sources Protect Huron River water quality by: • minimizing runoff from impervious surfaces *(outperform uncontrolled surfaces by 30 %)*, and • reducing the volume of land management chemicals used on campus by 40 %
Community awareness	We will pursue stakeholder engagement, education, and evaluation strategies toward a campus-wide ethic of sustainability	There is no goal recommendation for this theme. However, the report recommends investments in multiple actions to educate our community, track behavior, and report progress over time

[2] More information on the CSIA process, outcomes, and evaluation can be found at: http://graham.umich.edu/knowledge/ia/campus. Information on progress towards the 2025 Climate Action, Waste Prevention, and Healthy Environments goals can be found at: http://www.ocs.umich.edu/goals.html.

3 Changing Sustainability Culture

In 2011, U-M further elevated its commitment to sustainability in teaching, research, operations, and engagement following the goals emanating from the sustainability integrated assessment.[3] While many other universities have established sustainability goals related to climate action, waste prevention, and healthy environments, the U-M's efforts to raise levels of *community awareness and evaluate strategies toward an ethic or culture of sustainability* are unique.

The focus of the U-M's initial efforts to systematically raise levels of awareness, engage more of the students, staff, and faculty in sustainability activities, and move toward a sustainability culture on campus has been the Planet Blue Ambassadors Program.[4] The program, initiated in pilot form in 2011, has been evolving. To expand its coverage throughout the University, an on-line training tool consisting of five modules directly related to the U-M's sustainability goals was developed, widely publicized, and launched in 2013. The training involves an individual's commitment of 30–60 min and includes elements of gamification. Once completed the new "Ambassadors" have access to a personalized online dashboard to pledge actions, access resources, and track their personal progress. To date, over 1,300 people who have completed training and have collectively pledged over 30,000 actions, nearly 50 % of which have since been marked completed through their online dashboards. Perhaps the best measure of success is that over 80 % of the people who start a module complete the entire training.

Currently, efforts are underway to facilitate moving these online participants into a position of peer-to-peer sustainability leadership. For example, a monthly e-newsletter focuses on a targeted action such as using sustainable power settings on computers or increasing proper e-waste disposal. Important components of the e-newsletters are resources for encouraging others to take action, additional impact information, news, and special events or training opportunities. Additionally, a series of online and in-person events are hosted that strive to build community and skills, including a "Healthy Holidays" open house, local foods dinner, and a green computer webinar.

An important offshoot of the Planet Blue Ambassadors Program is the creation of Planet Blue Student Leaders (PBSLs). These students "eco-reps" live in residence halls and receive specialized training (via 1-credit seminars) in behavior change for sustainability. Acting as sustainability peer-advisors, they serve as a primary outreach and contact points to the over 8,000 U-M students who live in the residence halls and are establishing patterns relating to issues like recycling, food choices and energy conservation. Using the best practices in community-based social marketing, they hold contests among residents, initiate programs such as green dorm room certification and run events like skill shares. This student training

[3] For the complete text of President Coleman's 2011 sustainability initiative, see: http://sustainability.umich.edu/going-green-staying-blue-sustainability-michigan.

[4] See Marans and Shriberg (2012) for a description of the Planet Blue Ambassadors Program in its formative stage.

is part of sustainability leadership training programs at the U-M which span from first-year students to post-doctoral scholars. All are focused on integrating knowledge from different disciplines and applying this knowledge to the campus context as peer leaders.

For staff, a new program—Green Office Certification—was initiated in 2013. This points-based program—modeled after the U.S. Green Building Council's Leadership in Energy & Environmental Design (LEED)—certifies offices around the U-M campus as green based on their performance on a checklist. To date 89 offices comprising of over 3,900 staff members have been certified. Certification means important public recognition, but also personalized attention and consultation from the Office of Campus Sustainability. While the outcome is often changes relating to printing practices, energy management, recycling and composting, the certification program has also generated increased interest and engagement in sustainability. The program has an equivalent that is specifically designed for laboratories as well as components that reach out more broadly to the University. The Planet Blue Ambassadors Program, the on-line training and related activities and Green Office Certification are key elements in creating a culture of sustainability on the U-M campus.

4 Monitoring Change

For our purposes, "culture of sustainability" is defined as "a culture in which individuals are aware of major environmental (and social/economic) challenges, are behaving in sustainable ways, and are committed to a sustainable lifestyle for both the present and future" (Marans et al. 2010). U-M cultural change initiatives stem from the principles outlined under CSIA theme of Community Awareness. Among those principles is the "pursuit of evaluation strategies toward a campus-wide ethic of sustainability". This was articulated in President Coleman's 2011 address announcing the sustainability goals. Specifically, she stated that "we will scientifically measure and report our progress and behavior as a community… Institute for Social Research (ISR) researchers will measure the sustainability attitudes and activities of students, faculty and staff, as well as identify where we can improve."[5] The evaluation strategies involve a groundbreaking program for monitoring the U-M's progress in moving toward a culture of sustainability. Progress will be determined by tracking a set of sustainability cultural indicators over time.

5 The Approach

To create these indicators, a small group closely involved with the CSIA met for over a year working on what came to be known as the Sustainability Cultural Indicators Program (SCIP). The group examined the recommendations from the

[5] To read President Coleman's address and other information on the U-M's sustainability goals, please visit: http://sustainability.umich.edu/commitment.

CSIA culture team report (Marans et al. 2010), reviewed related literature, spoke to key national leaders working on similar efforts, ran focus groups with students and staff (see Schoolman et al. in submission) to determine current understandings of sustainability, and analyzed more than thirty existing campus surveys from numerous institutions (including the U-M) about topics such as recycling, transportation, etc.

One of the most useful resources for this work was the North American Association for Environmental Education's report "Developing a Framework for Assessing Environmental Literacy" (Hollweg et al. 2011). It provided a frame for developing questions under three categories; knowledge, dispositions or attitudes, and behavior. This went beyond many of the existing campus surveys which focus primarily on sustainability literacy or environmental literacy, or which focus exclusively on operational outcomes.

The design for the research was first proposed as part of the cultural team report and is described in Marans and Shriberg (2012). In sum, it proposed annual surveys of students, faculty and staff so as to track changes in levels of awareness, behaviors, degrees of commitment and dispositions. The sampling plan involved a repeated cross-sectional sample of faculty and staff and each cohort of students from freshmen to graduate students. In addition, panels from each undergraduate student cohort would be selected and queried each year during their tenure at the University so as to track individual change as well as cohort change. The initial proposal was to collect survey data annually for 6 years.

6 Questionnaires

Two separate web questionnaires were developed—one for staff and faculty, and one for students. While many of the questions were similar, different time frames and question sequences were used in the two versions. For example, the staff and faculty survey asked questions within a time frame of the past year while students were asked to answer questions based on their experience since the start of the fall semester. Also, students were asked several demographic questions at the start of the survey such as whether they live in campus housing or not in order to skip certain questions which did not apply to students living in campus housing while staff and faculty demographic questions were asked at the end of the survey.

Questionnaires were pretested with diverse groups of students, staff and faculty and reviewed by key operations staff members for additional input with the objective of writing questions which would be useful for campus efforts. Final versions of the two questionnaires were then programmed for administration as web surveys. Questionnaires covering the first year of SCIP were administered in fall 2012 to probability samples of students, staff and faculty. Responses were obtained from more than 4,000 students and over 2,000 staff and faculty. The overall response rate was 44 %.

7 Scope of Data Collected

A primary objective of SCIP was to work closely with the goals of the CSIA. Therefore, questionnaire modules were developed with questions focusing on travel and transportation, waste prevention and conservation, the natural environment, sustainable food, climate change, as well as U-M sustainability initiatives and respondent characteristics. Each module contained questions designed to measure levels of awareness, behaviors, and attitudes.[6]

8 Selected First Year Findings

Findings from the initial survey were reported in September 2013 (Marans and Callewaert 2013). They were organized around the four goals and presented in two ways. First, findings covering individual questions were presented in tabular and narrative form covering responses of each student cohort, faculty, and staff. Second, summaries of key findings were reported as *Sustainability Cultural Indicators* for each of these constituent groups. These indicators represent baseline measures against which indicators measured in subsequent surveys can be compared.

The sustainability cultural indicators are composite measures derived from two or more questions about a topic or theme. In a few instances, the indicator consisted of responses to a single question such as "Overall, how committed are you to sustainability?" and "How do you *most often* travel to and from campus?" Ten primary indicators dealing with *Climate Action, Waste Prevention, Healthy Environments*, and *Community Awareness* were developed whereas five secondary indicators not directly associated with U-M's goals but considered important indicators were also were reported. In order to compare indicators with one another and from year to year, all were converted to a common score on a 0–10 scale. The 15 indicators are noted along with summary scores for students, faculty, and staff in Table 2 and reveal several important findings.

First, there is considerable room for improvement with regard to the behaviors, levels of awareness, degrees of engagement and expressed commitment to sustainability among members of the U-M community.

Second, the behaviors of students are far more in tune with the goal of greenhouse gas reduction than the behaviors of staff and faculty. This is largely due to differences in the mode of travel to and from campus. Students are also likely to know more about transportation options available to them and are more engaged than either staff or faculty in sustainability activities on campus.

Third, compared to students and staff, faculty tend to act in a more sustainable manner with respect to conserving energy, preventing waste, purchasing food, and more generally, engaging in pro-environmental activities outside of the University.

[6] A full description of SCIP and the questionnaires administered during the first year are available at: http://graham.umich.edu/leadership/scip.

Faculty members also express a higher level of commitment to sustainability than others on campus.

Finally, students tend to be less knowledgeable than staff or faculty about protecting the natural environment, preventing waste, and sustainable foods. But they are more aware than faculty about what is happening at U-M with regard to sustainability. Nonetheless, members of the staff are most knowledgeable about the range of U-M's sustainability initiatives.

These selected findings representing the first year of SCIP are the baseline measures against which data collected in subsequent years will be compared. They have been shared widely among operational units representing plant operations, parking and transportation, student affairs, waste and recycling, and coordinators of the U-M's sustainability educational program. For example, the survey revealed that nearly 80 % of the faculty and staff who said they were "very committed" or "somewhat committed" to sustainability were not involved in a U-M organization addressing sustainability issues. These findings were instrumental in the decision to develop the Planet Blue Ambassadors training tool to engage a wider audience of University employees in sustainability activities. Additionally, data showing

Table 2 Summary sustainability cultural indicators for students, staff, and faculty (mean scores)

2012	Students	Staff	Faculty
Primary			
Climate action			
Conservation behavior	6.1	6.6	6.9
Travel behavior	8.2	1.6	2.2
Waste prevention			
Waste prevention behavior	6.6	7.0	7.3
Healthy environments			
Sustainable food purchases	5.5	5.7	6.3
Protecting the Natural Environment	8.6	6.5	6.1
Community awareness			
Sustainable travel and transportation	4.4	3.0	3.4
Waste prevention	4.0	5.0	5.1
Natural environment protection	3.1	4.1	4.3
Sustainable foods	4.3	4.7	5.6
U-M sustainability initiatives	5.1	5.4	4.9
Secondary			
Sustainability engagement at U-M	1.3	0.9	0.7
Sustainability engagement generally	1.9	1.9	3.0
Sustainability commitment	6.3	6.3	7.0
Sustainability disposition	3.5	2.9	5.3
Rating U-M sustainability initiatives	6.6	6.7	6.4

significant numbers of students, faculty, and staff were not using power saving settings on their computers and didn't know about recycling electronic waste led to the decision to address a recent e-newsletter and coordinated effort about these actions.

9 Current Status of SCIP

In fall 2013, a second set of questionnaires was administered to new samples of students, staff, and faculty. As in 2012, the sample plan targeted 1,000 first year students and 400 graduate students but fewer students from the sophomore, junior, and senior classes. The plan also targeted 1,500 staff members and faculty. Finally, the panel design aimed at surveying 500 sophomores (who responded as freshmen in 2012), 500 juniors (who responded as sophomores in 2012) and 500 seniors (who responded as juniors in 2012).

Data are currently being analyzed for nearly 3,200 responding students including more than 800 who participated in the 2012 survey, and over 1500 staff and faculty. Findings to date suggest that indicators and responses to individual questions show little if any change in behaviors, levels of awareness and commitments between 2012 and 2013. For example, when staff and faculty were asked how much they knew about the various transportation options available to them, there were no changes in the levels of awareness for 8 of the 9 available options. However, levels of awareness did increase among faculty and staff in response to "U-M's Green ride", a program designed to help members of the U-M community find partners to share rides to and from campus. In the period between the two surveys, a website called Greenride-Connect was launched by the Office of Transportation and Parking and widely publicized. The website was designed to make it easier for University employees to find rideshare partners or vanpools operating in communities around Ann Arbor. Faculty awareness of Greenride increased from 4 % in 2012 to 8 % in 2013 whereas the percent of staff who knew about Greenride increased from 8 to 17 % over the year.

Similarly, as a result of the University's outreach efforts, behavioral responses to the use of motion sensor power strips increased over the year. Only one in ten staff members reported using motion sensors in 2012 whereas in 2013, the proportion using motion sensors increased to one in four. Faculty too reported an increase in motion sensor use but the increase was not as dramatic (14 % in 2013 compared to 9 % in 2012). Further increases in use are anticipated for 2014 and beyond.

In addition to continuing analysis of the 2013 data and exploring cohort and individual differences between 2012 and 2013, on-going discussions of data use are occurring with personnel from U-M's operational units and educational programs. These discussions cover current and future uses of findings and how they might inform policy. For instance, concurrent with the collection of survey data is the collection of contextual environmental data for individual buildings located in different geographic regions of the University's wide-spread campus. Environmental information such as energy use per square foot, weight of recycled material, and the amount of waste sent to landfills has been collected and reported annually by the U-M's Office of Campus Sustainability.[7] In subsequent years when longitudinal survey data become available, it will be possible to examine the degree to which changes in environmental conditions impact changes in behaviors and vice versa. Such work can contribute to the development of predictive models demonstrating how changes in behavior impact operating expenditures.

Another area for exploration is the use of trials or experiments involving new sustainability initiatives that could reduce energy costs, raise awareness about recycling or protecting the environment, or increase sustainable behavior among students and University employees. Experiments are possible because of the large sample of staff and faculty respondents who work in buildings that are geographically distributed throughout campus. Similarly the classroom and other buildings (i.e. libraries, student unions, etc.) where student respondents spend time are also scattered throughout the University. These large numbers enable us to examine survey responses and index scores from year to year in U-M's different campuses, regions, and sub-regions. These geographic areas are shown in Fig. 1 on the next page.

As an example of a trial or experiment, University officials could initiate an experimental program or marketing strategy for energy conservation in a few sub-regions (or regions) but not in others. Examination of the behavioral responses and levels of awareness from a subsequent survey as well as energy costs in the two types of regions (experimental and non-experimental) could determine the effectiveness and success of the experiment. Assuming the program or marketing strategy were successful, it would then be launched throughout the entire University.

10 Potential for Replication

The Sustainability Cultural Indicators Program is intended to measure progress in changing sustainability culture at U-M's Ann Arbor campus. While other universities and colleges have been innovative in efforts to bring about change through operational and educational programs, we are unaware of any schools that have systematically measured progress in bringing about cultural change. Indeed, universities and colleges in the U.S. and elsewhere have expressed interest in replicating SCIP and we are eager to share our work and experiences with them. We are doing this in several ways. First, material covering the first year of SCIP including its questionnaires and the process of building indicators is available on the program website.[8] Second, we have discussed our work at professional and academic meetings during the past few years and will continue to do so as the program expands. Since its inception, presentations about SCIP have been made at

[7] The most recent report showing sustainability progress at U-M can be found at: http://sustainability.umich.edu/report/2013/.

[8] http://graham.umich.edu/leadership/scip.

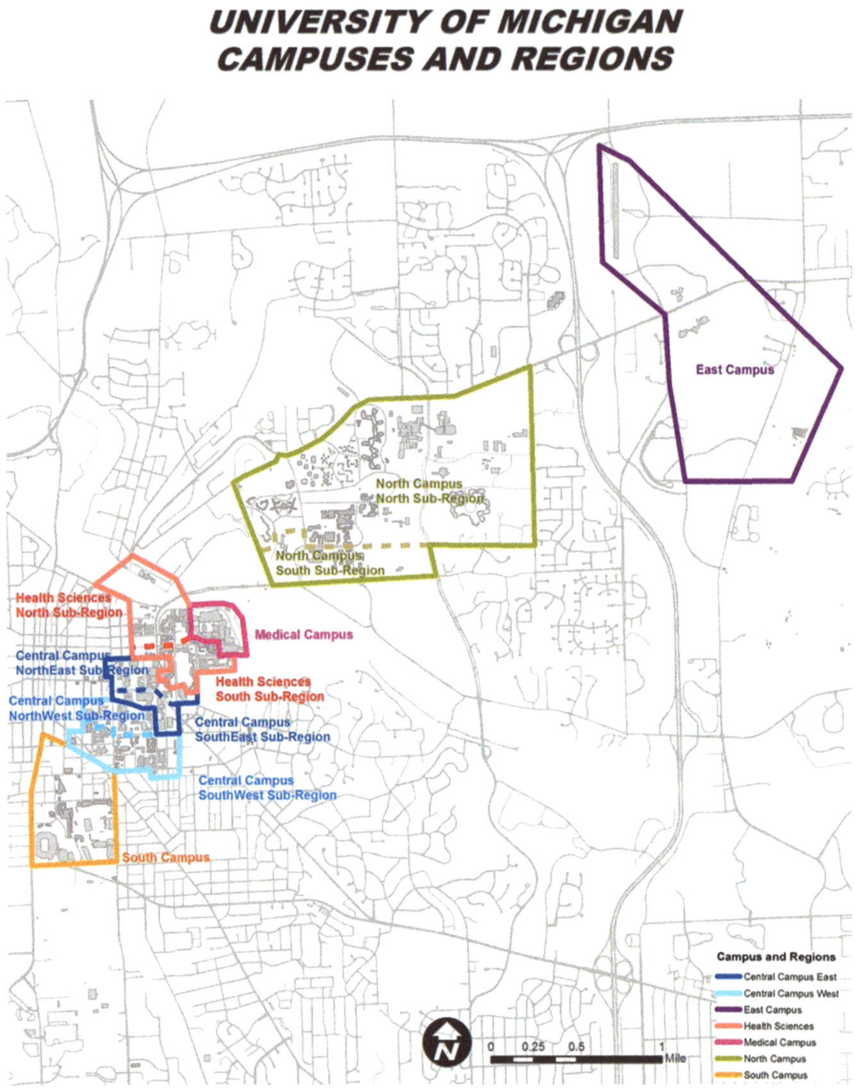

Fig. 1 Map of University of Michigan's Campuses, Regions, and Sub-Regions

conferences and meetings in the U.S., Brazil, India, Ireland and Taiwan. Furthermore, surveys designed to measure sustainability culture are being planned at universities in China and Brazil. Such surveys will be patterned after U-M's SCIP but will address issues of particular relevance to each setting. Finally, we believe that programs aimed at changing sustainability culture within public and private organizations as well as within cities and neighborhoods can be designed, implemented, and monitored. Discussions within these settings are just beginning.

11 Conclusions

This paper has argued that society must undergo a paradigm shift from a consumer culture to a sustainability culture as issues of climate change and resource depletion become more pressing. Colleges and universities such as the University of Michigan have been playing a leading role in fostering this shift and undoubtedly will continue to do so in the years ahead. A key question raised in the paper is whether there is any hard evidence indicating that a shift is taking place and if so, in what direction and by how much. Following a review of several recent initiatives designed to improve the culture of sustainability on the U-M campus, the Sustainability Cultural Indicators Program (SCIP) a process for measuring and monitoring sustainability culture for 6 years has been presented. That process involves annual web surveys of U-M's students, faculty, and staff. Key findings from the first year of data collection (2012) were summarized while preliminary data from the second year (2013) suggest that, overall, little change has occurred in cultural indicators scores. Nonetheless, modest but significant changes were found for a few individual items used to create indices. These items—awareness of the Greenride program and use of motion sensors—reflect specific University initiatives launched after the initial data collection. Finally, uses of the findings were discussed along with the potential for replicating SCIP at other universities and in other settings. Replication offers opportunities for comparative research on numerous topics related to the human dimension of sustainability including creating a sustainability culture.

References

Assadourian E (2013) Re-engineering cultures to create a sustainable civilization in Worldwatch Institute. State of the world 2013: is sustainability still possible? Island Press, London

De Young R (2013) Transitioning to a new normal: how ecopsychology can help society prepare for the harder times ahead. Ecopsychology 5(4):237–239

Friedman T (2009) Hot, flat, and crowded release 2.0: why we need a green revolution and how it can renew America. Picdor, Farrar, Straus, and Giroux, New York

Hollweg KS, Taylor JR, Bybee RW, Marcinkowski TJ, McBeth WC, Zoido P (2011) Developing a framework for assessing environmental literacy, North American Association for Environmental Education. http://www.naaee.net. Accessed 15 July 2012

Kunstler JH (2005) The long emergency: surviving the end of oil, climate change, and other converging catastrophes of the twenty-first century. Atlantic Monthly Press, New York

Marans RW, Shriberg M (2012) Creating and assessing a campus culture of sustainability: The University of Michigan experience. In: Filho WF (ed) Sustainable development at universities: New Horizons. Peter Lang, Frankfurt

Marans RW, Callewaert J (2013) Monitoring the culture of sustainability at The University of Michigan: fall 2012. http://graham.umich.edu/media/files/SCIP-Year-1-Report.pdf. Accessed 17 April 2014

Marans RW, Levy B, Bridges B, Keeler K, Avrahami T, Bennett J, Davidson K, Goodman L, Holdstein B, Smith R (2010) Campus sustainability integrated assessment: culture team phase I report. http://graham.umich.edu/knowledge/ia/campus. Accessed 10 Jan 2012

Schoolman ED, Shriberg M, Schwimmer S, Tysman M (in submission) Scratching the green surface: do higher education sustainability initiatives matter beyond the campus bubble? J Environ Stud Sci

Author Biographies

Robert W. Marans, Ph.D. is Research Professor at the University of Michigan's Institute for Social Research and Professor Emeritus of Architecture and Urban Planning in the university's Taubman College of Architecture and Urban Planning. For more than three decades, he has conducted evaluative studies and research dealing with various aspects of communities, neighborhoods, housing, and parks and recreational facilities. His research has focused on user requirements and the manner in which attributes of the physical and sociocultural environments influence individual and group behavior and the quality of community life. Dr. Marans' most recent book, *Investigating Quality of Urban Life: Theory, Methods, and Empirical Research* was published by Springer (2011). His current research considers the impact of the built and natural environments on quality of life, the role of neighborhood in the health of Detroit residents, and issues of sustainability and energy conservation in buildings and institutional settings. Dr. Marans is a licensed architect, a fellow of the American Institute of Certified Planners, and serves on the boards of several boards including the Detroit Riverfront Conservancy and the Huron–Clinton Metroparks, the regional parks authority serving southeast Michigan.

John Callewaert, Ph.D. is Integrated Assessment Program Director at the Graham Environmental Sustainability Institute and a Lecturer in the College of Literature, Science, and the Arts at the University of Michigan. As Integrated Assessment Program Director of the Graham Institute, Dr. Callewaert is responsible for designing, implementing and overseeing the day-to-day operations for the Integrated Assessment (IA) Program—the Graham Institute's engaged problem solving/translational research initiative. Dr. Callewaert came to the Graham Institute in 2009 after serving for 2 years as the Director of the University of Michigan–Flint Office of Research. From 2000 to 2007, Callewaert was the founding director of the Institute for Community and Environment at Colby-Sawyer College and an Associate Professor of Social Sciences and Education. Between 1984 and 1987 he served as an agricultural volunteer with Peace Corps Nepal working on improving seed supplies in remote hill villages. Dr. Callewaert serves as an associate editor for the Journal of Environmental Studies and Sciences and as an advisory board member for The Integrated Assessment Society. Dr. Callewaert earned a B.S. in Agricultural Engineering Technology from Michigan State University and his Ph.D. from the School of Natural Resources and Environment at University of Michigan–Ann Arbor.

Michael Shriberg, Ph.D. is Education Director at the Graham Environmental Sustainability Institute and Lecturer in the Program in the Environment at the University of Michigan. In addition to teaching courses such as "Sustainability and the Campus" and "Sustainability Leadership Development", he leads the Graham Undergraduate Sustainability Scholars Program, Graham Sustainability Doctoral Fellows Program and efforts to integrate sustainability across the curriculum at the University of Michigan. Dr. Shriberg is the North American Editor of the International Journal of Sustainability in Higher Education. He recently returned to academia after serving as Director of Environment Michigan and Policy Director at Ecology Center, leading policy efforts on Great Lakes protection, climate and energy, and as well as environmental health and toxics.

Previously, he served as Program Director and Assistant Professor of Environmental Studies and Interim Director of the Rachel Carson Institute at Chatham University in Pittsburgh, PA, as well as an adjunct Lecturer at Carnegie Mellon University. Dr. Shriberg earned an M.S. and Ph.D. in Resource Policy and Behavior from the University of Michigan's School of Natural Resources and Environment and a B.S. in Biology and Society from Cornell University. His research and practice focuses on organizational change and sustainability leadership, with an emphasis on higher education institutions.

Walking the Dead: The Systemic Premises Explaining Missed Opportunity for Sustainability Initiatives in Canadian Higher Education Organizations

Phil Cook and Anshuman Khare

Abstract

This paper reveals the "systemic" nature of the chronic missed opportunities for wide spread, meaningful, multistakeholder sustainability initiatives at Canadian Colleges and Universities. Prima Facia views of the opportunity for Higher Education (HE) organizations to engage in large scale collaborations with community and Industry elude the majority of Canadian HE organizations as if by design. This paper examines the structure of Canadian HE organizations and the systemic processes that mandate the impediment of large-scale multistakeholder involvement in Sustainability initiatives. While the majority of Canadian HE organizations have managed to place a priority on Sustainability in curriculum, on-campus initiatives, new building construction and certain community outreach programs, the analysis produced in this paper demonstrates that both internal objectives, rewards and perspectives and the "environmental coupling" of the organizations provide a "systemic" self-limiting structure to more than superficial sustainability initiatives. The research, in illustrating the dynamics at play responsible for the "why not", provides the reader with data analysis valuable in the assessment of situations where the "how to" and the "why" of sustainability initiatives are to be pursued. The practitioner pursuing initiation of large scale sustainability initiatives will benefit from an initial understanding of the potential road blocks and thus produce a strategy influenced by this prerequisite information.

P. Cook · A. Khare (✉)
Faculty of Business, Athabasca University, 301-22 Sir Winston Churchill Avenue, St. Albert, AB T8N 1B4, Canada
e-mail: anshuman.khare@fb.athabascau.ca

W. Leal Filho (ed.), *Transformative Approaches to Sustainable Development at Universities*, World Sustainability Series, DOI 10.1007/978-3-319-08837-2_13

Keywords
Canadian Higher Education (HE) · Association for the Advancement of Sustainability in Higher Education (AASHE) · American College and University Presidents' Climate Commitment (ACUPCC) · University and College Presidents' Climate Change Statement of Action for Canada (UCPCCSAC)

1 Introduction

The title of this paper pays homage to the Bowie (2013) song "Where Are We Now?" In the song Bowie describes the Bornholmer Strasse checkpoint—a bridge—between East and West Germany where during the evening of November 9, 1989 citizens of the GDR were allowed to leave East Germany to the West but were forced to have their passports stamped as "invalid" by border police. The meticulous process continued until, at 9:20 p.m. that night, border police were overwhelmed and lifted the gates, discontinuing the process and in essence beginning the fall of the Berlin wall. "Walking the Dead" refers primarily to the memories of a time of courage and solidarity, of human accomplishment and concerted effort toward change.

Courage is something that many of the thought leaders in the sustainability domain speak of, not the least of which is Martha Piper a leading Canadian figure for innovation and a sustainability champion during her tenure as president of the University of British Columbia. Piper suggests that profound transformations are quite often the result of a few courageous individuals "flapping their wings". She is of course referring to sizeable changes occurring from miniscule systemic input fluctuations—the "butterfly effect" of complexity theory (Lewin 1999). The unfortunate characteristic of nonlinear systems is that at times they are sensitive to such miniscule inputs and yet at other times they are not—the metaphor collapses—for if it was a simple matter of courage and the flapping of a few "wings" then perhaps greater numbers of Higher Education (HE) organizations in Canada would be far more driven toward sustainability.

Nonlinear dynamical systems while appearing complex prima facia may well be generated by a small set of simple sub processes (Lewin 1999). A group of elements with certain institutionalized relational characteristics interact in such a manner as to produce a global "order" which in turn feeds back an institutionalizing reinforcement signal (see Fig. 1). When applying this theory, of emergent order from the interaction of relatively stable and simple subsystems, to the HE organizations in Canada, one is drawn to the conclusion that the system is a self-organizing self-perpetuating group of subsystems systemically programmed to sustain a quasi equilibrious state incapable of anything greater than a superfluous scope of change toward campus wide Sustainability initiatives.

This paper examines the interactions and actions of the autonomous elements within the domain of Canadian Higher Education Organizations in an effort to illustrate the systemic nature of the resistance to widespread Sustainability

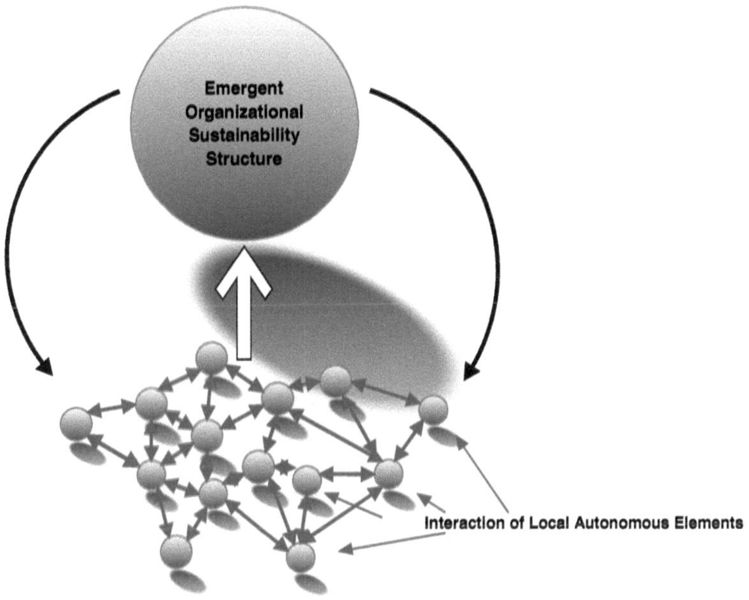

Fig. 1 Emergent sustainability structure resulting from local elemental interaction

initiatives—the emergent organizational structure that is a function of the elemental interactions. The results will be of benefit to those who might otherwise venture into this domain ill prepared to address the situation without this knowledge. Otherwise, it is feared, that in Canadian Higher Education organizations, Sustainability initiative swill wither to a state of mere memory—to the point of walking the dead?

2 North American Higher Education Sustainability Organizations

When one considers Sustainability in Higher Education in North America in general and in Canada specifically one is drawn to three main organizations; The Association for the Advancement of Sustainability in Higher Education (AASHE), the American College and University Presidents' Climate Commitment (ACUPCC), and the University and College Presidents' Climate Change Statement of Action for Canada (UCPCCSAC). The overall performance of these organizations and commitments while showing limited promise reveal a troubled movement when scrutinized longitudinally.

Founded in 2006 the AASHE was formed as the "first professional higher education association for the campus sustainability community" (AASHE n.d.). In 2009 the membership of AASHE increased by 137 institutional members from 797

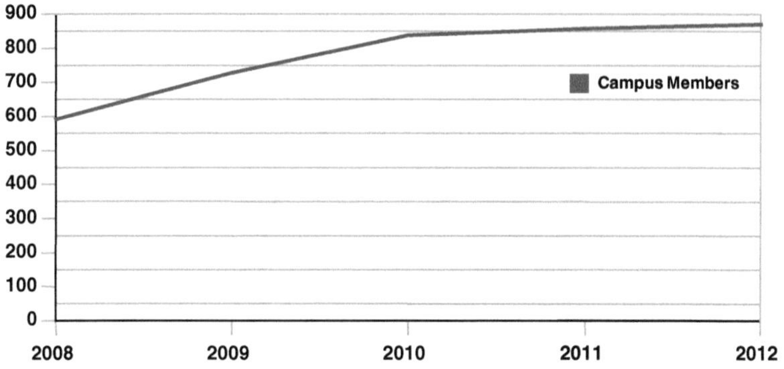

Fig. 2 AASHE campus membership growth 2008–2012

to 934. Of these, campus members grew from 591 at the beginning of 2009 to 728 at the end (AASHE 2009 annual report 2010). This represents remarkable growth in the 3 years between 2006 and 2009. However, it appears that this trajectory was not sustainable and may well have peaked in 2009. 2010 showed an increase of college and university membership from 728 to 839 (AASHE 2010a), 2011 saw campus membership grow by 19–858 by year end (AASHE 2011), 2012 saw this worsen with the end of the year arriving to report no more than 870 campus members (AASHE 2012b). Renewal rates in AASHE are also a disturbing negative rate—not exceeding 87 % for campus members since 2009. Figure 2 illustrates the plateauing of institutional AASHE members from 2008 through 2012.

Diving deeper into the AASHE statistics and singling out Canadian Colleges and Universities proves out a similar conclusion. 2009 saw Canadian membership at 54 with growth to 67 by 2012. While this represents a 20 % growth over the 3 year period, a closer look illustrates that with undulations this number has remained primarily static and ranges between these numbers (see Fig. 3).

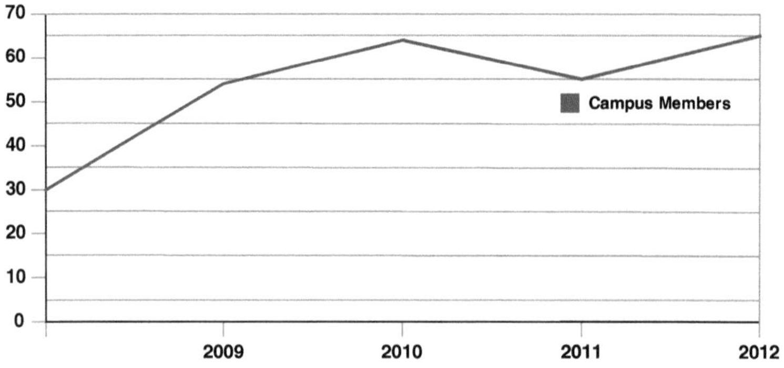

Fig. 3 AASHE Canadian campus membership growth 2008–2012

Stagnant and diminishing growth in one of the major organizations created to consolidate and represent the Sustainability movement on campuses is alarming. Considering that Colleges and Universities in the United States are approximately 5,000 in number (Digest of Education Statistics 2010 n.d.) and that Canadian Universities and Colleges number 250, the percentage of interested organizations of the continents leading organizations for representation and resources are an abysmal 16 and 26 % respectively of total possible participation. This means that in the United States of America 84 % of Higher Education organizations do not even have membership in the organization that represents the campus sustainability movement and that in Canada that number ranges between 79 and 74 %—in either case a large majority.

Unfortunately the lack lustre performance of Sustainability organizations does not end here. The American College and University Presidents' Climate Commitment (ACUPCC) formed in 2006 with twelve signatory University presidents participating (ACUPCC n.d.). By March 1, 2008 the ACUPCC had grown to over 500 Universities and Colleges signatory to the cause (ACUPCC 2007). This number grew to 605 Higher Education organizations by the end of 2008 and 665 by December 31, 2009. 2010 delivers up 676 members of this organization and a disappointing loss of 12 members to arrive at 664 to finish up 2012 (ACUPCC 2011). The dwindling membership is depicted in Fig. 4.

Of these organizations, roughly 30 % (664), have not met the organizations reporting deadlines and 22.5 % are in arrears with the organization. It is possible that the declining membership and lack of enthusiastic enrollment, not to mention delinquent payment and poor reporting performance, are based on the stringent and specific objectives set out by the organization. The ACUPCC have a well-defined "commitment" agreed to by the signatory at time of engagement. These commitments include:

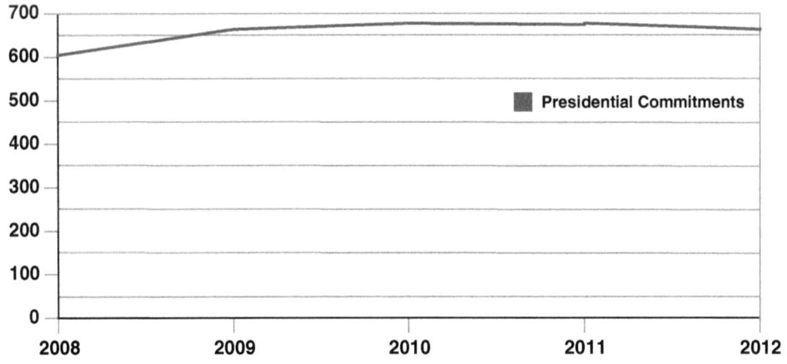

Fig. 4 ACUPCC membership growth 2008–2012

1. An annual greenhouse gas audit for their school
2. Specific organizational responsibility to coordinate and monitor the schools efforts to address climate change
3. Two or more "immediate actions" that demonstrate their commitment to results and to begin addressing the issue
4. Developing and implementing a plan with specific targets and timelines to achieve climate neutrality at a pace and in a manner that maximizes the opportunities for their school
5. Sharing their commitment plans and progress reports to facilitate and accelerate progress for their fellow institutions and society (ACUPCC 2007)

Prima Facie these obligations do not seem to be overly imposing given the enormity of the issue at hand—the sustainability of the species' interaction with itself and the planet. However, there must be some pressure imposed by this commitment to produce less than majority adoption of the concept. Or perhaps there is some deeper more complex reasoning underpinning this phenomenon?

While the ACUPCC is primarily a U.S. organization it does have affiliation and membership from several international members. In Canada one President has made the commitment—Patricia Lang President of Confederation College. Given the 250 Higher education organizations in Canada that one is involved in this organization represents a 0.4 % involvement in this initiative. This is a total disregard for the commitment to the ACUPCC. There is an affiliated group, however, on the Canadian side of this movement; The University and College Presidents' Climate Change Statement of Action for Canada (UCPCCSAC). This group formed in March of 2008 with six foundational Universities all from the province of British Columbia. The "statement" suggests the act to "expand" on the success of ACUPCC and yet the actions purported by the "Statement of Action" are less of a commitment and more of a wish list. The Action Statement reads out the following actions:

1. We will exercise leadership by reducing emissions of greenhouse gases in collaboration with our communities
2. We will develop measurable targets using research and science
3. We will develop achievable and practical plans to achieve reduction targets
4. We will put in place rigorous assessment and measurement procedures
5. We will fully disclose and be accountable for our actions (CUCPCCSAC n.d.)

While admirable in its intent this commitment carries with it no measurement of achievement and no deadline to accomplishment as does the ACUPCC. This is perhaps the reason that membership has grown since inception from 6 to an unimpressive 23 by 2012. If the Canadian case is any example, organizations are not shying away from the levity of the commitment as is perhaps suggested by the ACUPCC but rather there is merely a lack of interest and /or perceived value at the University and College level in North America with regard to the Sustainability movement on campus.

3 The Significance of Sustainability to Canadian Higher Education Administration

Drilling down a little further into the systemic nature of the stagnant performance in Sustainability initiatives amongst Canadian Higher Education organizations one must look no further than the Canadian Association of University Business Officers (CAUBO). Established in 1937 CAUBO is primarily an assembly of Administrators from a majority of the Universities in Canada. If nothing else they represent the "voice" and one would extrapolate the "thoughts" and value systems of this group. A review of the strategic plan from 2003 to 2009 uncovers no mention of Sustainability in its verbiage. A similar study of the 2009–2015 strategic plan reveals the term used three times once in a warning of "future" issues that may need to be addressed:

> ...the imperatives of climate change, reflected today in student activism and in public policy measures such as BC's recent legislation, will demand a step-change in universities' engagement with this issue and new levels of innovation from their administrators and leaders (CAUBO 2009b).

This statement is mixed in with other likely suspects such as the economic outlook, demographic shifts, Globalization and Public Expectations. One area though where Sustainability is conspicuously absent is that of the actual strategic goals of this organization through the year 2015.

Sustainability has a placeholder on the CAUBO website where one would find chat rooms or blogs. A single hygienic statement regarding the importance of Sustainability and the guarantee of disseminating information to members is all that exists in this area. However important this information is or urgently the membership needs it, it seems a rather low priority given is total absence ("Sustainability | CAUBO-ACPAU," n.d.).

A keyword search of the CAUBO (2009a) published Highlights and Achievements documents from the website produces six mentions of the term Sustainability in the 2008–2009 issue and six in the 2009–2010 issue. There is no occurrence of the term from 2010 to 2012. It would seem that if Sustainability had any heyday at all with the members of CAUBO it was during a similar time as those of the organizations AASHE and the ACUPCC. It would appear the University Business officers—those who hold and direct the purse strings of the Higher Education organizations—are not speaking or thinking of Sustainability as a current and strategic issue.

All of this lack of attention from the group who have not only reported on the multibillion dollar Accumulated Deferred Maintenance issue in Higher Education Facilities across Canada (CAUBO 2000)—claiming of the $5.1 Billion in deferred maintenance $2.4 billion is a health and safety concern—but also are ultimately entrusted with finding a solution to the problem. This is further reinforcement of the systemic nature contributing to the slow progress toward Sustainability initiatives on Canadian campuses across the country. Feasible models with sound returns exist to address the Accumulated deferred Maintenance issue in Canadian Higher

Education facilities not the least of which the authors have explored and carried out field work to implement (Cook and Khare 2012).

At the administrative level we see examples of lack of concentration and in some cases what appears to be some disconnect between the issues at hand and the solutions readily available. Drilling down a bit further into the situation illustrates the current status and the "face" of Sustainability Champions on campuses.

4 The Face of Sustainability Champions on Campus

A very close look at the administration of Sustainability on HE campuses in Canada would surely be incomplete without a detailed account of who the Sustainability champions are in terms of power position, numbers, income, authority, and ethnicity among other characteristics. Included with many others, Sharp (2009), no stranger to the on campus HE Sustainability movement, called for new governance approaches and decision making processes that would require ownership and responsibility for sustainability to fall in the hands of Presidents and Senior Executives in order to move from first and second wave movements toward wholesale organizational change movements. Sustainability positions should then be widespread and of a hierarchical level of senior leadership within the organization.

According to AASHE (2010b) 80 % of Sustainability staff held positions at or below the director level in the organization with the balance holding director or Chief Sustainability Officer positions. The same survey in 2012 produced similar results with only 5 out of 462 respondents claiming the title of Chief Sustainability Officer and 24 % holding a directorship or higher (AASHE 2012a).

The same reports from AASHE, including the 2008 report, show that the majority of sustainability champions in North American Higher Education organizations are part of the Facilities groups. In 2008 this number was 32 % followed closely by the Sustainability office at 31 %. In 2010 the figures were similar at 34 and 23 % respectively and in 2012 the numbers had increased for facilities to 43 %. Newport (n.d.) reminds us that while the facilities groups in these organizations do embark on important work toward Sustainability—social justice, economic equity and curriculum are generally far from the central focus of the facilities offices in HE organizations in North America.

Given that the Sustainability champions reside primarily in the facilities offices and are generally not of a senior executive status within the HE organizations is not the full picture of course. The previously mentioned reports also allude to the income levels, racial makeup, position status and shear lack of numbers of these representatives on HE campuses in North America. In 2010, for example, of the small number, 74 % of Sustainability champions were dedicated full time in their positions and by 2012 this percentage had grown to 85 %.

While it is somewhat difficult to place an exact figure on the average salaries of the sustainability champions in 2010 the average ranged between $81,361.00 and $43,714.00 annually and by 2012 had changed minimally to a high of $88,727.00 and a low of $40,770.00. To place these salaries in perspective we look to average

Dean's salaries of between $90,425.00 and $437,040.00 ("Median Salaries of Senior College Administrators by Job Category and Type of Institution, 2010–2011—Administration—The Chronicle of Higher Education," n.d.).

When one observes the numbers of staff available to manage the broad and encompassing endeavours of Sustainability offices throughout HE organizations the results are as telling. Sharp (2009) warns that HE organizations need to bolster staff tasked with enabling entire universities toward massive structural and organizational change and not merely leave this to a few or even one employee, generally ill equipped to manage change of this magnitude, reporting up through the facilities department. The AASHE data analysis do not provide a very promising illustration of this phenomenon occurring. In 2012 the AASHE report illustrated how severely understaffed the sustainability offices in North American HE organizations are with a report of 84 % between 1 and 5 Full time Equivalents and a meager 7 % at more than 11 FTE and 9 % at 6–10 FTE. It would appear that Ms. Sharp's pleas have fallen on deaf ears.

The feeble numbers are likely to be the primary reason for the poor reporting numbers and inconsistent quality of the reports across North America. Lozano (2011) cites thee studies suggesting the reporting standards and proliferation of corporations far out reaches those of Higher Education organizations.

The systemic interplay does not stop here though. The lack of diversity among Sustainability champions has remained heavily skewed as well. In 2010 AASHE reported 92 % white in response to the race and ethnicity question put to Sustainability champions and by 2012 this percentage did not change. This figure is in stark contrast with the growth of high school graduates that are enrolling in Universities by race ("Colleges Straining to Restore Diversity—WSJ.com," n.d.). Can this lack of diversity be expected to influence an already skewed concentration on enviro-centric Sustainability initiatives toward a more holistic approach that encourages consideration of social elements such as diversity and poverty (Zeegers and Clark 2014)? In his keynote to the AASHE conference Agyeman (2010a) calls for a "Just Sustainability" that rethinks the dominant image of Sustainability on campus as "greening the campus" and seeks one that models equality and diversity for all within the limits of supporting ecosystems. Higher Education organizations who do not model their sustainability champions after the socially just diversity that they wish to achieve in their communities are breeching what Agyeman (2010b) calls for as "being the change they wish to see".

5 The State of Sustainability on Campus: What Is and What Could Be

The authors have looked at the performance and position of many of the elements constituting this nonlinear dynamical system called Sustainability in Canadian Higher Education and specifically some of the quantitative and qualitative analysis of the interplay of these elements with the Sustainability movement. One begins to come to an understanding of the emergent order produced by the interactions and relations of the elements within the systems. Occasionally, one witnesses the extraordinary results of a few "wings flapping" and fantastic projects and accomplishments result. However, more often (certainly the case supported by the preceding analysis) opportunities that are seemingly straight forward and obvious paths to large scale Sustainability initiatives on Canadian campuses are simply not pursued and some that are pursued are done so in a shallow and somewhat misrepresented manner. Such claims require substantiation.

It is not perhaps enough to make observations regarding Universities in Canada and their lack of performance in general but perhaps to focus on a few key instances that are truly absurd denial of the pursuit of the potential that is virtually "lying in their laps". One Canadian University that has in place a "Presidents Advisory Committee on Sustainability", is a member of AASHE and does extensive research in the sequestration of carbon has not taken advantage of its greatest sustainable asset—this campus is placed geographically on a known source of geothermal energy capable of heating more than the entire campus—for eternity (MacPherson Engineering Inc and Helix Geological Consultants 2008). Since 2008 and before, this organization has had knowledge that they sat on a geothermal source of energy that given a 12 month utilization scenario could take advantage of an initial capital cost payback at current energy costs of 7 years. How could this be? How could an outside observer take this information and process it knowing fully that very highly skilled and knowledgeable individuals hold the controlling positions of this organization? The authors believe that the systemic nature—the emergent institutional structure—produced by the interaction of the elements involved also produce this sort of absurd reluctance to pursue what is seemingly an obvious and simple route toward a large scale Sustainability initiative.

In another case a major Canadian University has proclaimed a great reduction in carbon dioxide production in their "Sustainability Report" (Armstrong et al. 2011) through the replacement of grid based electricity and natural gas boilers with a Combined Heating and Power (CHP) plant otherwise known as cogeneration. What the University has done is to replace a grid electricity source of energy—which frankly could have been the conduit to green grid power (hydroelectric, wind, solar, etc.)—with a generation system that cannot take advantage of any type of renewable resource and is tied strictly to a non-renewable fossil fuel source—natural gas. A general and underlying theme of Sustainability is to take from the earth's crust at a rate in which that resource might be replenished naturally in the earth's crust. Burning fossil fuels is not sustainable no matter what minor increase in efficiency, 50–75 % in this case, can be realized. The verbiage used by the University to justify this decision is interestingly devious as well:

> The existing equipment is now at the end of its service life and is undersized to meet the requirements of the growing campus footprint ("Cogeneration | News & Events | University of Calgary" n.d.)

The observer is served up with a justification of replacement of an existing plant with a new one based on a growth fetish—one that can be presumed could not be managed through conservation and end of tail pipe solutions. In addition the University states that:

> [Cogeneration] is a very popular technology and many campuses across Canada have adopted it. It's the smartest and most sustainable option right now. ("Cogeneration | News & Events | University of Calgary" n.d.)

The authors believe that there are likely to be a great number of professional opinions that disagree with the opinion that "cogeneration" is the most sustainable option right now. It is not sustainable, at all, and as such defies being defined as sustainable.

While the action of replacing the standard plant with a cogeneration plant could be seen to reduce carbon emissions the mere reduction of carbon emissions should not be considered a single most important move toward sustainability when in fact alternatives are readily available that would eliminate carbon production and provide veritable "free" energy as long as the sun remained in its current state.

In addition to the controversial content of some "Sustainability Reports", Lozano (2011) reveals that the state of reporting in Canadian Higher Education organizations is laggard even though it is of such critical importance to the movement for organizations. Of the organizations studied by Lozano fewer than 30 % reported Sustainability initiatives in any significantly different way than for profit organizations, reports of Sustainability in research and curriculum were elusive and the reports themselves were potentially misleading and of limited value.

6 Conclusions

The intricate interplay of the elements of the nonlinear dynamical system under scrutiny here appear to be cause for the rather lack lustre performance of Canadian Higher Education Sustainability initiatives toward the institutionalization of Sustainability at these campuses. Observation of the movements of organizations over time show that Canadian HE organizations are either not participating at all or if participating interest has plateaued since 2008–2009. Presidential commitment in Canada while growing remains at 10 % of the entire population of HE organizations in Canada and the commitment that has been developed has no quantifiable time mandated initiatives.

While, it is possible to hold up the general global economic slowdown, beginning in 2008, as a defence of the poor uptake and failing performance of Sustainability Initiatives in Higher Education organizations in Canada, the authors feel that this is of minor impact and would require further research. Certainly enrollment continued to increase in Canadian Universities with 2009–2010 enrollment numbers up 8.5 % over 2008–2009 numbers (Back to School n.d.) and 2010–2011 numbers finally cresting the 1,000,000 number (AUCC 2011). It is suspected that in oil-producing provinces the decline in economic condition would contribute to a tapering of decline in enrolment if not an increase in enrollment and that enrollment in the non-oil producing provinces would continue to increase (Statistics Canada 2014) thus forming a somewhat recession proof situation.

At the administrative element level the lack of discourse and strategic initiative demonstrates that Sustainability is not on the agenda for the majority of Canadian HE Administrators. This element, that controls the purse strings of the organization and determines its direction greatly, is not driving toward an agenda of Sustainability commitments or objectives.

HE Sustainability champions as a result of several elemental interactions are generally small in number (where they exist at all), placed in positions of managerial authority in facilities departments, are ill equipped or mandated to act as organizational change agents, are not paid in anyway similar to other key organizational leaders and are skewed drastically in terms of racial diversity.

We conclude that from the interaction of these elements a Global Organizational Sustainability Structure emerges that seemingly by design produces an institutional self-limiting nature in terms of actual Sustainability performance in Canadian HE organizations. The small number of organizations that are even paying attention to the movement are bound in a web of interacting elements with opposing objectives, misunderstood needs, inadequate authority and misallocated focus and resources leading to Sustainability initiatives that are far short of what will be necessary to change the next generation of the species from multi-person consumers into singular person consumers who can reside within the limits of the ecosystem.

Considering the current trajectory outlined in this paper it seems utopian to imagine a world such as the one envisioned by Sustainability. Lessons learned in this paper would see the following wholesale changes occur:

1. University and College Presidents stepping up their commitments to Sustainability through specific, measurable, achievable, realistic, and timely objectives
2. Mandatory, legislated, and comprehensive involvement in the movement throughout the Higher education domain
3. Higher Education administrators of every discipline actively pursuing and understanding the economic, ecological and social imperatives of Sustainability through strategic initiatives
4. Sustainability Champions having positions of authority in newly created cross campus Sustainability departments. The term Chief Sustainability Officer should appear on every College and University organization chart
5. Sustainability reporting should be auditable and should adhere to rigorous compilations standards that are readily available today
6. Sustainability should underpin every activity, both academically and operationally,
7. Sustainability initiatives should be those which are truly sustainable. Higher Education must be the change they wish to see in the world

Of course the list need not end here, however, these few steps could surely redirect the movement currently.

The potential or current Sustainability champion will be well served to understand the interactions and the systemic nature of the local elements producing the seeming resistance to large scale effective Sustainability initiatives on Canadian HE campuses. Each of these elemental interactions must be appreciated and strategies developed to turn these relations away from the current state. This is an easy statement to make and a much more difficult one to enact. However, armed with the knowledge of how systemic the resistance is the Sustainability champion of the future is liable to be more effective in implementing such change. Systemic change at the local elemental level will be necessary in order to shock the movement back to the trajectory enjoyed up until 2008–2009 otherwise we may all be walking the dead.

References

AASHE (n.d.) About AASHE | Association for the Advancement of Sustainability in Higher Education (AASHE). http://www.aashe.org/about. Accessed 2 March 2014

AASHE (2009) AASHE 2009 annual report. International Food Policy Research Institute. doi:10.2499/9780896297852

AASHE (2010a) AASHE 2010 annual report. http://www.aashe.org/files/2010_annualreport_final.pdf

AASHE (2010b) AASHE 2010 higher education sustainability staffing survey. Higher education. http://www.aashe.org/files/2010_staffing_survey_final.pdf

AASHE (2011) ASSHE 2011 annual report. http://www.aashe.org/files/2011_annualreport_aashe.pdf

AASHE (2012a) AASHE 2012 salaries & status. http://www.aashe.org/files/documents/programs/2012_staffsurvey-final.pdf

AASHE (2012b) ASSHE 2012 annual report. http://www.aashe.org/files/annualreport_aashe_2012.pdf

ACUPCC (n.d.) Mission and history | Presidents' climate commitment. http://www.presidentsclimatecommitment.org/about/mission-history. Accessed 3 March 2014

ACUPCC (2007) ACUPCC annual report 2007: climate leadership for America: progress and opportunities in addressing the challenge of our time. http://www2.presidentsclimatecommitment.org/reporting/documents/ACUPCC_AnnRep_2007.pdf

ACUPCC (2011) ACUPCC annual report 2011. http://www.presidentsclimatecommitment.org/files/documents/report_2011.pdf

Agyeman J (2010a) "Just" sustainability: reimagining (e)quality, living within environmental limits. http://www.aashe.org/files/documents/blog/AgyemanAASHE.2010.pdf. Accessed 22 March 2014

Agyeman J (2010b) "Just" sustainability: reimagining (e)quality, living within environmental limits. http://www.aashe.org/files/documents/blog/AgyemanAASHE.2010.pdf

Armstrong R, Bell S, Ellard B, Houle L, Perdue J, Beatty B (2011) Stepping up together. Calgary. http://www.ucalgary.ca/sustainability/files/sustainability/InstitutionalSustainabilityPlan_0.pdf

AUCC (2011) Undergraduate student enrolment surpasses million mark—AUCC. http://www.aucc.ca/media-room/news-and-commentary/undergraduate-student-enrolment-surpasses-million-mark/. Accessed 23 March 2014

Bowie D (2013) Where Are We Now? On The Next Day. ISO, New York

CAUBO (2000) A point of no return: the urgent need for infrastructure renewal at Canadian Universities | CAUBO-ACPAU

CUCPCCSAC (n.d.) University and college presidents' climate change statement of action for Canada. http://www.climatechangeaction.ca/pdfs/ClimateChangeActionStatement-English.pdf

CAUBO (2009a) CAUBO highlights and achievements 2008–2009: working for you. http://www.caubo.ca/sites/137.149.200.5.pilot/files/fedora/repository/caubo_495/OBJ/2008-2009_Highlights_and_Achievements.pdf

CAUBO (2009b) CAUBO strategic plan 2009–2015: building on success: looking forward to 2015. http://www.caubo.ca/sites/137.149.200.5.pilot/files/fedora/repository/caubo_949/OBJ/CAUBO_Strategic_Plan_2009-2015_final.pdf

CAUBO-ACPAU (n.d.) Sustainability. http://www.caubo.ca/community/sustainability. Accessed 4 March 2014

Digest of education statistics (2010) (n.d.) National Center for Education Statistics. http://nces.ed.gov/programs/digest/d10/tables/dt10_276.asp. Accessed 2 March 2014

Cook P, Khare A (2012) Academia–corporate on campus sustainability collaboration: an exploration of the construct. In: Madu CN, Kuei C (eds) Handbook of sustainable management. Imperial College Press, London

Lewin R (1999) Complexity: life at the edge of chaos. Macmillan Publishing Company, New York, p 234

Lozano R (2011) The state of sustainability reporting in universities. Int J Sustain High Educ 12 (1):67–78. doi:10.1108/14676371111098311

MacPherson Engineering Inc, Helix Geological Consultants (2008) University of regina deep well geothermal conceptual study. MacPherson Engineering Inc. & Helix Geological Consultants, Regina, Saskatchewan

Newport D (n.d.) The Department of Change: the death of campus sustainability. http://davenewportblog.blogspot.ca/2012/02/death-of-campus-sustainability.html. Accessed 30 May 2012

Sharp L (2009) Higher education: the quest for the sustainable campus. Sustain Sci Pract Policy 5 (1):1–8. http://sspp.proquest.com/archives/vol5iss1/editorial.sharp.html

Statistics Canada (2014) Study: wages, youth employment and school enrollment, 2001 to 2008. http://www.statcan.gc.ca/daily-quotidien/140113/dq140113a-eng.pdf

Statistics Canada (n.d.) Back to school… by the numbers, 2012. http://www42.statcan.ca/smr08/2012/smr08_167_2012-eng.htm. Accessed 23 March 2014

The Chronicle of Higher Education (n.d.) Median salaries of senior college administrators by job category and type of institution, 2010-11—Administration. http://chronicle.com/article/Median-Salaries-of-Senior/126455/. Accessed 9 March 2014

University of Calgary (n.d.) Cogeneration | News & Events. http://www.ucalgary.ca/news/uofcpublications/oncampus/november2008/power. Accessed 13 March 2014

WSJ.com (n.d.) Colleges straining to restore diversity. http://online.wsj.com/news/article_email/SB10001424052702304732804579421164124444876-lMyQjAxMTA0MDEwNjExNDYyWj. Accessed 16 March 2014

Zeegers Y, Clark IF (2014) Students' perceptions of education for sustainable development. Int J Sustain High Educ 15(2):242–253. doi:10.1108/IJSHE-09-2012-0079

Author Biographies

Phil Cook is a Doctor of Business Administration Student at Athabasca University as well as working full time in Business Development for EH Price. His interests include Sustainable Development in Higher Education, Collaboration and Community Engagement.

Anshuman Khare is Professor in Operations Management & Sustainable Development at Athabasca University, Canada. He is an Alexander von Humboldt Fellow. His research focuses on the impact of environmental/climate change regulations on society, technology, innovation and corporate strategy.

Maximising Sustainability Outcomes by Amalgamating Dimensions of Sustainability

Dianne Chambers

Abstract
There is no single way for a university to become an organisation where sustainability is a basis of all activities. Rather, there must be many threads that together weave a culture of sustainability: threads that come from many parts of the organisation, have diverse approaches, and engage many people over an extended period. Although these need to have a range of sources, skills, and people, clear goals need to be shared and communicated so that all efforts are aligned and so that each initiative reinforces all others. This paper explores how one large, research-intensive university is working toward a more sustainable future through combining dimensions of sustainability, including teaching, campus operations and research in sustainability initiatives so that each dimension supports the other dimensions for mutual success.

Keywords
Sustainability · Education · Campus operations · Research · Governance

1 Introduction

Becoming a 'sustainable organisation' is a process rather than an act, and no single act will achieve the desired goal. Sustainability goals should be determined and shared by the leadership team so that all participants know where they are heading and so that plans can be made and implemented, but these goals must also be

D. Chambers (✉)
Melbourne Graduate School of Education, The University of Melbourne,
Melbourne, VIC 3010, Australia
e-mail: d.chambers@unimelb.edu.au

W. Leal Filho (ed.), *Transformative Approaches to Sustainable Development at Universities*, World Sustainability Series, DOI 10.1007/978-3-319-08837-2_14

flexible so that they can respond to changes in conditions and priorities. A university is a community leader and community role model, its research output can influence or even revolutionise practices and, most importantly, its graduates are the researchers, the decision makers, and the business, community and political leaders of the future. Thus it is more important for a university than another organisation of a similar scale and environmental footprint to take actions to ensure that sustainability is part of all aspects of its activities. For a university to become more sustainable there is a need for big picture planning and goal setting, for careful and detailed planning, for identifying the many aspects that need to be considered, planned for and coordinated, and for a need for all parts of the team to have a shared vision of success and a desire to achieve that success.

Much has been written about initiatives for enhancing sustainability in the curriculum or increasing the sustainability of a university campus and much can be learned from the experiences of others. However, it is proposed that for the goal of a sustainable university to be achieved that each dimension needs to be considered both individually and alongside at least two other dimensions. There should be efforts (and funds) directed towards creating opportunities where three or more dimensions of sustainability are brought together and that, in doing so, each dimension will be strengthened more than if addressed only in isolation. This paper will look at how a research intensive university is addressing sustainability, and in particular, it will outline some sustainability initiatives that address multiple dimensions of sustainability and the benefits that are being reaped from this approach.

2 Context

The University of Melbourne is a large research-intensive university situated in the City of Melbourne in Victoria, Australia. It was established in 1853 and the University is ranked number one in Australia and number 34 in the world in the Times Higher Education World University Rankings 2013–2014. In the 2013 *Universitas Indonesia Green Metrics* ranking (see http://greenmetric.ui.ac.id/) the university was ranked ninth worldwide for urban campuses and ranked 18 for overall excellence in campus sustainability. The main campus covers 22.5 ha adjacent to Melbourne's central business district and there are also six other campuses, both urban and rural. The university has approximately 45,000 students and approximately 8,000 academic and professional staff.

Since it first established targets in 2008, the University has achieved its sustainability targets. For example, carbon emissions were reduced by 50 % from 2008 to 2010 through a combination of energy reduction projects and the voluntary purchase of green power and abatements. In 2014 the voluntary purchase of green power and energy abatements ceased, and funds (>AUD$1 m) that previously were used for this are being invested in campus projects, such as voltage optimisation, HVAC improvements, LED lighting and solar PV installations. These projects will deliver ongoing savings in energy use and emissions that, together with other energy saving and energy generating initiatives, will contribute to achieving 2015 energy and emissions targets and the University's aspirational target of carbon neutrality by 2030.

In 2010 a review of research and research training at the University was undertaken, with the outcome being the report *Research at Melbourne: Ensuring excellence and impact to 2025* (University of Melbourne 2012). From the extensive process in developing this paper came the exposition of three 'Grand Challenges' facing society that universities can contribute to, and that the University would have as centrepieces for its research endeavours in the coming decade. One of these Grand Challenges is 'Supporting sustainability and resilience'. Placing sustainability at the centre of research at this research-intensive university is recognition of previous work and a framework for future research and research training. This report estimated that over 1,300 researchers are working in this area and that around AUD$218 million annually was being expended in research across energy, water, carbon management and related sustainability domains at the University. This will be boosted by a share in the AUD$100 million dollars that the University is injecting to work related to the three 'Grand Challenge' areas.

3 Aspects of Sustainability

The seven key dimensions of university sustainability functions are defined by Calder and Clugston (2003) as:

- Curriculum
- Research and Scholarship
- Operations
- Faculty and Staff Hiring, Development and Rewards
- Outreach and Service
- Student Opportunities
- Institutional Mission, Structure and Planning

There are thus many aspects of sustainability that need to be planned for and addressed as part of a university's journey to becoming more sustainable. Each aspect needs to be considered both as a specific area for planning and action, and also in conjunction with all others, so that each initiative supports and reinforces each of the others. This paper will look at how one large research intensive university has addressed these aspects of sustainability, and in particular, ways that the University has strengthened its performance and outcomes in sustainability through initiatives that address a number of sustainability dimensions, so that one is reinforcing the other. This collectively and slowly contributes to cultural change to a culture where sustainability is one of the considerations in the decision-making process at all levels, which continues to be sought and worked towards.

4 Leadership and Governance: Setting the Goals and Funding

For any organisation to attain long term goals there needs to be commitment from the senior leadership, and this commitment needs to be instantiated in policies, commitments, and structures. The commitment to sustainability is needed in policy documents and accountabilities, as otherwise a change in staffing could mean that priorities can shift in unintended ways.

The latest *Environmental Policy* (approved April 2013) extended on earlier policies and includes the following Principles:

> 1.1. The University has extensive environmental impacts in its role as an international teaching and research based University and a large organisation.
> 1.2. The University is committed to improving its environmental performance through teaching, research, engagement, operations and providing community leadership.
> 1.3. The University influences the environment beyond its local boundaries, and embraces social justice through its respect for limited resources globally.
> 1.4. Heads of academic budget divisions are responsible for ensuring that education for sustainability is embedded in all curricula.
> 1.5. Heads of budget divisions are responsible for ensuring all staff, students and stakeholders are engaged, encouraged and supported in achieving a sustainable campus.
>
> (University of Melbourne 2013)

Flowing from this policy commitment and from a 2012 Sustainability Survey, which found that the University needed more collective and senior Sustainability leadership, the University's *Sustainability Executive* was established in early 2013 to complement the broader *Sustainability Forum*, which had been established in December 2009. The *Sustainability Executive* is composed of high level senior managers of the University and provides collective leadership that consolidates and integrates environmental, social and economic aspects of sustainability in the University.

The *Environment Policy* and consequent activities are in addition to public commitments that the University of Melbourne has made to becoming more sustainable and there are many external and internal drivers toward this outcome. These include public commitments that the University has made about its graduates and their understanding of sustainability, national and international agreements to which the University is a signatory, the Australian National Action Plan for Education (EfS) for Sustainability (DEWHA 2009) that has as an objective that EfS is to be integrated into all university courses/subject areas, and other international drivers.

Recently the Vice Chancellor, in an opinion piece for the national newspaper (Davis 2014) and an email to all staff, stated that: "Sustainability is a core graduate attribute" and this is backed up by a recent revision of the University's *Graduate Attributes* which, under the Active Citizenship section, includes that graduates will:

- have a high regard for human rights, social inclusion, ethics and the environment;
- lead change through advocacy and innovation for a sustainable future. Setting clear University goals, having accountability for goals at a senior level, and putting sustainability on the agenda of each of the University's business units is a necessary foundation for the other dimensions of sustainability to be built upon. That is, if the university leadership is explicitly committed to a sustainable future and campus operations are working towards increasing sustainability in more and less visible ways, then the message that the university, through its hidden curriculum (Orr 1992), is sending to its staff, its students and the wider community is that it values sustainability. This is not to diminish the importance of the other dimensions of sustainability, as each one contributes to the overall culture of sustainability of a university, and each can reinforce the other.

5 Campus Operations: Making Things Happen

Campus operations as a unit has led the way in sustainability at the University of Melbourne, and this includes working toward the ambitious target of carbon neutrality by 2030, which is on track to be achieved, impressive water savings and remarkable achievements across all aspects of operations, some of which are outlined in Table 1 and others that will be components of case studies later in this paper. The team at campus operations has been proactive, creative, and energetic in looking at campus operations and seeing how things can be done in a more sustainable way, and have not constrained themselves to purely operational matters, which has led to some exemplary projects that will be outlined below.

Table 1 Some achievements of campus operations at the University of Melbourne in achieving sustainability goals

Energy and emissions	• Net emissions reduced by 43 %, after green energy and abatements
Water	• Potable water consumption reduced by 30 %
Waste	• 41 % of solid waste is recycled
Campus buildings	• New buildings built to 5-Star-Green-Star design rating, with the newest building on track to gain 6-Star-Green-Star rating • Major building upgrades built to 4-Star-Green-Star rating
Transportation	• Fleet reduced by 25 vehicles • Fleet emissions offset through Greenfleet since 2008 • Increased bicycle parking on campus
Sustainable IT	• Energy considerations of configuration and usage of all IT-related equipment, particularly computer labs and workstations
Sustainable purchasing	• Inclusion of sustainability elements in locally negotiated product and service contracts, i.e. sustainable products, reduced/recyclable packaging, take-back schemes for electronic equipment, catering

6 Campus as a Laboratory: Sustainability Dimensions Supporting Each Other

This section will focus on some case studies where the broad concept of 'Campus as a Laboratory' has been applied. The idea of integrating aspects of sustainability with each other is not new or unique, and McMillin and Dyball (2009) of the Australian National University have outlined benefits of integrating sustainability into many aspects of the university, and Mike Shriberg of the University of Michigan (Shriberg and Harris 2012; Shriberg et al. 2013) has also done extensive work looking at the connections of sustainability and curriculum. The case studies presented here will not focus on curriculum connections, as McMillan and Dyball and Shriberg have done, but rather on connections between campus operations (underpinned by governance), research and research training, and the engagement and education of staff, students and the wider community.

7 Case Study 1: Recycling at the University of Melbourne: Operations, Research and Engagement

A shining example of breaking down boundaries between dimensions of sustainability is recycling. Typically recycling and waste are seen as an operational activity, with staff and student engagement and education provided only as necessary. In 2010 the University developed a Waste Management Plan, set a 2015 target of 50 % of waste to be recycled (from a baseline of 16 % recycled in 2009) and appointed a Waste and Recycling Coordinator to drive the changes necessary to achieve the ambitious target. The person appointed to this role came to the position not only with waste management experience, but also 30 years of experience in environmental education.

Some of the early actions to achieve 50 % of waste being recycled that engaged all staff was a shift from a waste bin in each office to a approximately 1 L desktop container for waste and a larger bin for recyclable materials. Education of staff accompanied the staged roll out of the office bin system. The size of the bins was a clear indication of the expectations of the amounts of landfill waste and recyclables that staff would produce. This connection between operations and staff engagement is not particularly remarkable, but a later stage in the process of achieving the University goal took the much less usual path of combining campus operations with research. A study, entitled *Build It and They Will Recycle: The critical importance of infrastructure in changing recycling behaviour* (Gilmour et al. 2013) used the campus as the site of investigation and the staff and students as the actors on the campus, whose behaviours were influenced through the arrangements of landfill and recycling bins provided at different locations and whose behaviours were measured through the data collected about the amounts and proportions of landfill and recyclable materials that ended up in the bins.

The Waste and Recycling Coordinator, who initiated and was coordinating investigator of the study, worked with a researcher from the Department of Infrastructure Engineering and a postgraduate student from the Faculty of Science. The study was funded by the *Melbourne Institute for a Sustainable Society*, a multidisciplinary research institute of the University of Melbourne. The study focussed on people—staff and students with waste to dispose of and cleaning staff who are responsible for correct disposal of waste—and on infrastructure—the availability and placement of appropriate bins.

The study commenced with a review of literature to inform the later stages, with the key findings of this stage being that effective systems need to be easy to use and infrastructure needs to be conveniently located. Bins were audited using a methodology that was developed for the study to determine the composition of waste within individual bins using digital photographs and counting items using a superimposed grid. The data from bin audits enable an understanding how location, labelling, and distance to the nearest recycling/landfill bin was influencing the behaviour of people disposing of waste.

Findings include:

- Inside buildings the absence of a recycling bin led to a landfill waste bin containing an average of 38 % recyclable material (by weight).
- If a recycling bin was visible, but more than 5 m away, then there was no significant difference in the amount of recyclable material in the landfill bin.
- If the recycling bin was visible and 1–4 m away, then the amount of recyclables in the landfill waste bin dropped to an average of 17 %.
- If the two bins were placed adjacent to each other then this dropped to 14 %.

This provides evidence of the critical importance of co-locating landfill and recycling bins and, while perhaps unsurprising, demonstrates that the provision of suitable infrastructure can lead to dramatic improvements in rates of recycling and less recyclable materials going to landfill. An outcome of the study, and associated works, has led to an increase in recycling rates across the University from 16 to 41 %, which is well on the way to achieving the 2015 goal of 50 %.

In addition to the demonstrable success of the project in achieving its goal of increased rates of recycling on the campus, the project is a rare example of campus operations being a lead player in university research and working directly with university researchers and gaining funding from a research institute of the University. The divide between the roles of academic and professional (that is, non-academic) university staff is generally fairly impermeable at research-intensive universities, but this project demonstrates that these barriers not only can be overcome, but that major benefits can occur when this happens. This blurring of boundaries between academic and professional staff is explored as an emerging trend in universities in the UK and Australia by Whitchurch (2008).

Although the project and its outcomes do not impact directly on the taught curriculum of the university, the change in bins on the campus, such that bins are now all paired and labelled, not only increases recycling rates, but also sends a clear message to staff, students, and community members who walk through the University. The priorities of the University are evident from the bins and this is one more strand that builds the cultural shift to one where sustainability is a priority.

8 Case Study 2: Green Roofs: Research, Teaching, Community Engagement and Education and Operations

Not far from the main campus of the University of Melbourne is a smaller campus, known as Burnley because of the suburb it is located in, that has been the site of horticulture education for over 100 years and of environmental horticulture for over 30 years. The campus is the site of the heritage-listed 8.2 ha Burnley Gardens. This campus is a component of the Melbourne School of Land and Environment (MSLE), which is located at a number of campuses, including the main campus. The Burnley Gardens are open to the public and sells seeds and plants, as well as exchanging seeds and plants with other botanic gardens and research institutions.

Research is being undertaken at the Burnley campus into the efficacy of green walls and green roofs as an emergent technology to adapt urban centres to climate change and to improve urban environmental quality (Melbourne School of Land and Environment 2011). This site is one that plays a leading role in community education as well as research leadership. This community engagement includes annual open days and a 'Friends of Burnley Gardens' community volunteer group. Weekly seminars held at Burnley are open to the public and it is a place where connections between the University and its community are very strong.

The Green Infrastructure Research Group (see http://thegirg.org/) of the MSLE is undertaking wide ranging research, including into green roofs and walls. As part of this research a number of green roofs and walls have been constructed at the Burnley campus, where they are used for research, teaching, and community engagement and education. Much of the research into green roofs has been undertaken in cooler and wetter climates, and so research in Australian conditions is needed. A small research green roof was established in 2008, which demonstrated a 48 % reduction in energy use during summer for cooling the building.

The *Burnley Green Roofs* were completed in 2012 and are Australia's first dedicated green roof demonstration, training, and research facility. This suite of green roofs comprises a demonstration roof of 166 m^2, a research green roof of 80 m^2, and a biodiversity green roof of 52 m^2. The demonstration roof is on the roof of the heritage-listed Burnley Campus Hall and has been designed so that there is access across the entire roof and it shows how different substrates and different depths of substrates function, as well as demonstrating different uses and plants. The roof has 14 planting zones and the design of the roof, with access via a connecting curving walkway, facilitates small group teaching. Over 200 plant taxa are represented on the roof, with over 3,000 individual plants. The plantings are designed to demonstrate the range and variability of plants that can be used successfully on a green roof. The demonstration roof has gained much publicity in the community through open days and through the media.

The research roof is on the Main Building staff room and is based around four quadrants; three green roofs with different depths of substrate and one quadrant without a green roof. Climate and hydrology data are captured with automated instrumentation. Research is underway to seek to quantify the hydrology and energy benefits of different substrate depths and plant combinations. The biodiversity green roof was designed to attract and to provide habitats for lizards, insects, and birds. This roof features indigenous plants that are representative of Victoria's endangered grasslands, it has a small ephemeral pond and shallow creek bed that flows during rain events using stormwater from an adjacent roof. Different substrates have been used to provide a range of habitats for invertebrates and specific habitat features, such as logs and sand for insect burrowing, hollow twigs for nests of native bees, and hard surfaces for lizards and insects to bask on.

These green roofs are exemplary in bringing together many dimensions of sustainability at the University—they are a site for research and they are used for formal teaching and for community engagement and education. Research into green roofs is being undertaken both by University researchers and postgraduate research students, classes are held on the demonstration roof, and a *Specialist Certificate in Green Roofs and Walls* is a new postgraduate coursework qualification that is now being offered by the University. The course is targeted at professionals in the built environment and landscape sectors, and will enable them to undertake such projects for their clients in the community. This intensive will be held on the main campus and sites to be developed as part of this course are being chosen on the campus for both physical suitability and also for social benefits. This will extend the impacts of green roofs to the main campus of the University.

Thus, this undertaking brings together research, teaching, the hidden curriculum, and community engagement and education in a way that has captured the imagination of the community, as well as the energy savings and benefits to campus operations that assist the University in meeting energy saving targets.

9 Case Study 3: Shallow Geothermal

Shallow geothermal (or ground source heat pump) installations make use of the constant temperature below ground (gained from solar radiation) to heat spaces in winter and cool them in summer. This technology is not common in Australia, despite being extensively used in other regions, and research is needed to investigate its efficacy in Australian conditions. A wider understanding of the technology in the Australian community is beneficial, as, once installed, the energy used to run the heating and cooling systems are approximately 20 % of that using traditional energy sources. In Victoria, the state where the University is sited, the primary source of electricity production is brown coal, which is the most polluting of all coals, and so reduction in electricity use is of particular benefit in Victoria.

Initially a small experimental shallow geothermal system was installed to heat and cool a weights room in the University gymnasium. This 14 kW system consisted of four 30 m deep bores with associated above ground heat pumps. This demonstration unit had some small operational benefits through energy savings and had some, though limited, visibility to staff, students and the wider community [For a technical description of the projects see Johnston et al. (2014)].

As a second stage a larger (25 kW) shallow geothermal system was installed at the newly developed *Campus Sustainability Centre*, which is centrally located on the main campus of the University and was established to showcase sustainability-related research being undertaken across our campuses and to inform and engage on all aspects of sustainability at the University. This shallow geothermal system has five 50 m deep boreholes and heats and cools the ground floor of the building where the *Campus Sustainability Centre* is housed. The project was completed in 2013 and academic staff and students from the Department of Infrastructure Engineering were involved, working with staff from campus operations and industry partners. This gave valuable experience for both staff and students in this technology. The system has now been functional for a year and contributes to the University meeting energy targets.

This demonstration unit is part of a larger research project of staff of the Department of Infrastructure Engineering that is funded by the government Department of Primary Industries, and the installation of this unit was a significant achievement and an important first step of the overall project. The on-going monitoring of the system's operations will contribute towards local design and installation guidelines and the system will contribute to raising awareness of this technology, which is little known in Australia. This geothermal project is also a component of a research student's Ph.D. studies. This student worked with several contractors to design and install the system, which will inform his research into design techniques and guidelines. The installed system will allow field experiments to be performed and incorporate local heating/cooling loads and local geological characteristics in his research.

Knowledge gained is being used to inform larger scale shallow geothermal systems being designed for major construction work elsewhere on campus, in particular a 90 kW system now being installed at the *Bio21 Science Sub-school* (see http://www.sciencesubschool.bio21.unimelb.edu.au). This science sub-school is planned to be a leading centre for science education for secondary school students and teachers and is the fruition of a partnership between the University of Melbourne, led by the Bio21 Institute, and the Melbourne Graduate School of Education, the Victorian Government (Department of Education and Early Childhood Development) and two local secondary schools. Its year 11 and 12 students will have a lived experience of the technology and will be able to use the building as a living experiment in their studies with direct access to the data generated.

The demonstration shallow geothermal installations thus brings together campus operations, through lower energy usage and informing the design for new buildings, research and research training, through the work of academic researchers and research students, as well as the engagement and educational opportunities for staff, students and the wider community.

10 Conclusions

Although there have been successes at the University of Melbourne, there is still a long way to go in achieving sustainability goals. Some of the more visible targets, such as energy use and waste production, which are in the realm of campus operations, are well on track; governance and leadership in the area has led to good policies being in place; and specific areas of research are world class. However, and despite the case studies presented here that demonstrate successes, there is still much work to be done in bringing together the various dimensions of sustainability so that even more can be achieved in the future. The goal that each decision being considered through a sustainability lens, as well as the more usual lenses of University decision-making, is still not usual practice, and there is thus much cultural change needed. However, it is proposed that one effective way to work toward this desired approach is to look for, and to fund, opportunities that enable a number of the dimensions of sustainability to become entwined and for each to strengthen the other. This approach of designing sustainability initiatives to address three or more of the dimensions of university sustainability is one that has delivered great results and it shows much promise as a way forward for universities to work at becoming more sustainable.

References

Calder W, Clugston RM (2003) Progress toward sustainability in higher education. Environ Law Report 2003(1):10003–10022

Davis G (2014) Does Melbourne invest in fossil fuels? The Australian, 29 March 2014. http://www.theaustralian.com.au/higher-education/opinion/does-melbourne-invest-in-fossil-fuels/story-e6frgcko-1226867939986. Last Accessed 30 March 2014

DEWHA (2009). Living sustainably: the Australian government's national action plan for education for sustainability. Australian Department of Environment, Water, Heritage & the Arts. http://www.environment.gov.au/education/publications/pubs/national-action-plan.pdf. Last Accessed 27 Feb 2014

Gilmour P, Alcorn J, Moore G (2013) Build it and they will recycle: the critical importance of infrastructure in changing recycling behaviour. A Melbourne sustainable society institute report. http://www.sustainable.unimelb.edu.au/files/mssi/UoM%20Waste%20Mangement%20Draft%20Report.pdf. Last Accessed 27 Feb 2014

Johnston I, Narsilio G, Colls S, Kivi AV, Payne D, Wearing-Smith M, Noonan G (2014) Direct geothermal energy demonstration projects for Victoria, Australia. IPENZ Trans 41:1–10

McMillin J, Dyball R (2009) Developing a whole-of-university approach to educating for sustainability linking curriculum, research and sustainable campus operations. J Educ Sustain Dev 3(1):55–64

Melbourne School of Land and Environment (2011) Demonstrating sustainability: Burnley campus, The University of Melbourne. http://sustainablecampus.unimelb.edu.au/pdf/Demonstrating-Sustainability-at-Burnley-Campus-Brochure-July11.pdf. Last Accessed 27 Feb 2014

Orr D (1992) Environmental literacy: education as if the earth mattered, twelfth annual E.F. Schumacher Lectures Oct 1992, Great Barrington, Massachusetts. http://sfsf.com.au/Education.As.If.The.Earth.Mattered.pdf. Last Accessed 27 Feb 2014

Shriberg M, Harris K (2012) Building sustainability change management and leadership skills in students: lessons learned from "Sustainability and the Campus" at the University of Michigan. J Environ Stud Sci 2(2):154–164
Shriberg M, Schwimmer S, MacDonald L (2013) University of Michigan's sustainability scholars program: empowering leaders. Sustain: J Record 6(5):259–264
University of Melbourne (2012) Research at Melbourne: ensuring excellence and impact to 2025. https://www.unimelb.edu.au/research/docs/research-paper-2012.pdf. Last Accessed 27 Feb 2014
University of Melbourne (2013) Environmental policy (MPF1245). University of Melbourne policy library, The University of Melbourne, Australia. https://policy.unimelb.edu.au/MPF1245. Last Accessed 27 Feb 2014
Whitchurch C (2008) Shifting identities and blurring boundaries: the emergence of third space professionals in UK higher education. High Educ Q 62(4):377–396

Author Biography

Dr. Dianne Chambers is Assistant Dean (Learning Technologies) and a Senior Lecturer at the Melbourne Graduate School of Education at the University of Melbourne and teaches in the fields of Education for Sustainability and ICT in Education. Her research areas include education for sustainability in higher education, the uses of technology in teacher education, and technology-enriched problem based learning. She holds a Ph.D., B.Sc. (Hons) and Graduate Certificate in University Management from the University of Melbourne and an M.Ed. (Hons) from the University of Wollongong.

Dianne completed her Ph.D. in 1990 in the field of forest ecology, looking at the changes in soil that promote growth of mountain ash (*Eucalyptus regnans*) after a bushfire, before moving into teacher education. Dianne has been involved in a number of multimedia projects that support learning in teacher education and has won a number of awards for her teaching, including the inaugural *Award for Excellence in Teaching* from the Melbourne Graduate School of Education at the University of Melbourne and a *Universitas 21 Fellowship.*

She would like to thank colleagues from across the University and particularly to colleagues in Property and Campus Services for assistance in the preparation of this paper.

Driving the Energy Transition at Maastricht University? Analysing the Transformative Potential of the Student-Driven and Staff-Supported Maastricht University Green Office

Felix Spira and Alexander Baker-Shelley

Abstract

Strategies on how to improve energy efficiency at universities as part of the global energy transition are barely understood. This study aims to contribute to this body of knowledge, by investigating the energy efficiency transition at Maastricht University. Using the Multi-Level Perspective of transition studies, this research investigates the landscape trends in relation to energy efficiency at Dutch universities, the energy efficiency regime of Maastricht University and the impact of the student-driven and staff-supported Green Office, conceptualized as organizational niche, on this regime. The results suggest that before the Green Office was established, Maastricht University was trapped in a lock-in. Any external push to advance energy efficiency was missing and the institution's organizational capacity to work on this issue was very limited, thus allowing only for incremental, technology-driven improvements of energy efficiency in buildings. After successful lobbying of students, the university introduced the Green Office as a student-led and staff-supported sustainability unit. The results suggest that the Green Office—at least temporarily—enabled the university to break through this lock-in by expanding its organizational capacity to work

F. Spira (✉)
Dutch Research Institute for Transitions (DRIFT), Erasmus University Rotterdam,
Postbus 1738, 3000 DR Rotterdam, The Netherlands
e-mail: spira@drift.eur.nl

A. Baker-Shelley
International Centre for Integrated Assessment and Sustainable Development (ICIS),
Maastricht University, P.O. Box 616, 6200 MD Maastricht, The Netherlands
e-mail: alex.baker-shelley@maastrichtuniversity.nl

W. Leal Filho (ed.), *Transformative Approaches to Sustainable Development at Universities*, World Sustainability Series, DOI 10.1007/978-3-319-08837-2_15

towards energy efficiency through the mobilization of students. As a result, the Green Office implemented several additional projects on energy efficiency in IT, buildings and strategy development that all effected dimensions of the regime.

Keywords
Sustainability in higher education · Energy efficiency · Organisational change · Student engagement · Multi-level perspective

1 Introduction

Radically increasing energy efficiency is an important component of the global energy transition from fossil fuels to renewable energies. As institutions of learning and research, universities can provide interesting 'living laboratories' for experimentation and social learning around different ways to promote energy efficiency (Schneidewind and Brodowski 2013). Despite this importance, case studies on the energy efficiency transition at universities are rare. This case study examines the impact of a student-driven and staff-supported sustainability unit, the Maastricht University Green Office, to promote energy efficiency at Maastricht University. The Green Office was established in 2010 after successful lobbying of a student initiative, which convinced the university to provide a mandate, resources, and support for a student-led and staff supported sustainability unit.

The Multi-Level Perspective serves as the analytical framework to conceptualize the impact of the Green Office. The Multi-Level Perspective consists of three levels—the landscape, regime and niche—that have been adapted to Maastricht University. The landscape represents broad and long-term developments affecting energy efficiency at Dutch universities. The regime is defined as the complex set of socio-economic and technological practices that constrain and enable energy efficiency at the university. The Green Office is defined as an organizational niche, a relatively unconstrained place in which experimentation with new practices can take place (Kemp et al. 1998; Raven 2012).

Four research questions have been developed to understand the Green Office case and learn from it:

1. What were the major landscape trends in the Netherlands that enabled or constrained energy efficiency at Maastricht University?
2. To what extent have these trends impacted the energy efficiency regime of Maastricht University?
3. What were the dimensions of the energy efficiency regime before the Green Office was established in September 2010?
4. To what extent has the Green Office impacted this regime?

The results of the case study are relevant for the academic field of transition studies, as they present the first application of the Multi-Level Perspective to better understand organisational change towards sustainability within a university context. The case study is also relevant for the general literature on sustainability in higher education, as it presents a first detailed analysis of the impacts of a student-led and staff-supported sustainability unit on energy efficiency.

Participatory Action Research presents the methodological approach. 29 interviews with 24 participants were conducted between May and July 2013. A qualitative research design has been chosen to answer the research questions, as this requires an inductive content analysis of this specific case (Elo and Kynga 2008), rather than insights regarding a generalizable phenomenon.

The presentation of findings unfolds in the following way. The first section provides the findings of a literature review on energy efficiency in higher education. The second section introduces the Multi-Level Perspective as the analytical framework of the study. The third section describes the case study and the methodology. Thereafter, the results of the study are presented and discussed.

2 Literature Review: The Role of Energy Efficiency at Universities

Energy efficiency is an important topic in the literature on sustainability, environmental management and climate change mitigation at universities. Energy efficiency is a key component of sustainability audits for universities (Beyer et al. 2008; Mitchell 2011), is a major environmental impact that needs to be measured, controlled and reduced (Jain and Pant 2010; Bero et al. 2011; Disterheft et al. 2012), and features prominently in climate action plans (Cleaves et al. 2009; Rauch and Newman 2009).

Despite this importance of energy efficiency, the literature suggests that the energy savings potential of most universities has not been fully exploited yet. For instance, at German universities, the reduction potential of behaviour change measures was estimated to range from 9 to 20 % for heating and constitute 18 % for electricity at some institutions (Kattenstein et al. 2002; Gillen et al. 2002; Matthies et al. 2011; Starzynski 2012). In this sense, universities—alongside other public institutions—experience an 'energy efficiency paradox', as savings potentials exist that are not exploited (DECC 2012).

Among the common factors enabling or constraining energy efficiency at universities are campus size, incentive structures, power relationships, data quality, competence and knowledge of employees, risk perceptions, awareness, building status, funding, cost-calculation methodologies, geography and leadership (Levy and Dilwali 2000; Sorrell et al. 2000; Pearce and Miller 2006; Ward et al. 2008; Altan 2010; Ghosh 2011).

Numerous organisational measures have been identified that can help institutions to address this energy efficiency paradox. These measures include among others, revolving loan funds, partnerships with energy service companies, and energy savings in laboratories, student dorms and offices through technical and behavioural measures, such as community-based social marketing campaigns (Levy and Dilwali 2000; Kahler 2003; Marcell et al. 2004; Woolliams et al. 2005; Pearce and Miller 2006; Petersen et al. 2007; Wesolowski et al. 2010; Feder et al. 2012; Liers and Person 2012). Despite these very specific measures, no literature could be identified that investigates a more holistic approach towards advancing energy efficiency including a simultaneous attention to changes in, for instance, organisational culture, learning, policy, actor-networks, as well as infrastructure and technology. For this reason, we turn to the Multi-Level Perspective as an analytical framework.

3 Analytical Framework: Multi-level Perspective

Compared to existing studies on energy efficiency at universities, transition studies provide the required analytical perspective that is process, long-term and system-oriented. Transition scholars have developed several concepts to study transition processes, among which the Multi-Level Perspective is one of the most prominent. The Multi-Level Perspective provides conceptual insights into how transition processes unfold between and within three levels of analysis (see Fig. 1).

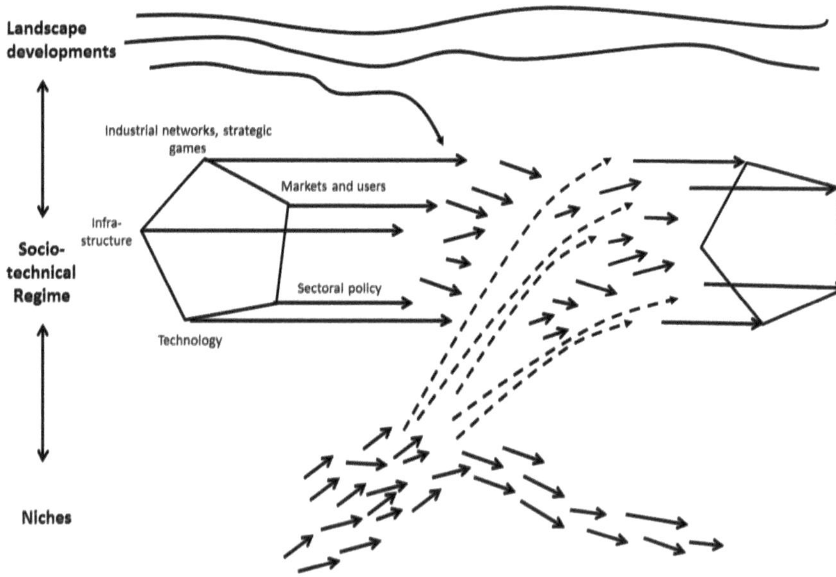

Fig. 1 Multi-level perspective (Adapted by authors from Geels 2002)

Figure 1 illustrates the three levels of the Multi-Level Perspective. They present a 'nested hierarchy', 'meaning that regimes are embedded within landscapes and niches within regimes' (Geels and Schot 2010; Loorbach 2010). The *landscape* constitutes the meso-level, such as privatization, demographics and globalization, describing long-term and global trends that regime actors cannot influence in the short run (Grin et al. 2010). *Socio-technical regimes* are nested within these landscapes. As illustrated in Fig. 1, Geels (2002) defines regimes with regards to different dimensions, including among others markets and users, sectoral policy, as well as technology and infrastructure. These elements weave into a complex and stable web of interdependencies, which constrain and enable certain practices and lock the regimes into a specific development pathway (Geels and Schot 2010). *Niches* are conceptualized as relatively unconstrained and 'protected spaces' in which experimentation with new practices can occur (Kemp et al. 1998; Raven 2012). It is the scaling of niche innovations into the regime—after the regime has been destabilized through landscape or regime pressures- that brings about energy transitions (Geels and Schot 2007).

The Multi-Level Perspective was adapted and operationalised to the specific requirements of Maastricht University. The landscape is conceptualized as the broad and long-term developments affecting energy efficiency at Dutch universities. The regime is defined as the complex set of socio-economic and technological practices that constrain and enable energy efficiency at the university. The Green Office is defined as an organizational niche, a relatively unconstrained place in which experimentation with new practices can take place. We thus assume that the Green Office can act as the catalyst to bring about changes within the energy efficiency regime.

4 Methodology

4.1 Case Description

After successful lobbying by a student initiative, the Green Office was established in 2010, as the student-driven and staff-supported sustainability office of Maastricht University. It is run by a team of seven to eight student employees- reimbursed on a part-time basis-, the university's environmental advisor and several student volunteers. The university provides the Green Office with an official mandate and an annual budget to pay for salaries, office space and project expenses. The Green Office is part of Facility Services and reports to its own supervisory board. From a governance perspective, the Green Office thus presents a structural mechanism to allow for a stronger integration of students into the university's sustainability efforts.

Since 2010, the Green Office has implemented several energy efficiency projects at Maastricht University. These projects have been grouped into three categories (Table 1): The *Green IT* project series is the overarching category for five projects

Table 1 Green Office projects under investigation

Green IT	Ctrl-Alt-Delete emissions business case (Jan–June 11)
	Policy proposal on PC power management (Oct 11–June12)
	Policy proposal on server virtualization (Oct 11–May 12)
	Policy proposal on data centres (Oct 11–on-going)
	Implementation of PC Power management (Sept 12–June 13)
Energy efficient buildings	Six student research projects on energy efficiency with regards to Maastricht University buildings (Jan 11–May 13)
	UCM Business Case (Oct 11–June 12)
	Partial implementation UCM Business Case (Dec 12–June 13)
Sustainability strategy	Climate action report (Feb 11–Nov 11)
	Sustainability policy 2012–2014 (Aug 11–March 12)
	Sustainability progress report 2011 (Jan–Nov 12)
	Sustainability progress report 2012 (Jan–July 13)
	Sustainability vision 2030 (Feb 12–Sept 12)
	Sustainability roadmap (Sept 2012–on-going)

to improve energy efficiency in the IT infrastructure. The project series *energy efficient building* is a mixture of six student research projects and a business case. The project series on *sustainability strategy* contains all governance projects in which energy efficiency plays an important element. In this sense, all of these projects comprise a multitude of measures, including student research, business case and strategy development.

4.2 Data Gathering and Analysis

The data was gathered through interviews and document analysis. Interviewees were selected through purposive sampling based on the interviewee's engagement with the Green Office projects, and snowball sampling through the recommendation of interviewees (Burns 2007). The sample included current and former Green Office student employees, volunteers and Maastricht University employees. Two external experts on Green IT and monumental buildings at Dutch universities were also interviewed. Six participants were interviewed at least twice and 14 provided feedback on research results either via email or in follow-up interviews. Ultimately, the most relevant stakeholders interviewed were chosen according to who worked on or was affected by the energy efficiency projects of the Green Office within the last three years. 29 semi-structured and open-ended interviews were conducted with 23 participants. The interviews were conducted in person between May and July 2013 at Maastricht University or online via Skype and lasted between 25 and 90 minutes. Interviews were recorded when interviewees agreed. Interviews were conducted in English, German or Dutch.

Table 2 Overview of documents gathered and analysed

Landscape	Agreement Dutch universities joining third Multiple-Year Agreement on Energy Efficiency (MJA3); MJA-bedrijfs- en brancherapportage 2009; MJA-Sectorrapport 2012 Wetenschappelijk onderwijs; ICT energy efficiency in higher education
Regime	Sustainable Building Document; Sustainability Policy 2008-2010; Energy Efficiency Plan 2009–2012; Energy Efficiency Plan 2013–2016
Niche	*Green IT*: Ctrl-Alt-Delete Business Case; Green IT meeting notes; Project descriptions by ICTS about data centre, server virtualization and PC PowerManagement; Presentations about SURF ICT Scan 2012 and PC Power Management *Energy efficient buildings*: UCM Business Case; UCM Business Case meeting notes; UCM Business Case Evaluation Report *Sustainability strategy*: Climate Action Report; Sustainability Strategy 2012-14; Sustainability Vision 2030; Sustainability Progress Report 2012 and 2013; Final draft version Sustainability Roadmap; Sustainability Roadmap meeting notes, planning documents and drafts

Documents were obtained through online research or by recommendations from interviewees in order to identify landscape trends, regime characteristics and niche activities. Table 2 depicts the documents that were gathered and analysed, including publications of external organizations, internal Green Office working documents, publications and policies. The insights gained from these documents were used to support or contrast the findings from the interviews.

The interviews and documents were analysed inductively following a Grounded Theory (GT) approach (Strauss and Corbin 1998). Inductive codes were developed from the interview transcriptions and the documents with regards to landscape trends, regime characteristics and impact of niche activities. These inductive codes were grouped into higher level codes where possible (Silverman 2010). Following this clustering, their qualitative relationships and relative importance of these codes towards each other was examined.

5 Results

5.1 Landscape Trends

With regards to the major landscape trends, four enabling and seven constraining trends were identified (Table 3). Among the seven constraining trends are, an increasing energy intensity of natural science and medical research, intensification of ICT usage and of building usage through longer opening hours, governmental funding cuts following the financial crisis, the lack of attention for energy efficiency in performance evaluations of universities, decreasing energy prices which raise the payback period for energy efficiency projects, and procurement regulations preventing universities from making ambitious demands for energy efficiency in appliances.

Table 3 Landscape trends constraining or enabling energy efficiency at Dutch universities

Impact	Landscape trends	Sources
Enabling	Tightening building and energy regulations, agreements and government policy	Project Manager FD 1 2013, pers. comm. 10 June; Environmental advisor 2013, pers. comm. 10 June
	Increasing societal awareness around energy, climate change and sustainability	Facility manager 2013, pers. comm. 11 June
	Efficiency improvements of equipment, technology and building materials	Project Manager FD 1 2013, pers. comm. 10 June
	More services and understanding to improve energy efficiency within organizations	Environmental advisor 2013, pers. comm. 10 June
Con-straining	Increasing energy intensity of natural science and medical research	Agentschap (2013)
	Intensification of ICT usage	ter Hofte (2011), Agentschap (2013), Ehnert (2013), Green IT researcher (2013), pers. comm. 12 July
	Intensification of building usage through longer opening hours of libraries and teaching facilities	Project Manager FD 1 2013, pers. comm. 10 June; Environmental advisor 2013, pers. comm. 10 June
	Budget cuts following the financial crisis	GO student coordinator 3 2013, pers. comm. 26 June; Energy coordinator 2013, pers. comm. 10 June
	Energy efficiency is not addressed in rankings, accreditation or other performance evaluations of universities	Environmental advisor 2013, pers. comm. 10 June
	Procurement regulations	Environmental advisor 2013, pers. comm. 10 June
	Decreasing energy prices	Environmental advisor 2013, pers. comm. 10 June

Only four trends were found that enable energy efficiency. All interviewees identified the third Multiple-Year Agreement on Energy Efficiency (MJA3) as the major enabling trend. Under the MJA3, Dutch universities commit themselves to improve energy efficiency by 30 % between 2005 and 2020 (VSNU 2008). One interviewee mentioned that societal awareness of energy efficiency was increasing, which might prompt stakeholders to take action on it. However, no official data could be found to support the observation. Finally, interviewees identified a trend towards more energy efficient equipment, technology and building materials, as well as energy efficiency related services as enabling trends.

The interview and document analysis suggests that all of these landscape trends have been found to affect Maastricht University. With respect to enabling trends, the university's accession to the MJA3 was the most important enabling trend that influenced Maastricht University, as the university amongst other measures, has to set up an Energy Efficiency Plan and formally commit to the MJA3 target (Green Office 2011). Similarly for constraining trends, the intensification of energy usage for research and IT, as well as low energy prices that increase payback periods are three issues Maastricht University struggles with.Also governmental budget cuts affected Maastricht University. Despite the fact that it was not possible to quantify the impact of individual trends, the data suggests that they mainly encouraged incremental, technology-driven improvements within the energy efficiency regime.

5.2 The Regime Before the Emergence of the Niche

Having described the landscape trends, this section zooms into Maastricht University, discussing the most important regime characteristics before the establishment of the Green Office in September 2010. Via the interviews and document analysis, five regime dimensions were developed inductively (Fig. 2).

Infrastructure and technology capture elements of the regime that directly determine energy consumption. The elements of this dimension are very diverse, including 30 buildings, the ICT system—including 6000 computers, 4,100 telephones, 860 network devices—and the 5,322 research apparatuses (ICTS 2011a, b, c).

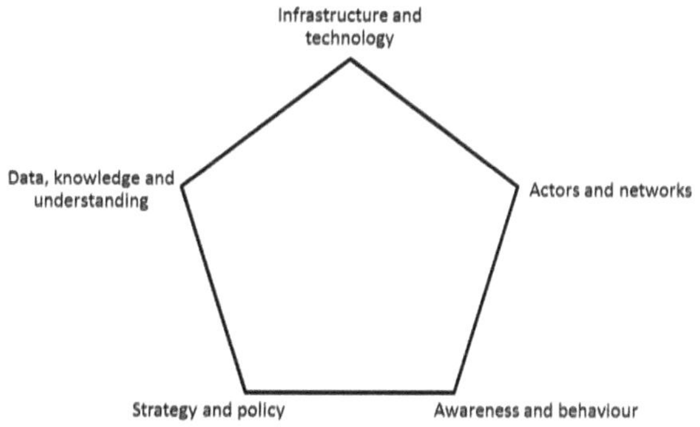

Fig. 2 Five dimensions of the energy efficiency regime

Before the Green Office was established, no energy efficiency measures had been implemented with regards to ICT and research equipment. Energy efficiency projects in buildings focused on technological measures with very limited user involvement.

Actors and networks describe the departments and employees determining usage, maintenance, policy and investment decisions with regards to energy efficiency. Facility Services formally manages the obligations of Maastricht University under the MJA3. In 2010, an energy group comprising six Facility Services staff met every 6 weeks to discuss progress on the Energy Efficiency Plan (Project Manager FD 1 2013, pers. comm. 10 June). On an operational level, the two energy coordinators were responsible for building optimization, monitoring energy consumption and maintenance of the basic infrastructure (Energy coordinator 2013, pers. comm. 10 June). The Executive Board was the highest governing body involved in formal investment decisions around buildings and energy efficiency. Before the Green Office was established no students, researchers or other staff—such as from the ICT department—were involved in formal decisions affecting energy efficiency.

The dimension of *awareness and behaviour* relates to the level of awareness of actors with regards to the importance of energy efficiency and subsequent behaviours to improve it. In 2010, several interviewees suggest that the overall awareness of the importance of energy efficiency for Maastricht University and acceptance of change seem to have been relatively low among students, staff, researchers, ICT departments and higher management. For instance, one Facility Service project manager explained that "the culture of the average university employee does not accept these issues—that is my experience" (Project Manager FD 1 2013, pers. comm. 10 June). Also no measures were implemented to raise awareness or promote behaviour change with regards to energy efficiency.

Strategy and policy describes the official targets, strategies and programmes of the university to improve energy efficiency. In 2010, the major focus of Maastricht University's energy efficiency strategy was adherence to the MJA3 (Maastricht University 2010). Technological fixes, such as building optimization and automation, that are largely invisible to students and staff, have been the strategic focus of the Energy Efficiency Plan 2009–2013. ICT, research and behaviour change were not part of the official energy efficiency strategy (Maastricht University 2009). However, despite its commitment to the MJA3, the efforts outlined in this Energy Efficiency Plan were insufficient to meet the targets: The proposed measures for the four years of the Energy Efficiency Plan aimed to achieve only 1.19 % efficiency improvements compared to the 8 % required by the MJA3 (Maastricht University 2009). The MJA3 was also not integrated into any strategic university planning documents, but rather managed as a voluntary add-on. Hence, we conclude that the strategic commitment of Maastricht University with regards to its own energy efficiency targets was insufficient.

The fifth dimension describes the *data, knowledge and understanding* that actors at Maastricht University had with regards to energy efficiency in 2010. The main sources of knowledge were business cases developed by Facility Services, consultancy reports, visits to energy fairs and biannual meetings with energy coordinators of other Dutch universities (Energy coordinator 2013, pers. comm. 10 June). With regards to user behaviour only one study was conducted by three researchers investigating the energy-related behaviours and attitudes of over 200 university employees (Lo et al. 2010). No projects followed from this study and it was the only research undertaken by academics concerning energy issues at the university. No investigation into the energy savings potential of ICT or research equipment had taken place. This is why we conclude that a major barrier to improving the energy efficiency of ICT infrastructure, research equipment and behaviour was thus the lack of knowledge about savings potentials and options.

The findings on the condition of the energy efficiency regime before the Green Office was established illustrate that the interaction between the five regime dimensions locked the university into a specific development pathway. Measures to advance energy efficiency through ICT, research equipment and behaviour change were not on the agenda for Maastricht University. Also, the ambitions outlined in the Energy Efficiency Plan were insufficient to meet energy efficiency targets. The organisational capacity for energy efficiency was constrained, as only a small group of Facility Service staff was involved for energy efficiency decisions. Accordingly, we assume that measures to improve energy efficiency remained far beyond a theoretical savings potential, as only technological improvements in buildings with minimal disturbances to users was implemented.

5.3 Impact of the Niche on the Regime

The interviews and document analysis suggests that the Green Office was able to break through the lock-in of the energy efficiency regime, at least temporarily. This was mainly achieved as the Green Office enabled students, as a new group of actors, to participate in the university's energy efficiency efforts, by formally entering the existing energy efficiency network. Two to three student employees and several student volunteers developed business cases together with the ICT department and Facility Services. This entrance of students significantly enhanced the university's organisational capacity to work on energy efficiency issues.

In addition to expanding capacity by mobilizing students, the Green Office also emerged as a network hub with connections to all major actors. For instance, the Green Office joined the Energy Efficiency Group of Facility Services, established connections to the higher management of the university, commissioned student research projects on energy efficiency, worked together with the ICT department on business cases and reached out to facility managers of buildings. In this sense, the Green Office was able to connect with and involve a much larger and diverse set of stakeholders in energy efficiency issues.

The three-project series of the Green Office also impacted the other dimensions of the energy efficiency regime. The projects had a small impact on *infrastructure and technology*, as one Green IT project is currently in the implementation phase and an LED lighting project failed. At this point, no projects focused on behaviour change or research equipment have been implemented. The findings suggest that the Green Office increased *awareness* for energy efficiency amongst the higher management and project managers involved in joint projects, but did not have an impact on the larger university community. The exact impact on *behaviour* of different actors could not be assessed. The Green Office influenced *strategy and policy*, as the energy efficiency goals were integrated into a larger framework of sustainability goals that include the promotion of renewable energies. The student research projects—though not widely communicated -, the business cases and the sustainability strategies provide improved *data, knowledge and understanding* of the energy efficiency potential for the university and options to reap this potential. These projects further enabled the Green Office to break through the lock-in of the regime.

The interviews also revealed two main challenges that the Green Office faced. First, whereas the Green Office improved the *data, knowledge and understanding* on energy efficiency potentials and measures through student research and business cases, this understanding has not yet translated into tangible energy reductions or savings. A project to install movement detectors and LED lighting in one building failed. Also the attempt to install PC Power Management on staff computers turned out to be more difficult than expected. Several reasons can be put forward why it has been difficult to implement these projects, yet the major conclusion remains that the impact on the *infrastructure and technology* dimension has been the weakest.

The second challenge relates to the mainstreaming of the energy efficiency agenda into the wider university community. A sustainability vision and roadmap have been developed, with ambitions to go beyond the MJA3 agenda. However, the reporting and implementation mechanisms for this roadmap have not yet been integrated into the strategic planning of the university. Also the Green Office has not yet achieved that other actors within the university themselves take up the energy efficiency agenda, but needs to keep pushing and driving projects that it implements together with other departments.

6 Discussion

A major conclusion with regards to the state of the energy regime of Maastricht University before the Green Office was a lock-in which prevented any radical steps forward. The majority of landscape trends were found to be constraining energy efficiency at Dutch universities, with a detrimental impact on Maastricht University. Also the different dimensions of the energy efficiency regime were combined in a way that kept the regime in a lock-in. A small group of relatively powerless employees were tasked with managing the energy efficiency strategy of the university. They focused on building maintenance and optimization with 'proven' technologies that did not cause any disruptions to students and staff. A lack of awareness and acceptance for energy efficiency measures penetrated different levels of the organization. The lack of organizational power to push for changes constrained the scope of action for Facility Services. The MJA3 presented the main commitment of the university to energy efficiency, which however was perceived as an add-on, voluntary commitment that was not integrated into a larger, binding policy framework. Finally, an understanding about the savings potential for ICT, research equipment and behaviour was missing. All these factors linked into each other which allowed for incremental improvements in building optimization, but prevented ambitious changes.

Figure 3 illustrates how the three Green Office project series affected all dimensions of the energy regime of Maastricht University. The Green IT project series was the most successful of the three project series. IT is an important determinant of energy usage and the Green Office managed to develop business cases together with ICTS that outline the reduction potential of measures. The energy efficient buildings project series was the least successful, as the student research projects did not have a tangible impact on projects and policies and since the UCM Business Case failed. These projects with no tangible impact on the regime are illustrated by the arrows bouncing back from the regime in Fig. 3. The sustainability strategy series had the largest impact on the strategy and policy, actors and networks and awareness and acceptance dimensions, because it included projects that enabled widespread participation from the university community. Despite the fact that the impact of the project series could have been improved, especially with regards to infrastructure, technology and behaviour, these findings suggest that the Green Office provided the internal push to break through the lock-in of the energy efficiency regime on some dimensions.

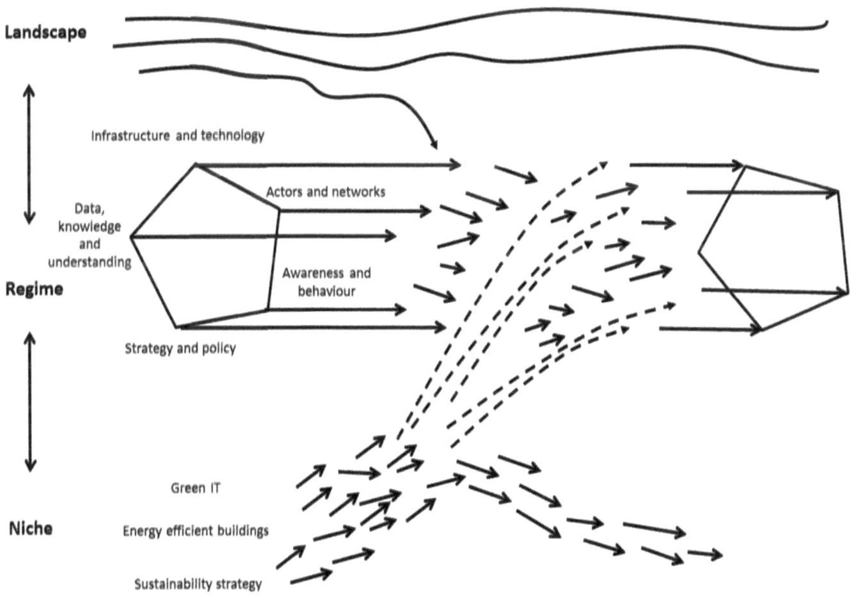

Fig. 3 Impact of the Green Office on the energy efficiency regime of Maastricht University

7 Limitations

This study faces five main limitations. First, precisely identifying landscape trends that affect the regime was difficult. To decrease the number of factors that could be included in the landscape as part of this research, its scope has been limited to enabling or constraining trends within the Netherlands that affect universities. Also the sample of interviewees was too small to make any reliable conclusions about energy trends in the Netherlands. Hence, the findings on the landscape trends only provide tentative insights into the landscape of energy efficiency at Dutch universities.

Secondly, the characteristics of the energy efficiency regime before the establishment of the Green Office could not be adequately assessed. The description of the regime in 2010 was based on interviews and document analysis that were conducted in retrospect in 2013. This made it especially difficult to trace more intangible developments, such as the network structure or levels of awareness. This shortcoming should be addressed in future research on student-driven Green Offices, by conducting a baseline analysis of the energy regime through qualitative and quantitative method in the founding year of the Green Office.

Thirdly, the relative importance of landscape trends, regime characteristics and impact of niche activities could not be established, due to the lack of quantitative data. This made it for instance difficult to appraise the level of awareness for energy efficiency among different university actors. Hence, only the argument which dimensions have been impacted by the niche could be made, but no conclusions about the depth of the impact could be drawn.

The final shortcoming relates to the choice of the Multi-Level Perspective as analytical framework. Transition processes occur on a long-term timescale of 30–40 years; however, this study focuses on a three-year period only. Accordingly, an important question that cannot be answered here is to what extent the Green Office has brought Maastricht University is on a lasting transition pathway. It might be that the Green Office brought about small changes before the regime then stabilizes in a new lock-in. In addition and as noted before, the amorphous nature of the Multi-Level Perspective makes it difficult to operationalize the three levels and apply them in an organizational context (Shove and Walker 2007). Nonetheless, the Multi-Level Perspective provides a great analytical tool to show how organisational change towards energy efficiency can progress along multiple dimensions.

8 Conclusion

The results suggest that before the Green Office was established, the energy efficiency regime of Maastricht University was trapped in a lock-in. Any external push to advance energy efficiency was missing and the institution's organizational capacity to work on this issue was very limited, thus allowing only for incremental, technology-driven improvements of energy efficiency in buildings. After successful lobbying of students, the university introduced the Green Office as a student-led and staff-supported sustainability unit. The results suggest that the Green Office—at least temporarily—enabled the university to break through this lock-in by expanding its organizational capacity to work towards energy efficiency through the mobilization of students. As a result, the Green Office implemented several additional projects on energy efficiency in IT, buildings and strategy development that all effected dimensions of the regime.

Despite the fact that the Green office has augmented a shift in specific dimensions of the regime, it has become clear that what is necessary is more widespread action on the part of staff and students themselves to prevent further lock-in scenarios from taking place. As mentioned in the results section, the second largest challenge in mainstreaming an energy efficiency agenda into the wider university community is the requirement for individuals to take up the energy efficiency agenda of their own volition. Because the Green Office needs to keep driving projects from the outset if this is not achieved, there is less time and resources available for other issues that might need tackling in the mid to long term in managing the sustainability portfolio of Maastricht University. A wider engagement of the university community in energy efficiency issues could be achieved by linking energy efficiency issues to education and research, for instance through academic research projects on how the institution could radically advance energy efficiency.

It is here recommended that further research on the effects of organizational niches in galvanizing a transition in the socio-technical regimes of energy efficiency take into account the historical processes and conditions leading up to the establishment of that niche. In addition, an assured and qualified base-line analysis of conditions at this point is recommended for any Green Office type niche elsewhere. This is essential to allow for the longer time frame required to more accurately encapsulate transition processes, and provides easy referral for each subsequent period of monitoring and reporting of progress. More specifically to this case and ultimately, the critical question had to be raised: to what extent is the Green Office now stuck in a new lock-in? Is current action enough for it to instigate a dynamic process, or is more required for the organizational transformation to gather steam? It is suggested, that future studies would do well to use these questions as a starting point.

References

Altan H (2010) Energy efficiency interventions in UK higher education institutions. Energy Policy 38(12):7722–7731

Agentschap NL (2013) MJA-Sectorrapport 2012 Wetenschappelijk onderwijs. Ministerie van Economische Zaken. http://vsnu.nl/files/documenten/Domeinen/Governance/MJA-Sectorrapport%20Wetenschappelijk%20onderwijs%202012.pdf. Accessed 02 Feb 2013

Bero BN, Doerry E, Middleton R, Meinhardt C (2011) Challenges in the development of environmental management systems on the modern university campus. Int J Sustain High Educ 13(2):133–149

Beyer M, Freund E, Grün N, Langer V, Kilburg M, Kirchgeorg T, Reuter R, Schmitt D, Wiese A, Winterstein M (2008) Klimaneutrale Universität—Studentischer Ergebnisbericht des Projektseminars "Klimaneutrale Leuphana Universität Lüneburg—Planung (KLIMA 2)im Wintersemester 2007/2008". Leuphana Universität Lüneburg, Lehrstuhl für Nachhaltigkeits-management, Lüneburg. www.leuphana.de/fileadmin/user_upload/uniprojekte/Nachhaltig keitsportal/files/Klimaneutrale_Leuphana__WS_0708.pdf. Accessed 03 Jan 2013

Burns D (2007) Systemic action research: a strategy for whole system change. The Policy Press, University of Bristol, Great Britain

Cleaves S, Pasinella B, Andrews J, Wake C (2009) Climate action planning at the University of New Hampshire. Int J Sustain High Educ 10(3):250–265

DECC (2012) What are the factors influencing energy behaviours and decision—making in the non—domestic sector? A Rapid Evidence Assessment, Department of Energy and Climate Change, UK Government, London. http://www.gov.uk/government/publications/factors-influencing-energy. Accessed 08 Jan 2013

Disterheft A, da Silva Caeiro S, Ramos M, Manuel U, Azeiteiro M (2012) Environmental management systems (EMS) implementation processes and practices in European higher education institutions: top-down versus participatory approaches. J Clean Prod 31:80–90

Ehnert H (2013) Problem statement: the implementation of Green IT at Dutch Higher Education—organizations: the case of Maastricht University. Unpublished Master Thesis, Maastricht University

Elo S, Kynga SH (2008) The qualitative content analysis process. J Adv Nurs 62(1):107–115

Feder E, Robinson J, Wakefield S (2012) Persistence of change: fume hood campaign lessons. Int J Sustain High Educ 13(4):338–353

Geels FW (2002) Technological transitions as evolutionary reconfiguration processes: a multi-level perspective and a case-study. Res Policy 31(8/9):1257–1274

Geels FW, Schot J (2007) Typology of sociotechnical transition pathways. Res Policy 36:399–418

Geels FW, Schot J (2010) The dynamics of transitions: a socio-technical perspective. In: Grin John, Rotmans Jan, Schot Johan (eds) Transitions to sustainable development: new directions in the study of long term transformative change. Routledge, New York, pp 9–101

Gillen T, Supp A, Scheibe A, Steinhäuser-Pigao J, Steiger V, Witzel C (2002) Projektbericht Energiesparen an Universitätsgebäuden am Beispiel des Psychologischen Instituts in Heidelberg. Unpublished student research project. Psychologisches Institut der Universität Heidelberg, Heidelberg. www.psychologie.uni-heidelberg.de/zentral/projekt_energiesparen/projektbericht0102.pdf. Accessed 15 Feb 2013

Grin J, Rotmans J, Schot J (2010) Transitions to sustainable development: new directions in the study of long term transformative change. Routledge, New York

Green Office (2011) Climate action report: establishing a baseline for Maastricht University on sustainable development. Maastricht University Green Office. http://greenofficemaastricht.skillcollector.netdna-cdn.com/wp-content/uploads/2012/03/Climate-Action-Report_Final-Version.pdf. Accessed 01 Apr 2013

Ghosh S (2011) Participation in the Green Power partnership: an analysis of higher education institutions as partners in the program. Int J Sustain High Educ 12(4):306–321

ICTS (2011a) Project Idee: PC policies. Unpublished internal documents, Maastricht University

ICTS (2011b) Project Idee: Serverhosting facilities UM. Unpublished internal documents, Maastricht University

ICTS (2011c) Project Idee: Server virtualisatie binnen de UM. Unpublished internal documents, Maastricht University

Jain S, Pant P (2010) Environmental management systems for educational institutions: a case study of TERI University, New Delhi'. Int J Sustain High Educ 11(3):236–249

Kattenstein T, Unger H, Wagner H-J (2002) Handlungskonzepte zur wirtschaftlichen Optimierung des Energiebedarfs und der Energieversorgung der Ruhr-Universität Bochum: Abschlussbe-richt zum Vorhaben: IV A4-20600298 optimierung der bestehenden Energieversorgung der

Ruhr-Universität Bochum. Ruhr-Universität Bochum, Lehrstuhl für Energiesysteme und Energiewirtschaft, Bochum

Kahler S (2003) The ripple effect: how one dorm room can affect a university's energy use. Int J Sustain High Educ 4(3):230–238

Kemp R, Schot J, Hoogma R (1998) Regime shifts to sustainability through processes of niche formation: the approach of strategic niche management. Technol Anal Strateg Manag 10 (2):175–198

Levy JI, Dilwali KM (2000) Economic incentives for sustainable resource consumption at a large university: past performance and future considerations. Int J Sustain High Educ 1(3):252–266

Liers J, Person R-D (2012) Energiemanagement in Hochschulen, Handbuch zur Unterstützung bei der Einführung eines Energiemanagements in Hochschulen, Hochschulinformationssystem. www.his.de/pdf/pub_fh/fh-201213.pdf. Accessed 13 Mar 2013

Lo SH, Peters GYJ, Kok G (2010) Universiteit Maastricht—Energie Gedragsonderzoek 2009–2010. Unpublished manuscript, Faculty of Psychology and Neuroscience, Maastricht University

Loorbach D (2010) Transition management for sustainable development: a prescriptive, complexity-based governance framework. Governance 23:161–183

Marcell K, Agyeman J, Rappaport A (2004) Cooling the campus: experiences from a pilot study to reduce electricity use at Tufts University, USA, using social marketing methods. Int J Sustain High Educ 5(2):169–189

Maastricht University (2009) Energy efficiency plan 2009–2013, facility services. Maastricht University, Maastricht

Maastricht University (2010) Sustainability policy 2008–2010. Maastricht University, Maastricht

Matthies E, Kastner I, Klesse A, Wagner H-J (2011) High reduction potentials for energy user behavior in public buildings: how much can psychology-based interventions achieve. J Environ Stud Sci 1:241–255

Mitchell RC (2011) Sustaining change on a Canadian campus: preparing Brock University for a sustainability audit. Int J Sustain High Educ 12(1):7–21

Pearce JM, Miller LL (2006) Energy service companies as a component of a comprehensive university sustainability strategy. Int J Sustain High Educ 7(1):16–33

Petersen JE, Shunturov V, Janda K, Platt K, Weinberger K (2007) Dormitory residents reduce electricity consumption when exposed to real-time visual feedback and incentives. Int J Sustain High Educ 8(1):16–33

Rauch J, Newman J (2009) Institutionalizing a greenhouse gas emission reduction target at Yale. Int J Sustain High Educ 10(4):390–400

Raven R (2012) Analysing emerging sustainable energy niches in Europe: a strategic niche management perspective. In: Verbong G, Loorbach D (eds) Governing the energy transition, pp 125–151. Routledge, New York

Schneidewind U, Singer-Brodowski M (2013) Transformative Wissenschaft: Klimawandel im deutschen Wissenschafts- und Hochschulsystem. Metropolis, Marburg

Shove E, Walker G (2007) CAUTION! Transitions ahead: politics, practice, and sustainable transition management. Environ Plan A 39(4):763–770

Silverman D (2010) Doing qualitative research: a practical handbook. SAGE, London

Sorrell S, Schleich J, Scott S, O'Malley E, Trace F, Boede Ul, Ostertag K, Radgen P (2000) Barriers to energy efficiency in public and private organisations: final report to the European Commission. University of Sussex, SPRU Environment and Energy

Strauss AL, Corbin JM (1998) Basics of qualitative research: techniques and procedures for developing grounded theory. Thousand Oaks, London

Starzynski S (2012) Verhalten ändern, Strom einsparen Die Kampagne "Next Step" am Geographischen Institut als Modellprojekt für die Universität Kiel. Unpublished Thesis, Universität Kiel

ter Hofte H (2011) ICT energy efficiency in higher education: continuous measurement and monitoring. Novay, SURFnet. www.surf.nl/binaries/content/assets/surf/en/knowledgebase/2011/

rapport_201111_ICT+energy+efficiency+in+higher+education+-+Continuous+measurement+and+monitoring+v1.1.pdf. Accessed 03 Mar 2013
VSNU (2008) Gemeenschappeljke verlkaring van de Minister voor Wonen, wijken en integratie, de HBO-raad en de VSNU over de toetreding van hogescholen en universiteiten tot de Meerjarenafspraak energie-efficiency (MJA3). VSNU, Utrecht
Ward I, Ogbonna A, Altan H (2008) Sector review of UK higher education energy consumption. Energy Policy 36(8):2929–2939
Wesolowski D, Olivetti E, Graham A, Lanou S, Cooper P, Doughty J, Wilk R, Glicksman L (2010) Use of feedback in lab energy conservation: fume hoods at MIT. Int J Sustain High Educ 11(3):217–235
Woolliams J, Lloyd M, Spengler JD (2005) The case for sustainable laboratories: first steps at Harvard University. Int J Sustain High Educ 6(4):363–382

Author Biographies

Felix Spira is a Ph.D. researcher at the Dutch Research Institute for Transitions (DRIFT) at Erasmus University Rotterdam in the Netherlands. In his PhD, he develops a conceptual framework to understand the process of scaling the impact of transition initiatives. Felix holds a master's degree in sustainable development and climate change from University College London. He is also one of the co-founders of Maastricht University Green Office and root ability, two organisations that aim to advance student-led and staff-supported change towards sustainability in higher education.

Alexander Baker-Shelley is a Ph.D. researcher at the International Centre for Integrated assessment and Sustainable development (ICIS) at Maastricht University in the Netherlands. He joined ICIS in December 2013 to commence his PhD research on the organisational transformation of higher education institutions towards sustainable development. He completed a master's programme in Sustainable Development at Uppsala University and the Swedish University of Agricultural Sciences, and worked as a sustainability consultant.

The Institutional Dimension of Sustainability: Creating an Enabling Environment for Sustainability Transformation at USM

Omar Osman, Kamarulazizi Ibrahim, Kanayathu Koshy and Marlinah Muslim

Abstract

Recognising the importance of strengthened institutional arrangements for the promotion of sustainability, Universiti Sains Malaysia has been taking steps to respond effectively to integrated approaches involving all the major stakeholders, especially the students. Over the years, USM has made strong commitments to internal and international efforts to prevent further irreversible environmental change and to promote sustainable development through a host of on-going activities such as Kampus Sejahtera, 2000, RCE-Penang, 2005, University in a Garden, 2006, Research University award, 2007, and the APEX University award, 2008. In order to create an enabling environment within which different sections of the university could implement sustainability depending on the demands of their discipline and special circumstances; USM has developed a new 'sustainability integration model', an innovative 'sustainability assessment methodology' and a new 'Policy on Sustainable Development 2013'. In order to facilitate smooth and effective operationalisation, the policy is accompanied by an implementation log frame and a user guide. While there are major barriers for

O. Osman
Chancellery office, Universiti Sains Malaysia, 11800, Penang, Malaysia
e-mail: omar_o@usm.my

K. Ibrahim · K. Koshy (✉) · M. Muslim
Level 5, Hamzah Sendut Library, Sustainability Studies, Universiti Sains Malaysia, 11800 USM, Penang, Malaysia
e-mail: kanayathu.koshy@gmail.com

K. Ibrahim
e-mail: kamarul@usm.my

M. Muslim
e-mail: marlinah_muslim@usm.my

W. Leal Filho (ed.), *Transformative Approaches to Sustainable Development at Universities*, World Sustainability Series, DOI 10.1007/978-3-319-08837-2_16

the enthusiastic integration of sustainability into the core activities of universities due mainly to the lack of awareness, understanding, workable models, difficulty in monitoring progress and above all, the apathy of senior level administration, the USM experience will show how in a modern university it can be practiced through committed and systematic approaches. What is innovative here is the use of a new SD-ESD concept model developed by USM, a new computerised methodology to assess sustainability, and a proactive management that promotes sustainability through targeted programs, new policies and procedures and focused financial support. The intention of this paper is to share USM's model and working experiences to a wider group of practitioners in the hope that it will contribute to a process of institutional learning for sustainability.

Keywords
Sustainability policy · Education for sustainable development · Higher education · Institutional dimension of sustainability · Sustainability initiatives

1 Introduction

As the current global growth paradigm continues to make it harder to set human development on a sustainable trajectory, University Sains Malaysia (USM) has reaffirmed the urgent need for its graduates to be fully literate in the knowledge necessary to chart a sustainable future for them. Numerous on-going activities at USM emphasize the need for development to be more equitable, less natural resource intensive, more protective of the environment and above all people centred. USM's teaching curriculum, research agenda, co-curricular and student-led activities have always been designed to put its graduates at the forefront of Malaysia's major development initiatives. However, by the turn of the century, the institutional dimension of USM's sustainability (used interchangeably with sustainable development) commitment took a consolidated expression with a number of focused initiatives designed to creating a 'making it happen' environment through strategies, roadmaps and policies that guide the "sustainability transformation" the university is seeking. This paper provides an insight into the institutional dimension of these multilateral and intergenerational sustainability initiatives, involving students, faculty and university administrators, which has been integral to establishing standards, policies and guidelines for sustainability at USM.

2 Background

Historically, it may be argued that USMs growth trajectory since its establishment in 1969 has involved three major waves of change: The first wave of change was the consolidation of teaching and training into a school-based interdisciplinary approach, implemented during the university's first two decades (Omar and Koshy 2012).

This was followed by a second wave of change that focused on achieving excellence in research, development and innovation, resulting in the university's recognition by the Ministry of Higher Education as a Research University in 2007. Then, in a third wave of change, beginning in the new millennium, USM embraced the concept of sustainability through a number of focused initiatives, four of which can now be seen to embody this emphasis in the university's life. These include the:

- Kampus Sejahtera (Campus well-being) Programme, 2000,
- USM as a Regional Centre of Expertise for ESD, 2005,
- University in a Garden scenario, 2006, and
- Award of Malaysia's Accelerated Programme for Excellence—APEX, 2008—status, which provides a special focus on sustainability education.

A brief description of *three* of the *four* initiatives and their role in the university's re-oriented directions given below, while the fourth is presented under Sect. 4. *APEX Phase I.*

3 USM Sustainability Transformation

3.1 Kampus Sejahtera Programme (2000)

The Kampus Sejahtera (Healthy Campus) Programme emerged in 2000 as a major initiative of the then Vice-chancellor Tan Sri Dzulkifli Abdul Razakfrom the realisation that students' ability to learn depends on their health and quality of life. The indigenous word *sejahtera* cuts across spiritual, social, physical, mental and environmental dimensions and in this sense it reaches beyond health to include ideas for sustainable development (Dzulkifli and Koshy 2010). As part of this programme, the USM Clinic (*USM Sejahtera Centre*) along with faculty and volunteer students, run annual anti-obesity and anti-tobacco clinics, recycling initiatives, and activities for 'differently-abled' students is an excellent example of intergenerational involvement of the university community to promote sustainability. This programme has been a major success in sensitizing the campus community to take ownership of their sustainable development and has spread to other countries as well.

3.2 USM as a Regional Centre of Expertise (RCE) for ESD (2005)

Following the launch of the Decade of Education for Sustainable Development by the United Nations in 2005, a global network initiative was launched by the United Nations University with the establishment of 7 foundation RCEs, RCE-Penang being one of them. Over the years, REC-Penang has carried out several programs together with the student community, such as the *White coffin campaign* (banned polystyrene-based containers on campus), *Campus-wide recycling* project, *Citizenship programme* (USM students working with schools to address sustainability problems), and *Going bananas* (handmade paper making technology using banana trunks) to name a few. (Zainal and Khelghat-Doost 2008; Frontiers 2005). For 2013/2014 RCE Penang, in collaboration with Yokohama National University, Japan has beeninvolved in a student Leadership Program in 'Sustainable Living with Environmental Risk' (SLER). This course involves eight universities in Thailand, Indonesia, Vietnam, Cambodia, Myanmar, Philippines, Kenya and USM. (http://www.rce-penang.usm.my/, accessing on 20th November 2013), (Frontiers 2005).

3.3 The University in a Garden Scenario (2006)

In order to ensure that the university remains relevant to the times and to the communities it is designed to serve, USM undertook an extensive 'scenario planning' exercise in 2005. From this consideration eventually emerged the *University in a Garden* scenario—a symbol for a sustainability-led university. (http://www.usm.my/index.php/en/info-gateway/2013-07-03-07-57-25/the-university-in-a-garden). This concept broadens the understanding of sustainable development using the garden as a metaphor for the role of a university. In an attempt to mirror the university in a garden idea into the mission areas, transdisciplinarity was incorporated in USM's activities which led to some of its departmental structures, especially those related to Research and Development, reoriented into a seamless cluster-based structure which still continues (USM 2007).

4 APEX Phase I

4.1 Laying the Foundation

Universiti Sains Malaysia's sustainability journey took an ambitious turn when it was chosen as the sole recipient of the new APEX award in 2008 by the then Ministry of Higher Education (MoHE). The APEX proposal to MoHE, 'Transforming Higher Education for a Sustainable Tomorrow', outlined ways to bring about a 'whole-system' sustainability transformation involving the top level management, faculty, students and support staff. This involved a two pronged approach —one which focused on the major global sustainability challenges and the other on campus sustainability. The former was designed to build capacity at the individual, institutional and systemic levels to produce graduates who are equipped to address the sustainability challenges facing their communities and the world at large while the latter will provide hands-on praxis to students and staff alike to 'walk the talk' and to lead by example (Dzulkifli and Koshy 2010).

Using the 'blue ocean strategy' that makes competition irrelevant through innovative leapfrogging as a major instrument for its sustainability transformation and its 'eliminate-reduce-raise-create' grid as an operational approach, USM has identified a number of areas in its core mission domains of teaching, research, community engagement and institutional arrangement that need to be transformed to be competitive in the current 'red ocean environment' (Kim and Mauborgne 2005). For this, it focused on what needs to be *eliminated, reduced* or *raised* while deciding at the same time what needs to be *created* to enter the blue ocean world of innovation and excellence. These ideas are presented in a strategy document, 'Transforming higher education for a sustainable tomorrow', nick named the 'black book' (USM 2008). In order to ensure that the institutional dimension of sustainability is aptly geared towards removing implementation barriers relating to resources, talents, governance, ombudsmen, quality systems and the overall global agenda, specific task forces were set up following the APEX award.

In addition to the task forces, USM embarked on a range of sustainability initiatives, the most important of which is the establishment of a *Centre for Global Sustainability Studies* (CGSS). The Centre is expected to promote sustainability within USM and outside. As part of this, a sustainability task force led by CGSS developed the USM-APEX Roadmap, the 'green book' for the implementation of the ideas in the black book (Zakri et al. 2009). Aligned to the global thinking such as World Summit on Sustainable Development—WSSD (2002), and the United Nations Conference on Sustainable Development, UNCSD or Rio+20, (2012), the national priorities and institutional capacity, the university selected water, energy, health, agriculture and biodiversity (WEHAB) as the five major sectors, and climate change-disaster risk management, production-consumption and population-poverty as the three cross-sectors as the major sustainability priorities for focused intervention—collectively called the 'WEHAB + 3' approach (Zakri et al. 2009).

4.2 The USM Sustainability Integration Model

The integration of sustainability into the core of a university's business requires a whole system enterprise that links major sustainability challenges on one hand with different educational approaches on the other. The sustainability challenges have to be selected very judiciously by each institution. Figure 1 illustrates such an integrated approach developed and followed by USM as part of its APEX implementation agenda. The interlocking circles on the left represent the sustainability pillars while the three circles on the right represent the common mission areas of Higher Educational Institutions. While the central box shows USMs WEHAB + 3 priority areas for sustainability studies, these may be replaced by priorities appropriate for other institutions as the case may be.

The model was applied following the guidelines of education for sustainable development, that emphasises not only knowledge and skills but also perspectives, value systems within which societal and development problems have to be seen and addressed. Education that follows both knowledge and ethical compass for direction produces graduates who think and act like sustainability champions and will be able

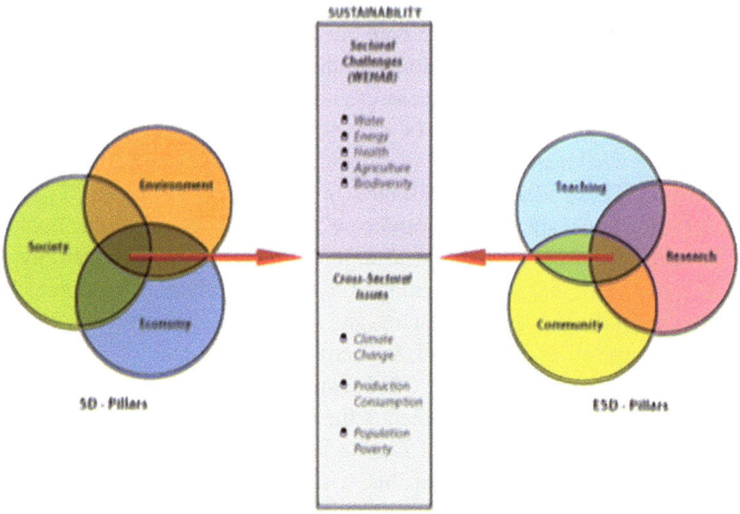

Fig. 1 USM sustainability model

to help shape knowledge societies, a subset of which is knowledge economies (Corcoran and Koshy 2012). Such graduates hopefully will be able to steer the world clear of developmental pathways that deplete natural resources, increase pollution, change climate, and widen the rich-poor gap. To them, protecting our planet, lifting people out of poverty and advancing economic growth are interconnected aspects of the same thing—sustainable development. This is what USM trusts to be true 'development with a difference', something worth sustaining, (Kamarulazizi et al. 2013). In order to train graduates of this caliber, USM had to make major changes in its work programs (USM 2013a).

From an institutional point of view this meant that USM had to negotiate with the government a new Constitution which was eventually approved and Gazettedon 1 July 2011 after receiving Royal Order under Section 26 of AUKU. This constitution provides fairly wide area of autonomy for the university in administrative, financial and policy exercise, supporting the mission of both APEX and Research University (RU) and created new structures for USM's Board of Governors, Senate, Ombudsman and Student Consultative Assembly (USM 2013b).

4.3 SAM for Monitoring Sustainability

A Sustainability Assessment Methodology (SAM), nested within the sustainability integration model, Fig. 1, was developed by the Centre for Global Sustainability Studies. SAM follows a three stage 'input, process, output' methodology (Fig. 2). The *input* data could be in the form of course synopsis, research report, project report, article, thesis, dissertation, book, manuscripts, policy documents, action plans etc. This information is analysed during the *processing* step, using 24 key

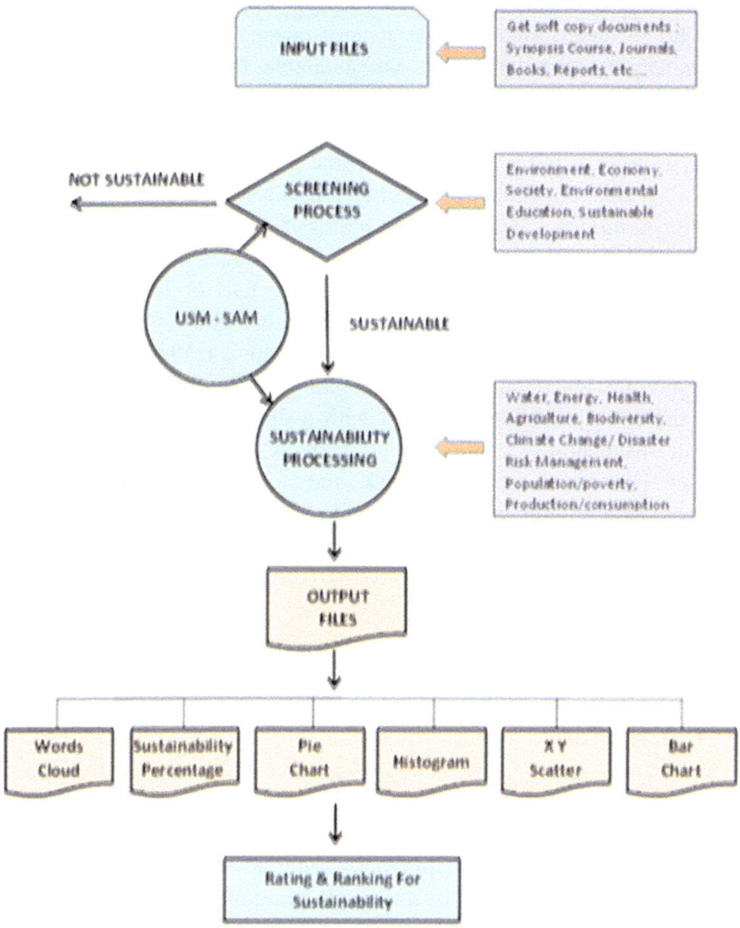

Fig. 2 SAM flowchart

sustainability criteria aligned to the content of the internationally agreed documentation such as, Stockholm Declaration (UN Conference on Human Environment 1972), Brundtland Commission Report (1987), Agenda 21(UN Conference on Environment and Development—Earth Summit 1992), Earth Charter (1994), Johannesburg Plan of Implementation (UN, World Summit on Sustainable Development, (WSSD 2002), The Future We Want (UN Conference on Sustainable Development—Rio+20, 2012) and, other relevant national reports from government and civil society reports on sustainability. In the final *output* step, the results are displayed in a variety of ways that appeal to a host of users such as academics, researchers, community or industry project implementers, middle and top level managers and policy makers. All courses offered by a School or Faculty will be analysed to know the overall coverage of sustainability themes, regardless of the number of courses used to cover the total sustainability content.

When we used SAM to do the USM Sustainability Audit, out of 2,671 courses examined, 80 % were found to have at least one element of the "Triple Bottom Line" sustainability. Similar results are available for research and community projects. Using these results and the sustainability worksheets in the USM Sustainability Roadmap, staff is able to reorient existing courses, design and manage new research projects and community-oriented programmes (Koshy et al. 2013).

Prompted by this assessment, CGSS has introduced a new sustainability course, *WSU 101: Sustainability—Issues, Challenges and Prospects*' in 2011 as an elective course with a view to making it mandatory for all degree students. In addition, an internationally networked master's program, '*Masters in Development Practice (MDP)* has been approved for offer from September 2014. On the research side, CGSS carried out a risk reduction to climate change related flood vulnerability project in a vulnerable community in Kedah. At a strategic level, CGSS has developed the concept of 'development with a difference' where it promotes the idea that all sustainability challenges come with a 'risk stage' first and then a disaster stage. Disaster being 'realised risk', the most prudent time to intervene is at the risk level when nothing adverse has in fact happened, just that the potential for harm is high. We have conducted trainings on this concept and published this approach recently (Kamarulazizi et al. 2013).

5 APEX Phase II

5.1 The Strategy

In 2013, the final year of APEX Phase I, USM has renewed its commitment to transform higher education for a sustainable tomorrow by pioneering sustainability mainstreaming in all mission areas to empower future talent and enable the bottom-billion towards achieving their overall well-being. In doing so, USM is fully cognizant of the fact that the inter-linked environmental, economic and social crises will make it difficult to provide the growing world population with food, water and energy as articulated in the Rio+20 outcome, The Future We Want (United Nations 2012).

Based on the considerable achievements of USM in laying a strong sustainability foundation during APEX Phase I, using the Blue Ocean Strategy, to transform governance, enhance efficiency of human and financial resource management, improve infrastructure, boost research and innovation output, and advances in academic reformations, USM is poised to enter an ecosphere of APEX Phase II where efforts will be accelerated to ensure our quest for excellence that will on the one hand generate human capital with first-class mentality and on the other help propel USM to become a sustainability-led university of world-class standing (USM 2013c). In this context, USM has already reaffirmed its conviction that for transformation, in addition to a bold vision, is needed concrete and practical measures. It is self-evident that for transformative outcomes we have to be patient with the process, as quantum leaps can only be realized with a broad alliance of students and staff working together to secure the sustainability future we want for all. With this in mind, the university established a high level *University Sustainability Council (USC)*, chaired by the Vice-Chancellor and a membership that includes the four Deputy Vice-Chancellors, Heads, Directors and Deans of sections responsible for sustainability mainstreaming at the institutional level. All sustainability initiatives of USM will be under the purview of USC. The Council has spearheaded the formulation of a USM Policy on Sustainable Development and a Sustainability Implementation Logframe during the latter half of 2013.

5.2 USM Policy on Sustainability 2013

Within the sustainability vision of the university and led by a mission to make tomorrow better than today, the goal of the USM policy is to mainstream sustainability into the core mission areas of the university to secure the future we want for all. USM aligns its sustainability policy (USM 2013d) with the national development goals as articulated in the 5-year Malaysia Plans, the New Economic Model (Koshy et al. 2010), the Higher Education Strategic Plan, the Rio+20 outcome—The Future We Want and other globally negotiated agreements such as Agenda 21, the Johannesburg Plan of Implementation and related Multilateral Environmental Agreements.

The sustainability foundations and priorities set during APEX Phase I guides USM sustainability policy by urging accelerated implementation for the transformation it seeks. The policy identifies the specific areas under WEHAB + 3 that need to form the continued focus for action so that it will be aligned to national development needs and global strategies for action. The policy stipulates that within this context, the principled approach should be to:

- Adopt sustainable development as a major guiding principle for the overall operations, including teaching, research, community engagements and institutional arrangements
- Educate our students with the right blend of knowledge on trans-disciplinary and inter-sectoral integration, skills, perspectives and value systems suitable for adaptive management, critical reflection and participatory approaches needed for sustainability (Corcoran and Koshy 2012)
- Ensure that research performed addresses environmental, social and economic problems and that efforts are made to translate research findings into policy and public knowledge
- Create an appropriate institutional culture by ‘walking the talk’ through campus sustainability efforts
- Establish alliances and partnership with relevant stakeholders in the public and private sectors to promote sustainable development
- Nurture and educate young people to be leaders and agents of change for a sustainable future.

The policy provides specific guidelines to address core issues and their symptoms in four domains as:

(i) Teaching (formal, non-formal and informal education)

Policy: Integrate sustainability into the academic curriculum at all levels using all modalities of delivery to ensure that necessary knowledge, skills, perspectives, value systems and issues are introduced and their relevance to the three pillars of sustainability emphasized fully through formal, non-formal and informal teaching.

(ii) Research and Innovation (process, fundamental/basic, action/applied and sustainability science research).

Policy: Champion sustainability research activities that are trans-disciplinary and promote advanced innovative thinking, new knowledge creation, and the ability to use and disseminate knowledge to find solutions to pressing sustainability challenges based on WEHAB + 3 priorities.

(iii) University-community engagement (industry, village, NGOs, civil society/non-state actors, business and policy communities)

Policy: Promote knowledge/skill transfer programs using education for sustainable development principles and practices to address community needs and challenges.

(iv) Institutional arrangement (utilities, infrastructure, transport, waste, and soft structures involving data, software and networks)

Policy: Integrate sustainability into the core governance system of the university to create an innovative enabling environment within which the other components (teaching, research and community engagement) operate efficiently.

Under each of the policy are descriptions as to what it should cover. Overall, when implemented it should help 'USM deliver as one' which is what the communities it serves will expect and has come to appreciate.

5.3 Sustainability Implementation Logframe

In order to implement the USM Policy on Sustainability, a detailed 'USM-APEX Phase II: Sustainability Implementation Logframe' (the logframe) has been developed for each domain such as teaching, research, community engagement and institutional arrangement, with specific goals, targets, and timelines. The interventions required and the sections involved have been identified with indicators and the impacts specified. This eight page landscape document together with an implementation handbook (CGSS 2013) is expected to trigger a host of activities across the university during APEX Phase II until 2025. The sustainability assessment methodology (SAM) developed by CGSS and other relevant sustainability indicators will be used to gauge the level of implementation of the policy on a project by project or unit basis. The policy and its logframe will be reviewed as necessary to ensure their relevance and effectiveness.

6 Conclusion

Can a modern-day university with all its pressures to excel in global competition take up sustainability as one of the core operational principles and practice it—'learning by doing'? If USM experience is anything to go by, the answer is 'yes' as USM believed that 'it is the most logical thing to do, like living a holistic life'. Starting with the Kampus Sejahtera program of the early new millennium championed by the then Vice-Chancellor and implemented across the campus by guided voluntarism, the programme grew to initiate student-led campaigns against plastic bags, Styrofoam containers, tobacco and other forms of substance abuse; the special activities of the campus health centre; the eco-hub group and a host of other sustainability oriented campus programmes involving the campus lakes, waste, and student halls. The champions and the best practice pilots of these early days became the reason for the international recognition of USM as one of the 7 foundation RCEs (Regional Centre of Expertise for UNDESD), which have now grown to 127 centres across the world (http://archive.ias.unu.edu/sub_page.aspx?catID=108&ddlID=183).

Convinced that the university's teaching and research must be relevant to knowledge pursuits as much as it is pertinent to finding solutions to pressing environmental, developmental and socio-economic problems, while helping to lift the poor from their plight, USM carried out an institution-wide 'scenario planning exercise' by the middle of the last decade. The planners finally settled on a university model that mimics nature for its own existence and the provision of goods and services. Like symbiosis that exists in ecosystems where everything is related to everything else, USM chose to be '*selectively relevant*' by ensuring a symbiotic relationship with the communities it was set up to serve. USM is symbiotically connected to everything having a stake in it—those whom it actions affect or influence and those whose actions affect or influence USM. From this consideration emerged eventually the University in a Garden scenario. This choice was an endorsement of the view that universities should be selectively relevant, as it is impossible to be all things to all people. This realization helped USM to focus and excel in a strategically chosen area that has wide implication for now and the future. It is this selectivity that eventually earned the APEX award for USM.

In higher education, one side of the sustainability coin represents the efforts of a university to be competitive to attract quality students, competent staff, adequate facilities and finances. Often, the focus here is on maintaining status quo, best practices and excellence. The other side of the coin represents the major global sustainability challenges, as highlighted by the Brundtland commission's report and its triple bottom line approach—environment, economy and society. Central to this approach is the realization that current developmental paradigms that deplete natural resources, increase pollution, change climate, and widen the rich-poor gap are simply not sustainable. Thus protecting our planet, lifting people out of poverty and advancing economic growth are interconnected aspects of the same thing—sustainability or sustainable development. Addressing these challenges is the realm of 'blue ocean strategy' in the context of USM.

Over a decade of USM's dedicated efforts to incorporate sustainability principles and practices into its activities, we have learned that unless these efforts move from the realm of voluntary participation by a few devoted staff and students to an institutionalised effort involving policies and procedures, the barriers—both perceived and real—in terms of staff awareness, attitudes, expertise and institutional commitment to accelerating the sustainability embedding processes will continue.

While it is true that unless the mind-set of people change first, implementation at the institutional level will be slower, it is also true that well planned and supported policies could catalyse an otherwise slow process as in the new environment the expectations and the recognitions are clearer and the resource allocation more transparent. To this end, USM needed to deliberately create an organisational means to ensure that this delicate balance between academic and administrative obligations toward sustainability integration is in place. Also it is understood that we need to continually re-examine our fundamentals as we strive to address pressing global challenges such as the implementation of the Millennium Development Goals and its successor development plan—the Sustainable Development Goals. It is the firm belief of USM that the new Policy on Sustainability 2013, its implementation logframe and the user guide, together with the strengthened institutional frameworks will go a long way in fast tracking sustainability across the university in the next decade and beyond.

References

Brundtland Commission Report (1987) Our Common Future. http://conspect.nl/pdf/Our_Common_Future-Brundtland_Report_1987.pdf

CGSS (2013) Implementation ideas: What can USM do to fast track sustainability. Center for Global Sustainability Studies USM, Penang. ISBN 978-967-394-159-9

Corcoran P, Koshy K et al (2012) The ethical dimension of sustainability in higher education: applying the principles of the Earth Charter in Malaysia and beyond. University Sains Malaysia, Penang. ISBN 978-967-394-040-0

Dzulkifli R, Koshy K (2010) Tomorrow Today, chapter transforming higher education for a sustainable tomorrow case of learning by doing at Universiti Sains Malaysia, Pub. Tudor Rose, UK on behalf of UNESCO, ISBN 0-9536140-8-5

Earth Charter (1994). http://www.earthcharterinaction.org/invent/images/uploads/echarter_english.pdf

Frontiers USM (2005) Bulletin for research and development, vol 4. Universiti Sains Malaysia, Penang

Kamarulazizi I, Koshy K, Asrar G (2013) Development with a difference: Neo-Disaster Risk Management for Sustainable Development. 'Opinion' paper, J Geomatics Nat Hazards Risk. http://dx.doi.org/10.1080/18475705.2013.82429. Taylor & Francis, England

Kim W, Mauborgne R (2005) Blue ocean strategy: how to create uncontested market space and make the competition irrelevant. Harvard Business School Press, USA, Cambridge

Koshy K, Norizan N et al (2013) An indicator based approach to sustainability monitoring and mainstreaming at Universiti Sains Malaysia, Sustainability assessment tools in higher education institutions. Book ID: 316255_1_En, Bk Ch No. 14, p 22, ISBN: 978-3-319-02374-8, Springer Pub, UK

Koshy K, Suzyrman S et al (2010) Implementation of the new economic model: the role of education for sustainable development in building knowledge base infrastructure. Centre for Global Sustainability Studies, USM, Penang. ISBN 978-967-5417-94-8

Omar Osman, Koshy K (2012) Universiti Sains Malaysia's sustainability journey: reflections on a knowledge base institution's vision for transformation. Future Perfect, Pub. Tudor Rose, UK on behalf of UNDESA, UK. ISBN 978-0-9568561-2-8

UN Conference on Human Environment (1972) - The Stockholm Declaration. http://www.unep.org/Documents.Multilingual/Default.aspdocumentid=97&articleid=1503

UN Conference on Environment and Development (1992) Agenda 21. http://sustainabledevelopment.un.org/content/documents/Agenda21.pdf

UN Conference on Sustainable Development (2012) The future we want. In: A/CONF.216/L.1, United Nations Conference on Sustainable Development outcome document, 20–22 June 2012, Rio de Janeiro, Brazil. http://sustainabledevelopment.un.org/futurewewant.html

United Nations (2012) The future we want. In: A/CONF.216/L.1, United Nations conference on sustainable development outcome document, 20–22 June 2012, Rio de Janeiro, Brazil

USM (2007) Constructing future higher education scenarios: insights from Universiti Sains Malaysia. ISBN 978-983-861-328-6, © Penerbit Universiti Sains Malaysia 2007

USM (2008) Transforming higher education for a sustainable tomorrow. The APEX University. Universiti Sains Malaysia. Reprinted, April 2008. ISBN 978-983-3986-30-9

USM (2013a) The APEX University: the state of the art, ISBN 978-967-394-126-1. Universiti Sains Malaysia, 2013

USM (2013b) The journey towards excellence. ©Universiti Sains Malaysia, 2013, Perpustakaan Negara Malaysia, ISBN: 978-967-394-122-3

USM (2013c) The Apex University second phase: 2014–2025, Version 1.0. A consultative document. Jan 2013. Universiti Sains Malaysia, Pulau Pinang

USM (2013d) USM policy on sustainability (2013). ©Universiti Sains Malaysia Pulau Pinang

WSSD (2002) Johannesburg Plan of Implementation. http://www.un.org/esa/sustdev/documents/WSSD_POI_PD/English/WSSD_PlanImpl.pdf

Zainal S, Khelghat-Doost H (2008) Regional centre of expertise as transformational platform for sustainability: a case study of Universiti Sains Malaysia, Penang. Int J Sustain High Educ 9(4): ISSN 1467–6370

Zakri A, Hamid, Koshy, K et al (2009) *USM-APEX Sustainability Roadmap: Aspiring to meet global challenges*, Centre for global sustainability studies, © Perpustakaan Negara Malaysia, Penang, ISBN 978-967-5417-59-7

Author Biographies

Professor Dato' Dr. Omar Osman Ph.D. He is currently the Vice-Chancellor of Universiti Sains Malaysia. He has served USM before as deputy vice-chancellor in charge of Student Affairs, and was the Director of Corporate and Sustainable Development Division. He is an ardent supporter of education for sustainable development, and in this field he has organised international workshops, promoted RCE-Penang, and has published and continues to network. He keeps his research and consultancy active in the areas of project management, ESD and higher education. He is a strategic planner and was involved in the strategizing and planning of the following major initiatives: USM as a Research University, the 9th Malaysia Plan and the APEX University programme. He has a keen interest in training and human resource development aimed mainly at students' development. Most of his publications have been in the area of sustainability in higher education, construction management, governance and economic issues in construction industry, education for sustainable development and human resource development. He is currently the President of Asia-Pacific University-Community Engagement Network (APUCEN) and the Chairman of the new, South East Asia Sustainability Network (SEASN), and travels widely for professional meetings and partnership building.

Professor Dr. Kamarulazizi Ibrahim Ph.D. He is the Director of the Centre for Global Sustainable Studies at USM and in involve heavily in sustainability research and policy matters at the local, national, regional and international levels, especially the UN system bodies. In addition, he is a Professor at the School of Physic in charge of Semiconductor Fabrication Processes and Characterization, Solar Cells, Clean room and Nanotechnology. He is also the Chairman for the Energy Research Program, Nano Optoelectronic Research and Technology Laboratory (NOR) and a Senate member of USM. He was the Coordinator for Energy Research Group USM, (1989–1992), Chairman for Pure Physics and Education, (1992–1997) and Master Trainer, Logical Frame Approach (LFA), USM (2002–2004). As the Director of the Centre for Education Training and Research Renewable Energy and Energy Efficiency (CETREE), USM (2000–2008) he was responsible for increasing knowledge and awareness of the role of renewable energy and energy efficiency for professionals, schools, tertiary institutions, schools and also to the public in Malaysia. He has travelled, lectured and published widely in Energy, Semiconductor Materials and Devices, Nanotechnology, and Sustainability field.

Professor Kanayathu C. Koshy Ph.D. He is the Professor of Sustainability at the Centre for Global Sustainability Studies (CGSS), Universiti Sains Malaysia, Penang. He was the former Professor of Environmental Chemistry at The University of the South Pacific in Fiji and the Foundation Director of the USP's Pacific Centre for Environment and Sustainable Development. Dr. Koshy was also the Director of the Pacific Centres of the global change research networks—START, the SysTem for the Analysis, Research and Training, Washington D.C. and APN, Asia Pacific Network for global change research, Kobe. Before joining The University of the South Pacific, he was a Faculty Member of the University of the West Indies, Jamaica, and the State University of New York at Buffalo. His research involves kinetics and mechanism of reactions; environmental chemistry; science, technology and innovation; education for sustainable development; and policy matters. He has served as Coordinator of the United Nations University Regional Centre of Expertise for Oceania (RCE-Oce), and the UNESCO-ACCU Centre of Excellence for Education for Sustainable Development in the Pacific region. He was the principal investigator of the Asia-Pacific Cultural Centre for UNESCO, European Union, Global Environmental Facility, and AusAID sponsored climate change projects for the Pacific. His current responsibilities involve capacity building, research, and professional networking at the national regional and international level. He was a Fulbright Fellow in USA and has lectured, published and travelled widely.

Mrs. Marlinah Muslim B.Sc. Marlinah is currently a Research Officer for Centre of Global Sustainable Studies at Universiti Sains Malaysia, Penang. She was a Research Executive at Sabah Forest Industries Sdn. Bhd. under Ballapur Industries Ltd. (BILT) India paper mill Company in Forestry and Tree Breeding field before and has developed a Vegetative Propagation and Nursery Working Protocol and was in-charge the nursery working area. She is also a former member of the law and policy organisation 'Safety and Health' of Malaysia and has been involved in several Standard Operation Procedure (SOP) and policy development for forestry workers at Sabah Forest Industries. Her current responsibilities involve research, publication, and networking and are in charge of correspondence for the newly established South East Asia Sustainability Network (SEASN).

Development, Structure and Impact of a Ten Year Outreach and Study Abroad Program in Sustainability and Environmental Health Disparities: University of Northern Iowa, the Iuliu Hatieganu University of Medicine and Pharmacy and the Romani of Pata Rat with a Report on a Cross-Sectional, Romani Environmental Footprint Study

Catherine L. Zeman, Jennifer Hall, Diane Depken and Ioan Bocsan

C.L. Zeman (✉)
HPE/HPELS and RRTTC/CHAS, University of Northern Iowa, WRC 207, Cedar Falls, IA 50614-0241, USA
e-mail: Catherine.zeman@uni.edu

J. Hall
Healthier Workforce Center for Excellence, University of Iowa College of Public Health, UI Research Park, IREH #102, Iowa City, IA 52242, USA
e-mail: Jennifer-1-hall@uiowa.edu

D. Depken
HPE/HPELS, University of Northern Iowa, WRC 223, Cedar Falls, IA 50614-0241, USA
e-mail: Diane.depken@uni.edu

I. Bocsan
Epidemiology and Primary Health Care, Iuliu Hatieganu University of Medicine and Pharmacy, UMF 8, Victor Babes Street, 400012, Cluj-Napoca, Romania
e-mail: bocsanioanstelian@gmail.com

W. Leal Filho (ed.), *Transformative Approaches to Sustainable Development at Universities*, World Sustainability Series, DOI 10.1007/978-3-319-08837-2_17

Abstract

A study abroad program which incorporates service learning and research into an undergraduate capstone experience focusing on systems analysis of complex environmental, health, and social justice issues (environmental health disparities and sustainability) has been on-going for 10 years. This program's development and execution is described and research developed from the collaborative project is reported here. This applied research will be used in the on-going educational program and to further inform audiences about the struggles faced by the Roma of Central Eastern Europe. As part of this study abroad program student and faculty applied research opportunities are incorporated. Recent research involved a cross sectional survey of Romani family groups at Pata Rat, Romania who primarily made their living in secondary materials recovery (recycling/scavenging) and which was completed in 2009 and 2011. The findings of this survey which related to dietary security, energy use, and work productivity are presented. The challenges facing Romani scavenger communities in regard to environmental and occupational health concerns are also discussed. Recommendations for on-going community engagement, education and community development are suggested by this work.

Keywords

Sustainability education · Study abroad · Romani · Ecological footprint · Environmental health disparities

1 Introduction and Background

This paper details a unique sustainability education program for undergraduate and graduate students offered each year for 4 weeks involving community engagement research and service learning. Current, applied research which is a product of this innovative sustainability education course is also reported. Since 2004 there has been a cooperative agreement for research, education and outreach activities in general between the Iuliu Hatieganu University of Medicine and Pharmacy (UMF), Cluj Napoca, Romania and the University of Northern Iowa, (UNI), Cedar Falls, IA, USA. This effort was founded by Dr. Zeman of UNI and Dr. Bocsan of the UMF with joint education and research activities being conducted since that time. This has included outreach activities to the Roma community of Pata Rat.

The Roma people are linguistically and culturally distinct peoples of Indo-Aryan origin, having left northern India around 1500 AD (i.e. gypsy) (Fonseca 1995; Beniamin 2008). Roma groups are found throughout Central and Eastern Europe, but are found in higher concentrations in certain countries such as Bulgaria and Romania ("Rom diaspora," n.d.). Beginning in 2006, the collaborative program Professors' research, outreach, and education efforts began to focus on the Roma populations of Pata Rat residing outside of Cluj Napoca and scavenging an active landfill site at the location. The collaborators efforts included undergraduate outreach/education programs focusing on environmental and occupational health concerns and graduate student research and involvement. In 2009 and 2011, cross-sectional interviews of Romani family units were conducted to identify the occupational health risks and dietary risks that could contribute to increased absorption of environmental toxins and to examine the beneficial contributions of the Romani community to local/regional waste management. In that regard, efforts were made to characterize and quantify the solid waste recovery and overall energy footprint of one Roma community living near the active landfill. This was done to illustrate both the risks the community faces and the considerable strengths and contributions of that community to the bioregion while providing case studies and applied research experiences for the Capstone study abroad experience in environmental health disparities. This innovative, systems oriented approach to the sustainability focused study abroad experience enriches the program for students, provides services to the Roma community, and enriches subsequent offerings of the program.

Roma Populations in Eastern Europe and across the world face a myriad of social and environmental challenges that contribute to health challenges. The social challenges include: poverty, low literacy rates (70–80 % illiterate), low status and educational attainment of women and girls, increased risk of infectious/communicable disease, high rates of chronic disease/injury, decreased life-expectancy, and widespread discrimination and bias (Zeman et al. 2003; Haijioff and McKee 2000; Idzerda et al. 2011; Hicken et al. 2012). Specific environmental health challenges include nutritionally deficient diets, shelter related inadequacies, rodent and insect exposures, toxic exposures at work and home, and work-related injuries (Cointreau 2006; Carpenter et al. 1996; Hunt 1996; Zheng et al. 2013).

These inter-related factors lead to the documented health related challenges facing many Roma populations and premature morbidity and mortality related to malnutrition and infectious diseases that are a constant fact of life in these communities (i.e. tuberculosis, skin infections, respiratory infections, STDs) (Zeman et al. 2003; Haijioff and McKee 2000). The stress of high poverty, lack of services and discrimination in employment, housing, and medical care, complicates and exacerbates the challenges cited; undoubtedly increasing the stereotype threat response (where stress and perceived discrimination drive a chronic stress or chronic allostatic load state) as has been documented in minority populations in the U.S. (Hicken et al. 2012).

Much of the research on Roma health attempts to document health risks and identify correlates, causes, and possible solutions for the "plight" of the Roma. The tendency to focus on the challenges that Roma scavenger communities face in regard to health and working conditions has left a gap in regard to the Roma's strengths and especially in regard to local solid waste and carbon cycle management as many impoverished Roma groups work in secondary materials recovery.

The Romani community of Pata Rat makes their living in materials recovery at an active landfill outside of a large metropolitan area in Romania (Fig. 1). The work is hard and hazardous. Similar to other scavenger communities around the world, they face increased musculoskeletal injury from heavy, repetitive work, the increased possibility of rodent bite, puncture injury, and infection (Arpinte et al.

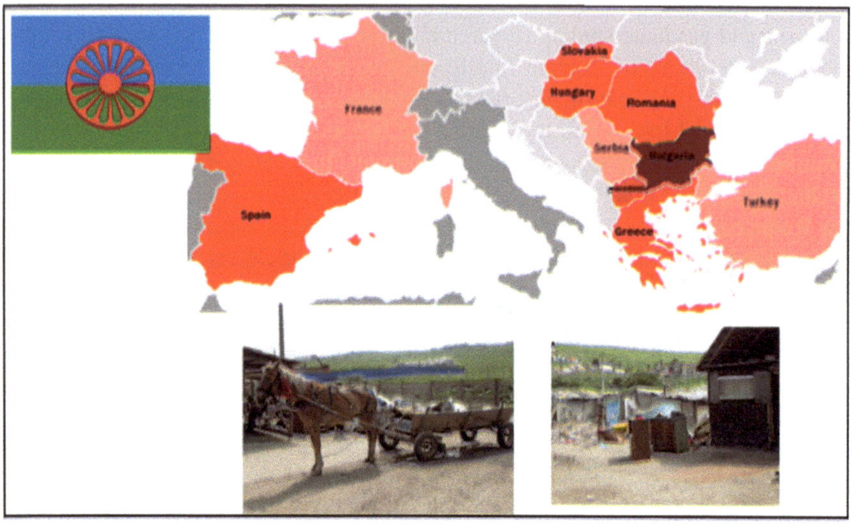

Fig. 1 Romani flag, Recyclers, and population distribution. *flag and population density of Roma by country ("Romani Flag," n.d., "Rom diaspora," n.d.)

2002; Thiraattanasunthon et al. 2012). And high exposures to a variety of contaminants including heavy metals and environmental toxins which are not only immediate dangers but can impact learning and development (Zheng et al. 2013; Carpenter et al. 1996). Additionally, most communities have modest incomes and are challenged to provide adequate food for a proper nutritional status, which increases their susceptibility to exposure sequel from their work. The Romani in this community are at increased risk in all of these areas.

2 Methods

2.1 Overall Study Abroad Effort

In 2004 an initial agreement for cooperative educational programs was signed between the International Programs offices of both UMF and UNI. From that point on yearly student/faculty visits and exchanges began. After the initial year, which was used to establish relationships between the institutions and with NGOs working in Roma communities, Drs. Bocsan and Zeman established a four week long study abroad program experience based on a senior undergraduate Capstone requirement which was part of UNI's liberal arts core requirements (UNI students must have attained junior/senior status before enrolling). The class is jointly offered at both institutions, so that students from UMF attend the class as a workshop with the students from UNI who receive the Capstone credit. The UNI students pay course tuition and associated travel fees through the study abroad program. Because of the reciprocal agreements with the partner institutions, students pay only student-dormitory rates for housing and meals, keeping the cost of the program reasonable. The study abroad course is capped at 25 American and 25 Romanian students in order to manage outreach activities in Roma communities effectively and maintain the quality of the experience. The numbers of American students have ranged between 16 and 25 while the Romanian student numbers have ranged between 5 and 15.

Dr. Zeman had a long-standing relationship with the UMF and the Romanian Institutes of Public Health, IPH after completing doctoral work with the IPH in rural Romania focusing on methemoglobinemia (1999) and a Fulbright in Romania prior to the summer of 2004. As a result, Drs. Bocsan and Zeman had a history of collaboration prior to developing the study abroad program. Benefiting from Dr. Bocsan's insights, discussions with the Romanian International Programs Office and with the NGOs that the program works with, the collaborators chose to focus their course efforts on a topical area of importance to Romania and the region of Transylvania– Romani environmental health issues. This would allow Romanian UMF students and American Capstone students to come together and focus on a topic that involved health, environmental sustainability, and social justice issues.

The Capstone experience is structured in such a way that the American students begin their preparation for the program one semester prior to the study abroad course which is held in May of each year. American students are required to take a one-hour, spring semester pre-departure seminar course. During this seminar, UNI students are required to learn about the countries they will be visiting in their study abroad, learn about the Romanian and Romani peoples and a little of their language, study the environmental, health, and social justice challenges that face scavenger populations, and spend time developing educational outreach campaigns and fundraising in order to provide environmental health education and giveaways to the Romani community. The outreach education is based on skits and interactive activities for adults and children, topics being chosen in conjunction with Romani staffed NGO groups in the Pata Rat community. This method of outreach was chosen due to the lower levels of literacy in the Romani community (less than 50 %) and the challenges in overcoming language barriers with complex and technical topics. The skits are translated into Romanian during the outreach education activities by colleagues with the NGO group, The Institute for the Child and Family, ICF. ICF is a Christian non-profit organization which was established by Pastor Rufus Whynot to provide outreach to Romani communities on a variety of levels including health services and education opportunities.

The in-country class consists of half-day class sessions followed by outreach activities and/or cultural opportunities. The class must read the text, *All Our Relations* by Winnona LaDuke (1999), excerpts from the text *Environmental Health and Racial Equity in the United States: Building Environmentally Just, Sustainable, and Livable Communities* by Bulalrd et al. (2011) and excerpts from the text *Thinking in Systems: A Primer* by Donella Meadows (2008). The Meadow's text and the topic of systems thinking, including the development of systems diagrams, is the first topic covered in the course as systems thinking is used as part of the overall course assignment. This course assignment involves students working

Table 2 Daily dietary intake in servings, major food groups

Food groups	Avg. servings/ household[a], range	Avg. servings/ individual[a]	percentage reporting < recommended servings[b]
Milk (all categories)	1.2, 0–6	1.2–1	51–83
Fruit (all categories)	0.95, 0–6	0.57–0.47	47–89
Vegetable (all categories)	1.44, 0–4	0.86–0.72	100

[a] Averaged across all reporting households and all individual age grades
[b] Based on actual percentages of reporting households

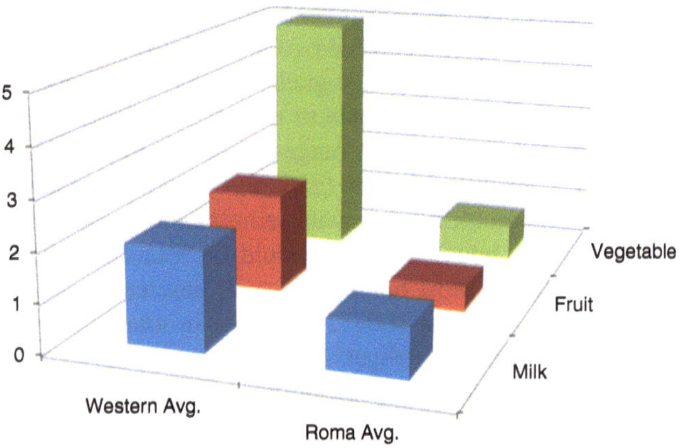

Fig. 2 Servings by group USRDA versus Roma Daily Average

manual labor and living in marginal housing as their work and living environments increase environmental exposures to lead and PCBs from the burning of plastics to recover metals and to use in cooking fires. The average servings per day of milk, fruit, and vegetable groups indicated that calcium, fiber and antioxidant/mineral intake is half or less of USDA recommended daily servings (Table 2 and Fig. 2). Based on an average household composition derived from the demographics the "average" individual of any household or age in the Roma group in Pata are consuming only 1–1.2 servings per day of milk versus the 2–3 suggested by dietary guidelines, of fruit the Pata consumption was 0.47–0.57 servings per individual per day versus the recommended 2–4, and of vegetables the consumption was 0.72–0.86 per individual per day versus the recommended 3–5 (*Pamphlet: U.S. Department of Agriculture*, 1990). Indirect studies of Roma mothers' nutritional status through cord blood analysis corroborate the poor general nutritional status of Roma groups (Rambouskova et al. 2009)

Romani reported collecting glass, metal (steel), plastic, copper, and aluminum as the primary materials recovered from their active landfill site. Metals and plastic were the top two materials recovered by weight with men recovering 38 kilos of metal (steel) per day, women 27 kilos, and children 10 kilos while men reported recovering 37 kilos of plastic, women 24 kilos of plastic, and children slightly less than one kilo (0.96 kilos) per day. When the total weights of all materials recovered per day are considered across all categories it was reported that men recovered 81.92 kilos per day, women 56.46, and children 11.59. Thus, based on the demographics data indicating that the "typical" household in Pata consisted on average of one male, one female and two children, a total of 160.84 kilograms per family per day is actively recovered at Pata. The U.S. EPA estimates that the average U.S. household generates 7.95 kilograms of waste per day (1.99 kg/capita/day by 4 individuals) and that on average U.S. households recycle about 34.5 % of that or 2.74 kgs/family/day (0.685 kg/capita/day by 4 individuals) ("Municipal solid waste" 2014). Thus, recycling activities as reported for men, women, and youth indicate that Romani recyclers recover 98 % more materials on a daily basis than the average American (Table 3 and Fig. 3).

When asked about their energy use patterns in the Pata community most households reported having a single incandescent, light bulb (84 %), and a CRT screen TV (68 %). Since only 26 % of households reported having a refrigerator, the use of a refrigerator in order to calculate energy use in the estimated, "average" Romani household was not considered. Individuals reporting that they had a light bulb indicated use ranging from less than one hour to 24 h per day (as a means of discouraging rodents in the home) with a mean of 9 h per day of use. Television usage ranged from less than one hour per day to 24 h per day (again mentioned as a means of discouraging rodents in the home) with a mean report of 7 h per day of use. Assuming a 60 W bulb and about 200 W for the standard crt television (the households did not have flat screens or LCD TVs) over a 30 day period, the average household would use 16.2 kWh per month to light their small homes and 42 kWh per month to operate the crt television for a total average energy use of 58.2 Kwh per month (Lunau 2013; Santiago et al. 2013). The US Energy Information

Table 3 Recycling Activity by Demographic and Material in kilograms per average day

Material by average and range	Glass	Metal	Plastic	Copper	Aluminum	Total avg./day
Men	4.8, 0–100	38.8, 0–200	37.3, 0–400	0.37, 0–5	0.65, 0–5	81.92
Women	3.2, 0–100	27.1, 0–100	24.4, 0–100	0.26, 0–3	1.5, 0–35	56.46
Children/ Youth	0, 0	10.3, 0–300	0.96, 0–30	0.03, 0–1	0.3, 0–7	11.59
Total/ household	Demographics for the 2009 data set indicate on avg. 1 male, 1 female, and 2 children per household for a recycling rate of 81.2 + 56.46 + 11.59*2 = 160.84 kilograms/day					

Fig. 3 Visual Comparison of Romani of Pata Rat versus US Recycling Rates, kilos/person/day, Romani recycling at Pata per capita/day is 98 % greater

Table 4 Energy use activities, Roma of Pata Rat

Percentage of households with a Light-bulb (N = 31)	Hours of light-bulb use per day average, range	Percentage of households with a TV	Hours of TV use per day	Percentage of households with a refrigerator
84 %	9, 0–24	68 %	6.8, 0–24	26 %
Average kWh use per month	9 h/day[a]0.06 kWh/hour[a] 30 days/mo = 16.2 kWh/mo		7 h/day[a]0.200 kWh/hour[a]30 days/mo = 42 hWh/mo	Less than 50 % of households not calculated
Total kWh use per month Romani of Pata Rat	58.2 kWh/mo (DOE Data, 2012 indicates average US home used 958 kWh/month)			

[a] (eia http://www.eia.gov/)

Association estimates that the average US home uses 958 kWh per month of energy (http://www.eia.gov/). Overall, calculated energy use is only 6 % of the typical, average American household's energy use per month (Table 4 and Fig. 4). It is also interesting to note that while the average Romani uses very little energy per month/year the average Romanian home (having many of the basic conveniences found in American homes uses only 1.1 ton of oil equivalents, (toe), per year in energy per dwelling while the average American uses 8x that amount or 8.35 toe per year in energy per dwelling (Lapillonne and Pollier 2014).

4 Outcomes from Overall Study Abroad Effort to Date

The 2014 environmental health disparities capstone will be the tenth year of this collaboration. It is clear to the author's that the Fulbright which Dr. Zeman had in 2003 to Romania was instrumental in providing the time to develop meaningful

Fig. 4 Visual comparison of US versus Romani of Pata Rat Energy Use, kWh/mo, US energy use is 94 % greater

collaboration and build joint projects. But the other bedrocks of this effort outside of the collaborators were the Directors and Associate Directors of the International Programs offices of the collaborating institutions, in particular, Dr. Yana Cornish, Director of the UNI, Study Abroad program and Ms. Rodica Marcu, Associate Director of International Programs at UMF. Both individuals were essential collaborators and consummate professionals in helping to arrange the logistics of the joint study abroad, manage student assistants working with the program and run every aspect of scheduling/payment, safety/liability, and cultural exchange. No professor who seeks to duplicate such a program should do so without the support and assistance of their International Programs or equivalent office and the professionals that staff it.

As of this tenth year of the program over 200 American and over 50 Romanian students have participated in the program. The American program can accommodate 25 participants and in recent years the program has been at capacity, on average 13–16 is more representative of participant numbers of Americans and 5-6 Romanian student participants (Fig. 5). Over 36 distinct curricular programs have been delivered which included practical skills classes for Roma participants. Electronic manuscripts of the Roma outreach curriculum, developed by UNI students, are provided, translated into Romanian and retained by the NGOs for reuse during yearlong programming. Topics covered have included hygiene, dental hygiene, water hygiene, water borne diseases, occupational hazards in scavenging operations, toxins in the environment, home hygiene, methods to counter diarrhea disease along with practical skills workshops such as making "tippy taps" or field hand washing stations, hand washing technique, and how to make a rehydration solution. To date over $5,000.00 in direct aid donations and cash have been provided to the community during the giveaway portion of the outreach program which is designed to support the theme of the program, for example, work gloves and work boots accompany programs focusing on occupational health risks and protection for scavenger communities.

Fig. 5 Romanian and US students attending class and outreach activities

Student evaluations of the program are consistently high to very high (90th percentile) for program quality, impact, and enjoyability. Select quotes from the outreach experience as relayed through reflective writing assignments by the American students speak to the broader impacts of the program for students:

> Our trip to Pata Rat was an amazing experience. There were so many environmental health issues that related to what we discussed in class…

> Visiting Pata Rat was a very eye opening and humbling experience for me in many ways…

These types of comments are typical for 85–90 % of students on average on the experience at Pata Rat. Additionally students are asked (in a final reflection piece) to comment on their experience with Romania and Romanian students. In 90+ % (93–96) of the comments the following excerpted comments are typical:

> I loved Romania! When I first signed-up for this study abroad, I didn't even know where Romania was! I made such good friends and experienced so many amazing things…

> Romania is awesome! I love the people I have met here who I know will be Facebook friends for life…

> Romania surprised me…it was the best thing I have ever experienced internationally…

Based on the comments consistently received about this sustainability and environmental health disparities experience it is obvious that is has a tremendous impact on the student participants.

5 Discussion and Future Efforts

The community based cross-sectional study provides poignant information that has been returned to the classroom portion of the sustainability and study abroad program as a research case study that can be built upon in future years. It is clear that the Romani are a contributing part of their community and that these contributions are significant, often unconsidered/measured and go unacknowledged (Goodwin and Buijs 2013). The contributions of scavenger-based communities in general and Romani populations specifically, to recycling and reusing solid waste streams, and the carbon offsets embodied in that, needs further study before generalizations can be made. If the contributions are similar and some recent studies in Brazil and Greece indicate that this is true, then exploration of ways to acknowledge and support these risky activities is warranted (King and Gutberlet 2013; Papaoikonomoua et al. 2008).

Additionally, although this study documented the very low household energy use and carbon footprint, in this community, this cannot be generalized until broader research is undertaken. While low resources and income may play a role in this frugal energy use, cultural norms are also likely a factor. The impact of Romani recycling from the perspective of a beneficial net good for the environment and society needs to be acknowledged as does the risk they face in doing this work. Services for occupational health and food security should be extended to the Romani as they provide a net environmental good to the entire community. In the United States there is currently a vogue discussion about "makers and takers"; this case study illustrates that unstudied ideas about "makers vs. takers" can be an unexamined, biased supposition.

Contributions to consciousness raising and multi-cultural understanding for American and Romanian students is also a major benefit of this type of experiential, systems analysis based approach to sustainability education in a cross-cultural context. The authors hope to institute a long-term summative evaluation of the effort to add to the current formative information. Sustainability education offered in experiential environments and structured through a systems analysis lens has proven to be a very popular structure for participating students and it offers the potential to have real, meaningful impacts for communities when maintained over the long-term.

6 Conclusions

The outcomes of this ten your effort in sustainability education focusing on environmental health disparities have been gratifying for organizers, the impacted communities, and students. To summarize:

1. The structure of the program includes all of the important elements of good sustainability education:

    i. A systems analysis approach that is both formally taught as a method of analyzing complex technology, society, and environmental issues; and,
    ii. A systems approach applied to the structure of teaching activities in the course. In that regard, students are involved in learning both through classroom work and analysis using systems thinking and by being engaged with a Roma community through service, outreach, and applied research.

   Perfecting this structure has provided organizers with an opportunity to apply in the field different pedagogical approaches to produce a workable mix of activities and learning with the additional intention of generating real attention for and understanding of the Roma and their struggles for development, economic security, and environmental and health equity.
2. The Roma community looks forward to the visits each year of Romanian and American students in this program. The consistency of presence has created a foundation for mutual respect and understanding which builds good will for all of the parties involved in the project. American and Romanian students have consistently rated the learning experience highly, through both formal course evaluations and as evidenced by comments made during reflective writing assignments.
3. Applied research completed in the Roma community during the course sessions by faculty and graduate students have provided information to enrich the course sessions and practical information for the Roma community, for instance, identifying community resources and environmental stressors in a community mapping project, identifying sources of toxics contamination and evaluating community water quality, and most recently (as reported here) documenting the positive environmental impacts of the Roma recycling activities while also highlighting their struggles for food security and adequate housing.

The collaborators have expanded their group to include a professor of performance communication and plan to continue working to develop environmental health literacy projects that will both inform the Roma community and provide them with opportunities for economic and educational development. In developing this sustainability education program, collaborators have worked toward a win-win for American and Romanian students as well as Roma community members. In the process knowledge, understanding and friendship has blossomed. All involved look forward to the next 10 years.

References

Arpinte D, Baican E, Ile S, Marin F, Moisa F, Preoteasa AM, Surdu M (2002) Roma working children and their families: socio-cultural characteristics and living conditions. RO Media Publishing House, Romania

Beniamin J (2008) Patterns in Roma migration. Hemispheres 31:6–16. http://dx.doi.org/

Bulalrd RD, Johnson GS, Torres AO (2011) Environmental health and racial equity in the United States: building environmental just, sustainable, and livable communities. American Public Health Association, Washington

Carpenter DO, Suk WA, Blaha K, Cikrt M (1996) Hazardous wastes in Eastern and Central Europe. Environ Health Perspect 104:244–248. http://dx.doi.org/

Cointreau S (2006) Occupational and environmental health issues of solid waste management: special emphasis on middle and lower-income countries. The World Bank Group, Washington

Flag of the Romani People. (n.d.). In: Wikipedia. http://en.wikipedia.org/wiki/Flag_of_the_Romani_people. Accessed 28 March 2014

Fonseca I (1995) Bury me standing: the Gypsies and their journey. First Vintage Edition, New York

Goodwin M, Buijs R (2013) Making good European citizens of the Roma: a closer look at the EU framework for national Roma integration strategies. German Law Journal 14(10):2041–2056

Haijioff S, McKee M (2000) The health of the Roma people: a review of the published literature. J Epidemiol Health 54:864–869

Hicken MT, Gee GC, Morenoff J, Connell CM, Snow RC, Hu H (2012) A novel look at racial health disparities: the interaction between social disadvantage and environmental health. Am J Public Health 103(12):2344–2351

Hunt C (1996) Child waste pickers in India: the occupation and its health risks. Environ Urban 8 (2):111–118

Idzerda L, Adams O, Patrick J, Schrecker T, Tugwell P (2011) Access to primary healthcare services for the Roma population in Serbia: a secondary data analysis (11:10). BMC International Health and Human Rights

King MF, Gutberlet J (2013) Contribution of cooperative sector recycling to greenhouse gas emissions reduction: a case study of Ribeirao Pires, Brazil. Waste Manage 33:2771–2780

LaDuke W (1999) All our relations: native struggles for land and life. South End Press, Cambridge

Lapillonne B, Pollier K (2014) Energy efficiency trends for households in the EU. http://www.odyssee-mure.eu/publications/efficiency-by-sector/household/household-eu.pdf. Accessed ODYSSEE-MURE

Lunau K (2013) The incandescent light bulb. Maclean's 136(51/52):122. http://eds.b.ebscohost.com/eds/detail?vid=9&sid=000d5f34-39ef-49dd-a65d-000aa0d1dcfa%40sessionmgr111&hid=109&bdata=JnNpdGU9ZWRzLWxpdmU%3d#db=crh&AN=92947239

Meadows DH (2008) Thinking in systems: a primer. Chelsea Green Publishing, United States

Municipal Solid Waste Charts (2014). http://www.epa.gov/wastes/facts-text.htm

Murphy SP (2001) Chapter 48: nutrition guidelines to maintain health. In: Nutrition in the prevention and treatment of disease (pp 753–771). [ScienceDirect]. http://dx.doi.org/10.1016/B978-01219315501/50050-7

Nutrition and your health: dietary guidelines for Americans [A pamplet] (1990). http://eds.a.ebscohost.com.proxy.lib.uni.edu/eds/detail?vid=3&sid=e4b5bc95-9a3b-4544-bb90-bb9e63eae26b@sessionmgr4004&hid=4102&bdata=JnNpdGU9ZWRzLWxpdmU=#db=edsgao&AN=edsgcl.15591293>

Papaoikonomoua K, Kipourosa S, Kungolosa A, Somakosb L, Aravossisc K, Antonopoulosd I, Karagjannidisd A (2008) Marginalised social groups in contemporary weee management within social enterprises investments: a study in Greece [Chapter in conference proceedings book]. Retrieved from Elsevier Science. http://eds.a.ebscohost.com/eds/detail?vid=9&sid=e4b5bc95-9a3b-4544-bb90-bb9e63eae26b%40sessionmgr4004&hid=4102&bdata=JnNpdGU9ZWRzLWxpdmU%3d#db=edselp&AN=S0956053X08004017

Rambouskova J, Dlouhy P, Krizoza E, Prochaska B, Hrncirova D, Andel M (2009) Health behaviors, nutritional status, and anthropometric parameters of Roma and non-Roma mothers and their infants in the Czech Repulic. J Nutr Educ Behav 41(1):58–64
Romani diapora. (n.d.). In: Wikipedia. http://en.wikipedia.org/wiki/Romani_diaspora. Accessed 30 March 2014
Santiago I, Lopez-Rodriguez MA, Gil-de-Castro A, Moreno-Munoz A, Luna-Rodriquez JJ (2013) Energy consumption of audiovisual devices in the residential sector. Energy. http://dx.doi.org/10.1016/j.energy.2013.08.018
Thiraattanasunthon P, Siriwong W, Robson M, Borjan B (2012) Health risk reduction behaviors model for scavengers exposed to solid waste in municipal dump sites in Nakhon Ratchasima Province, Thailand. Risk Manage Healthc Policy 5:97–104
Zeman CL, Depken DE, Senchina DS (2003) Roma health issues: a review of the literature and discussion. Ethn Health 8(3):223–249
Zheng N, Bakker HW, Tufts M, Raymond RE, Salihu H, Elliot MR (2013) Early childhood lead exposure and academic achievement: evidence from Detroit public schools, 2008-2010. Am J Public Health 103(3):72–77

Author Biographies

Dr. Zemana Professor of environmental health, HPE Division, UNI, has worked on sustainability issues in a multitude of capacities for the last 24 years through community organizing (1990–1992), as a consultant (1992–1998) and as a professor (since 2000). Dr. Zeman received her Ph.D. in Preventive Medicine with emphasis on Environmental and Occupational Health from the University of Iowa. Her research interests include nitrates in the environment and their impact on human health with focus on children's health, understanding the health issues of underserved populations, and environmental sustainability. Dr. Zeman was a Fulbright Scholar to Romania on environmental health issues during the Spring/Summer of 2003. Dr. Zeman is also the Director of the Recycling and Reuse Technology Transfer Center, at the University of Northern Iowa. She teaches classes in sustainability and environmental health disparities, epidemiology, human diseases, environmental health, and environmental and occupational health regulations.

Dr. Jennifer Hall currently serves as the Associate Director for Outreach for the University of Iowa's College of Public Health Healthier Workforce Center for Excellence (HWCE), one of four National Institute for Occupational Safety and Health's Total Worker Health™ Centers for a Healthier Workforce. In this role, Jenny conducts and translates research, and promotes best practices in integrated employee health, wellness and safety programs. Prior to her work with the Center, Dr. Hall taught courses in the Division of Health Promotion and Education at the University of Northern Iowa (UNI) and managed an international nonprofit organization to empower impoverished populations through accessible health and education services. Jenny holds a doctorate in Community Health Education with an emphasis in Environmental Health from UNI and has over 10 years of experience in planning, implementing and evaluating health protection, health promotion and education programs in diverse settings around the globe.

Dr. Diane Depken received her doctoral degree from Temple University in the area of Women's Health/Health Promotion. She also completed post-graduate work in epidemiology at the University Of Buffalo. Additional research and community service interests include; food systems, ecological theory, and applied research and programming related to community health; especially gender, health determinants, food access issues, and stress/wellbeing linkages. Diane also has a life-long interest in the sport of fencing.

Dr. Ioan Bocsan is Professor of Epidemiology and Primary Health Care at the Iuliu Hatieganu University of Medicine and Pharmacy in Cluj-Napoca, Romania. Founder (World Bank grant) of the first accredited School of Public Health in Romania (2001). National Expert at European Commission, Directorate of Research and Innovation for Health (2006–2013). Member of the International Advisory Board of EURIPA.

Perspectivaction: A New Educational Framework to Include Sustainability in Responsible Management Education

Norman de Paula Arruda Filho

Abstract

The Higher Institute of Administration and Economics (ISAE) is a business school located in the south of Brazil that has more than 5000 students a year in post-graduate courses. This article intends to present the Perspectivaction educational framework, whose goal is to complement the traditional scope of MBA courses, and highlight its contribution to the education of globally responsible leaders. Perspectivaction is a systemic educational framework with 14 interrelated activities. It is based on educational theories such as experiential education, transdisciplinarity, transversality and self-education. It offers students the opportunity to experience large companies' best practices in corporate sustainability, develop skills for producing articles and scientific research, plan their personal and professional development with the help of a coach, and participate in experiential challenges (such as rafting and flying in a balloon) which encourage them to advance team work focused on values. To validate Perspectivaction as an important tool in training globally responsible professionals, the face validity technique was applied to a sample of 50 professors. Results show that 92 % agree that Perspectivaction complies with the objective of educating globally responsible leaders.

Keywords

Responsible management education · Sustainability · Educational framework · Social responsibility · Globally responsible leaders

N. de Paula Arruda Filho (✉)
Higher Institute of Administration and Economics—ISAE/FGV,
Av. Visconde de Guarapuava, 2943, Curitiba, PR, Brazil
e-mail: norman@isaebrasil.com.br

W. Leal Filho (ed.), *Transformative Approaches to Sustainable Development at Universities*, World Sustainability Series, DOI 10.1007/978-3-319-08837-2_18

1 Introduction

Higher Institute of Administration and Economics (ISAE) is a business school that offers post-graduate, master's and in-company courses (short and medium-term), with headquarters in Curitiba, southern Brazil. It was created in 1995 by Getulio Vargas Foundation (FGV)—one of the most renowned educational institutions at global level since 1944.

This case study aims to describe the Perspectivaction educational framework, which was developed by ISAE as an important method for forming leaders capable of facing adversities in public and private companies and in society in general, seeking to locally overcome challenges towards sustainable global development.

Since its foundation, the institute has operated in an innovative and cross-sectional way, seeking to mobilize and help the market in respect to responsible practices that generate sustainable development. In 2004, Perspectivaction was created (with a perspective view to the future and integrated action in the present) to complement the traditional scope of Getulio Vargas Foundation's MBAs. Today, the framework is also applied in short and medium-term courses and is a great competitive differential for the institution in the local educational market.

The relevance of a study is evidenced by the significance, convenience and contemporaneity of its object. Perspectivaction—which completes 10 years on the date of publication of this article—is objectively focused on the four axes below:

- Contributing to develop executive education and the new role of business schools within the paradigm of responsible management education;
- Contributing to academia and its role in the transformation of society by developing reference frameworks that allow new visions and actions, offering students more effective learning processes;
- Contributing to the social agenda within the Brazilian reality, which demands strategies for training globally responsible leaders;
- Contributing to the companies' reflection towards new global educational policies, increasing the development of new frameworks for corporate education.

The application of this case study in executive education to develop managers and the new role of business schools (by means of the four types of contributions mentioned) mainly resides in Perspectivaction's originality and novelty—at least in the context of Brazilian education.

This educational framework is special for proposing a set of educational strategies based on the application and extension of competences (in addition to the traditional focus on the teaching-learning process). The adopted strategies and pedagogical tools complement the student's education with the integration of non-traditional activities in the curriculum, seeking to develop new skills and promote integrated learning processes. Another original contribution of this work is the inclusion of transversal themes or guiding concepts in the context of management education.

This article firstly presents the contextualization of how the sustainability theme is embedded in the case study. Then it describes the theoretical conceptual support, with reflections of renowned authors on education and the Principles for Responsible Management Education (PRME)—which support and found the development of ISAE's educational framework. Next, it details the Perspectivaction framework and its activities. Finally, the face validity methodology is described, validating Perspectivaction as an important tool to develop globally responsible leaders.

2 Contextualization

The challenges of sustainable development—protecting the environment, stabilizing world population growth, reducing differences between rich and poor and eradicating extreme poverty—require new forms of global cooperation. According to Sachs (2008), "in the twenty first century our global society will flourish or perish according to our ability to find common ground across the world on a set of shared objectives and on the practical means to achieve them".

It is only through education that leadership can meet the demands and expectations of this new reality, not only acting locally, but also on a global scale, for there is a greater comprehension of connectivity and complexity towards common good. "Education survives systems and, if in one system it serves to reproduce inequality and disseminate ideas that legitimize oppression, in another it can create equality between men and promote freedom" (Brandão 1986, p. 99).

In 1972, the Club of Rome—one of the most influential non-governmental organizations in the world, composed of heads of state, scientists, industrialists, political leaders, bankers etc.—published a report produced by a team of researchers from the renowned Massachusetts Institute of Technology (MIT) entitled "The limits of growth". By means of mathematical models, researchers concluded that, if the rate of population growth and the use of natural resources were maintained, by the end of the twenty first century humanity would be at serious risk of survival (Meadows et al. 1972). Thus, it was necessary to recognize that natural resources were, in fact, finite.

The United Nations declares that to reverse this situation it is necessary to invest 2 % of the world GDP in some key sectors—to start the so-called Green Economy (UNEP 2011). In "The great green technological transformation" (UNDESA 2011), the following sectors are listed: logistics, tourism, water, waste, production, energy, agriculture, buildings, fishing, forests.

The time in which we live inspires care and immediate search for solutions to build the future on a sustainable basis, focused on ethical values, greater transparency and reasonable use of scarce natural resources. This scenario indicates that it is time to innovate and break paradigms in all sectors—including education, for the educational environment (which has traditionally been linear, dominating and content-based) must be cooperative and hypertextual, promote freedom of thought and advance individual and group research (Mustaro 2003).

3 Education as a Vector for Sustainability

Resource attainment is essential for the process of change into a new economic model. But there is another equally important factor for overcoming this challenge: the setting of a new educational paradigm to advance the development of professionals capable of leading changes towards a wider future context.

Currently, the concepts of education, training, guidance and development cannot be regarded as absolute, but relative, once they reflect ideas and thoughts dominant in each epoch and must be adapted to new contexts.

This scenario claims for an education linked to the present and future global context, in which the formal framework of postgraduate courses for executives requires a thorough review. More than technical and traditional expertise in the world of management, the educational sector has to focus on a constant search for innovation in methods and content. This change of positioning has to widely promote collective values and individual characteristics.

"The Four Pillars of Education" are an important milestone in this new way of understanding education. The report containing the four pillars—published in the book "Education: A Treasure to Discover" (Delors 2006) was produced for UNESCO by the International Commission on Education for the twenty first century, headed by Jacques Delors. According to this report, the four pillars of education are: *learning to know, learning to do, learning to live together* and *learning to be*. Nowadays education is almost exclusively focused on *learning to know* and, on a smaller scale, *learning to do*. However comprehension and execution cannot be fully achieved without the other learning areas. *Learning to live together* and *learning to be* set up one of the greatest challenges for educators, because they involve feelings, beliefs, virtues, spirituality, values…—variables that cannot be objectively quantified or measured due to their subjective nature. Subsequently, a fifth pillar was added to the report: *learning to transform oneself and society*—which recognizes that each one of us can change the world, acting individually and collectively, and states that quality education provides the tools to change society (UNESCO 2005).

This demand for values is not only perceived in academic environments, but also in the market. According to a survey of the Association to Advance Collegiate Schools of Business (AACSB)—published in "Rethinking the MBA: Business Education at a Crossroads" (Srikant et al. 2011)—education based on concepts (such as ethics and corporate responsibility) and on decision-making processes in high complexity environments hasn't been deeply achieved by management education institutions in general.

In the same direction, the UN Global Compact has led the Principles for Responsible Management Education since 2007, which aims at mobilizing schools to adopt educational principles focused on sustainability when defining purposes, emphasizing new values, developing innovative methods, building partnerships and intensifying dialogue (PRME 2012).

To incorporate the effective commitment to the principles of a performance towards global sustainable management (within the paradigm of responsible management education), ISAE has signed the UN Global Compact and the Principles for Responsible Management Education (PRME). Nowadays, the institute is part of the board of the Global Compact Brazilian Committee and has directed the PRME Brazil Chapter since 2013. It also integrates the PRME Champions leadership group (which gathers the 30 most active PRME institutions in the world) and is a PRME Advisory School—working in a group that provides strategic advice and guidance to the Global PRME management committee.

4 Principles for Responsible Management Education (PRME)

Facing the context of a new educational paradigm for developing professionals, since 2007 the UN Global Compact has led a movement in favor of education—the Principles for Responsible Management Education, whose premise is to mobilize schools to adopt educational principles focused on sustainability, with respect to definition of purposes, emphasis on new values, development of innovative methods, partnerships and dialogue with society (PRME 2012).

The PRME initiative is composed of six principles that guide the management of signatory academic institutions, corporate universities and support organizations (such as regional councils, federations and business associations). The six principles are:

(1) Purpose: We will develop the capabilities of students to be future generators of sustainable value for business and society at large and to work for an inclusive and sustainable global economy.
(2) Values: We will incorporate into our academic activities and curricula the values of global social responsibility as portrayed in international initiatives such as the United Nations Global Compact.
(3) Method: We will create educational frameworks, materials, processes and environments that enable effective learning experiences for responsible leadership.
(4) Research: We will engage in conceptual and empirical research that advances our understanding about the role, dynamics, and impact of corporations in the creation of sustainable social, environmental and economic value.
(5) Partnership: We will interact with managers of business corporations to extend our knowledge of their challenges in meeting social and environmental responsibilities and to explore jointly effective approaches to meeting these challenges.
(6) Dialogue: We will facilitate and support dialog and debate among educators, students, business, government, consumers, media, civil society organizations and other interested groups and stakeholders on critical issues related to global social responsibility and sustainability.

5 Theoretical-Conceptual Support

This case study is based on theories and concepts of experiential education, transdisciplinarity, transversality and self-education, emphasizing that the educational phenomenon, for its very nature, is not a finished reality and cannot be conceived in a unique and precise way in its multiple aspects.

Experiential education is a conception of education that focuses on the transactional process between professor and student, involving direct experience with the learning environment and content. For the Association for Experiential Education (AEE), "experiential education is a philosophy and methodology in which educators purposefully engage with learners in direct experience and focused reflection in order to increase knowledge, develop skills, clarify values and prepare students for both careers and citizenship" (AEE 2009).

In this educational conception, the apprentice's experience has a central place in all considerations of teaching and learning. These experiences may include events from the apprentice's life and also those originated from activities implemented by professors/facilitators. A key element of experiential approach is that apprentices should perform the analysis of experience, by reflecting, evaluating and reconstructing it, in order to perceive meanings based on previous experiences (Andresen et al. 2000).

For an experience to be educational, John Dewey indicates that certain parameters must be present—continuity and interaction are the most important. Continuity refers to the idea that experience comes from experiences and leads to other experiences, driving the person to learn more and more. On the other hand, interaction occurs when experience reaches a person's internal needs or objectives (Dewey 1971).

Equally important is the concept of transdisciplinarity. According to Nicolescu (2000), the prefix "trans" concerns what is at the same time among disciplines, through different disciplines and beyond any discipline. Its goal is to comprehend the present world, in which the unity of knowledge is an imperative. Therefore, it seeks to comprehend complexity.

According to the Brazilian Interactive Dictionary of Education (DIEB), transdisciplinarity is the "theoretical principle that seeks intercommunication between disciplines, effectively referring to a common theme (transversal). In other words, in transdisciplinarity there are no borders between disciplines" (Menezes and Santos 2002).

Perspectivaction educational framework uses the concept of transversality in curricula. The practice of transversality in education consists in the insertion of transversal themes—guiding concepts—that can influence the process of changing society, in addition to traditional curricular content (Busquets et al. 1999).

Inspired by the PRME, the Higher Institute of Administration and Economics has guiding concepts (leadership, governance, sustainability, innovation, ethics, entrepreneurship) that direct all activities with stakeholders in respect to responsible education. The sustainability concept seeks to develop theories, methods and techniques towards a balanced rise in the value of the company, society and environment—aiming to achieve the environmental, social, economic, cultural, spatial and political dimensions of sustainability as a whole.

Transversal themes were established in 1999 and are part of the National Curriculum Parameters (PCNs/Brazil), based on the National Education Plan (PNE/Brazil). They are proposals to which school secretaries and units may refer in order to elaborate their own education plans. The PCNs suggest the elaboration of projects for implementing the proposal of transversal themes—one of the ways to organize the didactic work and integrate different ways for curriculum framework.

The concept chosen to complete the new vision of curricular practice is self-education, whose principles stated by Morin (2000) presented a model that provides not only technical learning, but also an education towards the search for the meaning of life. Self-education emerges as one of the action frameworks in the context of current education trends for adults, for it represents educational autonomy—or the ownership of each one's education power (Pineau 1999). Morin (2002) declares that educational institutions must contribute to individuals' self-education, by teaching them how to assume the human condition, how to live—thus teaching them how to become citizens.

The mentioned studies are consistent with the paradigm of the principles of responsible education, which, for their amplitude and contemporaneity, have remained as benchmarks in the education sector for many decades. Therefore, these educational processes are synergistic to ISAE's exclusive educational framework.

6 Perspectivaction Educational Framework

ISAE has assumed a new paradigm of academic-organizational expansion to develop globally responsible leaders, based on the ethics of relations, corporate sustainability, entrepreneurship and innovation.

To fulfill its mission and achieve its objective, ISAE has been adapting the status quo and proposing the construction of new solutions and educational methods to underpin this change. Therefore, Perspectivaction has been frequently renewed and updated, having the PRME as reference for continuous improvement.

Perspectivaction is composed by a set of mandatory and optional activities that permeate the student experience from the beginning to end of the course, with main focus on faculty training. With professors acting as promoters and stimulators of the educational process, students go through modules, workshops, courses and experiences related to the theme of sustainability.

The framework is focused on the integral education of the human being, leader, agent that changes reality into multiple perspectives, responsible for self-education. It is composed of interdependent activities that, taken as a whole, represent a new way to accomplish the education of leaders and managers.

This set of activities is distributed in four curriculum modules (*contextualization, self-development, theory and practice, applicability*), structured to accommodate the multiple strategies of learning, as shown in Fig. 1.

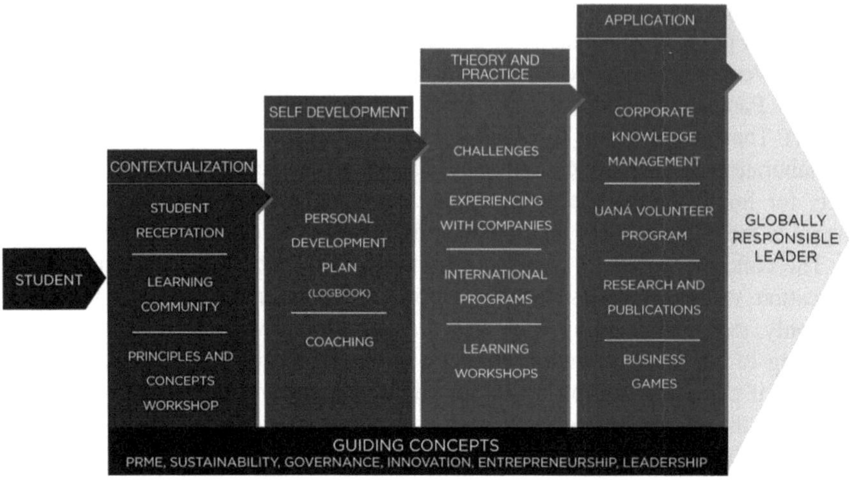

Fig. 1 Self-development module

Contextualization begins with the maturation of the decision of the student in relation to the course. It allows decision-making to be actually based on information. It emphatically seeks to integrate the student into the business school context. Students are guided through the most appropriate courses for their career and life.

Self-development offers support for the student to effectively advance self-knowledge and feel as part of a greater whole, integrated through the Personal Development Plan, Coaching and Mentoring.

In *Theory and Practice*, the set of activities prioritizes the experiences of the student in workshops, simulations and visits to national and international companies, aiming to develop knowledge and skills for the implementation of a theoretical basis in the company's daily activities.

Applicability focuses on the student's achievement of effective actions of "authorship" and "cooperation" in the creation and development of sustainable enterprises with colleagues and other partners. Opportunities arise such as incubation of projects, business games, voluntary consultancy, research projects, real cases and business plans for startups.

Each block, despite having a different emphasis, is closely related to the others, and altogether they converge to the desired profile: the globally responsible leader.

7 Perspectivaction's Main Activities

Student Guidance: At ISAE, guidance is the student's first contact with the institution, happening before enrollment. The objective is to align expectations according to the student's interests and needs.

Student Reception: It is an academic approach about all stages of the course and the main rules between student and institution. In addition, the PRME initiative is presented as an advantage for the students, who have access to contents focused on sustainability and corporate responsibility through Perspectivaction's activities.

Learning Community: The learning community is a dynamic process where students commit to share their experiences and knowledge, aggregating best practices to the learning platforms. The objective is to seek integration and alignment with ISAE's educational principles, besides promoting the living contract, with a participation approach focused on a systemic vision for the set of course activities, and an approach of the educational framework and Delors' learning principles (2010).

Personal Development Plan: It is a structured process that helps develop individuals' skills to reflect on their own learning, achievements, education and career planning. It focuses on the development of the person as a professional, citizen and global leader.

Coaching: A powerful tool for supporting people in their processes of growth, helping them to maximize their potential and performance. The coaching process in a business school aims to assist students in the implementation of their Personal Development Plan, with emphasis on the professional axis, but also considering them as citizens and their participation in society. Issues related to corporate social responsibility are also covered, valuing and empowering aspects of the economic, social and environmental commitment, the social management of human resources, solidarity, sustainable and responsible management, among others.

Corporate Knowledge Management: The main purpose of this activity is to progressively build new knowledge, skills and attitudes necessary for the scientific production of business, technical, human and cultural solutions. This goal is fully fulfilled with the production and presentation of course conclusion works.

Learning with the Company: Through technical visits, it promotes strategies of the teaching-learning process that allow the connection of knowledge developed in classroom with real life, the community and innovative practices of the business world.

These visits are held in large and medium-size companies and contribute to connect theory and practice, academia and society. It helps in the process of reviewing concepts when assessing business practices identified by students and alumni. It specially favors the understanding that business challenges in the current scenario have to be thought in a global context, extending the view of corporate purposes far beyond financial results, by assuming ethics and corporate social responsibility as strategic values in organizations.

Business Games: Business games simulate corporate management situations and processes, generally in the form of a business. They seek to maintain the players' attention through the exact simulation of real life events by using algorithms, and also to create a close bond among players who make decisions during actions, suffering consequences and results. ISAE has included special modules of business games in MBA courses, aiming to reproduce reality with simulations of real situations faced by a company and the market, combining the students' knowledge on business management and encouraging experience exchange in all disciplines.

International Programs: An international program provides a comprehensive view of economic, financial, corporate, educational and managerial elements that have been composing the process of advanced international fields in the world market, offering conditions for business managers to assess and improve their company's performance standard during ongoing changes. In addition, participants have the opportunity to evaluate the internal and external situation—regarding the current stage of integration and interaction of the Brazilian economy in the global scenario—and to identify the organizational behavior of internationalized companies within a macro and microeconomic scenario. International program's main activities include: promoting scientific exchange among faculty and students in the technological, cultural and philosophical fields, between ISAE and other national and international institutions; negotiating cooperative projects for the development of national and international projects; implementing and evaluating international cooperative projects; disseminating academic research and information regarding national and international educational opportunities among the institution's community; realizing events focused on responsible management education in partnership with national and international institutions.

Learning Workshops: The workshops are living areas to experience and reflect on current knowledge, necessary for everyone. They facilitate self-knowledge, stimulate individual and collective reflection and the acquisition of complementary skills for professional development. The main objective is to expand personal, professional and career skills guided by ethics, responsibility and sustainability, and to promote networking between students and companies. They are expected to understand that changes in management frameworks reveal the challenge to think and act in a global context. Companies' sustainable and ethical values have to be balanced with financial results. Methodology emphasizes knowledge construction through participation, questioning and, above all, based on the reality of situations, facts and life stories related to business management.

Uaná Program—Volunteering: In order to create a field to strengthen theoretical, practical and professional knowledge acquired by students in graduate, postgraduate and extension courses, in 2002 ISAE created the Uaná Program of Voluntary Consultancy in Management, inspired by a similar program developed by the Massachusetts Institute of Technology (MIT). Uaná, in *Tupi Guarani* indigenous language, means "firefly"—a being that illuminates paths with its own light. This is how the institution sees the volunteers that participate in the project. The overall objective of the program is to facilitate participation of the institution's community in voluntary consultancy projects, contributing to learning, social participation, sustainable development and poverty eradication. Its specific objectives are: to provide consultancy in management and promote the culture of competent entrepreneurship, stimulating the increase of income generation and jobs in the country; to promote research, development of course conclusion works, and other scientific production involving subjects related to the program; to offer students, alumni, graduates and internal community the opportunity to contribute to the eradication of world poverty through volunteering programs in management; and to encourage the participation of students in the project Voluntary Professional Consultancy in Business Management, which focuses on micro entrepreneurs and the third sector.

New Experiential Challenges: Concepts like ethics, respect and responsibility with people and the environment are included in activities, which are deeply experiential. The initiative aims to place the student in situations outside their comfort zone (either physical or emotional), strengthening cognitive, social and psychomotor abilities—factors of extreme importance for high performance teams in pursuit of common objectives. Experiential challenges include ballooning, treetop, rafting and theatrical dramaturgy. These activities encourage participants to develop collaborative leadership, interpersonal communication, planning, team work, collaborative spirit, decision making process and self-knowledge.

8 Method Used to Validate Perspectivaction

As an approach to validate the Perspectivaction educational framework as an important tool to form globally responsible professionals, the face validity technique was used with a sample of 50 professors from the FGV network.

Performing their functions, participants of the sample kept in touch with ISAE students through the formal course curricula and the supplementary process proposed in the educational framework. It is important to highlight that many of them also work for other educational institutions that are not part of the FGV network.

Results show that 92 % agree that the framework complies with the objective of forming globally responsible leaders, who are aware of sustainability in the corporate environment (Fig. 2).

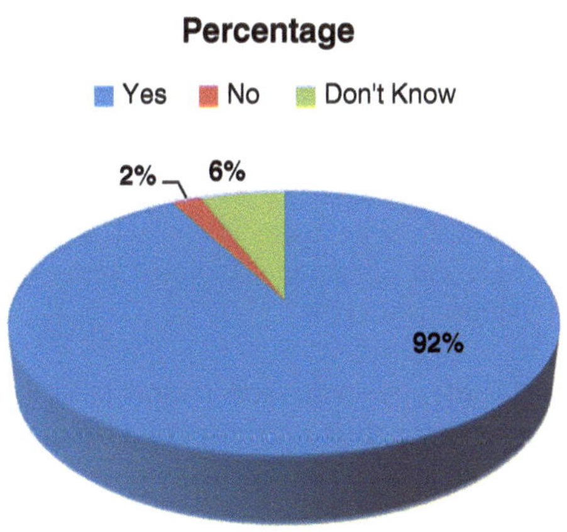

Fig. 2 Face validity for perspectivaction in respect of advancing the education of globally responsible leaders (*Source* ISAE)

9 Conclusion

By means of the theoretical-conceptual support and the face validity technique to evaluate the presented educational framework, this study concluded that it is appropriate and congruent in its elaboration and application as a complementary system for learning and embedding the sustainability subject in the curricula of courses (MBA, postgraduate, short and medium term).

It also concluded that the Perspectivaction framework recognizes diversity and individuality, not only seeking the awakening of the student for new habits, abilities and actions, but also promoting the integral development of the person by improving individual skills. In the long term, its intention is to develop in students the 5th Pillar of Learning, identified by Jacques Delors as the ability to *learn to transform oneself and society* (Delors 2006). Thus, as an expected result, this new personal and professional positioning shall gain scale through dissemination by the impacted individual.

This assessment is considered by the author as a management tool to measure and promote the educational framework created and implemented by ISAE in 2004. Extrapolating the local institutional context, the study can contribute to society and science by:

1. Offering a theoretical-practical reference on the evaluation of an educational model for teaching institutions, including PRME signatories;
2. Promoting the importance of global policies and the possible adherence by the formal educational environment, developing integration, convergence and global inter-institutional cooperation;
3. Offering examples to model courses and educational proposals combined between formal, non-formal and informal contexts, in order to disseminate the application of pedagogical engineering for ensuring consistency and appropriateness;
4. Disseminating a new system of learning for adults, following an experiential, transdisciplinary, transversal and self-educational conception;
5. Disseminating examples of practices that can be adopted by educational institutions responsible for training managers, so that they can promote sustainability, social responsibility and governance;
6. Founding a movement to seek innovative solutions for a more efficient and effective education, in order to advance the training of globally responsible leaders through extracurricular activities within education systems and companies;
7. Contribute to technical and scientific literature in the field of responsible management education and global leadership, with individual and collective actions, values and skills.

Finally, Perspectivaction aims at an improvement of the performance of managers within their organizations, by providing tools and knowledge for them to exert a transformative, collaborative, sustainable leadership, thus encouraging them as agents of change to inspire the necessary improvements in economy, society and the environment—and to advance the construction of a new civilization framework.

References

AEE (2009) A community of progressive educators and practitioner. Association for Experiential Education. http://www.aee.org/about/whatIsEE. Accessed 2 Sept 2009

Andresen L, Boudand D, Cohen R (2000) Experience-based learning. In: Foley G (ed) Understanding adult education and training, 2nd edn. Allen & Unwin, Sydney

Brandão CR (1986) O que é educação, 17th edn. Brasiliense, São Paulo

Busquets MD et al (1999) Temas transversais em educação: bases para uma formação integral. Ática, São Paulo

Dewey J (1971) Experiência e educação. Nacional, São Paulo

Delors J et al (2006) Educação: um tesouro a descobrir. UNESCO, MEC, 10th edn. Cortez, São Paulo

Meadows DH, Meadows D, Randers J, Behrens WW (1972) Limites do crescimento. Perspectiva, São Paulo

Menezes ET, Santos TH (2002) Dicionário interativo da educação brasileira. EducaBrasil, São Paulo, Midiamix. http://www.educabrasil.com.br/eb/dic/dicionario.asp?id=50. Accessed 12 Mar 2009

Morin E (2000) Os sete saberes necessários à educação do futuro, 3rd edn. Brasília Cortez, São Paulo (DF, UNESCO)

Morin E (2002) O método II: a vida da vida. Tradução de marina lobo, 2nd edn. Sulina, Porto Alegre

Mustaro PN (2003) Tecnologia crítica: pensando na integração dos professores em uma nova ação docente mediada pelas tecnologias. http://www.rau-tu.unicamp.br/nou-rau/ead/document/?view=23. Accessed 12 Mar 2009

Nicolescu B (2000) Um novo tipo de conhecimento: transdisciplinaridade. In: Educação e transdisciplinaridade, vol 1. Guarujá (SP, UNESCO)

Pineau GO (1999) Experiências de aprendizagem e histórias de vida. In: Carré P, Caspar P (eds) Tratado das ciências e das técnicas da formação. Tradução de Pedro Seixas. Instituto Piaget, Lisboa (Coleção horizontes pedagógicos)

PRME (2012) Inspirational guide for the implementation of PRME: placing sustainability at the heart of management education. GSE Research Limited, Leeds

United Nations Environment Program (UNEP) (2011) Caminhos para o desenvolvimento sustentável e a erradicação da pobreza—síntese para tomadores de decisão. http://www.unep.org/greeneconomy. Accessed 2 Oct 2014

Sachs J (2008) A riqueza de todos. Nova Fronteira, Rio de Janeiro

SEF—Secretaria de Educação Fundamental (1997) Parâmetros curriculares nacionais: apresentação dos temas transversais, 8th edn. MEC/SEF, Brasília

Srikant D, Garvin DA, Cullen PG (2011) Rethinking the MBA: business education at a crossroads. http://www.aacsb.edu/publications/enewsline/archives/2010/vol9-issue3-rethinking-mba.pdf. Accessed 24 Feb 2012

UNESCO—United Nations Educational, Scientific and Cultural Organization (2005) Década das nações unidas para o desenvolvimento sustentável—2005–2014. UNESCO, Brasília

UNDESA—United Nations Department of Economic and Social Affairs (2011) World economic and social survey 2011: the great green technological transformation. http://sustainabledevelopment.un.org/index.php?page=view&type=400&nr=208&menu=35. Accessed 21 Jan 2014

Author Biography

Norman de Paula Arruda Filho is an architect (1976) and Specialist in Urban Planning (1980) from Federal University of Rio Grande do Sul. Specialist in Public Administration from Getulio Vargas Foundation (1993). Master's Degree in Corporate and Public Management from the Brazilian School of Business and Public Administration, Getulio Vargas Foundation (EBAPE 2002). Ph.D. in Applied Business Management from the Higher Institute of Labor and Company Sciences, Portugal (ISCTE 2012). Current President of the Higher Institute of Administration and Economics (ISAE), licensed by Getulio Vargas Foundation. Board member and Coordinator of the Education Task Force of the Global Compact Brazilian Committee (CBPG). During his professional trajectory, he was Regional Coordinator of Urban Development Programs for Banco Nacional de Habitação (BNH—*National Housing Bank*), Superintendent of Planning for the Southern Region Superintendence of Development (SUDESUL), CEO of the Sanitation Company of Paraná, Director-General of the Paraná State Secretary of Urban Development, Regional Superintendent and Special Advisor for the Presidency of Caixa Econômica Federal, among other directing positions. A responsible-management-education enthusiast in Brazil, he heads the Brazil Chapter Secretariat from UN Principles for Responsible Management Education (PRME)—which since 2007 have guided the administration of more than 500 business schools around the world, including ISAE.

Student Perceptions on the First Masters in Education for Sustainable Development in Malta

Mark C. Mifsud

Abstract

The study is an in depth analysis of student perceptions on the first Master Course in Education for Sustainable Development in Malta. The course was carried out mainly through the use of lectures, fieldwork sessions and partly through an online virtual learning environment. The main aim of the course was to improve education for sustainable development knowledge, skills and values in educators and it mainly addressed local students from Malta through its contextualization. The study involved the use of mixed methodology, with questionnaires that focused on gathering the general picture and an in depth student evaluation written over a period of 2 weeks, with all the students to get deeper meaning and understanding. The reponses were categorised and a number of themes emerged. The research identifies a number of different areas which were the prime motivators for carrying out the masters course, while it also highlights areas which can be improved, changed or realigned in order to make the course more effective and operative. A 4D MESD model is presented as a means of visualizing the various polarised dimensions of the findings. Considering the focus on student perceptions on an ESD master degree, this paper will be valuable to students and lecturers that are interested in innovative methodologies and in developing or evaluating Education for sustainable development courses.

Keywords

Master in education for sustainable development • Mixed methods • 4D MESD model • Perceptions • Course analysis

M.C. Mifsud (✉)
Centre for Environmental Education and Research, University of Malta,
Msida MSD 1252, Malta
e-mail: mark.c.mifsud@um.edu.mt

W. Leal Filho (ed.), *Transformative Approaches to Sustainable Development at Universities*, World Sustainability Series, DOI 10.1007/978-3-319-08837-2_19

1 Introduction

The effectiveness or otherwise of education for sustainability courses at post-graduate level has never been studied in the Maltese context. This study is an in depth analysis of student perceptions on the first such course in Malta. The course was carried out mainly through the use of lectures, fieldwork sessions and partly through an online virtual learning environment. The main aim of the course was to improve education for sustainable development knowledge, skills and values in educators and it mainly addressed local students from Malta through its contextualization. The perception of environmental issues, attitude and knowledge of students following such a course is of great importance as knowing what students think about environmental issues will encourage pro-environmental education (Pawlowski 1996) and also because the involvement of students in decision-making and the implementation of environmental and developmental programmes is critical to the long-term success of Agenda 21 (UNCED 1992).

2 Background to the Study

2.1 The Maltese Environment

The Maltese islands are a small archipelago located in the centre of the Mediterranean. The Maltese archipelago comprises three inhabited islands—Malta, Gozo and Comino—with a number of uninhabited smaller islands. The most serious environmental problems arise from the fact that Malta is one of the smallest states in the world and one of the most densely populated. The high population density is augmented further by high tourist arrivals of about 1.2 million yearly (Mallia et al. 2002). The main environmental issues on the island include waste production and management, reliance on fossil fuels for energy production, high private motor usage and freshwater production requiring a high energy input.

2.2 Local Environmental Education Development

Environmental education in Malta started off with environmental NGOs in the 1960s, mainly in response to the major environmental issues on the island. The awareness raising campaign was a very long affair, hindered in its development by a number of factors that lead to the slow growth of environmental education in the islands (Mifsud 2009). These factors include:

- The highly competitive educational system;
- The non-committal policy of the government;
- The colonial mentality.

In the formal education sector one of the major problems is the dominant educational ideology that suppresses creativity and rewards rehearsed words in examinations. Individual teachers who are interested in the environment may attempt to highlight environmental education, but the majority of teachers are hampered with little timetable time and lack of locally produced resources. Some materials currently used in schools were produced for use in other countries, and their transferability in the local context should be cautiously studied (Mifsud 2012).

With regards to the local government, it is apparent that the Maltese government mainly funds projects that have short term goals. Additionally, the lack of a clear national policy on environmental education has resulted in a waste of human resources and the shelving of long-term environmental education initiatives. Due to the fact that Malta was a colony for a long time, a number of anthropologists have indicated that the Maltese people seem to have a problem realising they own the island itself and its environment (Boissevain 1990). Subsequently, they do not attach much importance to the outside environment. Furthermore, although the Maltese native language is taught and understood by everyone, the English language is still considered to be more prestigious. The situation started to improve mainly in response to the requirements imposed on the country by the European Union. The government is now realising that the environmental education is an effective and long-term solution to ensure environmental sustainability. Many government officials speak of their commitment towards sustainability, but their concern is mainly short-term (Mifsud 2011).

2.3 The University of Malta and Environmental Education

The Centre for Environmental Education Research (CEER) was set up in 2004 with the intention of acting as a centre of excellence for EE research in the Mediterranean. The Centre for Environmental Education and Research seeks to catalyse change towards a sustainable society by providing opportunities for environmental education that empower citizens, irrespective of age, gender and socio-economic status, to actively participate in environmental decision making fora and in initiatives that promote a good quality of life.

CEER has been promised funds and a building, but, these took a long time to be forthcoming, mainly due to the 'bureaucratic government system' which is prevalent on the island. Nonetheless, CEER has now managed to launch the first masters in education for sustainable development in Malta. This 3 year course targets teachers and education experts and aims to provide different perspectives of sustainable development, derived from the interaction of different environmental, societal and economic concerns (Mifsud 2011). This course and its first participants are the focus the present study.

2.4 The MESD Structure

The Masters in Education for Sustainable Development aims to present students with: different perspectives of (i) the environment, (ii) environmental education/ education for sustainable development, (iii) the interaction between the environment and society, and (iv) sustainable development. The course is structured (see Fig. 1) through a philosophy that will enable students to study issues related to ESD in practice and 'in situ' in different environmental realities, and to experience different environmental, social, cultural, political and educational perspectives. Additionally, the masters helps students to access and critically evaluate ESD research and to develop the skills and attitudes necessary to promote sustainable lifestyles.

3 The Study

The study involved the use of mixed methodology, with questionnaires that focused on gathering the general socio-demographic student picture and an in depth student evaluation written over a period of 2 weeks, with all the students to get deeper meaning and understanding. The reponses were categoried and a number of themes emerged. The questionnaire and in depth questions were constructed following the examination of other instruments in the literature (e.g., Mifsud 2010; Barrett and Kuroda 2002; Eagles and Demare 1999; Gambro and Switzky 1996, 1999; Hodgkinson and Innes 2001; Kuhlemeier et al. 1999; Makki et al. 2003; Mogenson and Nielsen 2001; Fien et al. 2000; Pawlowski 1996). The research identifies a number of different areas which were the prime motivators for carrying out the masters course, while it also highlights areas which can be improved, changed or realigned in order to make the course more effective and operative.

4 Outline of the Methodology

There were eight participants in this study which represent the whole cohort of the first Master in Education for Sustainable Development course in Malta. The participants were forwarded a simple questionnaire to locate their socio-geographical characteristics, and eventually were given in depth student evaluation diary in order to write their feelings, perceptions and ideas about the course over a period of 2 weeks. This amount of time gave the students time to reflect on any changes of thought or behaviour that happened during the timescale of the course. A number of short questions were inserted inside the diary to act as catalysers for the students. The in depth evaluation was given to the students towards the end of the course, to ensure that they were exposed to all the course units. The structure of the personal questions was a follows:

Study Units in the Programme:

Year 1

- *Different Paradigms of Development*
- *Principles of Education for Sustainable Development*
- *Transformative Pedagogies*
- Humans, Society and the Environment
- *Education for Sustainable Development* – Law and Policy

Year 2

- Local Agenda 21 Strategies
- Research Methods in Education *for Sustainable Development*
- Critical Thinking and Problem Solving Skills in Education *for Sustainable Development*
- Education *for Sustainable Development for* Adults, Youth and Community
- Environmental Education and Communication
- Education *for Sustainable Development* Research Seminars

Year 3

- Dissertation

Fig. 1 The structure of the MESD degree

1. Why did I do the course? What were my motivations?
2. Which were the main strengths and weaknesses of the course?
3. What is my overall opinion of the course?
4. Was the course structure and methodology suitable?
5. Did the course effect my knowledge, attitudes or behaviour?
6. Did I learn any skills during the course?
7. Would I change anything from the course?

The analysis of the in depth student evaluation diary was carried out following the 2 weeks afforded to students for data completion. The qualitative data analysis was based on an adaptation of the approach identified by Vaughn et al. (1996) who suggested four processes:

1. Identifying the big ideas—the participants' words and ideas and the intensity with which the participants responded provide an initial framework;
2. Unitising the data—identifying units of information from the text that will later become the basis for forming themes;
3. Categorising the units—bring together the units of data identified above that are related to the same content;
4. Identifying the themes—after considering the big ideas and the categorisation of the units the main themes are identified and refined.

The selection of the 'quotes' to include was informed by a professional judgement of what was important and significant for the context of the study. This, in turn, was based on the considerations of trustworthiness and transparency in qualitative studies, insights gained from the literature and the existing knowledge of environmental education in Malta and the Maltese environmental-political-educational context. One of the limitations of the study includes the very small sample size, that render the results non generalizable but rather only applicable to this case study. Other limitations include the issue of time management and access to participants due to the study being carried out after the course finished.

Table 1 illustrates the MESD students' backgrounds. A number of trends emerge:

- The majority of students have a non-science background.
- The main entry qualification is a first degree related to education.
- Most student's mothers were housewives.
- The students geographic origin was spread across the islands, but a peak was registered in the rather more polluted South region.

Table 1 Socio demographic characteristics of MESD students

Student	Age	Gender	Religion	Qualifications	Father's work	Mother's work	Status	Geographic location
1	40	Female	Catholic	B.Ed	Teacher	Housewife	Married	North
2	38	Male	Agnostic	BA. Sociology and PGCE	Project manager	Teacher	Single	Centre
3	26	Male	Catholic	BA., Geography	Retired	Housewife	Single	Centre
4	38	Male	None	B.Sc. Physics/maths	Secondary Education level	Secondary education level	Married	North
5	31	Female	None	B.Ed	Assistant director	Housewife	Single	South
6	38	Female	Catholic	B.Com/PGCE	Pensioner	Housewife	Married	South
7	26	Female	Catholic	B.A/Maltese PGCE	Soldier	Housewife	Single	Gozo
8	42	Female	None	B.Ed	N/A	Housewife	Single	South

5 The Findings

5.1 Results of the Findings from the Quantitative Analyses

No previous study had ever been made on students following a Master in Education for Sustainable Development in Malta. The present study was of a mixed design. A short questionnaire with a number of course parameters was designed in order to give an overview of the course and to act as a precursor to the more extensive qualitative analysis. The quantitative results show a generally positive outlook from students in most course dimensions which were examined. Nonetheless, it is clear that students did find that the course required to much effort (87.5 %) and that group-work especially was not very effective (37.5 %). This was a clear weakness in the course which was also corroborated in the qualitative analysis. Conversely, students appear to be very satisfied with both feedback (87.5 %) and tutor interaction throughout the course (87.5 %). Other dimensions had an average response throughout the continuum, such as; course difficulty, supervision and the suitability of teaching environment. An illustrative set of quantitative findings from the study are found in Table 2.

5.2 Results of the Findings from the Qualitative Analyses

The present study has brought to the surface quite a range of views regarding the course, its effects and the Maltese environment in general. With no direct influence from the researcher, the participants introduced several broad themes, the main points of which are summarised below.

Main Themes

- Motivation for doing the course
- Personal Changes
- Behavior and attitude modification
- Course Dynamics
- Significant Life Experience
- Strengths and Weaknesses
- The Future.

5.2.1 Motivations for Doing the Course

Participants had a number of different motivations for doing the course, both from a personal, work related perspective and also from a more beneficial community wide perspective:

> First and foremost, I did the course because I wanted to further enhance my education. At the same time I had been looking to start a Masters but most did not interest me. The Masters in ESD on the other hand caught my attention because it was something that interested me and something that I wanted to learn about (Student 1).

> It was a time when I wanted to continue studying at the master level. Since the course is closely related to my first degree, I thought it would help me in my career and my main expectations were to strengthen my views as regards the environment and how to approach it. Obviously, my expectations once this course will finish is that I apply more the knowledge learned throughout the course whenever the need arises (Student 3).

One student mentioned that the subject has been an interest to her from a very young age and it was only a matter of time until she applied for the course. It appears that her decision was made long before.

Table 2 Quantitative results

	Strongly agree (%)	Agree (%)	Not certain (%)	Disagree (%)	Strongly disagree (%)
The course covers the outlined scope, aims and objectives	75	12.5	12.5	0	0
The course as a whole was well-structured	75	25	0	0	0
The intellectual demands made on you were appropriate for the level of the course	62.5	25	12	0	0
Were you satisfied with your own level of preparation and participation	75	25	0	0	0
The course materials were adequate	62.5	12	12.5	0	0
The feedback was adequate	87.5	12.5	0	0	0
Teaching methods were adequate	62.5	12.5	12.5	12.5	0
The availability of reading materials listed in the bibliography was reasonable accessible	62.5	12.5	12.5	12.5	0
The course was interesting	75	25	0	0	0
The course was difficult	50	12.5	12.5	12.5	12.5
The course required too much effort	87.5	12.5	0	0	0
Adequate support was provided	75	25	0	0	0
The Teaching environment was adequate	62.5	12.5	12.5	12.5	0
Assignment supervision was adequate.	62.5	12.5	12.5	12.5	0
Tutor interaction during feedback was adequate	87.5	12.5	0	0	0
Group work was effective	37.5	25	12.5	12.5	12.5
Individual work was effective	62.5	12.5	12.5	12.5	0

> From a very young age I have been captivated by my surrounding environment. This inner curiosity led me to explore it through my eyes as a child throughout the scholastic years. As I grew older I found myself searching on the web to understand certain topics that I had read about. Thus, it was an easy decision to opt for this course as it was a subject of interest to me (Student 7).

5.2.2 Personal Changes

Many students reported a variety of different personal changes in their belief systems that occurred throughout the course and because of the course. Some students were very vague in and did not manage to be concrete in their discussions while others were very focused and named a number of changes individually.

> I definitely think that there has been a change in me. My way of thinking has definitely gone through a big change, and I feel that now I can be more objective on certain environmental issues (Student 1).

> I believe that the course helped me in being more empathic with people and in respecting their opinion even if this might be very different from mine and totally against sustainable development principles (Student 4).

> I think that the course helped me to improve my communication skills and in expressing my feeling and concern to others (Student 6).

5.2.3 Behavior and Attitude Modification

Most students reported a number of modifications that occurred to their attitudes and behaviours throughout the course. The majority also acknowledged a positive increase in environmental knowledge.

> The course certainly had an effect on my behaviour towards a lot of current environmental, social, economic and cultural issues. There were things that I used to do previously at the start of the course, which I used to do automatically (Student 1).

> Though most of my attitudes were there before the course, the course helped in strengthening and possibly provide deeper reason to why they are there (Student 5).

> First of all I started to make an effort to Comprehend all perspectives brought forward regarding SD issues and acknowledge their value. When reflecting on certain topics I began to analyze them in a holistic way and now I am aware that for SD to take place one has to take into account the 4 SD pillars. These led me to acknowledge my responsibility both as a Maltese citizen and as a global citizen and also made me believe that I can be an agent of change even with my limitations (Student 7).

5.2.4 Course Dynamics

A significant issue raised by the participants in their in depth writings was the examination and critical analysis of the course structure and its dynamics. There were a number of different viewpoints ranging from the underlying course philosophy to the actual deliver of individual units.

> I believe that formative assessment is useful as on looking back I find myself bewildered by the work I managed to produce and the knowledge acquired throughout its built up. However I believe that in order to assess students more justly, transformative assessment should also be included (Student 7).

> When it comes to course structure: in my opinion, the sequence of some units could have been different. There were some units that could have been done earlier, in the beginning of the course, rather than later on—for example research methodologies (Student 1).

> Best methodologies employed was when the lecturers fostered debates which led us to reflect on our own experiences and make sense out of them. The challenging of assumptions lead to set out the learning process (Student 5).

5.2.5 Significant Life Experiences

All students gave a synoptic overall personal lived experience of the course, both in terms of its delivery, its effects and its shortcomings. Some students also made a number of suggestions based on their experience as postgraduate students of the said course.

> My overall opinion of the whole course is definitely a good one. The course is certainly an interesting one and should definitely be continued and maybe be more promoted to encourage more further student, because it is certainly a course that would be of interest to a wide spectrum of students coming from all walks of life (Student 1).

> Sometimes I find myself trying to convince people or educate them in different ways—even by just exposing them to realities they never had thought of. I feel part of a cause (Student 2).

> It reinforced my attitudes towards the environment—more respect towards the environment. This course instilled in me more social responsibility towards the environment through little things that we do every day (Student 3).

> Though it was very tough and demanding, I am very happy with my experience and the quality of the course in general. I really value the fact that it expanded my horizon and my world view on a lot of issues. The only grey cloud was that the majority of us did the same job for a living, thus slightly limiting the diversity of backgrounds from where we came (Student 5).

5.2.6 Strengths and Weaknesses

Students mentioned a large number of areas which were considered to be strengths and others that were considered to be weaknesses. The following are extracts that illustrate the wide ranging weak and strong areas as identified by the participants:

Weak Areas

- Sustainable Development as a subject is too vast;
- Some overlapping content in some units;
- Not enough time for debate;
- Methodology used during lectures was not always engaging;
- Overall there were no irrelevant study units – but some units seemed to overlap in content;
- Lack of group-work in assigned work;
- More outdoor experiences.

Strong Areas

- Course coordinators very helpful and understanding;
- Great team where it comes to both the students and the lecturers;
- Course material very interesting;
- Lecturers availability;
- Small groups of students;
- Lecturers different and practical experiences of ESD;
- A number of units were thought provoking and challenged some pre conceived ideas that I had.

5.2.7 The Future

When discussing their opinion on the future some students focused on discussing the course future, others discussed the future of the area as an academic subject while others discussed the impact of such a degree and the expertise it brings to the wider community.

> I hope that CEER plans to carry out seminars and refresher courses for us students in order to keep up to date and in touch on what is going on in ESD. This could prove to be a valuable opportunity for us ESD educators to meet and share experiences, good practices and concerns that we might have (Student 4).

> In the future, it is of utmost importance of keeping a balance between economic, social, environmental and cultural development (Student 5).

> Sustainable development is a lengthy process and one cannot think that it can be achieved overnight. Such thinking usually leads to people being disheartened to loose faith in what they are doing (Student 6).

> I suggest that local issues should be explored more deeply throughout the course by giving students the opportunities to have more practical research sessions (Student 7).

5.3 The 4D MESD Perspectives Model

Informed through the findings of the present qualitative research, a model of the course perspectives, the 4D MESD Perspectives Model is being proposed. It has to be understood that this model is not being proposed as a universal model but a localised model on a particular cohort of people that is: Students following the MESd course. From the examination of the in depth student writings, four distinct axes of perspectives started to emerge with regard to the course standing:

- Extrinsic Motivation – Intrinsic Motivation
- Dynamic Path – Static Path
- Collective Benefit – Personal Benefit
- Attitudinal Change – Behavioral Change

Extrinsic Motivation: Intrinsic Motivation Some participants explained that the main reason for doing the MESD was for the possibility of getting a better job in the future, while others said that there main interest was in the subject of sustainability. Other student motivations included both intrinsic and extrinsic motivators, and a continuum line can be designed to accommodate all the student responses from the purely intrinsic up to the purely extrinsic motivators.

> I did the course mostly to learn about areas that I was already reading about. I expected to become more knowledgeable in these areas and now I'd like to use them for the best, preferably by finding a related job (Student 2).

> I expected that through this course I would expand my knowledge about sustainable development issues ad challenges which, in turn will help me to be a better citizen who acts accordingly in the best interest for a sustainable world (Student 4).

> I was motivated to do the course due to a personal interest in ecology and preserving the natural environment. Moreover, I wanted to pursue my studies is something that interested me personally rather than just for better job prospects since I like my present job. However, obviously if job opportunity in line with ESD arises in the future I will highly consider it (Student 5).

Dynamic Path: Static Path Participants appear to be interested to various degrees in change. Some have a genuine interest in each other and to further improve all aspects of activities, and have a genuine interest in 'saving the planet'. These participants appear to be active or even pro-active with regard to the events unfolding before them and argued in favour of their own possible contribution to increase environmental awareness in the family, at school and at work in the foreseeable future. On the other hand, some participants think that 'things are not that wrong' as they are currently, make vague remarks to maintain the status quo', and they appear to be somewhat passive to the events unfolding before them.

> I believe that I had the experience to study with a very good group of colleagues. Most of the time we shared documentation and paper needed for credits in a true spirit of a collaborative approach. If group dynamics weren't so good, I feel I would have missed out of some of the benefits of this experience (Student 4).

> For a sustainable world everyone has a role to play and thus education for all is crucial for reaching sustainability (Student 7).

> When I applied for the this course, I wanted to learn more about ESD, an to grow on a personal level, both spiritually and academically. I don't know what to expect now really... I feel more confident in the subject and im able to pass on the values, skills and knowledge, I gathered throughout this course to others and to my pupils at school. One of my main concerns will be passing on the values in an unofficial way to my colleagues at work since I feel they lack the skills, knowledge and beliefs to handle ESD (Student 5).

Collective Benefit: Personal Benefit Some of the participants, to various degrees, view the course as endowing a number of personal benefits to the individual, such as an improved critical mind and improved job prospects. Nonetheless, a number of students have a much wider view of the course benefits and put the community as the main beneficiary of the activity.

> Yes, because I really enjoyed myself doing the new job and basically instead of using the course to move forward in terms of promotion I am using what I have learnt for self-improvement and to be in a better position to offer a better service where I am needed (Student 4).

> The course equipped me with a critical mind in tackling sustainable development issues (Student 7).

Attitudinal Change: Behavioral Change Although all students stated that the course increased their knowledge in a number of different environmental dimensions, it is clear that not all students recorded changes in behavior. Actually some students reported changes in their attitudes only, while others reported changes in behavior only. Only a few students reported changes in both dimensions.

> Nowadays I reflect much longer and consider more things before making any type of decision. I feel more responsibility in my behaviour (Student 2).

> I now believe that small changes can bring large changes in the future (Student 8).

The majority of perspectives appear to occupy various points created by the intersection of these four continuum lines, around a transition area between the main polar views. No attempt has been made in this research to statistically analyse the responses in order to pinpoint various locations on the model that would represent groups of students, but rather the model represents the confines of all the perceptions that where discovered (Fig. 2).

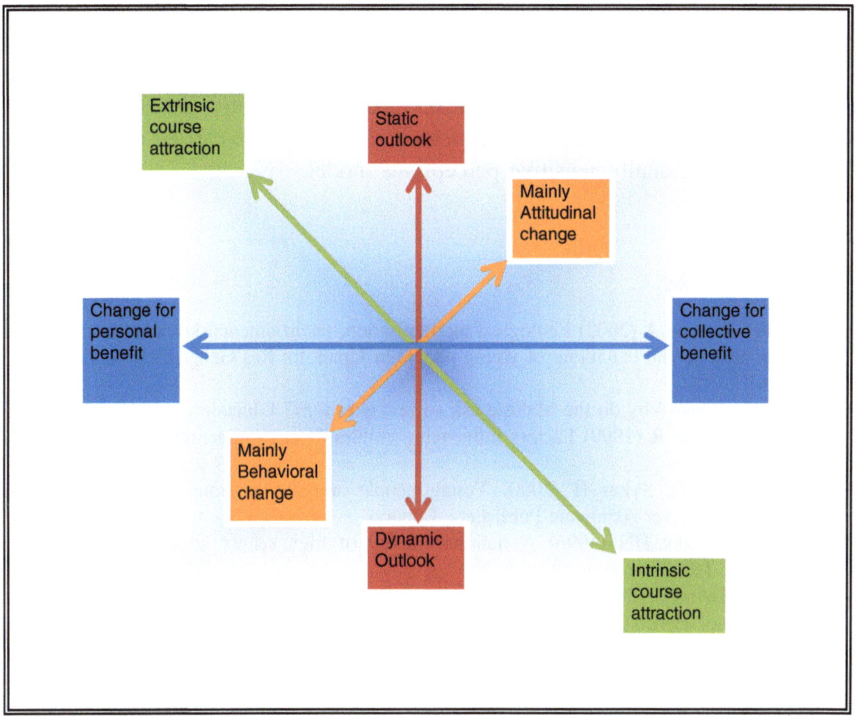

Fig. 2 The 4D MESD perspectives model

6 Conclusions: A Way Forward

The research has identified a number different areas of student perceptions with regard to the master course. These areas include; 'Motivation for doing the course', 'Behavior and Attitude Modification' and 'Significant Life Experience' amongst others. Additionally a number of socio-demographic trends were identified from the quantitative analysis, such as that the majority of students had a non-science background and that a majority of students' geographic origin was from the rather more polluted South region. Following further analysis, the findings were combined in a 4D MESD model as a means of visualizing the various polarised dimensions of the findings. Through the study a number of suggestions for further course improvement emerged. The suggestions arise from the examination of the data and the feedback given by the students and include the 'Increase in group work activities in the course', 'the sensitisation on the importance of collective behavior through improved course structure', the facilitatation of interdisciplinary work and the exchange of expertise on sustainable development both within the course and in the local community, and by means of outreach to other communities, nationally and internationally' and 'the production of educational materials specific to sustainable development as individuals and as groups'. All these suggestions should examined for incorporation in the next course in order to increase effectiveness. It is suggested that more research is carried out on future student cohorts of the MESD to test and further improve the 4D MESD model. Comparative research with other similar masters in the Mediterranean region should also be carried out in order to build a more regionally sensitive perceptions model.

References

Barrett BFD, Kuroda A (2002) Ecological modernisation, environmental knowledge and societal change: attitude and behaviour of young people in Japan. Int Res Geogr Environ Educ (11)3: 237–261

Boissevain J (1990) Why do the Maltese ask so few questions? Education 3(4):16–18

Eagles PFJ, Demare R (1999) Factors influencing children's environmental attitudes. J Environ Educ 30:33

Fien J, Yencken D, Sykes H (2000) Young people and the environment: an Asia Pacific Perspective. Kluwer Academic Publishers, London

Gambro JS, Switzky HN (1996) A national survey of high school students' environmental knowledge. J Environ Educ (27)3:6,28

Gambro JS, Switzky HN (1999) Variables associated with American high school students knowledge of environmental issues related to energy and pollution. J Environ Educ (30)2: 15–22

Hodgkinson SP, Innes JM (2001) The attitudinal influence of career orientation in 1st year university students: environmental attitudes as a function of degree choice. J Environ Educ (32) 3:37–40

Kuhlemeier H, Huub VDB, Nijs L (1999) Environmental knowledge, attitudes and behaviour in Dutch secondary education. J Environ Educ (30)2:4, 11

Makki MH, Abd-El-Khalick F, Boujaoude S (2003) Lebanese secondary school students' environmental knowledge and attitudes. Environ Educ Res (9)121–33

Mallia A, Briguglio M, Ellul AE, Formosa S (2002) Physical background, demography, tourism, mineral resources and land use. In: State of the environment report for Malta, 2002. Ministry for Home Affairs and the Environment, Malta

Mifsud M (2012) A contextual study of the events that have shaped the development of environmental education in Malta. J Teacher Educ Sustain 12(2)110–128. ISBN: 978-960-466-060-5, ISSN: 1790-3859

Mifsud M (2011) Factors influencing environmental knowledge of Maltese youth. Environ Sci 20 (2). ISSN-09668411

Mifsud M (2010) Focus groups effectiveness in studying youth environmental behaviour. In: Proceedings of the 6th international conference on education. ISBN: 978-960-466-060-5, ISSN: 1790-3859

Mifsud M (2009) A critical review of the global events that have shaped the development of education for sustainable development. Environ Sci 18(1). ISSN-09668411

Mogenson F, Nielsen K (2001) Students' knowledge about environmental matters and their belief in their own action possibilities. J Environ Educ 33(1)3, 33

Pawlowski A (1996) Perception of environmental problems by young people in Poland. Environ Educ Res 2(3):279–285

UNCED (1992) Agenda 21: programme of action for sustainable development. Rio Declaration on Environment and Development. NY, United Nations
Vaughn S, Schumm J, Sinagub J (1996) Focus group interviews in education and psychology. SAGE Publications, Thousand Oaks

Author Biography

Mark C. Mifsud is a Lecturer at the Centre for Environmental Education Research at the University of Malta. He is a Chartered Biologist, a Chartered Environmental Scientist, a Chartered Science Teacher, a Certified Wildlife Photographer and a Prince II Certified Practitioner with more than 20 years experience in Environmental Sciences, Environmental Education and Sustainable Development. Mark is presently the coordinator of the Masters in Education for Sustainable Development and his main areas of research include youth and the modelling of environmental knowledge.

Part II
Integrative Approaches in Teaching and Learning

eLearning and the Sustainable Campus

Brian Stewart and Anshuman Khare

Abstract

eLearning has been commonly accepted as a comparatively cost effective and environmentally friendly educational delivery technology. On the other hand, however, agreement as to its efficacy, quality and appropriateness for the social development and maturing of young adults, among other issues do not yield to a similar consensus. Following Brundtland's model of sustainable development, this paper employs the Sustainability Circle Framework developed by the Global Compact Cities Programme and applies it to eLearning to provide novel and beneficial insights into the critical factors for the on-going sustainability of eLearning. The paper analyses eLearning with regard to the four domains of ecology, economy, culture and politics, providing a comprehensive and rounded perspective of the critical factors that enable eLearning to be sustainable. In addition the findings will provide input into on-going discussions regarding the adoption of eLearning by different educational sectors, within differing disciplines and across differing economic and cultural regions. Achieving an improved understanding of educational sustainability drivers has potential to not only reduce the ecological footprint of educational institutions but also to ensure that such reductions become systemic and endemic into the fabric of the institutions.

B. Stewart
University of Alberta, 323 General Services Building, Edmonton, AB T6G 2H1, Canada

A. Khare (✉)
Faculty of Business, Athabasca University, 301-22 Sir Winston Churchill Avenue, St. Albert, AB T8N 1B4, Canada
e-mail: anshuman.khare@fb.athabascau.ca

W. Leal Filho (ed.), *Transformative Approaches to Sustainable Development at Universities*, World Sustainability Series, DOI 10.1007/978-3-319-08837-2_20

Keywords
eLearning · Sustainability circle framework · Sustainable campus

1 Introduction and Background

1.1 eLearning and Sustainability

The increased interest in eLearning is reflective of the increasing use of technology in society and the affordances it provides to teaching and learning. This has been overlaid with the apparent systemic unsustainability of the Higher Education sector, which is being revealed at an increasingly rapid rate (Christensen and Horn 2008; Christensen and Eyring 2011). The initial response of increasing revenues through expanded enrolment efforts, increasing tuition rates, private donations and government funding have provided a respite but not a solution to higher educational sustainability. eLearning is now seen as a technological solution to at least part of this problem, as it has grown to become a recognized component of formal education provision mainly represented in hybrid or blended courses (Mayadad et al. 2009).

With respect to the ecological aspect eLearning has been shown to provide a significant reduction in carbon emissions as compared to traditional campus based delivery. The study of the Open University UK, Roy et al. (2008) demonstrate the size of the differential of between 60 and 80; which was also found in operational research at Athabasca University in Canada where CO_2 emissions were estimated to be approximately 30 % of campus based institutions (Stewart and Khare 2012). While the exact value of CO_2 emissions is open to question and further elucidation, the order of magnitude is such that further research will be unlikely to erode the differential to any significant degree. This being the case, the adoption of eLearning as a mode of broad based education can be held to represent a potentially significant tool in the reduction of the ecological impacts of education.

This paper does not pursue the argument of eLearning as being equal to or superior to campus based education, nor does it suggest that eLearning should be adopted as a singular mode of educational delivery. Rather is takes the position that eLearning is a proven effective mode for teaching and learning and that it offers significant improvements to the sustainability of education as currently considered in the triple bottom line (Hendrique and Richardson 2004; Adams 2006). The paper instead focuses on understanding the nature of the sustainability of eLearning itself, and seeks through the application of an existing framework into a different context, to determine a comprehensive method of analysis that provides critical insights into the sustainability of eLearning initiatives. Following the concept developed by Suhonen and Suitenen (2014) of expanding Brundtland's (1987) three perspectives by the addition of an ethical dimension, we seek to evaluate the framework of Circles of Sustainability (CS) developed and used for cities, to assess its applicability and whether it offers a useful conceptual framework to the eLearning context (UN Global Compact Cities Programme 2013). The benefit of such a framework wouldbe to significantly improve the development, implementation and operation of sustainable eLearning initiatives by unifying existing disparate frameworks and knowledge sets. In addition it contributes to the growing bodies of knowledge around campus sustainability and the incorporation of eLearning into higher education.

1.2 Approaches to Understanding Sustainable eLearning

The use of the term eLearning has come to represent many varied interpretations as has been observed by Fournier (2006). Common understanding has also been heavily accented by its use in the distance and distributed learning fields, although this is lessening as blended or hybrid learning continues a steady adoption. Given its broad capability set and to provide as straightforward a description as possible for the purposes of this paper the following definition from JISC (2014) will be used "learning facilitated and supported through the use of information and communications technology."

1.3 The Need for an eLearning Framework

Conceptual frameworks consist of a set of wide-ranging ideas, principles and constructs that are relevant to a particular field of enquiry that are used to assist in the framing of research, analysis and presentation (Reichel and Ramey 1987). When used appropriately they provide structure and coherence to enable the understanding and interpretation of an accepted and expected knowledge domain, facilitating, communication, relevance, scope definition, and helps establish the context of research and analysis in an area.

In the more applied realm the need for a framework is exemplified by a major initiative to enhance the ecological sustainability of American campuses, reflective of the growing concern over climate change. The American University's and College President's Climate Commitment provides guidance and coordination to campuses to plan, execute and peer review their progress on climate initiatives. The site however does not provide a framework to facilitate a common approach for the participating institutions that would allow for reasonable comparison of activities, approaches and progress.

Bates (2010) suggests that eLearning provision of education is being seen as a partial remedy to the funding crisis facing third level institutions particularly in the US and Canada, exacerbated by the fiscal crisis of 2008. This is demonstrated by eLearning increasingly gaining new adherents over the last decade. In the U.S. the number of students enrolled at a degree-granting postsecondary institution in at least one online course has grown from 1.6 million in 2002 to 7.1 million in 2012. This is greater than a fourfold increase, and compares to an approximate 30 % increase in total enrolment over the same period (The Sloan Consortium 2013).

Thus the requirement for a conceptual framework is reflective of the lack of a coherent and complete approach, the growing use and importance of eLearning and of increased ecological awareness.

2 Developing a Framework

The Brundtland Commission's work for sustainable development initially proposed three perspectives to consider sustainability, Economic, Ecological and Social (Brundtland 1987). Although there still remains the requirement to develop a single universal definition of sustainability for eLearning, Stepanyan et al. (2013), Suhonen and Suitenen (2014) point out the lack of an ethical dimension in the three perspectives with respect to the sustainability of an interdisciplinary online PhD programme at the University of Eastern Finland. Stepanyan et al. (2013) conducted a scoping review to assess the range and nature of research on eLearning Sustainability. This study suggested three domains for categorising and mapping existing research articles in the field:

- Resource management (9 papers), focusing on the cost of eLearning, including studies of the strategies and approaches adopted by institutions to improve the effectiveness of human another resources.
- Educational attainment (12 papers), focusing on measures of student achievement, retention rates, skill acquisition, and personal development with emerging themes including evidence of benefits, perceptions of quality, usability of new technologies, and student performance.
- Professional development and innovation (11 papers), focused on strategies for adapting to change. Emerging themes within the domain include teacher training and development, institutional transformation, and educational leadership.

While economic, cultural and political elements are represented in the domains ecological elements are lacking as the studies focus on the institutional adoption, implementation, development and operation of eLearning services.

Robertson (2008) defines sustainable eLearning as eLearning that has become normative in meeting the needs of the present and future. Given the vulnerability of eLearning to disruption this definition may prove too restrictive in practice. This study used activity analysis, which provides a common lexicon to describe the organisational, technological and pedagogic perspectives in terms of subjects, tools, object and outcome, rules, community and division of labour. One of the strength of Activity Theory is the inclusiveness of its approach, particularly relevant in this emerging and expanding field of eLearning. Table 1 shows clear links with the CS framework. For example the three activity systems outlined organisational, technological and pedagogical system can be seen to be complimentary to the economic, political and social components of the CS framework. Indeed they would provide a methodology for further articulating these areas with regard to its

Table 1 Circles of sustainability—summary of the structure of the urban profile process

Domains	Perspective (or subdomains)	Possible issues
Ecology	• Materials and energy • Water and air • Emission and waste • Transit • Flora and fauna • Habitat and settlement • Embodiment and food	• Sources of energy including petroleum, food • Air quality; climate change and adaptation • Parks and gardens; tree coverage • Habitat destruction; land-use • Urban spatial development, housing • Physical human health, nutrition • Pollution, recycling and waste disposal
Economics	• Production and resourcing • Exchange and transfer • Accounting and regulation • Consumption and use • Technology and infrastructure • Labour and welfare • Wealth and distribution	• Industry and commerce; resources • Money; trade in goods and services • Regulatory systems • Consumption patterns; use of goods • Labour markets; economic provision • High-tech to low-level technologies • Poverty; unemployment; slums; inequality
Politics	• Organisation and governance • Law and justice • Ethics and accountability • Dialogue and reconciliation • Security and accord • Representation and negotiation • Communication and critique	• Legitimacy, current system of governance • Legal system; political justice and order • The press, media, news; dissent and protest • Participation by citizens, voting; civility • Political tensions; military presence • Customary rights; truth commissions • Corruption issues; public ethics
Culture	• Identify and engagement • Memory and projection • Belief and ideas • Enquiry and learning • Health and wellbeing • Creativity and recreation • Gender and generations	• Ethnicities; identities; public engagement • Celebrations; events and rituals, sport • Indigenous history; museums; monuments • Religions and spiritualties; ideologies • Gender relations; family life; generations • Education and training systems • Health and medical systems; mental health

UN Global Compact Cities Programme 2013

adaptation to eLearning. The ecological element is again not explicit, although it can be inferred in the identified tools instruments and mediating artefacts, being less impactful than other traditional types. Indeed it can be held that the literature on eLearning sustainability generally focuses on the social, political and economic domains and omits the ecological aspects of the mode, seeking to provide a conceptual framework within the context of the educational milieu, as opposed to the more holistic context of the entire operating ecosystem (Gunn 2010).

This is not unexpected and can be seen to be an entirely pragmatic approach as eLearning has generally been seen as an interesting experiment or supplement for specific populations, of inferior quality to the more accepted form of campus based delivery. For eLearning to be sustainable it must be effective at its core purpose, that of teaching and learning, it must also be accepted within all the communities it serves as such. Thus eLearning must offer a value proposition that is both sustainable and sufficiently desirable to overcome the initial barriers to adoption and change effort.

This value proposition has generally been seen to be based on economic or social rationale, through cost of provision reductions or access to underserviced populations. Stepanyan et al. (2013) argue that often the main purpose of eLearning initiatives is cost-effectiveness for reducing the costs of education or increasing the profits. It has yet to be broadly accepted as pedagogically relevant and equivalent to campus based delivery despite increasing evidence to the contrary (Mayadad et al. 2009; US Department of Education 2010).

2.1 Circles of Sustainability

The CS framework arose out of a joint project of the UN Global Compact, Cities Programme, and Metropolis/United Cities and Local Governments (UCLG) to develop a methodologically robust 'sustainability assessment process' for understanding urban regions. The method has been refined through projects in Melbourne and Milwaukee, and through an ARC-funded cross-disciplinary project that partnered with various organizations including Microsoft Australia, Fuji Xerox Australia, the City of Melbourne, World Vision, UN-Habitat and Metropolis. The process is outlined in James and Scerri (2011) and the Global Compact Cities Programme (UN Global Compact Cities Programme 2013).

The approach of CS provides a potential framework for developing deeper understanding of the factors of eLearning sustainability and in particular their interrelationship with the both micro and macro sustainability influences.

The framework consists of four domains Ecology, Economics, Political and Social.

- Ecology: The ecological domain is defined in terms of the intersection between the social and the natural, focussing on the important dimension of human engagement with and within nature.
- Economics: The economic domain is defined in terms of activities associated with the production, use, movement, and management of resources, where the concept of 'resources' is used in the broadest sense of that word.
- Culture: The cultural domain is defined in terms of practices, discourses, and material expressions, which, over time, express continuities and discontinuities, and commonalities and differentiations, of meaning.
- Politics: The political domain is defined in terms of practices of authorization, legitimation and regulation, where the parameters of this area extend beyond the conventional sense of politics as concerning the state to include not only issues of public and private governance but also basic issues of power.

Each of the four domains is broken into seven perspectives or subdomains and these are assessed through the identification of issues for consideration.

Table 1 is adapted from the CS Profile Process tool and shows the definition of the domains into subdomains and then into possible issues for evaluation (UN Global Compact Cities Programme 2013).

2.2 Using the Framework

The methodology of the framework is straightforward requiring the assessment of a subject's profile through the completion of a survey consisting of a series of seven questions for each of the issues, "by responding to the questions in the Urban Profile Question it is possible to generate a clear and simple graphic representation of the sustainability profile of that region" (UN Global Compact Cities Programme 2013). These are generally qualitative and subjective in nature. The questions provide a Likert scale numerical value from 1 being critical to 9 labelled vibrant, each number has a colour also associated with it to provide greater graphic clarity on the profile. These are then summed to provide a numerical value for representation on the profile chart.

The tools guide suggests that an assessment group be formed, comprising three to ten people with different and complementary expertise. The panel's efforts will be reflective of which of the four study types being undertaken. 2 h is optimal for a Rapid Assessment; 4 h to a day for an Aggregate Assessment, however, it will take significantly longer for an Annotated Assessment, and a comprehensive assessment can take from a few months to a year depending. The types of study are:

1. Rapid Assessment Profile: Responding to the single 'general question' under each 'perspective' by marking the 9-point scale.
2. Aggregate Assessment Profile: Responding to the 'particular questions' under each 'perspective' by marking the 9-point scale).
3. Annotated Assessment Profile: Completing the exercise at Level 2 and writing detailed annotations about how the points on the scale were derived.
4. Comprehensive Assessment Profile I: Completing the exercise at Level 3 and writing a major essay on the urban area using the questions to guide the writing. and/or
   Comprehensive Assessment Profile II: Completing the exercise at Level 3 and assigning metrics-based indicators to each point on the scale.

By responding to the questions in the Profile Questionnaire it is possible to generate a clear and simple graphic representation of the sustainability profile, as shown in Fig. 1.

3 Discussion

3.1 Fit with eLearning

At first the comparison may appear a little out of alignment as cities a sustainability model for cities would seem to pose little relevance to one for eLearning, as one is a technology, whereas the other is a physical entity. There are however many similarities between these apparent disparities, for example both cities and eLearning environments:

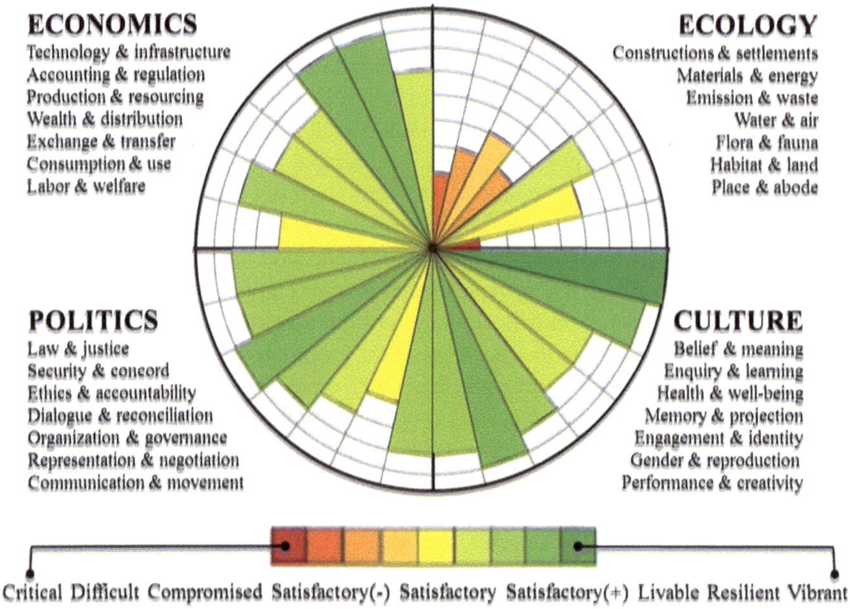

Fig. 1 Sustainability circle profile (Adapted from UN Global Compact Cities Programme 2013)

- Are designed for people to communicate, interrelate;
- Serve different populations with different aspirations and goals;
- Reflect dynamic complex ecosystems of complex sub systems;
- Are built on infrastructural layers that must interoperate;
- Are subject to the same forces operating within the four domains;
- Benefit from comparisons with other similar initiatives;
- Operate in strongly traditional cultures;
- Are perceived as identifiers of their respective populations;
- Operate within a given perceived quality paradigm;
- Offer non-homogenous service and product catalogues; and
- Operate within highly regulated political and regulatory environments.

The similarities serve to indicate that in relation to the sustainability of eLearning the Circle of Sustainability framework can be seen as a framework that has the potential to adapt very well to the needs of the eLearning field.

An initial view of the structure in Table 1 may lead to a rejection of the issue categories as not applicable to eLearning since they were developed for cities and the particular circumstances facing cities. However, while the specific application of the terms may have significantly different meanings in their application, the elements do provide appropriate issues to consider. For example under the economic domain the issue of poverty may not be seen as of deep consideration, however viewing it through the lens of the digital divide, the issue becomes very relevant to eLearning. Similarly with the ethnicities element in the Cultural domain, the domination of the English language in the computing and software industries filters through the eLearning environment and can be alienating if not excluding for non-English speaking populations. A sustainable provision of eLearning will need to allow for the context of ethnicity in its development and operation.

3.2 Adaption to eLearning

The adaption to eLearning would require the identification of the appropriate issues and the development of the survey questions and ranking to reflect the specific nuances of eLearning. For example the following issues pertaining to eLearning would require elucidation

- Pedagogy
- Academic Individualism
- Learning Science
- Reference Ware
- Assessment
- Adaptive Learning
- Gamification
- Learning Analytics
- Learning software
- Hardware
- Infrastructure
- Networks
- Access to Internet
- Privacy and security
- Systems integration
- Institutional cultures, traditions and reputations
- Student choice
- Faculty development
- Organisational structures
- Social missions
- Public policy

The identification of the specific attributes requires to be assessed for eLearning and the suitable questions to provide a basis for assessment are required before the profile can be used in any meaningful way in this context. Such will require significant research and consultation within the eLearning community to establish an agreed set for usage. In addition the dynamic nature of technology will require a review process also be established to review the agreed set on a periodic basis to ensure their continued relevance to the sustainability of eLearning.

3.3 A Note on Technology Adoption

A critical element to the sustainability of eLearning will be the on-going adoption of continuous innovations. The movement from the existing campus based instruction will move through a series of interim states before achieving a new steady state. This is due to technological adoption not generally being a smooth transition from one stage of development to a higher one. The challenge to pre-existing capital structures, financial, political, social, personal, and human, is non-trivial and creates a resistance that is commensurate with the degree of perceived disruption. This theme developed by Joseph Schumpeter (1942), whose theory of Capitalism critiqued the static neoclassical the framework, yielded the term "creative destruction", which describes the disruptive cycle of innovation that ultimately leads to the destruction of the existing order of value creation.

Models of technological adoption have been developed to provide a basis for understanding and prediction. Most commonly known of these is Rogers (2003), whose Diffusion of Innovations has become almost unchallenged in the business sector as the de facto description of innovation adoption. His model adopts a lifecycle approach describing the stages and determinants that characterise a technology's adoption curve. Several economic surveys have attempted an economic approach to technology diffusion, including Hall (2004) and Stoneman (2002). These have tended to view technology adoption in a narrower sense looking at first time usage of new technology. An Information Technology Adoption Model was developed and empirically tested by Davis (1989), whose Technology Acceptance Model hypothesizes a direct link between perceived usefulness and perceived ease of use, and that between a choice of two equally productive systems, the easiest to use will hold sway. Further empirical validation was obtained in Venkatesh and Davis (2000).

University education is highly tradition based, and has developed a culture that appears remarkably resistant to the adoption of technology; indeed it can be argued that it has successfully resisted most attempts to introduce technology into its canon since the inception of the form circa twelfth century, book technology notwithstanding. A hypothesis to understand the conflict of change that may be applicable in the higher education context was put forward in Perez (2002). Her thesis holds that an existing framework created to manage the growth of a given set of technologies will be unsuited to a disruptive replacement. This will lead to an initial mismatch between the existing techno-economic and socio-institutional spheres, of the respective old and new technologies, with the process of reconciliation and adoption being complex, protracted and painful. The adaption of a sustainability framework will need to address these econo-socio-political contexts and to provide an ability to assess the dynamics of a technology enabled service. This is further discussed below in the applying the framework section.

4 Applying the Framework

4.1 Open Source

A review of Open source sustainability helps to illuminate the motivations behind commitment to software development, which is of interest here as many educational software began as either university research or Open source initiatives that were later acquired by commercial entities. Xu et al. (2009) undertook a regression analysis to identify the factors leading to developer commitment, a critical element in the ongoing development of open source software. They identified Software needs reputation, enjoyment, leadership interpersonal relationships and ideology as the key variables, with software needs and interpersonal relationships proving to be the strongest influencers. The primacy of contribution to the development of Open source projects found by Gamalielsson and Lundell (2014) indicates the criticality of retaining the major committers to the effective sustainability of an Open source software initiative. The CS framework very effectively accommodates these critical factors in the cultural and political quadrants. Indeed they would be prompted in an a priori analysis of the sustainability of an Open source project, where initial focus generally falls on an initiatives economic sustainability.

4.2 Shared Services

A new emerging factor to potentially enable eLearning sustainability, and more broadly campus IT sustainability, is the move to adopt shared IT services. The primary reasons shared service initiatives are undertaken in Higher Education are very similar to those undertaken in other sectors (MCP 2013):

- Gain economies of scale through centralized transaction processing;
- Standardize practices, processes, and policies;
- Standardize and improve services to students;
- Leverage technology capabilities and investments;
- Uncover and utilize best practices among institutions;
- Foster increased inter-institutional collaboration;
- Focus campus staff on high-value activities;
- Reduce total unit cost to free up resources for reinvestment in institutional quality.

The adoption of shared services has been growing over the recent past with a number of initiatives across several countries. In Canada the most mature expression of this is taking place in British Columbia (BC) where BCNet and BC Campus are collectively delivering a range of shared IT infrastructure and application services to the BC Post-Secondary sector.

The implications for a sustainable campus are significant as the realisation of shared services can reduce both economic and ecological cost to each institution. Using shared services as a test of the framework can reveal the strengths and weaknesses inherent in both concepts.

4.3 Vulnerability to Change and the Impact of Dynamic Forces

While sustainability is desirable and a necessary condition to a service such as eLearning provision it is not a sufficient condition. The volatility of the developments in a technology enabled service make it vulnerable to transformative technologies through the invention and innovation of comparable but superior substitutable services. These will similar to the incumbent technology, face an adoption cycle and the necessity to overcome barriers of adoption, which, however their value proposition may be more than sufficient to overcome. The disintermediation of the physical media distribution in paper based publications and recorded music demonstrate the disruptive power of transformative innovations in a related field. More closer to this discussion the current shift to etexts is reflecting these previous disruptions in the educational course text book space.

The sustainability outcome matrix in Fig. 2 shows the relationship between the value of the service and its sustainability with regard to the source of disruption. In the context of eLearning the two main sources of disruption are from an inability to maintain a given service due to its inherent lack of sustainability and from an improved service due to its high relative perceived value introducing disenchantment with the existing service. Thus while sustainability is a necessary condition for ongoing provision, it is not sufficient. This condition will only be maintained if the value to users is perceived to be sufficient to meet their current and expected future requirements.

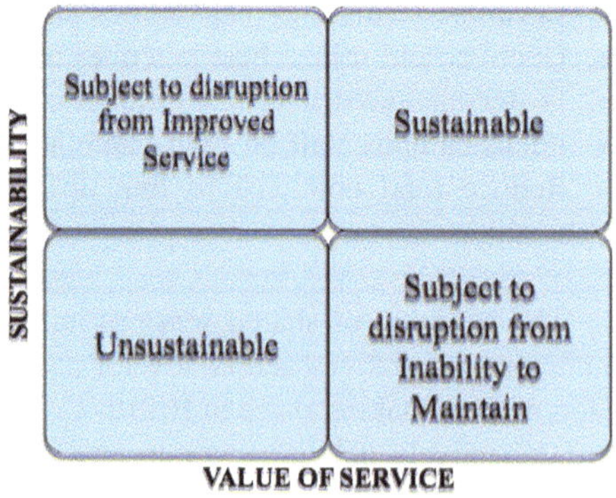

Fig. 2 Sustainability and destructive technologies

5 Conclusion

The analysis presented in this paper can be taken as an identification stage in the formation of an applicable conceptual framework for sustainable eLearning. The four quadrants of the CS framework were shown to have a comfortable fit with eLearning that can accommodate the criteria that will necessarily apply to its domains. It further provides a more unified and comprehensive framework that brings ecological sustainability into the eLearning domain while also expanding the accepted triple bottom line through the inclusion of power and ethical considerations. The application of the model to two use cases, open source and shared services, demonstrated its potential value in uncovering success factors not initially evident without the outlining of the impacts of technological change inherent in eLearning, creates a dynamic environment that disrupts existing paradigms in unpredictable ways, requiring a sustainability framework to allow for such in its construct, and thereby bringing the time element into the decisional milieu. In summary, the CS framework offers a potentially usable model that address many of the gaps in existing frameworks particularly through the inclusion of ecological perspectives and the development of the political and ethical dimensions.

The next step would be to move through a further identification, evaluation and validation of the framework initially through discussion within the eLearning community to determine an effective adaption of the CS framework, involving a development of an eLearning specific prototype of the framework. Upon completion of an accepted reformulated profile tool a trialing on current and past eLearning initiatives to assess its scope, fit and predictive-ness. Such could then be assessed through peer review to determine the use and value of the framework enabling greater awareness of its capabilities and its desirability of broad scale adoption and usage. These are nontrivial tasks and will require a degree of engagement and commitment similar to the establishment of the CS framework in the urban context. Nonetheless the potential value to guide eLearning development, lower its cost of adoption and enable its sustainable operation is of increasing importance. Particularly as resource strapped educational institutions and governments seek to adopt eLearning to address the challenges of the knowledge economy requiring the admission of a significantly enhanced proportion of the population to higher education. Without a coherent and relevant framework the likely outcome will be at best, a sub-optimal integration into current practices accompanied by an excessive use of resources that will both limit its effectiveness and undermine its sustainability.

References

Adams W (2006) The future of sustainability: re-thinking environment and development in the twenty-first century. http://cmsdata.iucn.org/downloads/iucn_future_of_sustanability.pdf. Accessed 30 Mar 2014

Bates T (2010) Innovate or die: a message for higher education institutions. eLearning and distance education resources by Tony Bates. http://www.tonybates.ca/2010/06/28/innovate-or-die-a-message-for-higher-education-institutions/. Accessed 30 Mar 2014

Brundtland G (1987) World commission on environment and development: our common future. Oxford University Press, Oxford
Christensen CM, Horn M (2008) Disrupting class: how disruptive innovation will change the way the world learns. McGraw-Hill, New York
Christensen CM, Eyring H (2011) The innovative university: changing the DNA of higher education from the inside out. Jossey-Bass, San Francisco
Davis FD (1989) Perceived usefulness, perceived ease of use, and user acceptance of information technology. MIS Q 13(3):319–340
Fournier H (2006) A review of the state of the field of adult learning: eLearning. Canadian Council on Learning, Ottawa (May)
Gamalielsson J, Lundell B (2014) Sustainability of open source software communities beyond a fork: how and why has the libre office project evolved? J Syst Softw 89(3):128–145
Gunn C (2010) Sustainability factors for eLearning initiatives. ALT-J Res Learn Technol 18 (2):89–103 (July)
Hall B (2004) Innovation and diffusion. Working paper no. 10212. National Bureau of Economic Research, Cambridge
Hendriques A, Richardson J (eds) (2004) The triple bottom line: does it all add up?. Earthscan, London
James P, Scerri A (2011) Auditing cities through circles of sustainability. In: Mark A, Toly NJ, Carney PL, Segbers K (eds) Cities and global governance. Ashgate, Farnham, pp 111–136
JISC (2014) "Key learning and teaching practice activities" learning and teaching practice. http://www.jisc.ac.uk/whatwedo/topics/learningteaching.aspx. Accessed 24 Mar 2014
Mayadad AF, Bourne J, Basich P (2009) Online education today. 323(5910):85–89 (January)
MCP (2013) Shared services opportunities in higher education, MCP insights, management consulting partners LLC. 2(1). http://www.mcpartnersllc.com/download/Shared%20Services%20Opportunities.pdf. 30 Mar 2014
Perez C (2002) Technological revolutions and financial capital: the dynamics of bubbles and golden ages. Edward Elgar, Cheltenham
Reichel M, Ramey MA (eds) (1987) Conceptual frameworks for bibliographic education: theory to practice. Libraries Unlimited Inc., Littleton
Robertson I (2008) Sustainable eLearning, activity theory and professional development. In: Paper presented at the Australiasian society for computers in learning in tertiary education 2008 (November) conference, Melbourne, Australia
Rogers EM (2003) Diffusion of innovations, 5th edn. Free Press, New York
Roy R, Potter S, Yarrow K (2008) Designing low carbon higher education systems: environmental impacts of campus and distance learning systems. Int J Sustain High Educ 9(2):116–130
Schumpeter Joseph A (1942) Capitalism, socialism and democracy. Routledge, London
Stewart B, Khare A (2012) Athabasca university reduces ICT carbon footprint. GSTF J Comput [serial online] 1(4):70–75 (January)
Stepanyan K, Littlejohn A, Margaryan A (2013) Sustainable eLearning: toward a coherent body of knowledge. Educ Technol Soc 16(2):91–102
Stoneman P (2002) The economics of technological diffusion. Blackwell, Oxford
Suhonen J, Sutinen E (2014) The four pillar model—analysing the sustainability of online doctoral programmes. Tech trends (in press)
The Sloan Consortium (2013) New study: over 6.7 million students learning online. http://sloanconsortium.org/publications/survey/grade-change-2013. Accessed 27 Mar 2014
UN Global Compact Cities Programme (2013) The circles of sustainability urban profile process. International Secretariat based at the Global Cities Institute, RMIT University in Melbourne, Australia. http://citiesprogramme.com/archives/resource/circles-of-sustainability-urban-profile-process. Accessed 29 Mar 2014
U.S. Department of Education (2010) Evaluation of evidence-based practices in online learning: a meta-analysis and review of online learning studies

Venkatesh V, Davis FD (2000) A theoretical extension of the technology acceptance model: four longitudinal field studies. Manage Sci 46(2):186–204
Xu B, Jones DR, Shao B (2009) Volunteers' involvement in online community based software development. Inf Manage 46(3):151–158 (April)

Author Biographies

Brian Stewart is Executive Programme Director at the University of Alberta. His research foci are on the *economics*, adoption and sustainability of eLearning, and the application of Information Technology in operational and organisational strategy.

Anshuman Khare is Professor in Operations Management and Sustainable Development at Athabasca University, Canada. He is an Alexander von Humboldt Fellow. His research focuses on the impact of environmental/climate change regulations on society, technology, innovation and corporate strategy.

Every End Is a New Beginning—a Realignment and Relaunch of an Environmental Management System at a German University Following a 15 Years Maintaining Period

Markus Will, Anke Zenker-Hoffmann, Jana Brauweiler, Bernd Delakowitz and Sebastian Riedel

Abstract

As the first Higher Education Institution (HEI) in Europe the University of Applied Sciences Zittau/Goerlitz implemented an Environmental Management System (EMS) according to EMAS (EMAS III 2009) already in 1999 and still is anxious to maintain, audit and improve this successful system on a high level. As most parts of this EMS are designed following Deming's PDCA cycle, the EMS aims at continuous improvement of performance. After such a long period the university's EMS reached a high degree of maturity. Recently, however, it seems that the envisaged continuous improvement process (CIP) has yet lost its dynamics. Obviously, there is still a lot of creativity how to improve the environmental performance. But dedication and motivation of both staff and students is slowly declining and low-hanging fruits, i.e. measurements that are cost-effective in the short term, have already been picked. On the other side, new challenges occur for HEI i.e. aspects of energy efficiency, occupational health and safety and sustainability which require a re-launch of the existing university's environmental management concerning the strategic focus and including improved processes specifically tailored to the needs of an academic institution. The paper presents a conceptual model of a new EMS for Institutions of Higher Education with a core alignment on strengthening and promoting initiatives of students, staff and stakeholders and on supporting small-scale but effective projects at a more operational level.

M. Will (✉) · A. Zenker-Hoffmann · J. Brauweiler · B. Delakowitz · S. Riedel
Environmental Management Team, University of Applied Sciences Zittau/Goerlitz,
Theodor-KoernerAllee 16, 02763 Zittau, Germany
e-mail: m.will@hszg.de

W. Leal Filho (ed.), *Transformative Approaches to Sustainable Development at Universities*, World Sustainability Series, DOI 10.1007/978-3-319-08837-2_21

Keywords
Environmental management at universities • Green campus

1 Introduction

Universities are recognized not merely as a place of research and education—or effectively managed and marked-oriented institutions as increasingly claimed—but an experimental laboratory for shaping societal progress (Mulder 2010). Impressively, more than 600 global HEI have committed themselves towards sustainability (cp. Leal 2011, p. 24). Due to this awareness, a growing number of university-level institutions are undertaking reviews or audits to assess their sustainability and environmental performance. An increasing number of caste studies and good practice examples are presented in the literature (Leal 2012). However, it appears too early to recognize a paradigm shift or sort of mainstreaming of sustainability in higher education (Wals 2014; Tilbury 2012; Disterheft et al. 2013). Many institutions are facing situations that hamper a full implementation (Raskin 2012; Lozano et al. 2013).

This paper is based on a 15 years process of successful and intensive EMS performance at the University of Applied Sciences Zittau/Goerlitz which according to the EMAS standard (known as "*Eco-Audit-Scheme*" by then) was the first HEI to implement, register and maintain this very system to date.

Through continuous improvement in the university's operation processes e.g. energy consumption, waste management, water treatment and procurement, a high grade of performance and acceptance has been reached. However, the necessary alignment of EMS at universities to focus more on indirect environmental aspects such as education, research, awareness and involvement of all target groups, particularly due to the frequency of incoming and outgoing students, has not been satisfactorily achieved. This paper presents explanations, experiences and analyses why a standardized and in parts little flexible EMS is actually not suitable to reflect the substantial challenges of an innovative EMS at HEI's. Based on the discourse of a maturity model the design of a new and more HEI focused EMS-approach is presented.

2 A Maturity Model of an Environmental Management System

Besides the international and widely accepted ISO 14001 EMS-standard (ISO 14001 2009) the complementary European EMAS-standard is considered more ambitious by integrating both (public) stakeholders and authorities. Based on a profound check of an organization's environmental performance both standards require to implement, to realize and maintain, to document and to constantly improve the EMS according to the Deming-Cycle (Deming 2000), i.e. regarding the PDCA sequence (plan-do-check-act). EMAS, in particular, demands the following aspects:

- Environmental policy
- Plan: identifying significant environmental aspects, legal requirements, defining objectives, targets, key performance indicators and programs
- Do: implementation and operation of EMS concerning qualification of responsible persons, internal communication, documentation, risk management
- Check: monitoring and measuring, corrective action, internal audits
- Act: Management review.

However, in addition to these largely general requirements, EMAS sets higher and more severe specifications regarding significant aspects:

- initial comprehensive analysis of current status relating to activities, products and services,
- proof of compliance with applicable legal requirements and permits,
- continual improvement of environmental performance, documented on the basis of key indicators,
- employee participation, involvement in the process of continual improvement and information,
- external communication with the public, interested parties, groups and clients based basis on an environmental report to be published annually.

It is the core of the idea of continuous improvement that an innovative EMS may be considered rather a journey or ongoing process than a destination. An EMS is, therefore, more than simply a score-keeper, a checklist-tick or award in contradiction to what business managers often see, when they receive a certificate to be displayed to the public for marketing and image purposes (cp. Enright 2012). Rather, an EMS should be considered as an effective tool to permanently improve environmental and also business performance and not merely to assess and rate it. An EMS itself can become a tool to provide effective, but gradual change. This is—in a nutshell—the idea of the CIP, which is based in Japanese lean management mind-set of KAIZEN (kai = “change”, zen = “good”). The principle of the CIP is an ongoing, thus never ending, effort to improve processes, structure and absolute performance (Brouwer and van Koppen 2008). According to Masaaki Imai, the organizational theorist and consultant who is considered as one of the godfathers of Kaizen, the purpose of CIP is a reflection of processes and the identification, reduction and elimination of suboptimal processes within an incremental and gradual rather than radical leaps (Imai 1996).

Most prominent in the US as well as in European industry is the PDCA, as a specific reading of Japanese Management philosophy. PDCA as an iterative four-step management method was developed and made popular by Walter Shewhart and William Edwards Deming (Deming 2000). Nowadays the PDCA is the key concept in management systems, where the cycle is to be conducted annually with external audits every 3 years. Maturity models as conceptual frameworks, may describe this development, i.e. the improvement of performance and sophistication of an area of interest, i.e. an EMS (Klimko 2001; Pigosso et al. 2013). Thus, the maturity model provides reflection and feedback considered as essential in CIP. In Table 1, different stages of sophistication and ambition are described. After sensitization and implementation phases, ideally a stabilization phase follows.

Shaping organizational routines, which correspond to stages three and four, got their pro's and con's. Clearly on the pro-side can be stated, that a reactive, trouble-shooting (problem-corrective-based) approach to get legal compliance is replaced by a more formal, systematic process, which aims at predictable improvements. However, the drawback of formalized routines and standardized operational processes is that a certain habituation as well as organizational blindness occurs, limiting the potential for innovative ideas and projects (Gastl 2005; Brouwer and van Koppen 2008). While effective and successful measures to improve environmental performance appear in the early years of the management system, this power slightly declines with increasing maturity. It becomes more and more demanding to keep dedication, enthusiasm and motivation of staff and students. This means that it happens to be more challenging to identify meaningful, reasonable, ecologically worthwhile and economically feasible ideas after the low-hanging fruits have been picked. This is especially true in the case, when long-term payback periods or tools such as *Total Cost of Ownership* are not permitted or supported by federal

Table 1 Maturity model according to ISO 9001, Müller-Christ (2013), Willard (2009)

Stage	Performance level	Stages	Level of ambition
1	No formal approach	Pre-compliance	Low ambition related to ecological aspects and sustainability
2	Reactive approach	Compliance	Compliance to legal regulation, corrective-based approach, trouble-shooting, only minimum data available
3	Stable formal system approach	Beyond compliance	Optimization conventional management + efficiency
4	Continual improvement emphasized	Integrated strategy	Optimization and investments conventional management + efficiency + preserving substance
5	Best-in-class performance	Purpose, passion and alignment with values and organizational culture	Optimization and investments conventional management + efficiency + preserving substance + responsibility

authorities (cp. Hermann and Acker 2011, p. 16). Effort for a more sophisticated and ambitious performance level and higher maturity ("best of class") will thus increase not only in terms of financial investments needed, but also in terms of purpose and passion, when underlying structures and organizational culture are challenged. If there will be a decision for more demanding engagement, i.e. no cost-effective measures in the short-term, the demand of organizational culture, sustainability oriented values and intrinsic motivation of top management (Gastl 2005, p. 242) still remain, particularly if there is no appealing market demand or legislative push.

3 Environmental Management at the University of Applied Sciences Zittau/Goerlitz

The University of Applied Sciences Zittau/Goerlitz offers around 40 Bachelor, Master and German Diploma programs in engineering sciences, natural sciences, social sciences and economics. More than 3,800 students are enrolled. Academic and technical staff consists of 122 professors and more than 360 research assistants.

It is worth mentioning that external R&D funds sum up to annually approx. 7, 5 Mio € of which another more than 100 researchers are employed. This is an exceptional high value displaying the university's high rated expertise and quality in specific research areas such as energy efficiency, renewable energies, natural sciences (e.g. molecular biotechnology, chemistry) and ecology/biodiversity and environmental protection.

The university has started implementation of its EMS already in 1998 and was successfully validated and registered according to EMAS in early 1999. Thus, the university is considered as one of the world's first HEI (Delakowitz and Hoffmann 2000). The EMS has been operated and improved since and was regularly re-validated by external audits. Selected examples will be given in this section of the paper.

3.1 Management of Resources

Here, environmental relevance is combined with large capacity for economizing, i.e. by reducing consumption of water, energy and greenhouse gas emissions.[1] More indirect environmental impacts, i.e. related public procurement, administration and decision making are considered, too. Environmental beneficial research, i.e. on more energy efficient technologies or environmentally sound energy conversation, is taken into account as well.

The University of Applied Sciences Zittau/Goerlitz—as a result of a self-assessment and with regard to the management procedures, not precisely in terms of

[1] All environmental data may be reviewed at http://www.hs-zigr.de/agumwelt/index.php (in German only).

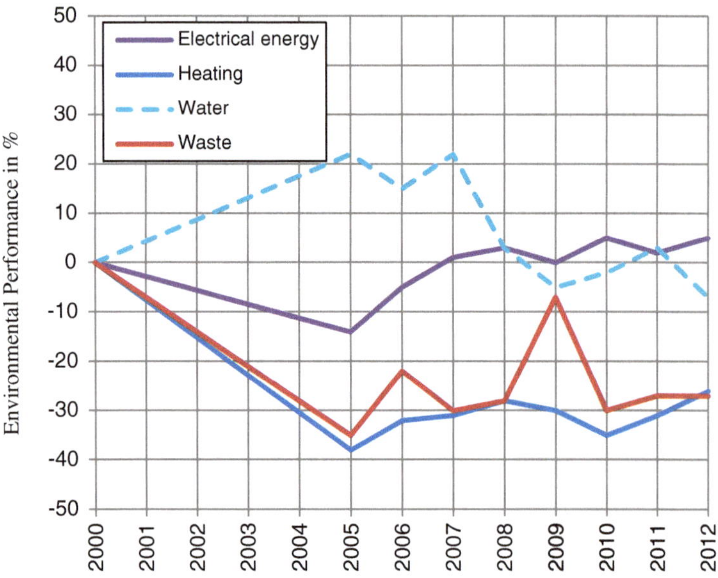

Fig. 1 Environmental performance from 2000 to 2012 relative to the number of students and staff, given in percentage indexed to the year 2000 (own elaboration)

absolute improvements—has reached stage 4 and remains at the frontier to stage 5 according to the maturity model presented in Table 1. There are measurable positive effects and a high external reputation of the EMS (Müller and Tegtmeyer 2012, p. 20) throughout the implementation and stabilization of EMAS at the university. Looking at the environmental performance over the period from the year 2000 to 2012, an improvement was reached in relative numbers, i.e. related to the number of students and staff (see Fig. 1)[2]:

- The demand in heat energy could be reduced constantly (−26 %), mainly to investments in insulation or new buildings.
- Energy demand is 5 % higher compared to the year 2000 which is also caused by growing energy demand in teaching facilities, laboratories and high-voltage field experiments mainly from the electrical engineering department.
- Waste generation peaked in the year 2005 and 2007 due to large scale construction and demolition activities. However, periodical peaking of waste accumulations is expected due to product life cycles and asset depreciation ranges, i.e. of IT equipment.
- Water demand was increasing in between, but could be reduced close to the level of 2000 (−7 %).

---

[2] Though, reduced water and energy electricity consumption cannot be proved in absolute numbers.

3.2 Education

There are several projects in education, research, information and communication that refer as good practice when it comes to foster multiplier functions of students and staff in order to create capabilities for change towards sustainability. Particularly, project-oriented work with students and staff is able to empower both. EMS measures thus become more relevant in a "learning by doing" manner, inducing a reflected learning process with an output desired, the continuous improvement of environmental performance and the EMS. Three examples are presented in Table 2.

3.3 Reason Why for a Re-launch

Despite of the level of internal and external acceptance and the results achieved a strategic realignment a re-launch of the EMS is considered as necessary because of the following reasons: Essentially, continual improvement of the university's EMS within the EMAS framework has become stuck after 15 years. Effective new options are missing and it seems that the strict EMAS framework is not fitting to this situation; instead it binds resources, needed for on-site projects. More specific reasons for a re-launch are:

- The effort to fulfill EMAS requirements, i.e. related to documentation of processes, data controlling and application of default key indicators is increasing.
- For two years now, the university's independence and influence on budgetary, property and facility management has been restricted due to restructuring measures within the State of Saxony. A state-owned centralized agency nowadays is responsible for all property of universities, which makes it challenging to follow environmental goals. Instead, following a continuous improvement agenda has become much more complicated. Thus, a lot of EMAS's requirements and processes and consequently their determination and documentation cannot be satisfactory fulfilled at the moment.
- A contextualized or systemic approach, tailored to the needs of an academic institution is not foreseen or required under EMAS. However, universities are faced with special conditions. For example, the major stakeholders, namely the students, do usually not stay longer than 3–5 years due to their study programs. It is, therefore, difficult to be compliant with the regulations of a standardized system, which relies on a CIP.
- Due to legitimated independence in research and education, statutory in the constitution, it is challenging to induce change from top-down, i.e. by passing instructions on contents of lectures, research topics or to create effective incentives.
- The increasing recurring expenses to fulfill validation requirements of EMAS are considered as disproportional to the possible positive effects. In addition, incentives such as deregulation in form of reduced monitoring and reporting

Table 2 Examples for projects in the area of educations, information and communication

Example 1: An internal project on operational procedures	
Target group	Mainly staff
Description	Together with seven other universities in Germany a project named "Change" was carried out that aimed at developing and implementing new and effective intervention strategies to foster efficient user behavior in organizations with the focus on energy use (Matthies and Wagner 2011). This project and related tools supported the university's own energy saving campaigns, i.e. by consistent information materials on saving options (e.g. flyer, brochures, posters). Projects to be implemented focused on minimal investments (e.g. switchable power strips), but at the same time clearly visible results. The project was based on elaborated expertise by environmental psychologists and change management experts
Benefits	The annual accounting of energy consumption showed slight savings on consumption. However, it is not straightforward to attribute improvements to specific measures. Benefits are undoubtedly seen in awareness raising and presentation of "real cases" with overall effects on motivation and engagement
Example 2: A student led internal environmental audits	
Target group	Selected students, staff
Description	It should be mentioned at this point that environmental management, life cycle assessment, environmental law and other environmental oriented issues have been addressed and integrated into the university's regular curriculum for more than 16 years. Since the starting phase of our EMS, students have been embedded into the implementation of the EMAS requirements. Every year, internal audits were carried out by students as an integral practical exercise within their study program
Benefits	Experiences with that have shown that it is very effective and efficient not only for the university but especially for students, who have the chance to practice and apply their theoretical knowledge within their own university. Another important result of these audits is to identify new topics for projects and activities in a dialog between staff and students
Example 3: A student led projects to develop innovative concepts	
Target group	Selected students, staff
Description	"Green purchase" of, for example, recycled paper and IT and communication equipment runs into several obstacles and psychological barriers in practice. It is thus necessary to assess compliance of the factual purchasing regime with internal procurement regulations. This is done annually by specific student task groups with the goal to issue guidelines for precise improvements, i.e. characteristics to be respected in purchasing such as content of recycled materials in office papers, maximal electricity consumption values in ICT relative to computing performance needed, etc
Benefits	As a result of this continual conceptual work, meanwhile a sound procurement policy was established. Preference for products with environmental labels (i.e. the well known German label "Blauer Engel") is regulated. In a follow-up student project it was found, that the established criteria are known, although there is partly a lack of acceptance and understanding of the necessity of this measure. To achieve an effective and continuous implementation process, this fact is respected and outlines the focus for next project works

obligations, faster licensing and permission procedure or reduced fees or taxes (Moosmayer et al. 2011)—valuable for industrial organizations - cannot be exploited by a university. The University of Applied Science Zittau/Goerlitz for instance could not use the advantage of lower regulations concerning reporting or monitoring, because there are no devices or facilities that have to be specifically monitored. Also a reductin of taxes is not applicable, because the relevant energy consumption is under the regulatory limit.
- The financial effort for the validation process and external audits amounts to more than 12,000 Euro over 3 years—money which cannot be made available for other environmental projects.
- A lack of general public interest has to be stated in many university's EMS that are validated under EMAS, especially in their published environmental reports. Furthermore, one of EMAS's benefits, to use its label (e. g. on letterheads, brochures, web page), is important but is not perceived by the stakeholders. There are no more than 2–3 inquiries of interested parties for the EMS of the University of Applied Science Zittau/Goerlitz per year. There is no verifiable relation between a growing numbers of students and the EMAS-EMS, for example in environmental oriented study courses.

The high maturity of the EMS on one hand and the observable decrease in motivation and engagement and low dynamic of the continual improvement on the other hand, supported the university's decision, not to continue with an external validated EMS according to EMAS. Instead, available limited resources will be assigned to develop and apply a renewed management structure with an indisputable goal and commitment to further increase environmental and sustainability performance.

4 Relaunching Environmental Management

In literature and practice there are different concepts and road maps of how to implement and maintain EMS—all of them having advantages and disadvantages. Full scale-EMS such as ISO 14001 or EMAS are internationally, at least European wide accepted and well known. Such general standards, however, remain in some way quite abstract as they are not specifically tailored to individual organization's needs and specific environmental impacts of branches. Contextualized tools like ready-to-use checklists are not available directly from these standards. Additionally, costs for certification/validation by external accredited bodies sum up to several thousand EUR per accreditation cycle. However, there are more flexible, less cost-intensive and contextualized EMS on the market, which may be considered as better applicable for Institutions of Higher Education compared to EMAS or ISO 14001. Selected requirements of EMS such as identification of environmental aspects, preparation of a policy documentation, stating vision goals as well as programs of actions to be taken are considered in more streamlined approaches while entry-barriers for implementation and follow-up costs can be kept low. Those EMS are often tailored to the needs of small businesses or organization or focused on a specific economic sector or industrial branch, i.e. chemical industry, touristic sector, for churches or educational institutions (Brauweiler and Sommer 2010, p. 302). Often, context-specific supportive tools which range from workshops to text books or special check lists with indicators are included. Some of those EMS are small-scale in terms of organizational effort, resource demand and documentation needs, but embrace sustainability aspects, especially in areas of social responsibility and philanthropy. Generally, costs for implementation, external auditing and certification and maintenance of the management system itself (related investments are hereby excluded) are considered to be significantly lower. The specific approach of those small-scale concepts, result in the fact that there is a variety of different concepts in Germany (ca. 16) and all over the world (ca. 40), with a quite small number of applicants. In Germany for instance there are over 7,000 organizations with ISO 14001 certified EMS whereas only about 1,000 organizations are validated according to EMAS. The number of organizations with small-scale EMS is obviously less, but not precisely known (Brauweiler and Sommer 2010, p. 319).

The University of Applied Sciences Zittau/Goerlitz recently decided to take a "pick-and- choose approach" to identify and employ advantages of various approaches to re-launch its EMS (see Fig. 2):

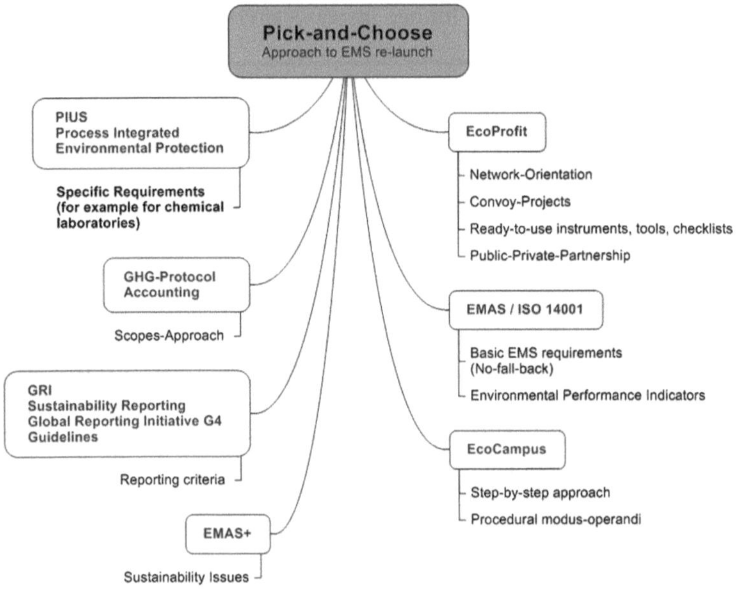

Fig. 2 Illustration of the "pick-and-choose" approach to select best practice from existing management systems and instruments (own elaboration)

It is not considered necessary to reinvent the wheel, though, the re-launched and modified future EMS should run smoothly. The vision statement may be summarized as:

- The EMS has to be pursued at the university but with an enhanced effectiveness, dedication and motivation.
- The re-launched EMS aims at reconquering the position of an internationally leading HEI in sustainability and environmental management and intends to achieve "best of class"-level.
- A new system should to be developed and established with a new focus on:
  - promotion and support of project initiatives,
  - taking advantage of opportunities on a more operational level,
  - adapting to particular requirements of educational institutions,
  - an organizational and cost-effective implementation model,
  - exploiting cluster-expertise in an appropriate network of educational institutions,
  - peer-to-peer audits in cooperation with competent bodies of other HEI instead of certification by external bodies.
- Distinct sustainability aspects such as energy transition, social issues/well-being responsibility will be integrated into the system.
- More concrete and specific activities for and with the university's stakeholders are planned, but under less formalities, bureaucracy and administrative demand as partly required in EMAS.

As Harvard Professors Kaplan and Norton stated, enhancing engagement and boosting the CIP means to create alignment, as "*Most organizations attempt to create synergy, but in a fragmented, uncoordinated way*" (cit. Kaplan and Norton 2006, p. 245). After 15 years running an EMS at a university quite a lot of projects have been induced by different actors. However, to reach a considerable higher level of engagement and efficiency, a alignment with management, staff and students following the vision stated above has to be created.

Considering the existing longtime experience, various strategy areas for operative sustainable management and projects have been identified in a brainstorming-process amongst the university's EM team (see Fig. 3):

- Operation,
- Procurement,
- Communication,
- Co-Operation,
- System, i.e. the management system,
- Education,
- Research.

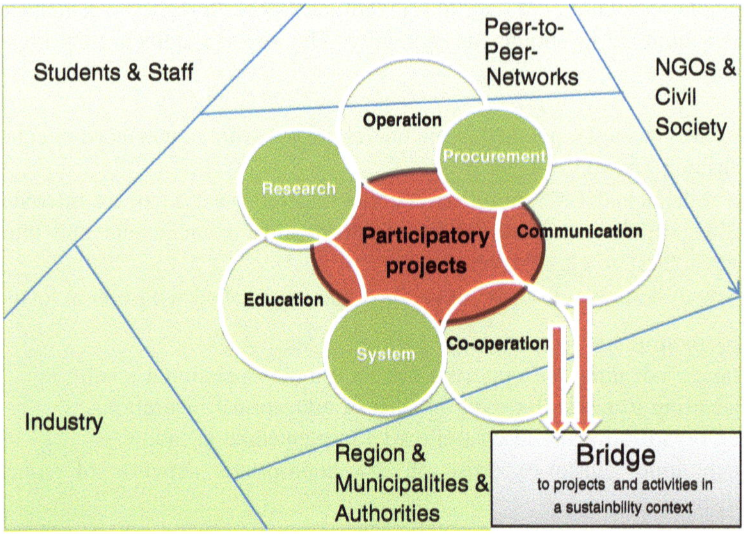

Fig. 3 Priorities and scope of action (own elaboration)

Participatory projects, focusing on clear environmental targets, will be the central element of the new EMS. So far, improvement processes have been triggered in a sector specific approach. As stated above, auditing and management, reviews and subsequent follow-up measures have reached a high degree of maturity, but simultaneously have lost their dynamics in terms of continual improvement and motivation of involved participants.

Thus, future activities will focus on supporting small-scale and short-term but effective projects on a precise operational level that involve both students and staff. Green campus—"join-in-projects" ("Mitmach-Projekte") aiming at participation and dedication from all stakeholders are highly desired. They may focus on wellness- and health management, work-life-balance and family-friendliness, healthy food and beverages (for examples at vending machines), take-back-, exchange and sharing campaigns (IT, chemicals, mobile phones, clothes, books etc.). Selected scopes of action, which are shown in larger bubbles in Fig. 3, are summarized in the following tables (Table 3).

Table 3 a Measures to be taken focusing on operational, b Measures to be taken focusing on education, c Measures to be taken focusing on communication, d Measures to be taken focusing on cooperation

Scope	Operational aspects
Background	Of course, operational aspects of the EMS in terms of optimization of processes and improvement of environmental aspects and performance will be carried on with even more ambition
Target groups	Faculties, research departments, institutes as well as administration, students and employees in academic and administrative sectors
Methods and instruments	Hot spot-analysis based on an initial comprehensive analysis of current status relating to activities, products and services, using the ecomapping-method and checklists (e.g. Günther et al. 2003)
Goal	Environmental aspects will be concentrated to hot spots (e.g. taking into account feasible ways to influence current praxis and consumption in a down-to-earth manner). Sectors to be considered are consumption of energy and water, waste generation and management structure
Scope	Education
Background	A central goal of Education for Sustainable Development is not to add the one or the other sustainability lecture or module into study courses, but to integrate this important topic into existing programs and modules. There are various approached to that issue, i.e. a lecture series that is mandatory for all students as integral part of their individual curricula. Since 1994 the University of Applied Sciences Zittau/Goerlitz has been offering a former 4 years Diploma and since 2011 a 7 semester Bachelor program on "Ecology and Environmental Protection". The philosophy of the course is to provide both theory and applied science that is not reduced to classical biotic ecology and natural conservation but that is holistic by integrating a variety of additional technical and economically oriented environmental protection issues as well as sustainability principles. The course's all-over intention is a preventative rather than an end-of-pipe approach in various kinds of organizational and industrial processes and business. Still, the program is quite unique all over Germany and, thus, highly attractive. A part-time MBA-program "Integrative Sustainability Management" was implemented in 2011 and is now executed in cooperation with external partners
Target groups	Faculties, students, lectures and professors, vice-rectors and vice-deans for academic and international affairs, Centre for Knowledge- and Technology Transfer (ZWT)
Methods and instruments	Desk research of module descriptions, interviews with lectures and professors. The evaluation is based on Education of Sustainable Development (ESD)-characteristics according to UNESCO (Wals 2012; Tilbury 2009). Finally a matrix will highlight conformities and gaps to requirements

(continued)

Table 3 (continued)

Scope	Education
Goal	Beside of specialized courses that are successfully implemented, aspects of sustainability should be integrated in other curricula and in advanced vocational training and professional development programs. Basically, an upgraded and integrated ESD program at all levels of education is planned. Prior to this, an evaluation will be made to assess the current status related to the integration of sustainability in education
Scope	**Communication**
Background	Obviously, communication is essential in the facilitation of environmental management particularly for the involvement of internal and external actors and stakeholders, the distribution of monitored performance data, presentation of successful projects, organizing collective actions, constructive mediation of (target-) conflicts
Target groups	Students, faculties, research departments, institutes as well as administration and employees in academic and administrative sectors
Methods and instruments	Regular reporting in the university's internal magazine ("Einblick") and on environmental data web portal, activation of the already existing online "green blog", organization of commemorative events and conferences, continuation of successful event formats such as the annually well-known university's "Earth Day"
Goal	To develop a communicative strategy and to implement a mid-term communication process for internal and external actors
Scope	**Cooperation with internal and external actors**
Background	A variety of approaches to foster sustainability already exist at the university which should be re-aligned and/or restructured from time to time. Additionally, the EMS-team is necessarily linked to different actors in the organization, such as the university' senate, university council, the rector and vice rectors as well as the university administration and the quality management group. Trough research and consulting projects, the university also achieves beneficial external effects, i.e. cooperation with municipalities and administrative districts within the European Energy Award (eea) to support local sustainable energy policies and rural development. Also municipal utilities, business and industrial enterprises benefit from the advice of experts or students in effective public-private-partnerships, following the triple-helix-approach (Etzkowitz and Leydesdorff 2000)
Target groups	Internal and external stakeholders
Methods and instruments	Survey, actors mapping, organizations of joint events, conferences and workshops
Goal	A large number of formal and informal contacts and networks already exists which perhaps provides valuable sustainability projects. However, in order to make use of synergies and to provide an overview of the recent state, some sort of coordination is needed. Unconditionally, the contacts to the mentioned public service provider in charge for facility management of universities in Saxony, need to be strengthened in order to broaden the scope of action and influence by the EMS task force itself

5 Conclusion

Based on a 15 years EMAS experience it can be concluded that neither EMAS nor ISO 14001 consider sufficiently specific challenges of HEI's which are:

(a) The focus must be on indirect and HEI-relevant environmental aspects which are difficult to measure.
(b) Involvement and integration of main target groups such as students and staff need to be more intensively addressed and endeavored.
(c) Integration of sustainability aspects and requirements into an EMS is considered essential.
(d) A more project-orientated approach is regarded preconditionary to realize (a) to (c) and to result more beneficiary for all stakeholders.

In the current phase of the project further consolidation and specification of various areas within the proposed EMS-model is essential. In pendency of these priorities, the design of the new EMS-model has to be further developed based on strategies and processes, organizational structures, internal and external audit procedures, communication instruments as well as focusing on the operational implementation. Some of the ideas based on the requirements for external evaluation (according to ISO 14001 and EMAS) which are considered focal elements for compliance and improvement consist in fostering transparency, reliability and independency.

In the frame of the re-aligned EMS a peer-to-peer review by competent bodies, i.e. environmental managers or auditors from other HEI, is believed to contribute additional benefit to the CIP of the university's environmental performance. For this purpose, the Environmental Management Team of the university intends to initiate a regional and education-oriented cooperation network which allows participating institutions to assess, compare and encourage each other in a constructive and fruitful benchmarking process. Partners on both national and international level are most welcome. The University of Applied Science Zittau/Goerlitz intends and happily is willing to accept again the role of guiding and leading the specific higher education EMS approach proposed in this paper as it has been starting the first European higher education EMAS some 15 years before.

References

Brauweiler J, Sommer P (2010) Niederschwellige Umweltmanagementansätze. In: Kramer Matthias (ed) Integratives Umweltmanagement. Gabler-Verlag, Wiesbaden, pp 301–320

Brouwer MAC, van Koppen Kris (2008) The soul of the machine: continual improvement in ISO 14001. J Clean Prod 16:450–457

Delakowitz B, Hoffmann A (2000) The Hochschule Zittau/Görlitz: Germany's first registered environmental management (EMAS) at an institution of higher education. Int J Sustain High Educ 1(1):35–47

Deming WE (2000) Out of the crisis. The Mit Press, Reprint

Disterheft A, Caeiro S, Azeiteiro MU, Leal W (2013) Sustainability science and education for sustainable development in universities: a way for transition. In: Caeiro S et al (eds) Sustainability assessment tools in higher education institutions. doi:10.1007/978-3-319-02375-5_1

EMAS III (2009) Regulation (EC) No 1221/2009 of the European parliament and of the council of 25 November 2009 on the voluntary participation by organizations in a community eco-management and audit scheme (EMAS), repealing regulation (EC) No 761/2001 and commission decisions 2001/681/EC and 2006/193/EC

Enright N (2012) Maturity matrices in action. http://www.sustainsuccess.co.uk/maturity-matrices-in-action. Accessed 20 Feb 2012

Etzkowitz H, Leydesdorff L (2000) The dynamics of innovation: from national systems and "Mode 2" to a triple helix of university–industry–government relations. Res Policy 29:109–123

Gastl R (2005) Kontinuierliche Verbesserung von Umweltmanagementsystemen und Umweltleistung—Die KVP-Forderung der ISO 14001 in Theorie und Unternehmenspraxis. Dissertation Universität St. Gallen, Hochschule für Wirtschafts-, Rechts- und Sozialwissenschaften (HSG)

Günther E, Fröhlich J, Winkler U, Klauke I, Brömmer Kathrin (2003) Checklisten für die Durchführung der Umweltprüfung/Umweltbetriebsprüfung an Hochschulen im Rahmen der Einführung von Umweltmanagementsystemen. Schriftenreihe der Kommission Umwelt, TU Dresden, Dresden

Herrmann A, Acker H (2011). Regelungen der Bundesländer auf dem Gebiet der umweltfreundlichen Beschaffung. UmweltbundesamtTexte 52/2011

Imai M (1996) KAIZEN—the key to Japan's competitive success. The KAIZEN Institute Ltd., McGraw-Hill Publishing Company, New York

ISO 14001 (2009) Environmental management systems—requirements with guidance for use, ISO 14001:2009

Kaplan R, Norton DP (2006) Alignment: using the balanced scorecard to create corporate synergies. Harvard Business School Publishing Corporation, Boston

Klimko G (2001). Knowledge management and maturity models: building common understanding. In: Second European conference on knowledge management, Slovenia, pp 269–278

Leal W (2011). About the role of universities and their contribution to sustainable development. High Educ Policy 24:427–438 (Palgrave)

Leal W (ed) (2012) Sustainable development at universities: new horizons. In: Environmental education, communication and sustainability, vol 34. Peter Lang Internationaler Verlag der Wissenschaften, Bern

Lozano R, Lozano F, Mulder K, Huisingh D, Waas T (2013) Advancing higher education for sustainable development: international insights and critical reflections. J Clean Prod 48 (2013):3–9

Matthies E, Wagner HJ (eds) (2011) Change—Veränderung nachhaltigkeitsrelevanter Routinen in Organisationen. LIT Verlag Dr. W. Hopf, Berlin

Moosmayer V, Zippel E, Lodigiani M, Koch C (2011) Fördermöglichkeiten und Privilegierungen für EMAS-Organisationen

Mulder K (2010) Don't preach. Practice! value laden statements in academic sustainability education. Int J Sustain High Educ 11(1):74–85 (Emerald Group Publishing Limited). doi:10.1108/14676371011010066

Müller J, Tegtmeyer R (2012). Green Campus: Vom Umweltschutz zur nachhaltigen Entwicklung. Wissenschaftsmanagement—Zeitschrift für Innovation. 18, Jahrgang, Heft 6, November/Dezember 2012

Müller-Christ G (2013) Ambitionsniveaus eines nachhaltigen Managements. In: Schäfer H/Krummerich K (Ed) Handbuch Nachhaltigkeit, Wiesbaden, pp 45–61

Pigosso D, Rozenfeld H, McAloone TC (2013) Ecodesign maturity model: a management framework to support ecodesign implementation into manufacturing companies. J Clean Prod 59(2013):160–173

Raskin PD (2012) Higher education in an unsettled century: Handmaiden or pathmaker? In: GUNI (ed) Higher education in the world 4. Higher education's commitment to sustainability: from understanding to action. GUNI series on the social commitment of universities 4, vol 1. Palgrave Macmillan, Basingstoke, pp 18–28
Tilbury D (2009) Tracking our progress: a global monitoring and evaluation framework for the UN DESD. J Educ Sustain Dev 3(2):189–193
Tilbury D (2012) Higher education for sustainability. A global review of commitment and progress. In: GUNI (ed) Higher education in the world 4. Higher education's commitment to sustainability: from understanding to action. GUNI series on the social commitment of universities 4, vol 1. Palgrave Macmillan, Basingstoke, pp 18–22
Wals AJ (2012) Shaping the education of tomorrow: 2012 full-length report on the UN decade of education for sustainable development. Commissioned by UNESCO
Wals AEJ (2014) Sustainability in higher education in the context of the UN DESD: a review of learning and institutionalization processes. J Clean Prod 62(2014):8–15
Willard B (2009) The sustainability champion's guidebook: how to transform your company. New Society Publishers, Gabriola Island

Author Biographies

Markus Will Dipl.-Ing. (FH), works as lecturer at the University of Applied Sciences Zittau/Goerlitz. His academic interests lie in life cycle assessment, technology assessment and innovation theory. Recently, he deals with methodological questions of greenhouse gas accounting for regions. He regularly works as facilitator and consultant, trying to support the development of sustainability strategies in organizations and municipalities. Despite his deep examination with the topic, his environmental footprint is still larger as it would be adequate in a fair and sustainable world.

Anke Zenker-Hoffmann Diplom Kauffrau, works as research associate and lecturer at the University of Applied Sciences Zittau/Goerlitz, Germany. She is an expert in environmental oriented business management with consulting, auditing and training and teaching experience in the field of management systems on environment, occupational health and safety, energy and their integration in the terms of sustainable management. She is a member of the task group on EMS at the University of Applied Sciences Zittau/Goerlitz since the beginning in 1999. Anke Zenker-Hoffmann is also involved in different processes to develop and implement study courses/programs in her field of work. She has conducted and managed several research and applied projects with national and foreign universities as well as industry partners.

Dr. Jana Brauweiler is a professor of "Integrated Management Systems" in the Faculty of Mathematics and Natural Sciences at the University of Applied Science Zittau/Goerlitz. Her teaching and researches focuses on management systems and project management. She has conducted research and practical projects with national and foreign university and industry partners and was visiting lecturer at universities abroad (e. g. Kazakhstan and the Czech Republic). Dr. Jana Brauweiler is qualified as an Internal Environmental Auditor (TÜV academy) and certified Project Management Professional (GPM), IPMA Level D®. She is freelancing in the fields of environmental management as a consultant and trainer, for instance for the programs "Environmental Management Specialist" or "Specialist for Occupational Health and Safety Management" in cooperation with TÜV Rheinland academy®. In addition she is a lecturer at the International University Zittau, a central scientific institution of the Technical University of Dresden.

Bernd Delakowitz holds a diploma degree and a Ph. D. in geosciences and geochemistry of the Ludwig-Maximilians-University, Munich. Since 1996 he is full professor for environmental management, environmental law and life cycle assessment (LCA). Bernd Delakowitz is an environmental expert with approximately 30 years of professional expertise in the fields of nuclear waste disposal, lecturing on higher education level, research & development, training of innovation management systems (EMS, QMS, OH&S) as well as in higher education administration. For nearly 11 years he has been dean of the faculty of Mathematics and Natural Sciences at the University of Applied Sciences Zittau/Goerlitz. In addition he is the representative for the university's EMS and the representative for the Baltic University Program (BUP) network. Furthermore, Bernd Delakowitz holds the position of the director of the univerity's *Institute for Ecology and Environmental Protection*. Prof. Delakowitz is a visiting professor and frequent guest lecturer in numerous universities such as in Mexico, Brazil, Lithuania, Latvia, Russia, Sweden, Poland, Kazakhstan, Scotland, England, Austria and the USA.

Sebastian Riedel graduated in Ecology and Environmental Protection at the University of Applied Science Zittau/Goerlitz in 2009. Since then he has been a lecturer and contributed to various projects. His main areas are the German environmental law and the implementation of Integrated Management Systems in companies.

Toward Temporal Conceptions of Education *as* Sustainability: Contributions from Complexity and Process Thinking

Sarah Chave

Abstract

Conceptions of education *as* sustainability (for example Bonnett 1999, 2002; Foster 2001, 2011; Sterling 2003, 2009, 2010; Vare and Scott 2007) locate education *as* sustainability within systems theory which emphasises the nested, interdependent spatial relationship between education, society, the economy and the ecosphere. This paper, however, builds on temporal conceptions of education *as* sustainability. Temporal conceptions place emphasis on the temporal continuity of a system, rather than focussing mainly (or only) on spatial continuity. Temporal conceptions are hard for people to engage with and understand. The anthropologist Graeber (2001) suggests that this stems from a worldview based on the Parmenidian approach prevalent in Western culture. People see the world as constructed from static objects and reflections of ideal forms: even relationships and feelings are reduced to objects. He suggests that science and philosophy are now recognising a more Heraclitean worldview in which everything is in flux, where new forms and structures are emerging: a world of becoming and potentiality. This paper explores what education as sustainability could look like if people can embrace such a worldview and why this is important. It concludes with a short case study of engaging with such an approach in a University Teacher Education module on Education for Sustainable Development (ESD).

Keywords

Systems thinking · Temporality · Complexity · Sustainability · Subjectification

---

S. Chave (✉)
Graduate School of Education, University of Exeter, Heavitree Rd., Exeter EX1 2LU, UK
e-mail: ssc203@exeter.ac.uk

W. Leal Filho (ed.), *Transformative Approaches to Sustainable Development at Universities*, World Sustainability Series, DOI 10.1007/978-3-319-08837-2_22

1 Introduction

This paper proposes that to engage with education *as* sustainability it is necessary to develop understandings of the temporal continuity of systems (the life of the system over time), including the role of emergence of the new. Since existing understandings of education *as* sustainability focus mainly on spatial continuity aspects of systems (how what can be seen as separate parts of the world actually fit together as interdependent parts of a whole) this paper suggests both an innovative focus and methodology. It is, however, important to note that temporal conceptions are intended to enhance not replace current spatial conceptions. This paper engages with what Biesta (2010), drawing on the ideas of the philosophers Arendt (1977 [1958]) and Levinas (1998 [1974]), calls a 'pedagogy of interruption'. This paper includes a short case study of using this approach with in-service trainee-teachers undertaking an ESD module. This paper also considers how a 'pedagogy of interruption' supports the sustainability values identified by the United Nations General Assembly (2000) of responsibility, freedom and equality.

Why is it relevant in today's world to explore temporal conceptions of systems? When society speaks about sustainability it is generally talking about living things —things that might die or disappear or come to harm if society does not care for them in a certain way. However, to sustain life, the living thing must grow (emerge). If it does not it dies (or is already dead). Therefore sustainability must enable emergence. The paradox is that to 'sustain' life it must be allowed to 'grow' into something different than it is, for example an acorn into an oak tree, a human baby into a ballerina. Moreover, it is not knowable in advance what that 'something' will be—a space of possibility needs to be created in which novelty can emerge. Temporal, emergent understandings allow exploration of this issue and consideration of how education can enhance this process. The process philosopher Whitehead (1977 [1929]) uses the phrase 'the fallacy of misplaced concreteness' to describe an opposite approach—one which understands the world as fundamentally a static 'thing', which complies with universal forms and theoretical invariants. Graeber (2001) suggests that this stems from a worldview based on the Parmenidian approach prevalent in Western culture. He proposes that science and philosophy are now recognising a more Heraclitean worldview in which everything is in flux, where new forms and structures are emerging. This Heraclitean conception is also echoed in the work of the philosopher Bergson (1998 [1911]) who saw the world as something which both endures and grows—it is an incomplete universe.

'Misplaced concreteness' also hinders people's ability to change or believe that change is possible or worthwhile. This paper proposes that developing temporal understanding of systems is highly relevant since it can enable people to feel that they *can act and make a difference*—a change in attitude which is greatly needed in the world today.

2 Education *as* Sustainability

Systems thinking has opened up new ways to understand the world and this has been influential in education *about*, *for* and *as* sustainability (Foster 2001; Sterling 2003, 2009, 2010). Systems theory emphasises the nested, interdependent spatial relationship between education, society, the economy and the ecosphere. This is a *relational* approach to sustainability and ESD in which the focus is on interconnectedness and mutuality. This contrasts with an atomised worldview in which 'the physical world is constituted by an indivisible number of invisible corpuscles moving randomly in an infinite void' (Taylor 1995, p. 64). This atomised idea of the biosphere has informed the dominant Western epistemology since the Renaissance.

Conceptions of education *about*, *for* and *as* sustainability have been explored by a wide variety of theorists including Bonnett (1999, 2002), Foster (2001, 2011), Orr (2004), Vare and Scott (2007) and Sterling (2003, 2009, 2010). Education *about* sustainable development is first order learning covering knowledge and skills relating to sustainability issues which, as Sterling (2009, p. 60) identifies, 'can be assimilated quite easily within the existing education paradigm'. Learning *for* sustainable development involves more reflective activities, encouraging the development of critical thinking skills, problem solving and trans-disciplinary thinking which many believe will be needed to solve sustainability issues in the future. However, as Sterling (2009, p. 61) points out:

> There is often an assumption that we know clearly what values, knowledge and skills 'are needed'.

Sterling (2009, p. 61) also recognises the spatial and to some extent temporal aspects of education *as* sustainability. He sees education *as* sustainability as:

> transformative, epistemic learning response by the educational paradigm, which is then increasingly able to facilitate a transformative learning experience. This position subsumes the first two responses but emphasises process and quality of learning… There is a keen sense of emergence and ability to work with ambiguity and uncertainty. Space and time are valued, to allow creativity, imagination and cooperative learning to flourish.

3 Temporal Aspects of Systems, Emergence and Potentiality and What This Could Mean for Education

So what is meant by the concept of the temporal life of a system? To explore this it is necessary to consider different ways of conceptualising time. In Newtonian science time is conceptualised as a mechanical reversible process in which a system, irrespective of the number of changes it undergoes, can return to its original state. Now common sense suggests this is not so. A system cannot retrace its previous steps: raindrops cannot go back into clouds; people cannot become younger. As natural sciences developed, the understanding grew that the Newtonian approach is highly abstract. It works for motion and machines/mechanically organised systems, but not for all areas of the observable world. The world is not a clockwork mechanism. Rather, in the natural world 'open systems do interact with their environment, changing themselves and their environment in the process' (Osberg and Biesta 2007, p. 49). This introduces the concept of irreversible time in which the 'Arrow of Time' can only go in one direction.

In addition, research into complexity theory and open and closed[1] systems introduced novelty and potentiality into thinking about systems. A key theorist in this area is Prigogine who explored novelty and emergence in 'far from equilibrium' systems. These are systems where inputs and outputs of energy and matter are not evenly balanced. The discovery made by Prigogine (1983, 1997) is that instead of increasing disorder (entropy) we can see increasing order and the creation of what Prigogine calls dissipative structures. These developments clearly underline the Arrow of Time and the impossibility of going back. Systems soon create their own unique and non-predictable histories. Prigogine uses the term 'bifurcation point' (see Fig. 1) to describe a point at which a system has 'to decide whether to go one way or another and this decision cannot be predicted in advance' (Osberg and Biesta 2007, p. 32). What is more, a process is created whereby properties that have never existed before and, more importantly, are inconceivable from what has come before, are *created* or somehow *come into being* for the first time.

This paper proposes that these ideas of complexity and emergence are also important for the social sciences, including education. Education is highly complex, messy and unpredictable, operating in what could be called emergent open systems. In such an emergentist approach to education knowledge is neither representation of something more 'real' than itself, nor an object that can be transferred from one place to the next. Knowledge is understood, rather, to emerge as we human beings participate in the world. Furthermore:

> every meaning that emerges is uniquely new, something which has not been in the world before, and cannot be predicted from the ground from which it emerged (Osberg and Biesta 2008, p. 313).

Drawing on process philosophy, Nelson (2003, p. 54) makes a similar point about knowledge, stating:

> the question of knowledge must be reframed. Our knowledge is not just about reality, in process thought. Rather, it creatively *adds* to reality.

It then follows that if knowledge and education is seen in this way it sets up a question as to how we can best organise opportunities for this kind of emergent learning to take place. Education in general and ESD in particular are no longer about moving knowledge from one place to another (usually from the teacher to the student) but creating a setting where learners are active participants in the making of

---

[1] Osberg and Biesta (2007, p. 49) describe closed systems 'as those which do not interact or exchange information with their environments'.

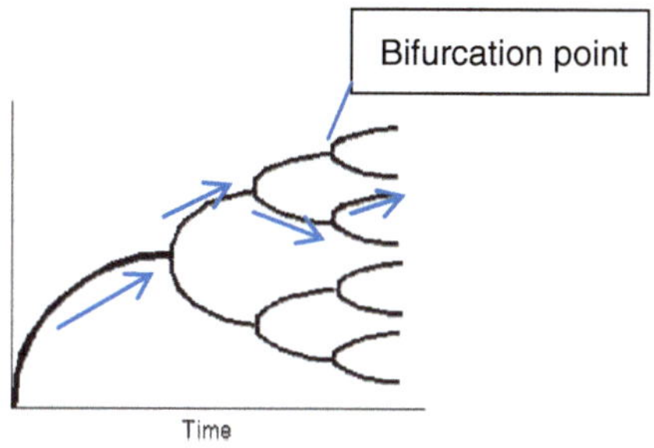

Fig. 1 The development of bifurcations in far from equilibrium systems—unique path followed by system

meaning. In this approach the intersubjective[2] *relationship* between the teacher and learner becomes important. The focus is on creating a space where learners can be response-able. This is what a temporal understanding of systems can contribute to education *as* sustainability. But what could this look like? This paper suggests that the Biesta idea of pedagogy of interruption has much to offer here.

4 Pedagogy of Interruption

Biesta (2010) begins his exploration of a pedagogy of interruption by asking 'what is education for'? He identifies three dimensions. Firstly, education is for qualification—learning skills and knowledge and also the dispositions and skills to do something. This *is* an important aspect of education. However, Biesta also identifies socialisation and subjectification. Socialisation is developing the social skills and attributes to be inserted into the existing social order. Subjectification, however, is about the process of becoming a subject—a person who acts in the world, and also is acted upon. Through the process of subjectification students become aware of who they are and their own 'beginnings' in the world. They also become aware of the 'beginnings' of others and how these impact upon them, and also the unpredictable ways that others respond to their 'beginnings'. To explain 'beginnings' Biesta (2010, p. 82) draws upon Arendt:

> For Arendt, to act first of all means to take initiative, that is to begin something new. Arendt characterises the human being as an *initium*: a beginning and a beginner (Arendt 1977, p. 170; emphasis added). She argues that what is unique in each of us is our potential to do something new, something that has not been done before.

For Arendt action is intimately connected with freedom, but this is not to be understood as a freedom of will but rather as the freedom to call into being something which was not present before. Such an exercise of freedom, since it is not an inner feeling, must necessarily require action in a 'public space'. This does not indicate a physical space but rather a 'quality of human interaction'. It requires plurality and Biesta draws on the ideas of Levinas (1998 [1974]) to explore this. Levinas asks questions about what it is to be, challenging the notion of the subject

[2] Intersubjectivity—the interaction between two subjects shaping and forming each other.

as a *cognito*, saying rather that the being is constituted in relationship with others. Biesta emphasises that, for Levinas, an individual is not representative of the rational community but rather a unique instance, who can speak outside the rational community. Much of education is about developing the rational community where one can speak as a representative example. There is nothing wrong with this—it is an important aim of education. But, there are moments when spaces occur where individuals have the opportunity to ask questions, to speak as unique instances, to explore their intersubjective relationship. These are not separate moments, ones that can be programmed into learning, but rather ones that emerge within the educational process. These moments interrupt the work of the rational community, unsettle it. For these moments to occur it is necessary to engage with the idea that education can produce human subjectivity—if this can be done temporal spaces might open up for uniqueness to come into the world. All these ideas seem to have something very important to say for education *as* sustainability. Can teachers and students create, through education, such temporal spaces of possibility where humans can act, can do something new, rather than being socialised into the existing order; a space where they can respond to their own and each other's beginnings? The following short case study provides some attempts to explore this.

5 Case Study

This case study explores work undertaken on a PGCE/Certificate in Education Module with a group of nine in-service teachers undertaking their initial teacher education alongside a post in the lifelong learning sector.[3] Teachers were employed in various settings including further education colleges, charities working with young people and also those with addiction. The trainee-teachers taught at introductory to degree level. The module was designed using Heron and Reason's (2001) cooperative enquiry approach to researching *with* people rather than *on* people. It is:

- a way of working with other people who have similar concerns and interests to yourself, in order to:
- Understand your world, make sense of your life and develop new and creative ways of looking at things.
- Learn how to act to change things you may want to change and find out how to do things better (Heron and Reason 2001, p. 179).

The module adapted this to a cooperative approach to teaching, working with the trainee-teachers as active agents, allowing ideas to emerge from discussions and activities and contributions from all course members and the tutor.[4] Through an

[3] The Lifelong Learning Sector includes: Community learning and development, further education, higher education, libraries, archives and information services work based learning (UK Commission for Employment and Skills 2010).

[4] The tutor is also the author of this paper.

action research cycle these ideas would be explored in the teachers' own practice, again working with their own learners using a similar cooperative approach. As Biesta (2010) points out, a pedagogy of interruption is a 'soft pedagogy'—it is something which can occur in moments of practice, rather than an approach to adopt leading to a desired outcome. However, it seemed possible that a cooperative approach to teaching could encourage such moments to occur—to allow spaces for the temporal development of a system in which newness can emerge. This is recognised by Heron and Reason (2001, p. 185) who state that in a cooperative enquiry approach:

> a mental set is needed which allows for the interdependence of chaos and order, of nascence and knowing, an attitude which tolerates and undergoes, without premature closure, inquiry phases which are messy. These phases tend, in their own good time, to convert into new levels of order. But since there is no guarantee that they will do so, they are risky and edgy. Tidying them up prematurely out of anxiety leads to pseudo-knowledge. Of course, there can be no guarantee that chaos will occur; certainly one cannot plan it. But the group can be prepared for it, tolerate it, and wait until there is a real sense of creative resolution.

In the first session the trainee-teachers were wary of the topic of sustainability—seeing it as yet another demand on teachers, and potentially more 'preaching' to or at students on how to live their lives. Then, one trainee-teacher asked the question 'How can we expect students to engage with saving the planet when they don't even take responsibility for bringing a pen to their classes?' This question really made the group stop and think—an interruption into the 'rational community's' approach to developing knowledgeable world citizens. Rather than ignore the inconvenience of this question the group decided to use this as a starting point and consider approaching the module from the point of view of responsibility.

The trainee-teachers took the concept of responsibility back to their own learners and discussed projects that the learners would like to undertake. The ideas were very varied and included reducing cyber-bullying; taking responsibility for their own music equipment and helping others with theirs; exploring mechanical engineering aspects of a swing bridge that linked two areas of a harbour together enabling regeneration; taking responsibility for revision and writing of engineering formula sheets; taking responsibility for development of new values in addiction and substance abuse settings; environmental building methods and responsibility of those in the building trade to explore these with customers; taking responsibility both for our own studies and attending to needs of our fellow classmates; developing self as a healthy individual and music and literacy as ways for young people to have a voice. This is a very varied list and maybe not the traditional topics explored in an ESD module. The tutor consulted the module moderator to ensure that such topics would be accepted by the University. This matters as the module was engaged with qualification as well as subjectification—both are important. The moderator confirmed these topics were acceptable. She also provided the following UN definition of ESD as learning which aims to:

also helpful. Teachers need to plan in programme preparation time to follow up avenues decided on by the group and avoid the desire to 'spoon-feed'.

In the case study discussed in this paper the 'interrupting' question related to responsibility but this theme cannot just be transported to a new setting. Instead it is important to see what questions which interrupt our usual 'norm' emerge in different settings and then respond to these. This is a challenge to any simple transferability—it requires openness to 'interruption'. Adopting a cooperative enquiry approach to teaching was shown in the case study to assist this process and it offers exciting possibilities for the future.

Arendt (2006 [1961], p. 193) succinctly and effectively expressed many of the ideas explored in this paper in her essay.[5] *The Crisis in Education*:

> Education is the point at which we decide whether we love the world enough to assume responsibility for it and by the same token to save it from that ruin which, except for renewal, except for the coming of the new and young, would be inevitable. And education, too, is where we decide whether we love our children (*and young people—my addition*) enough not to expel them from our world and leave them to their own devices, nor to strike from their hands their chances of undertaking something new, something unforeseen by us.

An education which can embrace all of these aspects—one which sustains, enabling development and the emergence of the new—would indeed be education *as* sustainability.

References

Arendt H (1977 [1958]) The human condition. Chicago University Press, Chicago
Arendt H (2006 [1961]) Between past and future. Penguin, London
Bergson H (1998 [1911]) Creative evolution (Trans: Mitchell A). http://web.archive.org/web/20060415050425/, http://spartan.ac.brocku.ca/~lward/Berson/Bergson_1911a/Bergson_1911_01.html. Accessed 26 Mar 2014
Biesta G (2003) Learning from Levinas: a response. Stud Philos Educ 22(1):61–68
Biesta G (2006) Beyond learning: democratic education for a human future. Paradigm, Boulder
Biesta G (2010) Good education in an age of measurement: ethics, politics, democracy. Paradigm, Boulder
Bonnett M (1999) Education for sustainable development: a coherent philosophy for environmental education? Camb J Edu 29(3):313–324
Bonnett M (2002) Education for sustainability as a frame of mind. Environ Educ Res 8(1):9–20
Foster J (2001) Education as sustainability. Environ Edu Res 7(2):153–165
Foster J (2011) Sustainability and the learning virtues. J Curriculum Stud 43(3):383–402
Graeber D (2001) Toward an anthropological theory of value: The false coin of our own dreams. Palgrave, New York
Heron J, Reason P (2001) The practice of cooperative inquiry: research 'with' rather than 'on' people. In: Reason P, Bradbury H (eds) Handbook of action research: participative inquiry and practice. Sage, London
Jensen B, Schnack K (1997) The action competence approach in environmental education. Environ Educ Res 3(2):163–178

[5] Biesta (2006) also uses this quote in his final chapter.

Levinas E (1998 [1974]) Otherwise than being or beyond essence (Trans: Lingis A). Duquesne University Press, Pittsburgh
Milgram S (1963) Behavioural study of obedience. J Abnorm Soc Psychol 67(4):371–378
Miller WR, Rollnick S (2002) Motivational interviewing: preparing people to change, 2nd edn. Guilford Press, New York
Nelson J (2003) Confronting the science/value split: notes on feminist economics, institutionalism, pragmatism and process thought. Camb J Econ 27(1):49–64
Orr D (2004) Earth in mind: on education, environment, and the human prospect, Special edn. Island Press, Washington
Osberg D, Biesta G (2007) Beyond presence: epistemological and pedagogical implications of 'strong' emergence. Interchange 38(1):31–51.
Osberg D, Biesta G (2008) The emergent curriculum: navigating a complex course between unguided learning and planned enculturation. J Curriculum Stud 40(3):313–328
Prigogine I (1983) The rediscovery of time: a discourse originally prepared for the Isthmus Institute, presented to the American Academy of Religion. http://www.mountainman.com.au/ilyatime.htm. Accessed 23 Feb 2014
Prigogine I (1997) The end of certainty. Time, chaos and the new laws of nature. The Free Press, New York
Rokeach M (1973) The nature of human values. The Free Press, New York
Sterling S (2003) Whole systems thinking as a basis for paradigm change in education: explorations in the context of sustainability. http://www.bath.ac.uk/cree/sterling/sterlingthesis.pdf. Accessed 3 Feb 2012
Sterling S (2009) Sustainable education: re-visioning learning and change. In: Stibbe A (ed) Schumacher Briefing No. 6, 2nd edn. Green Books, Dartington
Sterling S (2010) Living in the earth: towards an education for our time. J Edu Sustain Dev 4 (2):213–218
Taylor C (1995) Physical atomism. In: Honderich T (ed) The Oxford companion to philosophy. Oxford University Press, Oxford
UK Commission for Employment and Skills (UKCES) (2010). Lifelong learning sector: sector qualifications strategy. UKCES, London. http://webarchive.nationalarchives.gov.uk/+/, http://www.ukces.org.uk/upload/pdf/LLUKSQS_edit.pdf. Accessed 21 Mar 2014
United Nations General Assembly (2000) United Nations millennium declaration: 2000. UN Department of Public Information, New York
United Nations (2005) World summit outcome document, 15 September 2005. http://www.who.int/hiv/universalaccess2010/worldsummit.pdf. Accessed 9 May 2011
Vare P, Scott W (2007) Learning for a change: exploring the relationship between education and sustainable development. J Edu Sustain Dev 1(2):191–198
Whitehead AN (1977, [1929]). Process and reality. An essay in cosmology. Gifford lectures delivered in the University of Edinburgh during the session 1927–1928. The Free Press, New York
World Wildlife Fund Scotland (2005) Linkingthinking: new perspectives on thinking and learning for sustainability. World Wildlife Fund Scotland, Aberfeldy. http://www.decd.sa.gov.au/efs/files/links/Linking_thinking_toolbox.pdf. Accessed 22 Mar 2014

Author Biography

Sarah Chave has taught for over 20 years in further and higher education, including at City College Plymouth and Plymouth University with particular interests in equality and diversity, education for sustainable development, teacher education, business education and heterodox economics. She is currently working on a conceptual Ph.D. at Exeter University to bring insights from philosophy to her practice areas. She is a director of the South West Learning for Sustainability Coalition. She thanks participants in the case study and her Ph.D. supervisors Dr. Alun Morgan and Dr. Deborah Osberg.

Multicultural Constructive Community Learning Course for Education in Sustainable Development Applying Backcasting

Jordi Segalàs and Gemma Tejedor

Abstract

The International Seminar on Sustainable Technology Innovation is a course offered in the framework of the master of Sustainability of the UPC-Barcelona Tech University and financed by the ERASMUS Intensive Program scheme. It introduces backcasting scenarios methodology in real sustainability problems. The learning environment is international, transdisciplinary, intergenerational and intercultural. It includes stakeholders' dialogues and discussions. The course is organized around current sustainability relevant topics that are analysed in study cases based in different contexts: going from developed to developing countries and from local to global cases. Students apply scenario methodologies to the study cases in order to set up the most contextualized sustainable strategies. The course is structured in four phases: first the local situation analysis where students analyse the topic in their own countries/regions; then students are grouped in international multidisciplinary teams and define the state of the art of the case studies; afterwards students, lecturers and stakeholders meet in Barcelona where the course is run for two weeks and finally students analyse their learning experience in terms of competences acquisition. The topics analysed in the course vary each year and are related to relevant sustainability challenges: urban solid waste management; food and drinks packaging waste; overfishing and marine ecosystem degradation; sustainable mobility, agro-ecology and sustainable community energy systems. In 2014 UPC Barcelona Tech is running the course for the seventh time and so far more than 170

J. Segalàs (✉) · G. Tejedor
University Research Institute for Sustainability Science and Technology,
Universitat Politècnica de Catalunya, Barcelona Tech, Av Victor Balaguer 1,
08800 Vilanova i la Geltrú, Barcelona, Spain
e-mail: jordi.segalas@upc.edu

W. Leal Filho (ed.), *Transformative Approaches to Sustainable Development at Universities*, World Sustainability Series, DOI 10.1007/978-3-319-08837-2_23

students, 40 lecturers and 60 stakeholders have participated in the course from 39 countries. This paper explains the learning environment and the challenges and lessons learnt when organizing such a course, and the learning results obtained by students.

Keywords
Engineering education · Transdisciplinary learning · Community learning · Constructivism

1 Introduction

One of the UN Millennium Development Goals (UN 2012) is directly directed to education. No one will dispute that education is a must for sustainable development, but so far there is no direct relation between educated societies with the highest rate of "educated" citizens and the highest sustainability (WWF 2012). Sustainability demands a specific kind of learning; quoting E.F. Schumacher (Schumacher 1973): "The volume of education… continues to increase, yet so do pollution, exhaustion of resources, and the dangers of ecological catastrophe. If still more education is to save us, it would have to be education of a different kind: an education that takes us into the depth of things". In addition, some authors call for a deep change in society to order to achieve more Sustainable Development (SD) (whatever it is). SD is not just a matter of acquiring some extra knowledge. Attitude is also important. Moreover, it is often necessary to change social structures (Segalas et al. 2006).

Stephen Sterling (Stirling 2005) maintains that the nature of sustainability requires a fundamental change of epistemology, and therefore, of education. He writes: "Sustainability is not just another issue to be added to an overcrowded curriculum, but a gateway to a different view of curriculum, of pedagogy, of organizational change, of policy and particularly of ethos. At the same time, the effect of patterns of unsustainability on our current and future prospects is so pressing that the response of higher education should not be predicated only on the 'integration of sustainability' into higher education, because this invites a limited, adaptive, response…. We need to see the relationship the other way around—that is, the necessary transformation of higher education towards the integrative and more whole state implied by a systemic view of sustainability in education and society".

In relation to technological education, the Barcelona Declaration (Barcelona Declaration 2004) approved during the Engineering Education in Sustainable Development (EESD) conference in 2004 declares that today's engineers must be able to:

- Understand how their work interacts with society and the environment, locally and globally, in order to identify potential challenges, risks and impacts.
- Understand the contribution of their work in different cultural, social and political contexts and take those differences into account.

- Work in multidisciplinary teams, in order to adapt current technology to the demands imposed by sustainable lifestyles, resource efficiency, pollution prevention and waste management.
- Apply a holistic and systemic approach to solving problems, and the ability to move beyond the tradition of breaking reality down into disconnected parts.
- Participate actively in the discussion and definition of economic, social and technological policies, to help redirect society towards more sustainable development.
- Apply professional knowledge according to deontological principles and universal values and ethics.
- Listen closely to the demands of citizens and other stakeholders and let them have a say in the development of new technologies and infrastructures.

The Universitat Politècnica de Catalunya (UPC Barcelona Tech), aware of the new competences that engineers should have, offers a master degree in Sustainability Science and Technology. The aim of the master's degree in Sustainability is to provide advanced training in sustainable human development that enables students to understand the complex interaction between society, technology, the economy and the environment, so that they can tackle the social and environmental challenges inherent to sustainability: climate change, the depletion of natural resources, North-South imbalances, environmental justice, etc. This master's degree prepares students to become entrepreneurial professionals and agents of change for sustainability who will, depending on their specialization, design and assess global, sustainable solutions for the uncertain, complex scenario in which we live. They will take an interdisciplinary approach and ensure scientific and technical rigor in the diverse cultural and professional contexts in which they work. It becomes a highly demanded opportunity of teamwork and to mix with students of different knowledge cultures (Fry 2003).

Within the Master course, there is the subject Sustainable Technology Innovation. These next sections explain the innovative learning environment and the challenges and lessons learnt when organizing such a course, and the learning results obtained by students.

The course is innovative in terms of applying backcasting to real case studies with involvement of real stakeholders where students from different cultures and backgrounds work together. This constructive community future oriented approach has shown to be very effective to open the minds of the students, making them aware of the relevance of transdisciplinarity in finding agreed sustainable solutions for the future.

2 ERASMUS Intensive Program in Sustainable Technology Development

The Sustainable Technology Innovation (STI) course is a subject within the Master of Sustainability Science and Technology offered by Barcelona Tech University. It is a 5 ECTS (European Credit Transfer System) course, which uses constructive and community oriented learning. Since 2012 the course is an ERASMUS Intensive

Program (http://ec.europa.eu/education/erasmus/ip_en.htm) financed by the European Commission Education and Training programme.

2.1 Goals and Learning Outcomes

The goals and learning outcomes of the course are in line with to the Barcelona Declaration (2004):

- To increase the understanding of SD in the long term and the role of technology therein embedded in systems.
- To increase the capability to apply foresighting, forecasting and backcasting.
- To contribute to the development of the scientific work competences of students.
- To increase the capability of teachers to teach the approach of future imaging, foresighting, forecasting and backcasting.
- To equip UPC with the insight and capacity to transfer knowledge on future imaging and backcasting for sustainable development in the Masters in SD and further afield in Catalonia and Spain.
- To become an experts' meeting point and create networking activities among different groups and institution.

2.2 Methodology

The course is organized around a real and 'hot' topic which is chosen every year to give context to the participants' work. The topic is analysed first at the national level, and afterwards through case studies. The case studies are chosen in order to cover the broad spectrum of sustainability problems: from local to global and from developed and developing countries perspectives (Table 1). The content of the lectures and the dialogues is defined by a transdisciplinary committee, so that the most innovative and updated lectures on each issue can be analysed. Students, in multicultural and multidisciplinary groups, apply backcasting methodologies to the case studies, proposing strategies that would improve their sustainability.

The course is structured into 3 phases (Fig. 1):

Phase 1: Virtual, distance-based. This phase is divided in two steps:

First step: Local situation analysis. From March to May, students analyse the topic in their own countries/regions. The goals will be to introduce students to the sustainable topic of the year in their national framework. To report the results of this task, they will have to develop a poster with the state of the situation (data and flows, system and technology solutions, regulation, social perception and so on), which will be presented on the first day of the third phase.

Second step: Case study analysis. In May, students are grouped into international multidisciplinary teams and define the state of affairs in the case studies.

Table 1 Topics and case studies analysed in the programme

2008	*Topic*: *Water sanitation and treatment*	
Case studies	Developing	Recovery of Morro de Moravia, a suburb of the city of Medellin established on a landfill, with serious sanitation problems and water shortages
	Developed	Water supply difficulties in the Barcelona metropolitan area, with a climate that is inexorably warming up and drying out
2009	*Topic*: *Urban solid waste management, food and drinks packaging waste*	
Case studies	Developing	Dematerialization in food and drink packaging, analysing Bulgaria
	Developed	Dematerialization in food and drink packaging, analysing Catalonia
2010	*Topic*: *Overfishing and marine ecosystem degradation*	
Case studies	Global	Atlantic bluefin tuna is in urgent need of extensive international cooperation, required in order to ensure appropriate management of the species
	Local/ Developed	Initiative to provide a figure of protection to an Environmental protected Area (3000 ha approx.), at coasts of Garraf (Mediterranean Sea), taking into account: the entire ecosystem, the involvement of all stakeholders in the area linked to the sea and fishing
2011	*Topic*: *Mobility and global transport*	
Case studies	Global/ Developing	The Guatemala's technological corridor project intends to build mega infrastructures to connect overland the Pacific to the Atlantic Ocean, with the target to convey goods disembarked from merchant ships
	Local/ Developed	Mobility-related university activity affects the metropolitan area of Barcelona environmentally, socially and economically, with a lack of coordination between the different levels of territorial planning
2012	*Topic*: *Agroecology and consumption model*	
Case studies	Global/ Developing	The COMPETE (competence platform on energy crop and agroforestry systems for arid and Semi-arid Ecosystems in Africa) project, addressed the interrelated problems of low quality of life, limited energy access, and lack of livelihood opportunities in rural Africa, that lead to a degradation of arid and semi-arid ecosystems through unsustainable land-use practices
	Local/ Developed	*Salveml'Ortoll* is a civil society movement that tries to safeguard a peri-urban space in Vilanovai la Geltrú (Catalonia), which is instate of deterioration. The gravity of the situation can cause this area is irreversibly damaged, leaving urbanization as its only possible future
2013	*Topic*: *Sustainable energy systems: community participation*	
Case studies	Local/ Developed	The association of forest owners Massís del Garraf. An association of forestry owners of the Massís del Garraf (APF Massís del Garraf) was created in order to promote biomass extraction in a collective way and to try to reduce the costs of extraction and biomass collection
	Global/ Developed	The European energy legislation framework. The European Union has introduced its 20/20/20 targets to reduce climate change impacts and energy consumption, with the final target being to achieve a low carbon economy that promotes green growth and jobs to strengthen Europe's competitiveness

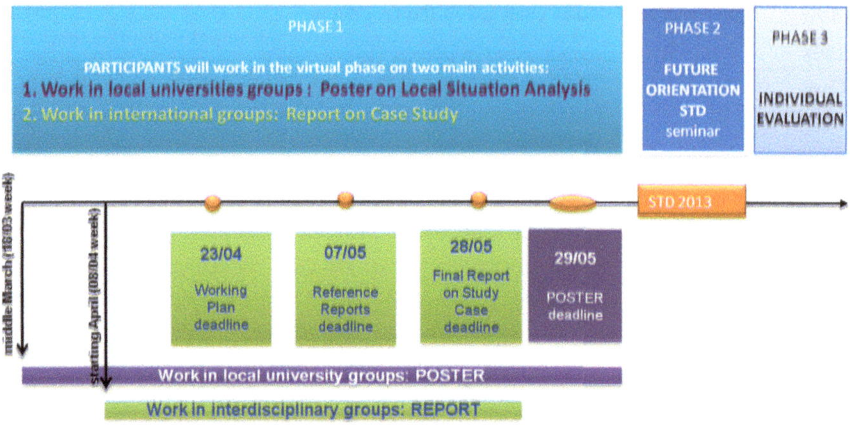

Fig. 1 Phases and schedule of the programme

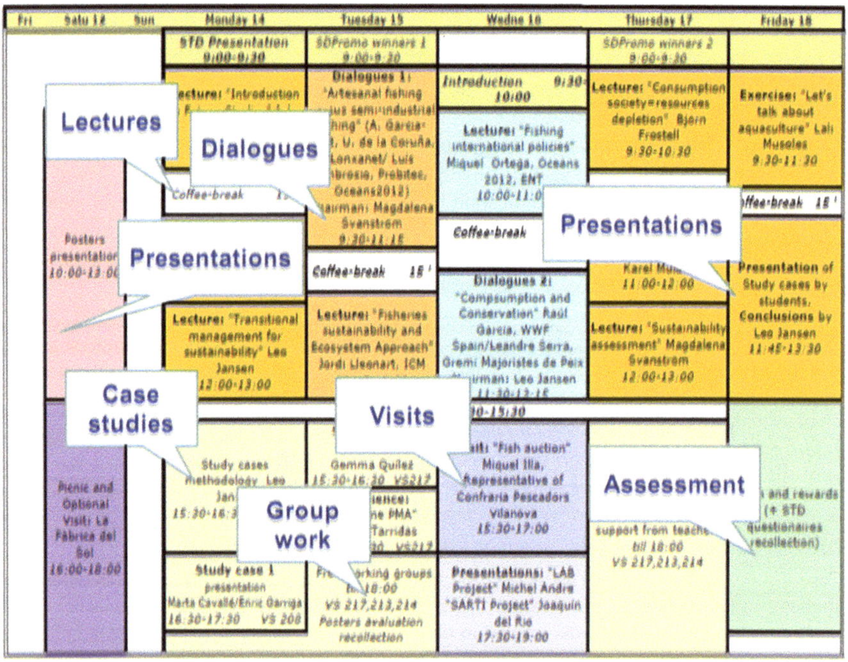

Fig. 2 Learning methodologies that are used in the presential phase of the programme

Phase 2: Seminar at UPC. In June, students, lecturers and stakeholders meet in Barcelona where the course is run for two weeks in a presential framework. There are two weeks of different types of learning methodologies (Fig. 2): Poster presentation by students with peer review; lectures from expert professors; practical visits; interaction with stakeholders in debates; and group working sessions on the case studies, applying backcasting scenario analysis methodologies and assessment sessions. Practical visits draw the connection between theory and practice, related to the case studies. Finally, the students set up a final presentation and a report on applying backcasting to the case studies, in order to propose feasible transitions towards sustainable solutions.

Phase 3: Evaluation of the course. Students evaluate their learning experience in terms of competences acquisition.

2.3 Five Successful Years of the Programme

The course has been run every year since 2008. During those six years, the topics and case studies have been based in the *five challenges* of the *Pla UPC Sostenible 2015* (UPC 2009), chosen to concentrate efforts on the key issues that society demands: *Construction, energy and climate change; Integrated water management; Socially responsible technology; Land use planning, mobility and logistics; and Material cycles, eco-design, waste management*. Table 1 shows the topics analysed and the cases studies worked out by the participants (Figs. 3 and 4).

During these 6 years, more than 160 students, 25 lecturers and 50 stakeholders have participated in the seminars.

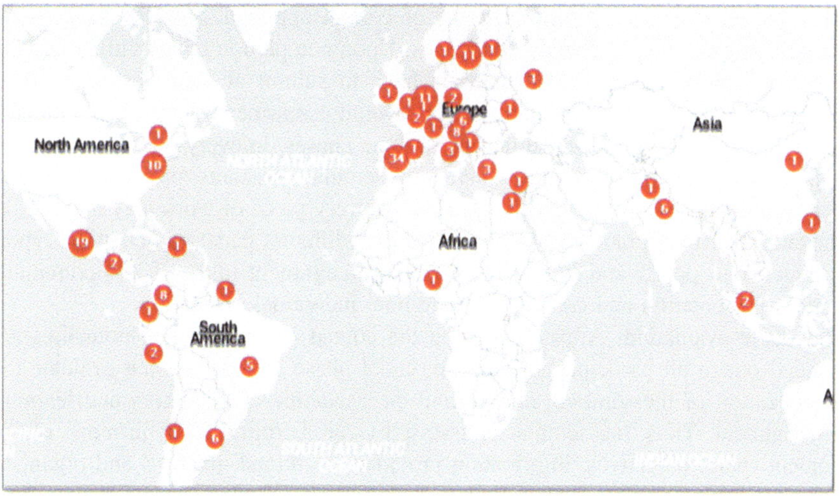

Fig. 3 Multicultural atmosphere: students from 39 nationalities. (number of students by country)

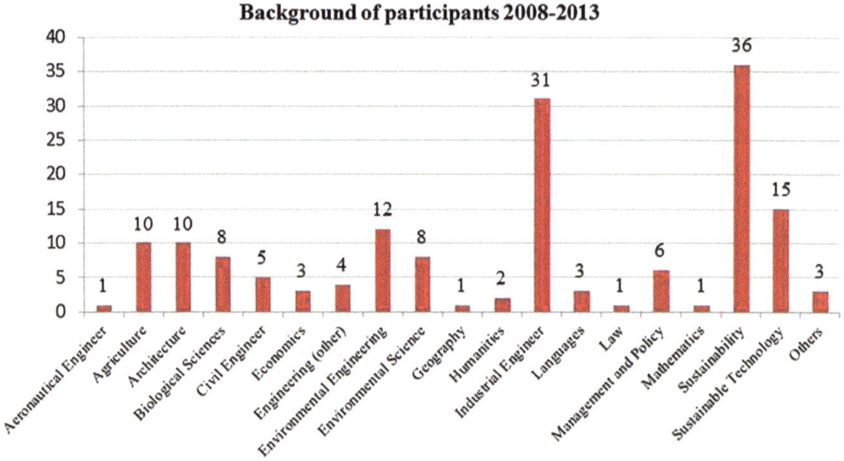

Fig. 4 Transdisciplinary atmosphere: Background of participants

2.4 Programme Assessment

During the course two types of assessment are carried out:

Students evaluation: Students are evaluated by their communication skills in terms of poster design and presentation, and the report communication value; their capacity to work in multidisciplinary and multicultural atmosphere in terms of team-building competence; the cognitive thinking contained with the content of the final presentation and report, in terms of correct application of the backcasting methodology to the case study challenges. Upon completion of the virtual and the presential phase, students and teachers have to submit an essay assessing their individual experiences, expectations, and overall assessment of learning achieved. Overall, students are evaluated during all three phases. In every phase students are supervised either by their home university, seminar assistants, project staff or the academic panel. They will receive grades for every piece of work they hand in. In the end, their final grade will be composed of 4 different parts: participation, report (divided into poster and virtual report), final assignment (divided into communication, presentation and content) and the final individual evaluation.

Course evaluation: After completing the course, evaluation questionnaires are handed out to all participants at the presential phase in order to first evaluate the organization of the seminar, and second the academic competences and learning environment. Thus, the seminar is assessed (scored from 1 to 5) in terms of the content (plan, objectives, information provided, workload, interest) and organization (facilities, logistics, accommodation, organizers involvement). Lecturers and external collaborators are evaluated in terms of their communication skills, introduction of interesting elements for the discussion and reflection, and support for the students' work. Finally all the pedagogical methodologies (pre-work, students' presentations, lectures, dialogues, visits, study cases) are evaluated.

3 Results

The course is very successful among students of masters related to sustainability, as every year the number of students and universities participating in the course increases.

The course evaluation shows a high satisfaction of the students in relation to the content and organizational issues and also to the pedagogical methodologies applied. See Figs. 5 and 6.

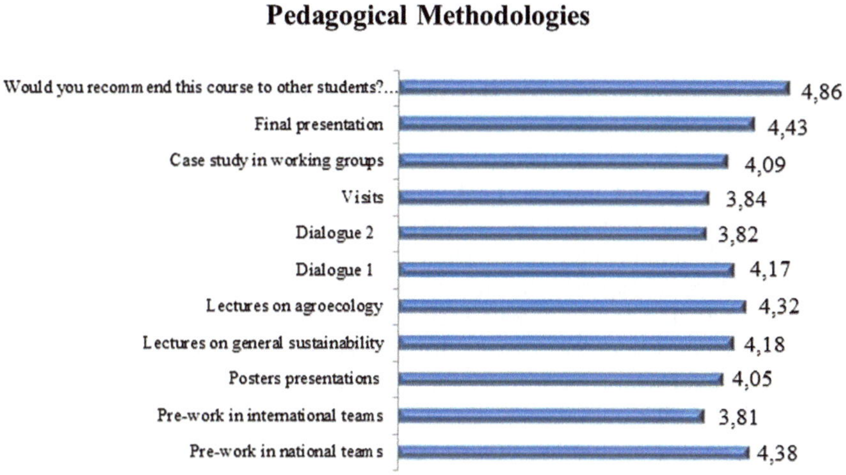

Fig. 5 Assessment of pedagogical methodologies. (1 absolutely disagree, …, 5 absolutely agree). Year 2012

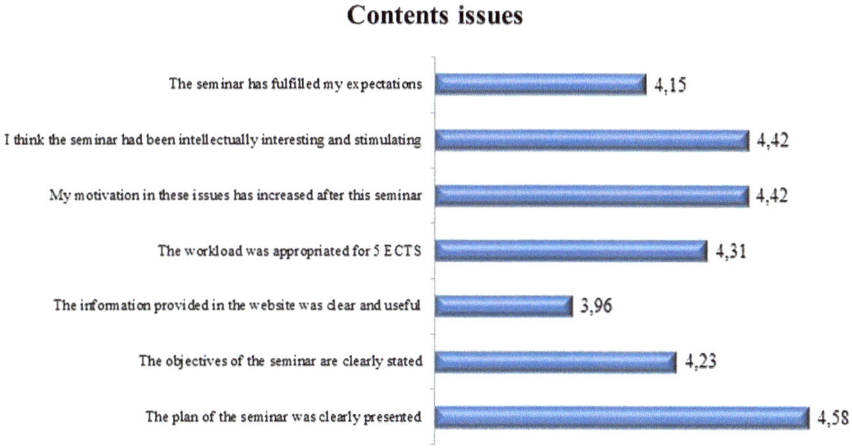

Fig. 6 Assessment of the programme. (1 absolutely disagree, …, 5 absolutely agree). Year 2012

4 Conclusions

After 6 years, the Seminar has achieved much more than a purely cognitive theoretical type of learning. The community-oriented case studies, together with the stakeholders and social movement dialogues, bring together academia and society. The course has been shown to be an area with enormous potential to bring social needs to the world of ideas, beyond a learning space.

Originally the Natural Step (Holmberg and Robert 2000) backcasting approach was used, but currently students are introduce also to participatory backcasting (Quist 2007) developed in The Netherlands. This allowed students to better understanding of the backcasting approach as they learn two different methodologies. Moreover each case study is analysed by four groups of students applying both backcasting methodologies. This allows comparing the results between groups and between methodologies.

During these 6 years of experience some aspects has shown to be crucial in order to assure the learning process. First, students need sufficient time to apply the methodology with enough time to discussions where interdisciplinary learning takes place (20 h in the face to face phase); second, working group sessions should be tutorized by a experienced supervisor in both workshops and backcasting methodology; and finally, students need to have sharp guidelines and milestones during the seminar in order to assure that they have enough time to go through all the steps of the backcasting methodology.

In relation to the objectives of the course, it uses cross-disciplinary and transformative learning approaches needed for sustainable development. The different nationalities, cultures and background of the students have shown to be very relevant to enable these learning to take place. The path to transdisciplinarity starts when stakeholders participation is added in the process and mutual learning appears.

In order for students to become change agents for sustainability, it is crucial that stakeholder involvement and participatory processes are contained within the learning process including problem definition, analysis (current situation, vision) and the advancement of possible.

Acknowledgments We thank the support of the European Commission Education and Training Erasmus program, whose funding has enabled to run the seminar since 2012. Also with these lines, we pay tribute to Leo Jansen, who died in August 2012, having participated actively and enthusiastically in the course, from its creation.

References

Barcelona Declaration (2004). Engineering education in sustainable development conference barcelona. http://www.upc.edu/eesd-observatory/who/declaration-of-barcelona. Accessed 14 Apr 2014

Fry G (2003) Training needs for interdisciplinary research. In: Tress B, Tress G, van der Valk A, Fry G (eds.) Interdisciplinary and transdisciplinary landscape studies: potential and limitations, DELTA Series 2. Wageningen, pp 118–123

Holmberg J, Robèrt KH (2000) Backcasting: a framework for strategic planning. Int J Sustain Dev World Ecol 7:291–308
Quist J (2007) Backcasting for a sustainable future. The impact after 10 years. Ph.D. Dissertation, Technological University of Delft. Eburon academic publishers, Delft
Schumacher EF (1973) Small is beautiful. economics as if people mattered. Blond & Briggs Ltd., London
Segalas J, Mulder K, Ferrer-Balas D (2006) Embedding sustainability in engineering education. Experiences from Dutch and Spanish technological universities. In: Conference: higher education for sustainable development: new challenges from a global perspective, Luneburg
Sterling S (2005) Higher education, sustainability, and the role of systemic learning. In: Corcoran PB, Wals AEJ (eds) Higher education and the challenge of sustainability: problematics, promise and practice, Kluwer, Boston, pp 49–70
UN (2012) The millennium development goals report. UN, New York
UPC (2009) Pla UPC sostenible 2015. http://www.upc.edu/sostenible2015/pla-upc-sostenible-2015. Accessed May 2013
World Wildlife Fund (2012) The living planet report: biodiversity, biocapacity and better choices. WWF International, Gland, Switzerland

Author Biographies

Jordi Segalàs (1967) works as associate professor and Deputy Director at the Research Institute of Sustainability Science and Technology in the Universitat Politècnica de Catalunya UPC-Barcelona Tech. He is Director of the Master in Sustainability Science and Technology at UPC-Barcelona Tech. He is the head of the Research Group on Sustainability Education and Technology in Higher Education. He has been the Director of the Catalan Network of Education for Sustainability. He obtained his PhD in Sustainability Education in Engineering from Barcelona Tech University. He has been working in curriculum greening policies and actions plans at the Barcelona Tech University since 2000. He is also working in TEMPUS (trans-European cooperation scheme for higher education) projects related to sustainable development in higher education. He has published more than 60 articles on higher education and sustainability.

Gemma Tejedor (1966) works as project manager and researcher at the Research Institute of Sustainability Science and Technology in the Universitat Politècnica de Catalunya UPC-Barcelona. Her PhD work is in transdisciplinarity learning for sustainability in engineering education. She holds a Master degree in Sustainability by UPC – Barcelona Tech and a Master in Forestry Engineering by University of Lleida. She has been working as a forestry engineer for about 15 years, both in public and private sector, as well as in cooperation projects in Latin America. She is currently working in TEMPUS (trans-European cooperation scheme for higher education) projects related to sustainable development in higher education.

Assisting Sustainability Through Achievement of an Integrated Health, Safety and Environmental Management System at Cardiff University

Lara Hopkinson and Julie Gwilliam

Abstract

Universities use a multitude of methods to assist in driving sustainability forward. As large organisations, it can often be difficult to instigate change covering all aspects of the institution, without using a systematic method for implementation. This paper investigates the approach used by Cardiff University over a 5-year period to implement an integrated Health, Safety and Environmental Management System. It will show the approach used to ensure coverage across every academic school and professional service, and discuss this in terms of being one of the first UK Universities to achieve such integrated certification. It will then investigate how this approach has assisted in the development of wider sustainability issues and programmes across the entire institution, including "Green Impact", through detailed investigation within three academic schools and a professional service. The problems and barriers encountered during the process will be discussed, along with approaches to mitigation of these. A brief analysis of Cardiff's progression towards sustainable development compared to other Universities in Wales is also considered. Finally, some recommendations for further improvement at Cardiff will be discussed.

Keywords

Sustainability · Environmental management · Institution approach

L. Hopkinson (✉) · J. Gwilliam
Welsh School of Architecture Cardiff University, Bute Building,
King Edward VII Avenue, Cardiff CF10 3NB UK
e-mail: HopkinsonL@cardiff.ac.uk

J. Gwilliam
e-mail: GwilliamJA@cardiff.ac.uk

W. Leal Filho (ed.), *Transformative Approaches to Sustainable Development at Universities*, World Sustainability Series, DOI 10.1007/978-3-319-08837-2_24

1 Introduction: Sustainability and Higher Education

Sustainable development is a well-established concept, having been written about and discussed in many circles from as far back as the 1960s. However, consideration of sustainability within the Higher Education sector is still a relatively new and emergent area of inquiry (Jones 2012). In terms of defining sustainable development, the most widely used and familiar definition is that published within the Brundtland Commissions report 'Our Common Future' as follows: *Development that meets the needs of the present without compromising the ability of future generations to meet their own needs* (WCED 1987, p. 43). This quotation has been rewritten and misrepresented countless times, and yet is still relied on by many as the descriptor of sustainable development, by Governments across the World.

Higher Education, by its very nature, provides significant opportunity for the application, dissemination and delivery of sustainability, with the potential to catalyse or accelerate a societal transition towards a more sustainable way of thinking, acting and living (Jones 2012). Within the education process itself, there is further potential to ensure students are more 'globally aware', employable, and empathetic towards their fellow humans, whilst still considering the need to be economically and socially secure. Higher Education prepares the world leaders and professionals of the future, and therefore has a clear moral responsibility to consider issues relating to sustainability (Cortese 2003). Education for Sustainable Development (ESD) enables people to consider the longer term impacts of their decisions (UNESCO 2013).

But universities, also have substantial operational impacts, and within this more pragmatic, 'business' aspect, that enables the educational practice and research to occur, there are further opportunities to respond to the sustainability agenda positively, providing leadership to other institutions and organisations. It has been noted in literature that both formal (hard) and informal (soft) systems are required in order to have effective sustainability strategies, but that all too often, this is not the case (Epstein and Buhovac 2010). In Higher Education, the focus to date has been to 'green' campus activities, working more towards 'environmental sustainability' (James and Card 2012) and to learn from lessons from such work (Dobson et al. 2010).

A clear way to achieve such environmental sustainability would be to utilise a formal environmental management approach such as the implementation of an Environmental Management System (EMS). Such a system places a formal framework within an organisation of any type, to enable continuous environmental improvements. There are a number of recognised EMS, including the International Standard ISO 14001, the European Community's Eco-Management and Audit Scheme (Pérez et al. 2007), and British Standard 8555. In essence, all follow a similar Plan-Do-Check-Act methodology, clearly identifying environmental areas of improvement, and undertaking an audit to ensure that progress has been achieved.

In terms of sustainability, an EMS can be seen as a framework for how a sustainability programme can be physically implemented (Clarke and Kouri 2009; Bero et al. 2012), helping to clearly identify the measures to be looked at, and enabling the collection of the appropriate evidence associated with this. Formalised systems (i.e. those that are accredited by an external auditor) then have a verification route through the requirement for an external audit process.

Limitations do arise within the sector. In general, sustainability (in whichever format the University decides to act upon) is centralised within estates departments (Jones 2012), as these control the areas of highest environmental impact including the university estate and waste management. Further, there is a tendency to use sustainability officers 'on the cheap' (Dobson et al. 2010), by passing responsibility to project officers who may have the drive but lack specific sustainability expertise. It is also important to note that sustainability should not just be seen as something relating solely to '*managerial efficiency*' (Bannerjee 2003, quoted in Jones 2012), which is perhaps how the adoption of an EMS is sometimes perceived.

A further driver for sustainability within the Higher Education sector is the People and Planet 'Green League'. This was first undertaken in 2007 (Jones 2012) and is an environment-sustainability league table for all Universities in the UK. It is carried out annually and scores establishments using the standard student degree categories (i.e. first, upper second, lower second, third, fail) based upon a number of criteria, including carbon, community, campus and curriculum (People and Planet 2012). The development of such a league table can outline sensitivities within university senior management. As Dobson et al. (2010) states "*University managers....rightly or wrongly believe that it makes a real difference to an institution's prospects whether it is near the top or bottom*". There is also the argument that universities can use such a league table to 'greenwash' their wider sustainability engagement agenda (Jones 2012). At present, there is little evidence that the Green League is changing university behaviour (Jones 2012), but this may well change with the proposed overhaul of the scoring system in 2014.

2 Specific Welsh Challenges and an Integrated Management System Approach

In 2008, the Higher Education Funding Council for Wales (HEFCW) issued a circular announcing funding to each institution to "*develop a suitable EMS and equivalent measures which cover their operations in order to minimise their overall resource use to sustainable levels*" (HEFCW 2008). The circular was not prescriptive and allowed institutions to choose an EMS that was appropriate to their needs.

As a result of this circular Cardiff University decided upon an integrated management system approach for Health, Safety and Environment. Literature suggests that an integrated approach provides more leverage for integrating sustainability issues into the mainstream (Asif et al. 2011), provides greater 'business excellence'

(Karapetrovic 2003) and ability for integration of audits due to their similarities (Karapetrovic and Willborn 1998). Summarising the relevant theoretical literature, Bernardo et al. (2009) state that integrated management systems are "*a process of linking different standardised management systems into a unique management system with common resources aiming to improve stakeholder's satisfaction*".

This paper aims to identify the extent to which the Cardiff University integrated management system approach has helped to drive sustainability more widely within the institution. As earlier considered, this paper aims to explore whether the integrated approach has assisted in enabling sustainability to be considered as mainstream in a diverse organisation, in this instance Cardiff University. The achievement of the integrated system is innovative for the sector, with Cardiff reporting that it was the first UK University to achieve such an integrated certification (People and Planet 2013). Thus, the research presented here, begins to establish a body of knowledge on the impact of integrated management systems relating to sustainability in the HE sector. The research has utilised an interview-based methodology and presents recommendations for further improvements.

3 Background: The Approach Used at Cardiff University

The starting point for the adoption of the integrated management system approach at Cardiff University was the development of an Environmental Management System (EMS) within the Cardiff and Vale NHS Trust, which included the University of Wales College of Medicine (UWCM) at the time. The NHS was given the challenge to adopt an EMS, with the Trust deciding on ISO 14001, achieving this in September 2003.

In August 2004, UWCM was formally merged with Cardiff University. This provided a focus for existing staff with responsibilities for the EMS at UWCM level to encourage the University to adopt this system as well. This coincided with the HEFCW circular, calling for Welsh HEs to adopt EMS. Cardiff therefore decided to build on the existing EMS experience in UWCM and implement the International Standard ISO 14001, thus enabling good practices to be shared internally. Further, as an international standard, it was also considered the appropriate choice for an aspirational world-leading University. Further discussions helped to highlight the similarities between ISO 14001 and 18001, the International Standard for Health and Safety management system, to formalise existing practice within the University, and as a result, it was agreed that an integrated approach would be beneficial. In audit terms, it reduced the burden of numbers of audits, and the systems also shared certain aspects of work.

One final driver to aid the adoption of an EMS was the People and Planet 'Green League', with points in the assessment awarded to institutions that had adopted an EMS. With the resulting comparison of institutions in the league table, this provided a further driver in terms of seeking senior management support.

As the HEFCW circular suggested using "*good practice examples*" (HEFCW 2008), Cardiff was keen to utilise existing structures and learn from lessons within UWCM. At the time of starting the implementation, there were a number of existing committees to which the initiative reported, including the Sustainability Group (providing approval on this, the University Travel Plan and the development of an Education for Sustainable Development assessment tool called STAUNCH). In addition, there was also a pre-existing environment working group, energy working group and steering group (chaired at the time by the Vice Chancellor). In order to streamline processes, these groups were merged and redefined to form the EMS Steering Group, which is still in operation today, chaired by the Deputy Vice Chancellor. The University also benefitted from the introduction of an eco-champions network, which was started by a senior member of staff within the Finance division. Eco-champions were typically self-selected, as people who were passionate about assisting with environmental improvements within their local area, although informally they received support from Heads of Schools or Divisional leads. This network changed in 2012, with the development of a more formal Environmental Compliance Officer (ECO) role. The major change was that at this time this became a formal role, with Heads of School/Divisional leads being invited to nominate people to undertake the role. The role is formally acknowledged in job descriptions, has a role profile of its own, but is still generally undertaken by those who are passionate to drive change in their individual area.

At the time of issue of the HEFCW circular, there was only one member of staff within the University who had responsibility for issues relating to implementation of the EMS. Indeed, the Professional Service division responsible for the integrated system consists of a small team, each with clearly defined roles. The funding was therefore used to appoint an EMS Officer who assisted with the process.

The design of the process was to ensure that all areas of the University participated in the integrated system. Beginning in 2009, a pilot was staged, involving a number of academic schools and one directorate. In this pilot, each participating area was given full support to undertake an environmental review, from which a report was written and recommendations made for further improvements. These recommendations then formed action plans that were agreed by each pilot area, and then implemented.

During the pilot phase, there was a great deal of promotional activity running simultaneously. Again, to utilise existing good practice, existing fora were used to promote the existence of the integrated system development. The pilot areas were used as examples of good practice (using case study development to facilitate work in other areas of the University during roll-out). The promotion and communication around the integrated system was also important, as it made the systems 'visible' to the rest of the staff body. It was very clear that the University was keen not to make this just another 'senior management' initiative.

Once the pilot phase was complete, a roll-out phase was launched. Due to the significant numbers of areas to be covered (28 academic schools and 9 major Directorates, further sub-divided into a number of divisions each), the roll-out was undertaken as a staged approach, using 3 phases. Due to staffing restrictions, the onus in the roll-out phases was to encourage areas to undertake work by themselves. Guidance and promotion was given to utilise within each area, and a team approach was advocated to enable work to be carried out in a timely fashion. Each area was asked to define a team leader—normally someone senior within the school or directorate to provide an appropriate level of gravitas to enable progress. Each team leader was provided an environmental review template; to facilitate this, meetings were staged initially to outline why the information was needed and some guidance to the best methods of collection. Deadlines were set for return of the information, which was then reviewed centrally. This review firstly checked whether the correct information had been provided, and then defined an associated action plan based upon the findings. At this stage, each area was sent the draft action plan to agree, along with detailed management plans on how the actions plans could be achieved and timescales.

At the same time, over-arching legal and environmental aspect registers were developed, as well as other associated management system documentation e.g. manual, associated management plans and audit protocols. It was viewed that the registers should be kept centrally, but reflects the legal requirements and environmental aspects identified within every area of the University. Whilst this was the case when the integrated systems were audited for the first time, it has been since recommended that bespoke versions of these documents are created for each School and Professional Service, and held at these levels. The manual was designed to be as simple as possible, whilst still reflecting the complexity of the University.

At accreditation audit, it was recognised that the University had more established strategic frameworks in place for Health and Safety, than those proposed for EMS and that these were more embedded within the institution. Accreditation of the integrated system was achieved in November 2011, and it was acknowledged that at the time, the University was the first in the UK to achieve such an integrated approach. This system currently covers all academic schools and Professional Services within the main Cathays campus; while, the medical schools based with the Local Health Board are still covered by the NHS system, due to issues surrounding facilities and shared resources. At present, the University is in the re-certification phase, and the auditor is challenging the University to become more sector leading in terms of the objectives and targets it sets.

4 Methodology: Assessing Sustainability Through the Integrated System

In order to assess how sustainability within the University has been assisted through the integrated management system, a questionnaire approach was utilised. It was decided to interview one school within each of the 3 Colleges of the University (College of Arts, Humanities and Social Sciences (12 Schools); College of Biomedical and Life Sciences (8 schools); and College of Physical Sciences and Engineering (7 schools)), as well as one of the 33 Professional Services. The areas selected for interview were chosen based on their progress with the integrated system since implementation and accreditation.

The questions aimed to establish:

1. the perceived importance of implementation of the integrated management system;
2. whether practices had changed due to implementation;
3. whether senior managers within the local area were sustainability-minded;
4. if there was a perception that the University senior management were sustainability-minded;
5. what the respondents would recommend for further change.

The questionnaire used was approved through the Welsh School of Architecture Ethics Committee in February 2014. Interviews were staged face to face in February and March 2014. Whilst the number cannot be considered to be representative, the findings represent a reasonable overview of the stage and effectiveness of implementation at this point in time. It is anticipated that that this process will be carried out within all academic areas of the University (i.e. all 27 schools), as well as every Professional Service area to gain a full picture, in the future, the benefits of which from one of the key recommendations of this paper. By staging interviews face to face, this enabled as open as possible a conversation around the subject, guided by the approved interview schedule, and aimed to enable deeper qualitative interpretation of the results.

In all cases, the interviewee was the Environmental Compliance Officer (ECO) for the area under question. As stated earlier, this role has formal recognition within the University, and is nominated through all Heads of Schools and Professionals Services; in all cases, this is a role that is carried out by staff undertaking other roles within the University, normally safety officers. The ECO role is normally undertaken by staff passionate about driving environmental change within their area; therefore these staff could be described as the 'converted', staff who already show a high degree of pro-environmental behaviour.

The survey aimed to assess the ECOs perceptions of sustainability performance within their own local area, and whether this was being driven by the implementation of the integrated management system.

5 Results: Analysis of Interviews and Identification of Sustainability Good Practices

It can clearly be seen that Cardiff University demonstrates good practice by having the ECO role within each area of the University. Such 'champions' can be instrumental in driving forward initiatives to assist in embedding sustainability (Stoughton and Ludema 2012). However, the responses show that support for this role is varied among the respondents, with some having clear line management and committee support, while others are just '*being left to get on with it*'. In all cases, support is acknowledged from Heads of Schools or Divisional leads, outlining the importance of the ECO role more widely. In all cases, the areas interviewed became involved with the integrated management system as part of its roll-out across the University, although the exact point of entry to this was not clear for any respondent. Again, in all cases, there was no issue with initial involvement either. There are a number of reasons for this—it was either seen as a '*positive process*', a process facilitated by existing policies already in place, or as '*part and parcel of what the staff wanted to do*'.

Respondents were then asked whether practices had changed as part of the integrated management system, and whether this change was a noticeable one. It is notable that here, the differences between academic schools and professional services becomes marked. In a positive light, changes that were highlighted included '*more awareness due to having formal frameworks in place*', '*improvement on waste management including better management of sharps waste*', '*a greater emphasis placed on environmental issues*' (where the respondent indicated that this had been less important before), and the chance to develop greater team-working within areas. In terms of whether the changes were noticeable, all areas agreed that this was the case. One area said that they have probably '*done more as a result of the integrated management system*', and that this was '*a systematic approach whereas before it was not planned internally with input from other University colleagues*'. There was emphasis on better planning and control, for example in terms of storing waste oils, chemicals and so on. One area indicated that the integrated management system has enabled the '*inclusion of environmental issues*', as these had not been considered as highly as Health and Safety issues in the past.

When queried whether the integrated management system had helped to boost sustainability within their area, again respondents were mixed. One noted that there is now a '*much stronger evidence base; there is now a coherent approach and internal plan which still enables an ad hoc approach driven by the grass roots, due to people being passionate about it*'. Another said that the '*success of the management system was down to the support to highlight what is needed*'. Conversely, one respondent saw the integrated management system as '*Safety, Health and environment with a small e*'. One area did not believe there to have been any change. One issue raised with this question was awareness amongst the wider body of staff. It was noted that not all staff (within this specific area in question) were aware that the University had an integrated management system, leading to discussion on auditing and whether practices could be improved to capture this sort of information; sustainability is not currently captured within the audit process. It was also noted that in some cases, the ambitions of the University could be seen to be at odds with improvement in sustainability performance. For example, within the University strategy 'The Way Forward', there is an ambition that a percentage of students will have '*studied, worked or volunteered abroad during their time at Cardiff* (Cardiff University 2012)'; but this could be viewed as increasing the potential for travel emissions.

Ways in which the respondents felt that sustainability was boosted due to the integrated management system included better links between the Health, Safety and Environmental issues affecting the University and the re-branding of committees to include all issues. One respondent mentioned some societal impacts that activities within their area related to (for example, waste products being used within Third World countries), but indicated that this was not the norm at present. One respondent outlined that they felt their area was still '*reactive*' when it came to dealing with issues.

For the areas that felt sustainability had not been boosted, they were asked what could be done to improve upon this. One respondent indicated time being a considerable issue—the time it takes to consider the new initiatives on top of everything else that needs to be done, especially when related to environmental issues. The other respondent suggested changes to induction for both staff and students, to help to highlight what the University does and what they can do to help. It was also suggested that senior management within academic areas need to take their responsibilities more seriously, otherwise it is not in their minds to consider it.

When asked whether sustainability was now considered at the forefront of decision making within their area, the respondents were clearly split between academic and professional service areas. All academic areas agreed that sustainability was not considered within the decision making process. Reasons for this included '*there is no incentive which is part of the problem*' and '*it is difficult for environmental/sustainability issues to gain traction when compared to teaching quality and REF*'. However the professional service area indicated that sustainability was one of their '*key criteria*' when considering projects and other issues. For example, the operational plan for the professional service contains sustainability as a key concern, and the respondent works closely with colleagues in other areas to help develop joint policies and plans at University level. When discussed in more detail with all respondents, the responses can be validated by the level of control each area considers themselves to have when considering sustainability issues. It is clear to see that a professional service (central to the University and able to control many significant sustainability related issues) would see themselves able to consider sustainability as part of their decision making process, as this enables practices to be cascaded down across all areas, including academic ones. Where as, academic areas have many pressures placed upon them, and perhaps fail to see the initial significance that utilising sustainability-based decisions can provide, let alone have to time to consider them.

Respondents were then asked whether there was senior management support for sustainability within their area. Responses were evenly split between yes and no. Those saying yes indicated that key senior staff were involved in relevant committees or had relevant research and teaching portfolios relating to a sustainability issue. They also noted that there was less clarity on '*whether this is implemented further down in day-to-day decision making*', and that issues relating to the REF constrained the ability to make decisions relating to sustainability issues e.g. when considering teaching staff, their travel away from work, and their ability to do this in a sustainable manner whilst visiting the places needed in the timeframes available.

Related to this, respondents were asked whether they thought the University senior management considered sustainability as part of its decision making process. All respondents were agreed that there were considerable good practices and good leadership examples at senior management level—one respondent noting that there was a '*small and good core centrally with the right attitudes and beliefs*'. Indeed, environmental and social sustainability is noted as one of the key guiding principles for the University in 'The Way Forward' (Cardiff University 2012). It was noted that more could be made on interlinking individual good practice to '*deliver on the challenge of being a sustainable institution*'.

Finally, all respondents were asked what one change they would recommend to further drive sustainability practices within the University. Most of the recommendations made were strategic in nature. The first was investment to reduce environmental impact from the University estate. It was then suggested to target senior management within academic settings, to evidence their commitment, highlight how they influence change, show them that they also have a journey to take, and that if they were willing to take this journey, others would follow. The final strategic recommendation was to outline that the University, Colleges and academic schools have to demonstrate environmental performance and be held accountable to this. The integrated management system is an ideal way of demonstrating this, and it was noted that if this was already the case, that this requirement was not communicated down to college level. There was also discussion around resources and capacity to aid this demonstration of performance. The final recommendation made was to promote and incentivise sustainability performance. For example, enable individual areas to purchase better performing equipment through subsidies and so on. It was also suggested that a strap line is needed to encourage people to be more sustainability minded—why do it? The University needs to be seen as a '*driving force for change*'.

6 Discussion: Improvement Opportunities for Cardiff University

As stated earlier in this paper, this research was carried out in only one academic area of each of the Colleges within the University, and one Professional Service area. In order to obtain a clearer view of whether sustainability is fully considered within Cardiff University due to the integrated management system, it is recommended that interviews are staged within each area. This may help to highlight further pockets of good practice that are known to exist across the University, but that are not being communicated to other staff. It will also help to identify other opportunities for change that have not been captured to date.

The interviews at College and Professional Services level highlight that although Cardiff has many good practices, and that the integrated management system is largely seen as a positive framework to enable improvement to occur, there are still improvement opportunities that the University can take advantage of.

Taking the recommendations suggested by the interviewees as a starting point, perhaps the most notable would be that related to investment to reduce environmental impact from the University estate. This is not a new recommendation; during the implementation of the integrated management system many staff involved in the process from across the entire University raised this issue. Due to the size and building typology common across the University estate, this is also not a recommendation that can be dealt with quickly or easily; for example, the estate has many ageing, historic and event protected listed buildings. However, where possible sustainable retrofit, would improve both the conditions within the buildings, but also increase resource efficiency and reduce carbon emissions (Carbon Trust 2008), in line with the University's Carbon Management Plan (which aims to achieve an absolute reduction in CO_2 emissions by 20 % by 2020). One clear improvement opportunity from this recommendation would be to communicate more clearly to staff what the anticipated refurbishment plans are along with timescales. A further opportunity is to utilise staff expertise within the field of sustainable refurbishment, as well as encourage student engagement within small, discrete projects to identify sustainability perspectives. This also enables students to enrich their learning experience, and assists with the Education for Sustainable Development (ESD) agenda. This issue will also be addressed within the Estates Strategic Review, due for completion by September 2014, which will enable input from staff across the University.

The next recommendation made relates to engagement of senior management within academic areas on sustainability issues. Literature suggests that senior managers who are '*responsive and engaged*' enable actions to occur quickly (Dobson et al. 2010). This has improved within Cardiff University with the appointment of a new Vice Chancellor in 2012, with sustainability, particularly within the curriculum, now featuring within the portfolio of the Deputy Vice Chancellor. However one positive practice (the Pro-Vice Chancellor led sustainability group) was disbanded in 2011, despite it being responsible for agreeing the integrated management system, travel plan and educational sustainability work. This was in answer to a rationalisation of University committees, especially as the EMS Steering Group reports into the University Health and Safety committee. It is noted that senior staff have considerable workloads, and perhaps this is detrimental to full acknowledgement of sustainability issues. One clear way to rectify this would be to appoint suitably expert staff at operational level for which sustainability is their key role. This model is used in a number of other Universities across the UK to great effect for example, the University of Gloucestershire has a sustainability team, Kingston University has a sustainability hub, while Exeter benefits from a Sustainability Manager. Another way to achieve this recommendation is to raise awareness more fully within the University senior staff about the importance of sustainability in the longer term. This is an issue where HEFCW could help, although HEFCW are bound by Welsh Government and the guidance they receive within their annual remit letter, the last of which mentioned sustainable development was 2012/13. Within this model, operational staff would become the method of guidance to senior managers, providing the link between sustainability and good University business practice.

The next recommendation relates to demonstrable environmental performance, and ensuring that someone is held accountable for such. When this arose in the interview process, discussion was held around whether the integrated management system should already achieve this. A way to rectify this particular issue involves greater communication about how the integrated management system relates to each individual area within the University. Greater emphasis could be placed upon achievement of individual action plans in line with the overall University objectives and targets, linking also to other important policy documents including the travel and carbon management plan. In addition, more communication needs to be provided to all staff to highlight that they all hold a responsibility for actions related to the integrated management system.

This leads into the final interviewee recommendation of promotion and incentivisation of sustainability performance. When discussed in the interview process, this was suggested relating to purchasing equipment. It is acknowledged that existing purchasing structures and financial constraints do play an important role in how items are procured. Where financial constraints are of uppermost importance, consideration of sustainability performance is increasingly pushed aside purely on cost measures. However, the University is attempting to re-dress the balance, and is currently looking at a central purchasing policy for recycled paper for the entire University, quality permitting. At individual area level, those with responsibility for purchasing should be given guidance on sustainability issues to consider, and all staff should be guided to actively seek more sustainable options when purchasing items for their courses, projects or groups. Promotion of this can happen using existing staff communication mechanisms, without further cluttering inboxes with more emails and examples are already being used across the University in fields such as inclusion within staff and student induction sessions, staff and other appropriate meetings, promotion through the ECO network and so on. Equally important is the 'why should I bother?' question. With so many issues being seen as central University issues, it can be difficult for staff to even consider becoming engaged, especially if previous efforts have not resulted in any significant or demonstrable change. The use of incentives was discussed; in interviews, it was suggested in terms of the University 'subsidising' the purchase of more sustainable equipment, therefore enabling areas to move beyond their financial constraints. In practice, this is unlikely to be possible across the board. However incentives such as competitions for example, a recent competition to win vouchers for participation in a sustainability skills survey could be used in the short term, and should be linked to potential savings that could be achieved through more sustainable purchasing practices. Incentives, if appropriately used, can therefore be seen to help to drive this cultural change, to enable staff to believe that their actions can contribute to a greater good, and therefore also help to drive change.

In terms of further opportunities for the University, there is also the potential to include curriculum within the audit process of ISO 14001. At present, the authors are aware of one other UK University that has achieved this within the past 12 months. More research would be needed to assess the significance of this, including the practical aspects behind adding curriculum into the audit process, the evidence that is needed and the scale of inclusion.

7 Conclusion: The Future for Cardiff

The key lessons from this short study are that practices in Cardiff University are improving relating to sustainability, and that the integrated management system is a useful framework from which further improvements can be driven. However, the impact of the integrated management system is still limited relating to sustainability performance. The main lessons learnt from this study are as follows:

Clearer communication is key to driving change, and in order to have impact, the message needs to come from the senior management team of the University, down through the new College structure into individual academic and professional services areas. The existing ECO network is a valuable mechanism to ensure information is disseminated at local level, although there may be disparity across the network on what is communicated and when. More leadership on communication will help to address this issue and ensure parity across the institution.

There needs to be acknowledgement of a clear senior management commitment to increasing sustainable performance across the campus; not only operationally but within formal and informal curricula as well. Utilising the integrated management system as a framework can assist directly in this, while implementing the recommendation to place a key member of operational staff, with sole responsibility for sustainable performance, will also help to drive understanding of achievements. Estate improvements are at the core of a further recommendation for action, where ensuring that forthcoming widespread estate improvements has the potential to provide significant revenue savings as a result of capital investment. Adjacent to this proposal would be mechanisms to enable acknowledgement of individual recommendations in this field, from across all institutional staff, helping to engender not only a sense of community, but a sense that individuals suggestions are listened to and actioned. Whilst the authors note that there is only a limited budget for estate improvements, consideration of whole life costing when exploring sustainable retrofit solutions in this area can lead to cost savings in the longer term, enabling more capital to be available for further improvements. Related to the integrated management system, this mechanism can directly assist with the continual improvements that are required as part of the audit process, as well as highlighting the institutional commitment to apportioning budget to the improvement of the estate.

Finally there needs to be consideration of curriculum issues, with Education for Sustainable Development as a powerful agenda that can help to transform students, educators and operational staff towards greater consideration of sustainability issues. Widespread adoption of the principles of ESD would enable students to become better leaders of the future, being more sustainability literate, understanding the issues affecting them and their fellow man. Educators can become more expressive through their curricula, and better able to confidently outline sustainability concerns within their field of specialism, meanwhile operational staff can become involved in projects, driving further improvements within the University as a direct consequence. In terms of the integrated management system, Cardiff might consider including curriculum within the scope of the system. This is still a fledgling concept within the UK Higher Education sector, and would need careful scoping, but has the potential to directly contribute to the ambition to become a sector leading institution.

These are all big issues; but Cardiff continues to work towards its challenges, and therefore strives to be the '*driving force for change*'.

Acknowledgments The authors wish to express their thanks to all colleagues within the University who assisted with this work. In particular, those who were interviewed and agreed for their quotes to be used, and to colleagues within the Occupational Safety, Health and Environmental Unit who provided background to the integrated management system development. The authors also wish to acknowledge the support of the European Social Fund (through the Welsh Government) to enable the development of this paper.

References

Asif M, Searcy C, Zutshi A, Ahmad N (2011) An integrated management systems approach to corporate sustainability. Eur Bus Rev 23(4):353–367

Bernardo M, Cadadesus M, Karapetrovic S, Heras I (2009) How integrated are environment, quality and other standardized management systems? an empirical study. J Clean Prod 17 (8):742–750

Bero BN, Doerry E, Middleton R, Meinhardt C (2012) Challenges in the development of environmental management systems on the modern university campus. Int J Sustain High Educ 13(2):133–149

Carbon Trust (2008) Low carbon refurbishment of buildings—a guide to achieving carbon savings from refurbishment on non-domestic buildings. London

Cardiff University (2012) The Way forward 2012-2017. http://sites.cardiff.ac.uk/thewayforward/. Accessed 20 March 2014

Clarke A, Kouri R (2009) Choosing an appropriate university or college environmental management system. J Clean Prod 17(11):971–984

Cortese AD (2003) The critical role of higher education in creating a sustainable future. Plan High Educ 31(3):15–22

Dobson A, Quilley S, Young W (2010) Sustainability as a competitive advantage in higher education in the UK. Int J Environ Sustain Dev 9(4):330–348

Estein MJ, Buhovac AR (2010) Solving the sustainability implementation challenge. Org Dyn 39:306–315

Higher Education Funding Council for Wales (2008) Circular: education for sustainable development and global citizenship. Reference W08/07HE, Cardiff, 12 Feb 2008

James M, Card K (2012) Factors contributing to institutions achieving environmental sustainability. Int J Sustain High Educ 13(2):166–176

Jones DR (2012) Looking through the "greenwashing glass cage" of the green league towards sustainability challenges for UK universities. J Organ Change Manag 25(4):630–647

Karapetrovic S, Willborn W (1998) Integrated audit of management systems. Int J Qual Reliab Manag 15(7):694–711

Karapetrovic S (2003) Musings on integrated management systems. Measuring Bus Excellence 7 (1):4–13
People and Planet (2012) Green league report 2012: driving UK universities transition to a fair and sustainable future. People and Planet, Oxford
People and Planet (2013) Cardiff university awarded 2:2 class in the people and planet Green League 2013. http://peopleandplanet.org/green-league-2013/tables?ggl13profile=9926&test=aab387. Accessed 14 Apr 2014)
Pérez EA, Ruiz CC, Carrasco-Fenech F (2007) Environmental management systems as an embedding mechanism: a research note. Acc Auditing Account J 20(3):403–422
Stoughton AM, Ludema J (2012) The driving forces of sustainability. J Organ Change Manag 25 (4):501–517
UNESCO (2013) Education for sustainable development (ESD) in the UK—Current status, best practice and opportunities for the future. Policy brief, 9 March 2013
WCED (1987) Our common future. Oxford University Press, Oxford

Author Biographies

Lara Hopkinson has been involved in environmental and sustainability research for the past 17 years. She spent several years researching and practically assisting in implementation of Environmental Management Systems for Small to Medium Sized Enterprises (SMEs) in Wales, devising a phased approach for implementation prior to the development of BS8555. She then assisted Cardiff University in the development of the integrated management system described in this paper. She is currently ESDGC Manager for the Built Environment Sustainability Training Programme (BEST), researching use of the agenda across the UK. She assists in the teaching of a core sustainability module for several Masters courses within the Welsh School of Architecture, and has also devised a sustainability awareness course for Built Environment professionals in Wales.

Dr Julie Gwilliam has spent the last 10 years teaching and researching at the Welsh School of Architecture, Cardiff University. She received her doctorate in Architecture also from Cardiff University in 2008, which was focused on the potential impacts of climate change on occupant health in housing in South Wales. Her main interests lie in the role of education and continuing professional development in the delivery sustainable development; the study and teaching of environmental design and sustainability as applied to the built environment. She is also interested in the social production of buildings, collaborative practices, the investigation of professional working practices in the delivery of built environment sustainability. She is the director of Post Graduate Teaching at the WSA as well as the Welsh Energy Sector Training Project (WEST) and the Built Environment Sustainability Training programme (BEST).

Creating Sustainable Homes and Environmental Minds at Staffordshire University—GreenPad: A Student's Green Fund Project

Katie L. Ferneyhough

Abstract

GreenPad was one of the successfully awarded projects as part of the NUS led Student's Green Fund. It is an innovative approach to driving sustainable development (SD) within higher education institutions (HEIs), being principally led by a Students' Union rather than the University itself. GreenPad will address key aspects of achieving SD; attitudinal and behavioural change, education and community outreach and act as an institutional driver to achieve SD across the University. It will create a sustainable housing stock for Staffordshire University students to rent from within the private rental sector, as well as encourage and reward positive environmental behaviour. It will culminate in the creation of a Students' Union led, sustainability focused, letting agency. This will be achieved by firstly uncoupling student rent from bills, moving away from the "all-inclusive" billing culture and incentivising students to use less energy, and secondly, by installing smart meters in GreenPad homes and offering rewards for living sustainably, students will become more aware of their energy usage and be incentivised to use less. Alongside this, landlord engagement events and incentives will ensure local landlord involvement, providing the housing stock which will be regenerated to create the energy efficient GreenPads. Houses will be assessed according to a unique assessment criteria to become certified GreenPads, carried out by the trained Student Auditor team. The positive effects of this scheme will reach far beyond the University, infiltrating the surrounding communities and providing a much needed regeneration of the Shelton area in particular which was recently named as the area with the highest number of cases of fuel poverty. Over the course of the project, GreenPad will save 140 tCO_2e

K.L. Ferneyhough (✉)
Staffordshire University Students' Union, Beaconside, Stafford, ST18 0AD, England, UK
e-mail: Katie.ferneyhough@staffs.ac.uk

W. Leal Filho (ed.), *Transformative Approaches to Sustainable Development at Universities*, World Sustainability Series, DOI 10.1007/978-3-319-08837-2_25

through GreenPad tenant lifestyle changes, and provide a reliable framework of a Student Union led letting agency for other organisations to follow. It will be a long-lasting legacy for both current, and future, students to be actively part of and learn from, and will be a key institutional driver of SD at Staffordshire University.

Keywords
Energy · Efficiency · Accommodation · Sustainability · Behavioural change

1 Introduction—Sustainability Within Higher Educational Institutions—the Global Agenda

We are currently facing arguably one of the most crucial and significant challenges in recent human history: addressing the global state of our natural environment. As we as human beings become increasingly industrialised and reliant upon finite resources to fuel our daily lives, we are both directly, and indirectly, contributing to inevitable loss, and irreversible damage, to our environment (Stern 2006; Ralph and Stubbs 2014). The resultant environmental crisis is multi-faceted, ranging from habitat loss and species extinction to unsustainable agriculture and loss of resources (Sharp 2002), many of which are inextricably linked to the global crisis of climate change as a result of our rising demands for energy and subsequent production of greenhouse gases (Thomas et al. 2004; Lombard et al. 2008; Perry et al. 2008; Cahill et al. 2013; Pfeiffer 2006).

As the current environmental state is more of a result of our own social behaviour than as a result of the environment itself (Wackernagel and Rees 2013), we must be our own solution. Higher educational institutions (HEIs hereafter) therefore form the perfect platform in which to embed sustainability within the very core of our future generations, creating sustainable societies as a result (Cortese 2003). Being multi-faceted organisations by their very nature, HEIs covernumerous aspects from community, research and academia (Oelreich 2004; Cortese 2003) through to lifestyle and behavioural influence (Sharp 2002; Ferrer-Balas et al. 2008; Lonzano et al. 2006; Arbuthnott 2008; Ralph and Stubbs 2014). As a result, they have the ability to drive sustainable development (SD hereafter; WCED 1987) within the broader society (Sharp 2002), and due to their capacity for research, "academic freedom" (Cortese 2003) and their financial influence (Ralph and Stubbs 2014) they are able to become industry leaders and set a clear example for others to follow (Kirwan 2010). Incorporating SD into HEIs is therefore paramount to achieving a sustainable future, becoming a key focus on the global agenda as a result (see Wals 2013 for a review). Some of the key ways HEIs can achieve SD is by driving behavioural change amongst their future graduates via attitudinal influence (Nickerson 2003; Arbuthnott 2008; McKenzie-Mohr 2008), education (Hopkinson et al. 2008) and community outreach (Cortese 2003). Less noted within literature, is the ability of their partnering Student Unions' to help drive SD within

HEIs. With the United Nations declaring 2005–2014 the Decade of Education for Sustainable Development, now is the time for universities and their partnering unions to achieve this change. This is also supported by the fact that students are calling for change themselves, acting on their own initiative to incorporate sustainability (Helferty and Clarke 2009), as well as the majority now believing that sustainability should be incorporated and promoted within Universities (Drayson et al. 2012).

Here, the author will outline an innovative project lead by Staffordshire University Student's Union (Staffs SU hereafter)—GreenPad—and how it addresses key aspects of SD via a sustainable housing initiative in order to become an institutional driver of sustainability.

2 The UK's Carbon Footprint—Sustainability Within the Housing Sector

One of the main contributors to the current environmental crisis is our production of greenhouse gases, stemming from our reliance and subsequent burning of, finite fossil fuels as an energy source (Pfeiffer 2006; Hook and Tsang 2013). Despite this, on a global scale, our demand for energy has only become greater in recent years, with CO_2 emissions rising by 49 % over the last two decades alone (Lombard et al. 2008). In response to this, the UK Government has set its own CO_2 emission reduction target, aiming for a 50 % reduction by 2050 with an initial target set for a 34 % reduction by 2020 (DECC 2008). Outlined as the key drivers to achieve this goal is an increase in the energy efficiency of buildings, both commercial and domestic, and a greater use of renewable energy sources (DECC 2008). As domestic energy production accounts for 31 % of the UK's total carbon Emissions (DECC 2010) it is a key target area to address in order to achieve our national targets and in turn reduce our global environmental impact.

With a significant proportion of students living within private rented housing for the majority of their time during University (NUS 2014), helping current and future students within UK based HEIs to manage and reduce their own carbon footprint, would be a significant contributor to meeting our national targets. As the secondary target outlined by the government was to ensure that no household is living within fuel poverty by 2016 (DECC 2008), enabling and encouraging private landlords to make efficiency improvements to their homes is a key to achieving this goal. Teaching students about energy efficient homes and what they should be looking for in a private rented house is also paramount to driving this change. Staffordshire University alone, for example, has over 24,300 students, 3,800 of which live within local rented accommodation. As Keele University is also close to the campus, this is a further 3,000 students living within local rented housing. If the GreenPad project is able to help influence only half of these 6,800 students into becoming energy aware citizens who appreciate their environment and actively want to reduce their carbon footprint, this would be a huge achievement. This is further emphasised due to the current problems for young people attaining home ownership. It is estimated

and environmental waste management. GreenPad also aims to address the rise in the all-inclusive billing culture, a system which results in students that are unaware of energy consumption and reliant upon unsustainable lifestyles. The project will also have strong academic and research links, being incorporated into both formal, and informal, education offered to Staffordshire University Students. As the project is centered around improving the energy efficiency of local student homes, the inevitable improvements to student housing will also provide a strong regenerative element to the project, improving the local community across both campus sites. Addressing these aspects of SD, GreenPad will be a key driver within the institution, achieving sustainable homes and environmental minds as a result.

5 GreenPad—Achieving Attitudinal and Behavioural Change

By introducing a smart metering system into all GreenPad homes, both students, and ourselves, will be able to monitor the energy usage of tenants, rewarding and encouraging positive environmental behaviour change. Each term, the GreenPad team will be calculating the energy usage of each household, per person, and rewarding the student household that is living most efficiently. Equally, it will enable the targeting of students who may need extra guidance on using less energy. This will be further supported by the hiring of a team of student GreenPad Ambassadors which will carry out home visits to students within GreenPads to advise them on how to live more sustainably. The team will also work closely with existing societies and events, helping at workshops which will run alongside the initiative to support the sustainability message. As each GreenPad home will also be fitted with a GreenPad handbook, students will be able to access simple tips at any time on how to live sustainability and reduce their environmental impact. Simple documents such as an instruction manual for their boiler and the properties EPC will all aid in students understanding of domestic energy use. Also included in the handbook will be everything from simple energy saving advice—e.g. boiling the kettle with only the water you need, not leaving electrical devices on standby—how to travel sustainably to University—e.g. local bus maps—how to recycle—e.g. who to ring to gain recycle bins, location of local recycle centres—and how to cut their food carbon footprint—a guide to seasonal veg and buying locally. It will enable students to think, and actively live, sustainability, successfully teaching them all of the multi-faceted elements which encompass a sustainable lifestyle. This will all be supported throughout the University calendar of events. For example, students choosing to travel sustainability to University will be able to receive free bike maintenance from regular visits from local organisations which deliver this service, as well on-going support from our University police officer who will provide access to free bike locks and safe travel packs. Staffs SU's annual housing fair will also provide a great opportunity for students to explore all aspects of housing, from discussing their housing rights, meeting local accredited GreenPad landlords right through to getting their housing contracts checked and taking away some free energy saving devices. All of these factors will contribute to not only educating students on the private rental sector, but also changing their attitude and giving them the power and the knowledge to demand more from their housing.

6 GreenPad as an Education Tool

One of the key factors for HEIs to address to fully incorporate SD is including sustainability within the curricula both formally (Cortese 2003; Sedlacek 2013) and informally (Hopkinson et al. 2008). Formally, GreenPad will help drive this change by acting as an educatory driving force, being incorporated into current curriculum options within different faculties at Staffordshire University. Within the Engineering faculty, final year students undertake a module based on the energy efficiency of University buildings. In line with the GreenPad project, in 2015 these students will be able to carry out these assessments on student homes, enabling both enhanced data collection on our housing stock but also enabling the students to see real life examples of energy efficiency and how they would improve these homes and save energy as prospective engineers. The behavioural change as a result of the GreenPad project will also provide a number of research projects for Psychology students, who are able to map the behavioural changes that students are making on a real-time basis. An example of this would be the investigation of the effects of smart meters on the energy use behaviour of student's tenants. This will be achieved by providing a set of housing stock with the smart metering systems, and an alternative set which are not given these devices. Psychology students will then be able to determine if these systems do help achieve a reduction in energy usage. Not only does this allow the quantitative mapping of the change the project is achieving, but it also incorporates sustainability into a subject which it is not formally associated with. The novel criteria against which houses are assessed according to their sustainability status to determine if they are of GreenPads standard was also devised with the help of various University academics, including current Ph.D. students. GreenPad has also enabled the formal training of the Student Auditor team. They carried out energy training by both an external organisation and also a local institution to give them a full insight into carrying out the assessments and advising other people on living energy efficiently and sustainably. As the team is made up of 12 students which are both undergraduate and postgraduate, various ages and from varying degree disciplines, GreenPad is helping to teach a student audience about sustainability that may otherwise not have known. Their strength as a team is delivering advice from student to student, and have found that students are much more likely to listen and learn from their fellow peers, who are actively living what they teach to others. This therefore forms an element of informal education, with the student team talking to their friends about the project and advising them how to save energy which in turn passes the sustainability message onto further people within that group and so on. The leading organisation itself is also formally learning from this experience. Staffs SU has already carried out a Housing Survey for students which addressed their attitudes towards both sustainability and accommodation and helped to shape the project from the beginning. The GreenPad project has also sought to address landlord's attitudes towards sustainability, as they will provide the housing stock that will be regenerated and will also be a key target audience in which sustainability practices can be embedded within. As a result, GreenPad have now released a landlord survey to local accredited individuals. The GreenPad project will enable both of these surveys to take place once again in 2015 to access attitudinal changes after the first year of the project has taken place.

Informally, the GreenPad web presence provides a source of relevant, easily accessible, information to students on how to live sustainably. Both or website, and social media platforms have been successful in helping to deliver the sustainability message to students. One of the key elements to informal education is also an element of active learning and space for creativity (Hopkinson et al. 2008). This has been achieved via the work of the Student Auditor team who are able to actively contribute to the role and show initiative relevant to sustainability issues. It essentially gets them to think for themselves, with a number of students coming back with ideas of what they would like to incorporate into both the criteria and the advice service they provide to the student homes they visit. As the GreenPad team is working closely with a local Housing Association, a number of their visits are carried out alongside trained professionals from which they are also able to learn from. In conjunction with the Student Advice centre at Staffs SU, the project will also play a role in the development of a new series of workshops which will provide a voluntary learning experience for students to attend, in which they can learn about the economic implications of inefficient homes.

7 GreenPad and the Community

The inevitable improvement to the local housing stock which the GreenPad project will encourage will lead to an overall improvement to the community as a result. This is particularly key at the Stoke-On-Trent campus, as the bordering Shelton area was recently named as having the highest cases of fuel poverty in the UK (DECC 2013). This improvement was also achieved by the partnering of GreenPad with Stoke-On-Trent council and the local Housing Association, which was the result of the council lead initiative to secure government ECO funding. The council's aim was to provide energy efficient regeneration to the Stoke-On-Trent area, and hired the Housing Association to carry out home visits to 600 properties within the Shelton area and gain energy efficiency data on each property. However, as 70 % of their 600 target properties were student let, they came across issues gaining access. Our student team were able to combat this, as students were much more receptive to allowing fellow students to visits their properties, and working together they were able to visit and carry out energy audits at over 600 homes in a 3 month period. The data collected will not only be used to better the GreenPad project, but will provide the evidence required to potentially receive the government funding. As students have attended numerous housing visits within the community, they have already become a familiar presence to local residents within the area, extending their advisory service for energy advice to residents as well as their own peers. Being united by the economic values which surround energy usage, both parties are both consciously, and unconsciously, reducing their carbon footprint.

8 The GreenPad Legacy

The GreenPad project will culminate in the creation of a Green Letting Agency, a not for profit social enterprise providing sustainable, good quality, student accommodation. It will form a tested model that other Student Unions' across the country can adopt; promoting sustainable lifestyles amongst students whilst ensuring their accommodation needs are accounted for by focusing exclusively on sustainable properties. GreenPad's legacy will not only be the sustainable letting agency created for current and future students to rely upon, but also the students who will graduate from this institution as environmentally aware, conscientious individuals, with the tools and knowledge required to reduce their environmental impact and live sustainably in their everyday lives. As a result of the formal, and informal, education elements of the project, students will become empowered to demand more of their accommodation and begin to recognise property energy efficiency as a key focus when searching for accommodation. Saving over 140 tonnes of CO_2 over the duration of the project, it will be significant contributor to both the Universities, and the nations, own carbon saving targets and, although not the complete answer to achieving SD within Staffordshire University, GreenPad will no doubt be a key driving force within the institution and provide an innovative first step in achieving institutional wide change.

9 Conclusions

There is no longer a question as to whether we can debate the inclusion of sustainability within the very essence of HEIs (Sharp 2002). Given our current environmental state, educating our future leaders and decision makers in the necessity of a sustainable lifestyle is the key to the future of our planet. We can longer exist in an unsustainable society and expect that our environment will experience no consequence, and as the current environmental state is more of a result of our social behaviour than as a result of the environment itself (Wackernagel and Rees 2013), we must be our own solution. The role of HEIs in developing graduates that are conscientious and environmentally aware is therefore greater than ever, and universities nationally, and globally, must adopt SD within to their daily practices in order to achieve this result. As Lonzano (2006) stated, "SD must be the golden thread throughout the university system". However, as the author has shown here, the partnering Unions' of Universities must also mirror this drive for sustainability, or be the drivers themselves in order to fully achieve this. Both need to work in partnership to achieve an overarching resolve of sustainability throughout every aspect of University life, as whilst the University has access to finance and academia, for example, the Students' Union has a unique insight into the lifestyle choices of students and has the ability to influence student behaviour arguably more so than the University itself. For this reason, the GreenPad project is able to influence how students live outside of their University timetables, rather than when they are physically on the campus, helping them to make changes in the way they live their daily lives within their homes. As behavioural changes made off-campus will inevitably filter into their on-campus activities and vice versa, Union led projects such as GreenPad are therefore strongest when incorporated with a similar resolve from their partnering University. Together, they are able to influence both on-campus, and off-campus, student activities and behaviours. Finally, whilst GreenPad is by no means a full approach to achieving SD, it forms a worked example of how Student Unions' can be the driving force, and potential leaders of, institutional change, as well as the importance of infiltrating sustainability into students' off-campus behaviour.

References

Arbuthnott KD (2008) Education for sustainable development beyond attitude change. Int J Sustain High Educ 10(2):152–163

Cahill AE, Aiello-Lammens ME, Fisher-Reid MC (2013) How does climate change cause extinction? In: Proceedings of the royal society of London, p 280

Cortese AD (2003) The critical role of higher education in creating a sustainable future. Plan High Educ 31(3):15–22

Department of Energy and Climate Change/DECC (2010) Digest of United Kingdom energy statistics 2010. http://www.decc.gov.uk/en/content/cms/statistics/publications/dukes/dukes.aspx. Accessed 29 Mar 14

Department of Energy and Climate Change/DECC (2008) Climate Change Act. Chapter 27. http://www.legislation.gov.uk/ukpga/2008/27/contents. Accessed 30 Mar 14

Department of Energy and Climate Change/DECC (2013) Sub-regional fuel poverty data: low income high costs indicator. https://www.gov.uk/government/collections/fuel-poverty-sub-regional-statistics. Accessed 30 Mar 14

Drayson R, Bone E, Agombar J (2012) Student attitudes towards and skills for sustainable development. National Union of Students and the Higher Education Academy, UK

Ferrer-Balas D, Adachi J, Banas S, Davidson CI, Hoshikoshi A, Mishra A et al (2008) An international comparative analysis of sustainability transformation across seven universities. Int J Sustain High Educ 9(3):295–316

Helferty A, Clarke A (2009) Student-led approaches to reducing greenhouse gas emissions at Canadian Colleges and Universities. Int J Sustain High Educ 10(3):287–300

Hook M, Tsang X (2013) Depletion of fossil fuels and anthropogenic climate change—a review. Energy Policy 52:797–809

Hopkinson P, Hughes P, Layer G (2008) Sustainable graduates: linking formal, informal and campus curricula toembed education for sustainable development in the student learning experience. Environ Educ Res 14(4):435–454

Joseph Rowntree Foundation (2012) Housing options and solutions for young people in 2020. Joseph Rowntree Foundation, New York

Kirwan WE (2010) The 21st century: The century of the American research university. Innov High Educ 35:101–111

Leal Filho W (2011) About the Role of Universities and Their contribution to sustainable development. High Educ Policy 24:427–438

Lombard LP, Ortiz J, Pout C (2008) A review on buildings energy consumption information. Energy Build 40(3):394–398

Lonzano R (2006) Incorporation and institutionalization of SD into universities: breaking through barriers to change. J Cleaner Produc 14:787–796

McKenzie-Mohr D (2008) Fostering sustainable behaviour: an introduction to community-based social marketing. Beyond Brochures 108–118

Nickerson RS (2003) Psychology and environmental change. Lawrence Erlbaum Associates, Mahwah, pp 4–18

National Union of Students/NUS and Unipol (2013) Accommodation Survey 2013/2014. http://www.nus.org.uk/Global/Campaigns/Accommodation%20Costs%20Survey%20V6%20WEB.pdf

National Union of Students/NUS (2014) Homes Fit for Study. http://www.nus.org.uk/Global/Homes%20Fit%20For%20Study/Housing%20research%20report_web.pdf

Oelreich K (2004) Environmental certification at Malardalen University. Int J Sustain High Educ 5 (2):133–14

Parkin S (2010) The positive deviant—sustainability leadership in a perverse world. Earthscan, London, pp 1–15

Perry S, Klemes J, Bulatov I (2008) Integrating waste and renewable energy to reduce the carbon footprint of locally integrated energy sectors. Energy 33:1489–1497

Pfeiffer DA (2006) Eating fossil fuels: oil. Food and the Coming Crisis in Agriculture. New Society Publishers, Canada, pp 1–23

Ralph M, Stubbs W (2014) Integrating environmental sustainability into universities. High Educ 67:71–90

Staffordshire University (2010) Carbon management plan. https://www.staffs.ac.uk/assets/Carbon%20Management%20Plan%20web%20ver_tcm44-36926.pdf. Accessed 30 Mar 14

Stern N (2006) Stern review on the economics of climate change. http://webarchive.nationalarchives.gov.uk/?/http://www.hm-treasury.gov.uk/independent_reviews/stern_review_economics_climate_change/sternreview_index.cfm. Accessed 29 Mar 14

Sedlacek S (2013) The role of universities in fostering sustainable development at theregional level. J Clean Prod 48:74–84

Sharp L (2002) Green campuses: the road from little victories to systemic transformation. Int J Sustain High Educ 3(2):128–145

Thomas CD, Cameron A, Green RE, Bakkenes M, Beaumont LJ, Collingham YC, Erasmus BFN, Ferreira de Siqueira M, Grainger A, Hannah L (2004) Extinction risk from climate change. Nature 427:145–148

Wackernagel M, Rees W (2013) Our ecological footprint: reducing human impact on the earth. New Society Publishers, Canada, pp 7–38

Wals EJA (2013) Sustainability in higher education in the context of the UN DESD: a review of learning and institutionalization processes. J Cleaner Produc http://dx.doi.org/10.1016/j.jclepro.2013.06.007

World Commission on Environment and Development/WCED (1987) Our common future. Oxford University Press, Oxford

Author Biography

Katie L. Ferneyhough has spent her life fascinated by the natural world and has always based her lifestyle and academia around her strong passion for conservation and sustainability. When choosing her degree course at University this translated rather naturally into her studying Zoology (B.Sc) at Liverpool University. During this time she was an avid volunteer with the RSPB and acted as an Educational Field Assistant to young adults passionate about the environment. In her final year project, she studied the sexual dimorphism of Lepidopterans, namely the Cabbage White *Pieris brassicae* in a novel study which outlined the potential evolutionary drivers behind their dimorphic colour traits. After graduating in July 2012, Katie chose to study a Masters of Research in Conservation and Resource Management at Liverpool University. During this time, she centered her two main research papers on the behaviour of seabirds in response to human induced pressures. She was keen to show that the camera trapping and tagging technique relied upon by the majority of seabird ecologists to study seabird conservation had detrimental effects on the breeding cycle of Black Legged Kittiwakes (*Rissatridactyla*) and affected the diving behaviour of European Shags (*Phalacrocoraxaristotelis*). The field research Katie collected on these topics will go on to provide the back bone of the research publication which will culminate in 2015. Before graduating from her MRes in December 2013 Katie was hired by Staffordshire University to be the project lead on their HEFCE funded Students Green Fund initiative led by NUS; GreenPad. Katie would like to thank her dedicated student auditor team and her incredible manager Lisa Hough for helping her make this project everything it has, and will, become.

The Role of the University in Promoting Photovoltaic Usage in Brazil

Alek Suni, Mario Corrêa de Sá e Benevides and José Baltazar de Andrade Guerra

Abstract

This paper examines Brazil's progress toward solar photovoltaic (PV) usage via university involvement in its propagation, and through a case study of the REGSA project's PV installation initiative at an elementary school in Rancho Queimado, Santa Catarina. This pilot project aims to add social and economic benefits to the local community and disseminate renewable energy knowhow. The experience is used to analyze a range of traditional policy approaches used in stimulating PV usage in the Brazilian context. The paper assesses the applicability of a feed-in tariff, renewable portfolio standards, tax incentives, and more favorable financing schemes for solar energy in Brazil. Finally, through the lens of REGSA's Rancho Queimado PV installation, the paper considers the above scenarios and the role of universities or other institutions in promoting renewable energy diversification and PV installation. Given the incipient nature of PV presence and familiarity in Brazil, universities are examined as a possible conduit for encouragement and PV suitability assessment among industries and developers in the Brazilian state of Santa Catarina. Comparisons of electricity consumption levels among major regional industries aim to determine which local energy consumers may be most predisposed to complement with individual generation.

A. Suni (✉) · M.C. de Sá e Benevides · J.B. de Andrade Guerra
REGSA Project, Universidade do Sul de Santa Catarina, Florianópolis, SC, Brazil
e-mail: alek.suni@fulbrightmail.org

M.C. de Sá e Benevides
e-mail: mariocsbenevides@gmail.com

J.B. de Andrade Guerra
e-mail: baltazar.guerra@unisul.br

W. Leal Filho (ed.), *Transformative Approaches to Sustainable Development at Universities*, World Sustainability Series, DOI 10.1007/978-3-319-08837-2_26

Keywords
Solar power • Renewable energy • Photovoltaics • Brazil

1 Introduction

This paper examines the role of the university in promoting solar photovoltaic installations in the greater Florianópolis area in southern Brazil. First, a brief summary of the underlying regulatory framework is presented, which represents one key determinant in the outcomes of strategic shifts in electricity generation. National-level shifts in electricity sources are generally gradual by virtue of the structural changes that must take place and the rigidity of some large-scale producers. Therefore, understanding the regulatory backdrop of a gradually privatizing and diversifying Brazilian electricity sector is crucial to establishing effective initiatives through universities and other institutions toward alternative energy sources. Background is also provided on the growing and changing nature of electricity demand in Brazil, particularly in the local context of the Florianópolis region, which is the primary focus of the university's initiatives. The activities of UNISUL and the international, university-affiliated REGSA Project are examined in this context, promoting alternative electricity sources through photovoltaic installations. The focus of the REGSA initiative in Brazil is a pilot photovoltaic installation on a local school, accompanied by efforts toward diffusing knowledge of alternative energy options and new policy conditions for residences and businesses. Data on local insolation presented alongside daily and seasonal electricity consumption patterns support the notion of complementarity of solar power use in the area.

2 Regulatory History of the Brazilian Electricity Sector

Before the Government of Fernando Henrique Cardoso, the Brazilian electricity sector was controlled by the Brazilian State. It was also vertical: some distribution companies (DISCOS) were also transmitting (TXCOS) and/or generators (GENCOS), and their scope was geographically defined by the Government. With the inflationary outburst that lasted until the deployment of the Real as the Brazilian currency, the expansion of Brazilian electricity generation had been practically stopped. The Cardoso Government, elected to a first term from 1995 to 1998, implemented its privatizing policy and started the electricity sector's reform by "deverticalizing", splitting, and privatizing some of the companies, mainly DISCOS and some GENCOS.

The National Electric System Operator (ONS in Portuguese) was then created as a body responsible for the operation of the national interconnected system (SIN in Portuguese). The ONS is maintained by the Electric Companies, both private and non-private. The privatization of GENCOS was performed based on a minimum price plus a premium: the state control of the concessionaire was transferred to the competitor that offered the best price, and the contracts included the obligation of the expansion of the generation capacity—besides the offer of project-finance by the National Bank for Economic and Social Development (BNDES in Portuguese). New projects, notably of hydropower generation, were auctioned by the same logic of the best price. The companies were no longer regional, and could now participate in power plants anywhere in the country. President Cardoso was re-elected for another four-year term. Later, the Cardoso Government's opposition, led by Luiz Inacio Lula da Silva (more commonly referred to as simply Lula), was elected and re-elected, serving as president of Brazil from 2003 to 2010, and then succeeded by his ally and current President Dilma Rousseff, to run for re-election in 2014.

The main modification introduced by the Lula administration was the lowest fare scheme, created in 2004: the winner of the auctions for the expansion of the national generator complex should be the proponent capable of offering the lowest quote for the energy generated. The market resumed its regulatory stability. However, this lasted only until September 2012, when the Rousseff Government launched the Provisional Measure no 579, later the Decree 7805. The Provisional Measure and the Decree established the reduction of some of the taxes included in the energy price and the possibility of anticipation of the renewal of concessions that would end between 2015 and 2017, but based on asset depreciation calculations and electricity rates solely based on operation and maintenance costs, which made some renovations unfeasible on the part of enterprises not belonging to Eletrobras. At the same time, it forced Eletrobras, as a company managed by the Federal Government, to accept the renewal of concessions, but affected its capacity to invest in its expansion. The purpose of the Provisional Measure and the Decree was to reduce the price of energy, which in practice has not occurred. Among other reasons is a drought at the end of 2012 that decreased the capacity of the reservoirs of hydroelectric plants, which led to new measures imposed by the government. These included increasing thermal generation even when at some moments this might not be necessary ("out-of-merit dispatch"), to preserve the capacity of reservoirs for the expected increase in energy demand, for instance due to the World Cup in Brazil (2014) and the Rio Olympic Games (2016). Another rule introduced without prior discussion was the sharing of the additional costs of thermal generation by all market players, including generators and consumers.

Such measures have created uncertainties for the expansion, compromising the reduction of tariffs and taxes that consumers should expect. In short, instead of proposing a structural tax reform being debated within the National Congress to reduce energy prices, the Rousseff Government imposed short-term solutions, without having prior discussions with the main stakeholders. This has shaken the electricity sector's regulatory stability, reducing market value, investment capacity and/or interest of some GENCOS, and potentially bringing negative impacts to the economy as a whole. It is expected that the market will adapt and normalize, and that possible new measures will be discussed in advance with the agents of the sector and society as a whole.

3 Rising Demand in a Stressed System of Centralized Power Sources

As a rapidly growing Brazilian middle class has taken shape in recent years, consumer demand has burgeoned for many goods previously out of reach to most of the population. Coupled with wider electrification and use of domestic appliances, this has led to rapidly increasing electricity consumption. Even as population growth has slowed dramatically to 0.83 % and fertility rates have stagnated at roughly 1.8 children per woman, electricity demand has continued to skyrocket in much the same way as in countries with rapid population growth (IBGE 2013). In all, national electricity demand is expected to increase by 4.1 % per year between 2012 and 2022, according to Brazil's Energy Research Company (EPE in Portuguese), an arm of the Ministry of Mines and Energy (EPE 2012).

This added demand has put increasingly problematic strains on the country's electricity sector. Brazil's conventional source of hydroelectric power has traditionally increased along with rising consumption to meet elevated demand. However, with heavy reliance on volatile river levels, blackouts have continued to threaten electricity supply. Environmental, indigenous, and other groups have long decried the large projects' negative externalities on regional communities. Therefore, although hydropower projects do provide renewable solutions to electricity shortages, diversification of renewable power sources is desirable to achieve a national energy mix that is both reliable and responsible.

In light of this concern, the federal government has introduced programs to increase the participation of alternative sources in the grid and lessen the electricity sector's dependence on volatile river levels supplying hydroelectric dams. One such program, PROINFA (Portuguese acronym for Program of Incentives for Alternative Renewable Electricity Sources), has focused on promoting wind power, biomass, and small-scale hydropower. Wind power has been a special emphasis at the nexus of competitive pricing and minimal negative externalities. To meet rising electricity demand, the Ministry of Mines and Energy (MME) plans to guide the wind sector toward 9.5 % of installed capacity by 2022 (Plano Decenal de Energia 2013).

Wind power's introduction has come in the form of utility-scale projects commissioned through energy auctions. Their potential for mitigating some of the controversy from hydropower-centered government approaches are undoubtable, and their competitive, commercial-scale pricing is proven (Bloomberg 2013). Meanwhile, solar installations[1] have the potential to address shortages at the individual level by providing alternative sources to residences and businesses directly. Commercial establishments and industrial facilities reach their daily peak electricity usage during daytime hours. Both would mitigate impacts of power failures by relying on autonomous sources of power during periods of highest demand.

[1] For purposes of simplicity, solar will refer only to solar PV in this paper.

Many barriers exist to addressing blackout risk via distributed solar PV installations. Despite having abundant solar incidence when compared to many northern countries with much more installed PV capacity, Brazilian consumers have long lacked the same government support and incentive schemes for PV. Together with added import tariffs of 12 % on imported panels–and additional state and industrial taxes on inverters–interested parties are faced with high costs for components. While the situation has not changed with regard to financial support from the government, an important regulation introduced in 2012 by the country's national electricity regulator (ANEEL) has paved the way for net metering on alternative renewable generation systems of under 1 MW.

The new net metering policy, enacted by ANEEL Law 482, allows consumers to earn credits for electricity generated in excess of their consumption (ANEEL 2012). The regulation allows for small generators of less than one MW to feed electricity into the grid to be later consumed in the same quantity within 36 months. Although the law is applicable to wind and small hydropower systems as well, the easy siting, more accessible components, and easily customizable sizing has favored PV as the leading generation means for those aiming to benefit from Brazil's net metering. Distributors were legally required to be compliant with requests to connect to the grid beginning in December 2012, but bureaucratic delays have at times impeded the integration of grid-tied PV systems in the regulation's early stages. With few examples spread across the vast Brazilian territory, many consumers are unaware of the regulatory change to mandated compensation that allows for more economically viable PV installations.

By creating an avenue for more affordable PV projects without providing direct subsidies or tax write-offs, the new policy of net metering has left consumers to assess independently the time horizons for their return on investment in a PV installation. A knowledge gap therefore exists for potential installers. Although market forces would be expected to eventually entice consumers toward PV installations' attraction of long-term savings, the diffusion of community familiarity with net metering could be slow.

In the absence of a government push or strong incentive structure, there exists the possibility of other institutions filling an intermediary role in familiarizing potential end-users of solar PV with net metering grid-tie procedures. One such institution can be universities, and university-affiliated research centers and initiatives. The following sections explore the experiences of the REGSA project, an international consortium of universities promoting renewable energy in South America, in its local solar energy initiatives in Brazil. REGSA's primary focus is a pilot project installing a photovoltaic system at an elementary school in the rural town of Rancho Queimado in the southern Brazilian state of Santa Catarina. The initiative's complementing approach is through the promotion of photovoltaic systems among commercial and industrial establishments in the greater Florianópolis area—also in Santa Catarina state—as a means of pushing electricity sector diversification and hedging blackout risk through individual PV system investment and distributed generation. The initiatives both aim to encourage alternative renewable electricity sources locally, but through different approaches and with different target audiences. Notably, both projects also promote the advantages of the recently introduced net metering policy, without which PV installations would remain much less financially viable.

4 The REGSA Pilot Project at a Public Elementary School in Rancho Queimado

One of the REGSA Project's stated objectives is the creation of pilot programs by which to generate interest in and examples for alternative renewable energy use. These local initiatives are meant to serve as catalysts for increased familiarity with emerging technologies that are not yet commonly installed in Latin America. REGSA's pilot project in Brazil was the installation of a 6.6 kW solar PV system at the Roberto Schütz public elementary school in the rural town of Rancho Queimado, Santa Catarina. The installation of the system was opened in November 2013 and has begun supplying all of the school's average annual power consumption of 4,632 kWh. Through REGSA's efforts school's classrooms were also renovated with high-efficiency light bulbs and repainted with lighter colors that gave the areas a brighter and more visually accessible appearance.

Once officially connected and registered by the distributor, CELESC, the school became one of the first three successfully grid-connected PV systems in Santa Catarina state, as well as the first non-residential case. Looking forward, the school's excess generation during peak sunlight hours and in summer months will continue to be logged as credits for future use. At an existing rate of 0.28455 Brazilian Reais (BRL) per kWh, a total of BRL1, 318 will be saved on a yearly basis. Notably, because the installation's expected annual generation of 8,784 kWh will consistently exceed the school's consumption, REGSA and CELESC are negotiating ways in which to direct credits toward other school facilities.

Given the nature of the project, the connection process to the grid via CELESC was faster than has been experienced by similar installations in Santa Catarina, as well as in other states. As Escola Roberto Schütz is a public elementary school, the initiative was able to engender multi-sector cooperation in forming a unique partnership between local government, the quasi-government electricity distributor, and REGSA.

Because the system was funded by an upfront grant from the REGSA budget, the 45,000 BRL system will provide immediate savings for the school. In the sense that the project serves as a model to spawn local PV interest, the pilot project does not aim to provide a financial example to be easily replicated by residential consumers. However, in the Florianópolis area a new not-for-profit financing scheme for small-scale PV installations called the *Fundo Solar,* managed by the Instituto Ideal, was launched in May 2013. The fund disburses direct grants of up to 5000 BRL to defray financing costs for residential consumers. Drawing on familiarity with these sources, REGSA's role as a conduit for interested residential community members includes connecting appropriate actors with one another.

As the project has become publicized, the audience of residential consumers potentially interested in installing their own PV systems has grown. With few existing examples of PV connections to the grid in Santa Catarina, the Roberto Schütz installation has gained attention in the region as the state's first school with the capacity to fulfill all of its electricity needs through solar power. The project has featured prominently in several Santa Catarina television news outlets, including coverage by Band News, RBS News, Ric Record News, and the TV UNISUL and TV Cultura university channel.

This publicity has highlighted the potential for small-scale consumers to enjoy long-term savings by connecting to the grid through net metering as is being demonstrated by Roberto Schütz School. In order to support the diffusion of PV knowledge in Santa Catarina communities, the REGSA Project remains available as an academic resource and bridge for continued knowledge transfer. Using the Rancho Queimado installation as a catalyst, the REGSA Project hopes to fill this role as a conduit between communities and alternative renewable electricity sources.

5 Encouraging Industrial and Commercial PV Applications

While the Roberto Schütz PV system serves as an example to the local community mainly for an audience of residential electricity consumers, the complementing portion of REGSA's initiative is to consider commercial and industrial consumers. This fits within another of the REGSA Project's objectives of dissemination activities and awareness raising. Disseminating new net metering allowances ought to also be directed at other sizable segments of Brazil's electricity usage. As Fig. 1 shows, these industrial and commercial segments represented 59 % of total Brazilian electricity consumption in 2012.

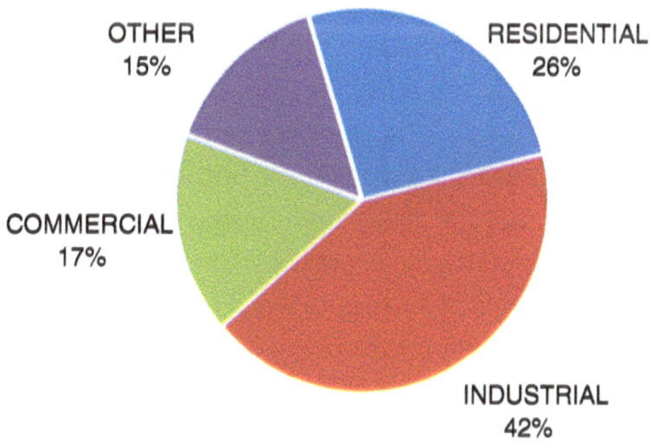

Fig. 1 Brazil's electricity consumption by sector, 2011. *Source* EPE

The majority of electricity demand evidently derives from commercial and industrial sources. Moreover, government projections for the southern region in which Santa Catarina is situated predict that the region's strongest annual demand growth will be seen in the commercial segment, which will increase an estimated 5.5 % per year through 2022 (EPE 2012). This represents the driving force behind total southern Brazilian electricity demand growth of only 3.8 % over that period.

The REGSA initiative therefore aims to complement its Pilot Project installation with an informational campaign among some of the region's larger electricity consumers, which comprise portions of the industrial and commercial segments. As one local study has highlighted, local industry and commerce are by and large willing to invest in alternative renewable energy options, with as many as 62 % of respondents in one survey or local business managers indicating they would invest in solar PV even if it incurred greater costs on the company (Instituto IDEAL 2011). The survey went on to indicate that of those that responded positively, 49 % would pay up to 10 % higher energy costs to install PV. However, with few pioneers in the area of individual business initiatives toward PV, these types of alternative renewable energies can still be approached with uncertainty and unfamiliarity. This information gap between willing consumers and alternative renewable sources is what the REGSA initiative aims to bridge.

Notably, while residential connections to the grid via net metering have been spreading since the regulation was enacted, industrial connection requests have been rare. Almost one full year after electricity distributors' mandatory compliance with net metering, only 3 of 107 imminent PV connection requests being processed by utilities as of September 2013 were by industrial consumers (CEMIG 2013). One interpretation could be to attribute this disparity to the fact that limited net metering policy favors small-scale producers and consumers. This a valid consideration, seeing as individual producers are limited to one MW in order to receive compensatory credits for their PV system's output. However, it is also worth noting that few of the installations connecting to the grid thus far have approached the cap, meaning larger consumers have not attempted to take full advantage of net metering's maximum installed capacity.

One of these few industrial installations is located in the northeastern state of Ceará, where B&Q Energia, a utility, has integrated a system at their corporate facilities in Fortaleza. Another industrial application of solar PV is coincidentally located near the REGSA pilot project in Santa Catarina state. This example is found at Agua Santa Rita, a water purification company that installed a 70 KW solar PV system atop the company's bottling facility.

Commercial end users, on the other hand, have shown more receptiveness to net metering changes, according to data from CEMIG, a Minas Gerais utility. Through September 2013, a reported 37 % of connection requests in Brazil were submitted by commercial consumers (CEMIG 2013). The benefit of seasonal complementarity is clear with commercial consumers, as this segment follows a similar pattern to residential electricity consumption, as seen in Figs. 2 and 3. Both segments ostensibly exhibit winter drops in electricity usage consistently, with demand falling significantly during southern hemisphere winter months and reaching its low

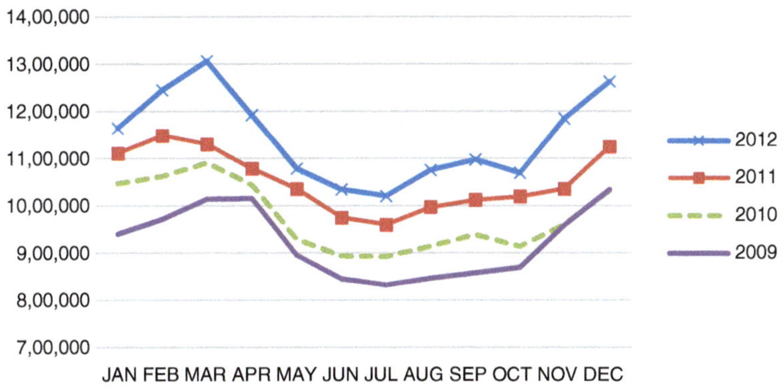

Fig. 2 Southern Brazil commercial electricity consumption, 2009–2012 (MWh). *Source* EPE

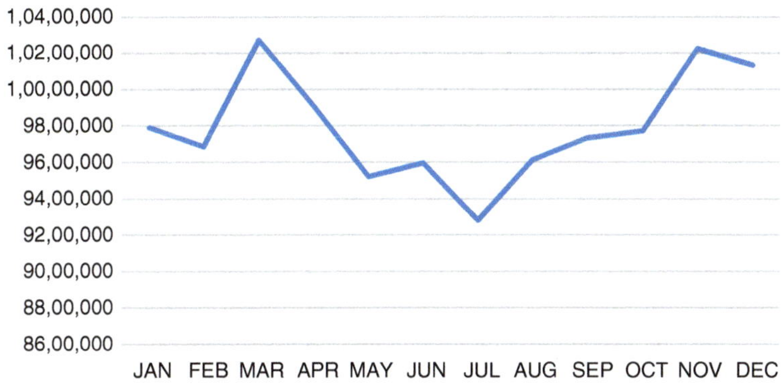

Fig. 3 Southern Brazil total residential electricity consumption, 2012 (MWh). *Source* EPE

point in July in 2009 through 2012. As temperatures in Brazil's southern region[2] fall in mid-year, so does electricity demand, as air conditioning units are not used. Since most homes and businesses lack indoor heating systems, falling temperatures do not cause heating-related demand increases.

In Fig. 4, the region's electricity demand behavior is overlaid with the seasonal changes in expected hours of sun per day. Using data from the Brazilian Solar Energy Atlas, solar incidence per month shows marked complementarity with electricity demand over the year, as both rise in the early and late months of each year, and wane during the colder mid-year months. Notably, this is only a locally observable phenomenon, as more tropical regions of Brazil, such as the Midwest,

[2] Brazil's south is considered to be comprised of the states of Paraná, Santa Catarina, and Rio Grande do Sul.

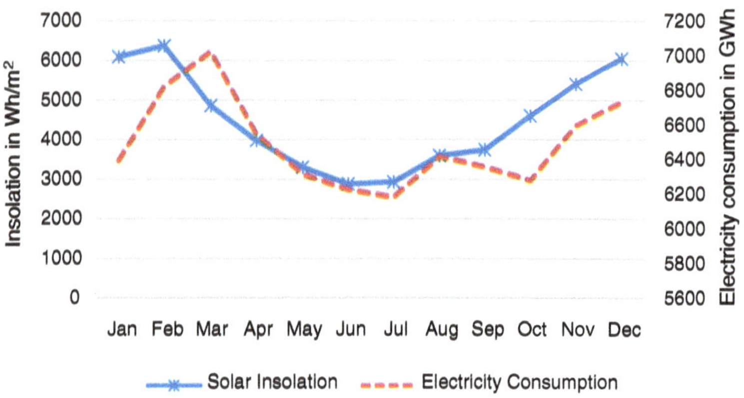

Fig. 4 Monthly insolation in Florianópolis and electricity consumed in Southern Brazil, 2012. *Source* EPE, Brazilian Solar Energy Atlas

the North and the Northeast remain reliant on indoor climate control-related electricity sources throughout the year, which results in consistently high demand from the commercial and residential segments.

6 Dissemination of Solar Energy Familiarity Among Businesses

The REGSA project's goal for encouraging PV installations in these two segments must consider the respective characteristics of each grouping in order to find effective paths to greater PV participation in the Brazilian grid. Through the project's informational sheet, the initiative targets commercial and industrial end users while taking into account each sector's profile and primary motivations for installing. Building on the Instituto Ideal's survey of Florianópolis businesses, the REGSA Project disseminates the electricity sector's new and more favorable policy of compensatory credit disbursement through net metering.

The informational sheet provides an overview of the improved regulatory climate for PV installations under net metering and was distributed with in-person meetings to 15 hotels and condominiums, and shopping malls in the Florianópolis area. More importantly for large-scale consumers such as the shopping malls, which may not benefit from net metering if their demand regularly surpasses 1 MW, information is included illustrating complementarity of solar incidence with peak commercial demand. Shopping centers can benefit from complementarity not only seasonally, as shown in Fig. 4 in demand patterns, but also over the course of a business day. As seen below in Fig. 5, central Florianópolis, which is an area with high-density of commercial consumers, displays expected increases in demand that closely match sunlight hours, both in summer and winter months. This advantageous complementarity is highlighted in distributed materials. The sheet also

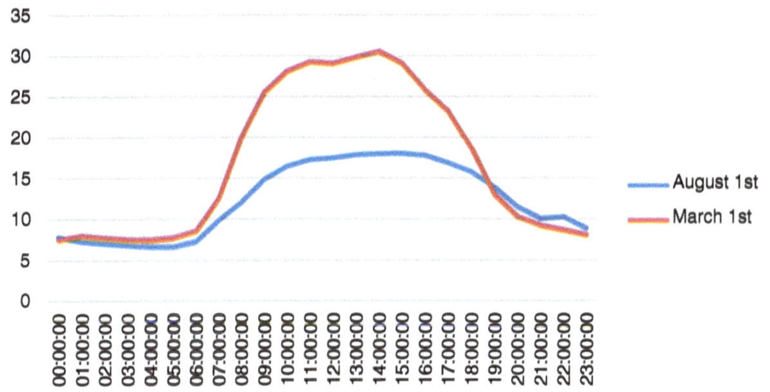

Fig. 5 Electricity Usage (MW) in Downtown Florianópolis, Brazil on March 1 and August 1, 2013. *Source* CELESC

includes estimates for economies of scale when dealing with larger PV projects in order to clarify potential savings for large commercial consumers such as shopping malls.

When considering large commercial and industrial consumers, one common obstacle to transitions toward PV solutions is a legal requirement (ANEEL Law 10.848/2004) of larger consumers to give 5 years notice when requesting to be reconnected to the regulated distributor's grid after purchasing directly in the open electricity market. Nonetheless, as central locations with high visibility, the malls would aid in diffusing solar PV knowledge among local residents. Ultimately, while it is important for larger consumers to be aware of policy changes for medium-term time horizons, consumers currently purchasing electricity from the distributor will be able to more quickly benefit from new net metering policy through new installations.

7 Conclusion

The REGSA Project's experiences in promoting solar PV installations in Santa Catarina, Brazil, are working in an incipient Brazilian solar environment. As the technology continues to become more affordable through falling prices and viable through accommodating regulations, REGSA's aim is to continue spreading familiarity and spurring more widespread PV usage. Working in conjunction with private and public entities to carry out the implementation of the solar school project, REGSA has sought to act as a bridge between sectors in delivering alternative renewable electricity to the community. This was successfully carried out as the role of REGSA and UNISUL was demonstrated in quickly delivering a connected PV system in a regulatory environment that has otherwise delayed in granting two-way metered connections.

References

ANEEL (2004) Agência Nacional de Energia Elétrica. Lei No 10.848, de 15 de Março de 2004. N. p., Web. http://www.aneel.gov.br/cedoc/blei200410848.pdf. Accessed 15 Mar 2004

ANEEL (2012) Agência Nacional de Energia Elétrica. Resolução Normativa No 481, de 17 de Abril de 2012. N.p., Web. http://www.aneel.gov.br/cedoc/ren2012481.pdf. Accessed 17 April 2012

CELESC (2013) Florianópolis daily electricity demand in 2013. N.d. Raw data. Celesc Internal Database, Florianópolis, Brazil

COGEN (2012) Associação da Indústria de Cogeração de Energia. Inserção da Energia Solar no Brasil. Rep. N.p., Web. http://www.cogen.com.br/Solar/Biblioteca_Virtual/Relatorio_Final_GT_COGEN_SOLAR.pdf

EPE (2012) Empresa De Pesquisa Energética. Projeção Da Demanda De Energia Elétrica. Rep. N. p., Web

EPE (2013) Empresa de Pesquisa Energética. Monthly and annual electricity consumption by region and subsystem. Raw data. http://www.epe.gov.br/, n.p.

IBGE (2013) Instituto Brasileiro De Geografia E Estatística. Population statistics. N.d. Raw data. http://www.ibge.gov.br, n.p.

Instituto IDEAL (2011) Avaliação Do Selo De Energia Solar Fotovoltaica. Rep. N.p., Web

Instituto IDEAL (2012) Condições De Importação De Equipamentos De Mini and Micro-Geração Distribuída Fotovoltaica No Brasil. Rep. Câmara De Comércio E Indústria Brasil-Alemanha, Web

Instituto IDEAL (2013) Fundo Solar. Amrica Do Sol. N.p., n.d. Web

Ministry of Mines and Energy (2013) Secretaria de Planejamento e Desenvolvimento Energético. Plano Decenal de Expansão de Energia 2022. N.p., Web. http://www.mme.gov.br/mme/menu/pde2022.html

Nielsen S (2013) Wind farms dominating Brazil power auction set for record year. Bloomberg News. N.p., Web. http://www.bloomberg.com/news/2013-11-18/wind-farms-dominating-brazil-power-auction-set-for-record-year.html. Accessed 18 Nov 2013

Ponte R (2013) Telephone interview. 12 Nov 2013. Interview with Coelce engineer

Sistema De Organização Nacional De Dados Ambientais (2013) Monthly solar insolation in greater Florianópolis. Raw data. http://sonda.ccst.inpe.br/publicacoes/atlas_solar.html, n.p.

Souza MEM (2013) Requisitos Para Conexão De Microgeradores às Redes De Baixa Tensão. IV Seminário O Mercado De Energia Solar Fotovoltaica No Brasil: Tendências E Oportunidades. Rio De Janeiro. 5 Sept 2013. Lecture

Author Biographies

Alek Suni is a returning U.S. Fulbright Fellow who was based at the Universidade do Sul de Santa Catarina in Brazil. He holds a B.A. in economics from the University of Pittsburgh and has previously contributed to REGSA and JELARE publications while in Florianópolis. His areas of research interest include sustainability, energy, and international economics.

Mario Corrêa de Sá e Benevides is an Electrical Engineer, graduated from Universidade do Estado do Rio de Janeiro, 1977, MBA and Post-MBA from Fundação Dom Cabral and Kellogg School of Management, 2005–2006, with more than 30 years of experience in different areas of industry, and in the last 10 years as a consultant in corporate sustainability, mainly in electrical power generation in Brazil.

José Baltazar Salgueirinho Osório de Andrade Guerra, Ph.D. is Dean and Professor at Unisul Business School (Universidade do Sul de Santa Catarina, UNISUL, Brazil). Director, at UNISUL, of two research projects: JELARE—Joint European-Latin American Universities Renewable Energies Project and REGSA—Promoting Renewable Electricity Generation in South America, both projects financed by the European Union (through ALFA III and the thematic program for environment and sustainable management of natural resources, including energy). Member of The Scientific Committees of the World Symposium on Sustainable Development at Universities (WSSD-U-2012 and WSSD-U-2014), a parallel event of Rio+20, and the Green Campus Summit 2013. Author and organizer of several books in the fields of Economics, International Relations Environment, Sustainability and Renewable Energy.

Sustainable Development as a Local and Regional Goal in Natural Science Education in Central Asia

Denise Galvin and Aibek Upenov

Abstract

Funding from the European Commission TACIS Tempus instrument has allowed many Central Asian universities to design or implement natural science degree programs that align with the Bologna Process. Inherent to the Bologna Process are curricula which are designed around competence based education, that is, degree programs that prepare students for entry to the labour market. Some of these new study programs, which are convergent across the Central Asian sub-region, aim to introduce 'Integrated Natural Resource Management' based on the concept of sustainable development. This is a totally new direction for both the managers and users of natural resources in the region. Nevertheless, the Central Asian countries are at different rates of progress in terms of incorporating Education for Sustainable Development into national legislation and policy. The paper identifies some of the issues that affect the design and accreditation of these new degree programs most of which are related to notions of what constitutes sustainable development at local, national and sub-regional levels.

D. Galvin (✉)
International Projects Office, Faculty of Economics and Business Science, Universidad de Alicante, Campus San Vicente Del Raspeig, Ap. 99, 03080 Alicante, Spain
e-mail: denise.galvin@ua.es

A. Upenov
Faculty of Natural Sciences and Pedagogics, Talas State University, K. Nurzhanov Alleya, 25, 724200 Talas, Kyrgyzstan
e-mail: a_upenov@yahoo.com

W. Leal Filho (ed.), *Transformative Approaches to Sustainable Development at Universities*, World Sustainability Series, DOI 10.1007/978-3-319-08837-2_27

Keywords
Sustainable development · Bologna process · Education for sustainable development · Integrated natural resources management · Natural science education in Central Asia · Competence based education

1 Introduction

Present understandings of sustainable development anticipate a balance between economic and social growth that has limited environmental impact. Central to the concept of sustainability is the notion of an improvement in the quality of life for both present and future generations (World Commission on Environment and Development 1987). These ideas are both comprehended and supported in the Central Asian countries which gained their independence in 1991 and wish to contribute to the realisation of sustainable development at the national, sub-regional and global level. Environmental problems in the Central Asian (CA) sub-region are closely connected with global issues, as well as directly connected with local economic and social problems; commonalities exist despite the cultural idiosyncrasies of each country. At the time of writing, the general environmental situation in CA is characterised by exceptional environmental degradation in some areas which has led to the depletion of major hydrological systems, excessive soil erosion, as well as pollution, and in some circumstances, deterioration in the quality of life for present generations.

The individual countries that make up the CA region, as members of UNECE (United Nations Economic Commission for Europe); accept that sustainable development is a complex issue that encompasses economic, environmental and social dimensions. In other words, development is essential to satisfy human needs and improve the quality of human life. Yet at the same time, development must be based on the efficient and responsible use of scarce resources—natural, human and economic (CAREC 2007). The CA countries share a vision for the future "…that embraces common values of solidarity, equality and mutual respect between people, countries and generations". The sub-region aspires to be identified with economic vitality, social justice and cohesion, as well as environmental protection based on the sustainable management of natural resources (CAREC 2007).

Major regional environmental problems are trans-boundary. Some of the better known issues are the Aral, as well as the Caspian Seas and, the efficient use of life support systems. All of which impact on the health of the population, affect interaction between national and sub-regional agricultural activity, as well as the energy sector, and, impact on the trajectory of future development. The CA countries and various donor organisations are tackling these issues from a united front. Such interaction could lead to the real development of sustainable national and sub-regional environmental, social and economic dialogues, as well as programs. However, these aspirations are inextricably linked to the integrated management of natural resources which can only emerge when Ministries of Education are willing to adopt teaching programs that embody Education for Sustainable Development (ESD).

To help achieve this goal the Central Asian Education ESD and Environmental Education (EE) Programmes were developed during 2003–2006 which identified that sub-regional problems were to be resolved by national educational sectors in the following way:

- Establish a sustainable connection for the exchange of experiences and a methodological base for EE between educational institutions in individual countries;
- Develop common approaches and principles for EE programs;
- Develop data bases on EE that are accessible to the public;
- Create a network of information, methodology and training centres in each country;
- Introduce modern textbooks, manuals and materials;
- Attract new investment to the region to assist with much needed funding for the implementation of programs and projects. In support of these objectives the Central Asian EE Program provides practical aid for educational institutions, ministries and NGOs through the mutual implementation of projects to develop new teaching materials, posters, CDs, videos on climate change, water resource problems and biodiversity, etc.; conducts training sessions for educators and holds annual sub-regional conferences for stakeholders (CAREC 2007).

To support of these objectives the Central Asian EE Program provides practical aid for educational institutions, ministries and NGOs through the mutual implementation of projects to develop new teaching materials, posters, CDs, videos on climate change, water resource problems and biodiversity, etc.; conducts training sessions for educators and holds annual sub-regional conferences for stakeholders (CAREC 2007).

To further assist this push for cohesion, the European Commission through instruments such as the Tempus Program has made funding available to improve the capacity of higher education institutions. Some of these Tempus projects have funded the design of new degree programs that aim to move individual CA countries from a fragmented approach to the study of the natural sciences to that of Integrated Natural Resource Management (INRM). In addition, and to ensure that local degree programs support the exchange of knowledge between CA countries and the European Union, new degree programs are designed to align with the Bologna Process and competence based education wherein specialisation begins in the second year of undergraduate work (Zasada et al. 2012). Nevertheless, the political history of the CA sub-region, as well as individual country physical, socio-economic and cultural contexts impede the accreditation of curricula designed around the types of Key and Field Specific Competences that twenty first century natural resource managers need to acquire. In addition, the percentage of ECTS (European Credit Transfer and Accumulation System) devoted to subjects that are obligatory, but not related to Lifelong Learning or INRM in both graduate and postgraduate programs, is relatively high across the sub-region. In some circumstances, these obligatory learning modules can take up to sixty per cent of available ECTS for a Bachelor or Master Degree program (Education, Audiovisual and Cultural Executive Agency 2011).

This situation, along with marked differences between the efforts of individual countries to assimilate ESD, limits the potential for graduates of convergent study programs to acquire the knowledge and skills required to manage trans-boundary environmental issues. In addition, cross roads of differences between cultures and their knowledge of what constitutes sustainability pose a challenge for those engaged in the design of degree programs: ministries responsible for education and natural resource management are required to collaborate on issues that in the past were addressed in isolation; teachers from CA countries need to find the way to bridge cultural differences, as well as knowledge gaps; and, students have to be able to deploy multi-disciplinary tools, models and methodologies that focus on an integrated approach to natural resource management.

Using the curricula developed from two Tempus projects that were designed and implemented around both the Bologna process and INRM as a point of reference, we are able to demonstrate that different understandings of what constitutes sustainability between government, educators and employers from different political, socio-economic and cultural context limit the potential for the trans-boundary management of natural resources. Even though political, historical, socio-economic and cultural anomalies between educational systems generally support the concept of sustainability at the local level, differences between professional transversal skills and field expertise at national level do not always support the management of issues when viewed from the sub-regional level.

2 The Concept of Sustainable Development and Education in Central Asia

The concept of 'sustainable development' is enshrined in global, trans-national, national, regional as well as local legislation and policy that directs how humans interact with nature. One of the difficulties in translating sustainable development at any level lies with the fact that definitions of development are different. Development can be understood as a condition or static state, or development can be a dynamic process. Sustainability is assumed to imply the conservation of an 'ecological balance' by avoiding the depletion of natural resources even though it does not represent an 'average' goal or carrying capacity or a grade of environmental degradation. The concept establishes what is acceptable or not in relation to the economy or the environment. In other words, the wider society now has an obligation and also an opportunity to participate in the striking of a balance between growth and conservation.

Because of its indefinite nature, the concept has become a social product or the object of discussion. In the words of Kofi Annan

> Our biggest challenge this new century is to take an idea that seems abstract – Sustainable Development and turn it into a reality for all the world's people United Nations (2001)

However, conflict has arisen over the control of the significance of the concept as it defines the elements of social practice because 'sustainable development' has become a dichotomy; that is, the push for economic development continues despite the need to reconcile the rate of economic growth with the physical limits of our ecosystems. Moreover, 'sustainability' has become a new social norm (Galvin et al. 2006) which because of its ambiguous nature is tricky for environmental and educational professionals to implement in a practical sense because the concept aims to connect what are generally considered to be three distinct aspects of contemporary life:

- Ecological: Protect natural resources; maintain the basis of production; reduce and avoid environmental degradation; conserve biodiversity; minimise destruction to ecological systems caused by agricultural production
- Economic: Contribute to the productivity of the whole economy; safeguard and improve employment opportunities in agriculture; food security and food quality
- Social Reality: Avoid and solve global environmental problems; international fairness in distribution of and access to natural resources; food security in the global context; access to international treaties and agreements on SD (Harris 2003). These three objectives can only be realised if greater attention is paid to how we educate, train and raise public awareness.

Accordingly, the UN General Assembly adopted the Resolution of the Decade ESD (2005–2014) in December, 2002 which supports the sustainable management of the environment, economy and society. UNESCO was nominated as the coordinating agency at the global level. The end of the decade of ESD coincided with another global development agenda: The Millennium Development Goals which is predicated on the success of national Poverty Reduction Strategies wherein the educational sector plays a central role. While noticeable achievements have been made in the context of ESD, significant challenges remain. Peoples from war-torn zones, remote communities, ethnic minorities and women still face difficulties accessing education. Although there has been a marked increase in literacy levels, this improvement does not meet the twenty first century needs of both Asian and Pacific countries where currently there are the largest number of illiterate adults of any region in the world (UNESCO 2012).

The CA countries, due to their geopolitical position have a unique opportunity to take part in two regional processes, i.e. in the region under regulation of the United Nations Economic Commission for Europe (UNECE), as well as in the Asia-Pacific region (APR) (UNESCO 2013). On this note, a group of environmentalists who were concerned that the developing nations of the southern hemisphere should incorporate sustainable development into educational programs can claim responsibility for the evolution of EE (Guíllern Rodriguez 1996). EE can have two strands of communicative learning—formal and informal. The informal approach is applied to describe the transfer of 'indigenous knowledge'—or traditional views of nature. That is, knowledge which is handed down from generation to generation in an informal manner. This kind of 'informal knowledge' has a role to play in natural resource management even though it is not *black and white* (Andersen 1999). The problem remains however, that because it lacks a well-defined epistemology, this knowledge and the manner in which it is transferred is often discounted by those who subscribe to the western scientific view Galvin and Sánchez (2008). In addition, a great deal of capacity building has to take place before local knowledge can be integrated into natural resource management (Commonwealth of Australia 2004). For these reasons, it is difficult to integrate indigenous—or local—knowledge into field specific competences.

It is also difficult to articulate an all-encompassing definition for ESD. For this reason we adopt the definition given by the SDE Network which is contained in a working paper prepared by The Environmental Association for Universities and Colleges, United Kingdom, June, 2009 and reads:

> Sustainable Development Education is the process of acquiring the knowledge, skills and attitudes needed to build local and global societies that are just, equitable and living within the environmental limits of our planet, both now and in the future.

ESD encompasses a much broader circle of competences and topics compared to EE. The UNECE Strategy on ESD (article 14) outlines that this holistic approach to education is still developing as a broad and comprehensive concept and addresses a wide range of development subjects, as well as targeted forms of education. At the present time, degree programs that embody EE over-ride those of ESD in the sub-region (CAREC 2009). Nevertheless, the goal of these EE programs is to equip students with the kind of knowledge and professional competences required to mitigate the negative consequences of economic growth (Boldjurova 2007). Individual countries have begun to implement their ESD objectives at national level in the hope that this will maintain the sub-regional goals that are to be supported through new EE programs. Yet, as the discussion in Sect. 3 reveals countries are at different stages with the required institutional and legislative changes (CAREC 2009) which has implications for the successful integrated management of natural resources at both national and trans-boundary levels.

The key principles of ESD such as value-orientation, a holistic approach, reflexivity and transformation provide a learning philosophy and a strategy relevant to emerging qualities of the postmodern society (Fadeeva and Galkute 2012) which also concurs with the value oriented, reflexive and holistic principles of INRM. INRM implies a conscious process of incorporating multiple aspects of natural resource use into a system of sustainable management, to meet explicit production goals of farmers and other land users (e.g., profitability, risk reduction), as well as goals of the wider community (e.g., cultural values, sustainability). These production and other goals will be different over a variety of physical, political, economic and social settings. It is at this point that educators from the natural sciences have to confront the dichotomy that exists between global aspirations for ESD and the reality of local environmental contexts.

In other words, local industrial and other standards may not be totally in synch with what is considered to be global best practice. Equally, the competences that 'qualified' environmental professionals need to attain can vary from place to place as is the case in the CA countries where it is the Ministries of Education that strictly regulate curricula (Zasada et al. 2012). However, ESD requires participatory teaching and learning methods that motivate and empower students to focus their behaviour and action towards sustainable development. This means that curricula designed for INRM should promote competencies like critical thinking, imagining future scenarios and collaborative decision making (UNESCO 2005–2014), skills which can only emerge in a multi-disciplinary context.

In addition study programs that embody ESD need to have inbuilt mechanisms that allow for a change of direction if and when required. It is in this kind of change scenario that both informal and formal knowledge can be incorporated to point the way forward. For all of these reasons there has to be far more collaboration between all stakeholders involved in the deployment of the concept of sustainable development. This kind of participatory approach has to be supported by new approaches to the crafting of institutions, as well as the development and implementation of new legislation that upholds both ESD and EE. At the time of writing this is not entirely the case in all of the countries that comprise the Central Asian sub-region.

3 Best Practices in Education for Sustainable Development and Environmental Education in Central Asia

In 2002, the CA countries began to develop and implement regional strategies on ESD objectives at the national level. Yet, up until quite recently natural resource management degree programs were not systematic across the sub-region because there still exists a profound misunderstanding of the connection between the environmental, social, as well as the economic processes at the policy, management, community and personal level. This is reflected in a review prepared by the Central Asian Regional Environmental Council (CAREC) of the status of ESD capacity in the CA countries within the frameworks of the UN Decade on ESD and the UNECE Strategy on ESD on the one hand, and the sub-regional—the CA Initiative on SD (CAI). This review illustrates that there are still political and other barriers that impede an integrated approach to the management of life support systems such as water, soil, forests and biodiversity. This situation continues to prevail despite the existence of regional organisations such as the Interstate Commission for Water Coordination for Central Asia.

Thus far, the fragmented management of two major waterways—the Amu Darya and Syr Darya Rivers—has contributed to conflict between both upstream and downstream countries regarding the types and regimes of use because no balance of interests acceptable to the involved countries has been struck. Future river flow fluctuations and decreases as a result of climate change are likely to lead to greater competition for use of water resources (UNECE 2011). This and similar sub-regional issues exist because the five countries included in this study: Kazakhstan, Kyrgyzstan, Tajikistan, Turkmenistan and Uzbekistan are at varying stages with the development of study programs framed around ESD.

Nevertheless, the member nations of the CA sub-region have an opportunity to co-opt existing human resources and experience accumulated for EE in order to further develop ESD and widen cooperation networks at all levels: formal and non-formal, pre-school, primary school, higher, as well as professional education. In this regard, all of the CA countries appointed ESD coordinators in 2005–2006 and many of them are members of the joint CAWG on ESD which is a body that works in conjunction with the UNECE Secretariat on ESD. In addition, the CA countries developed their first voluntary reports on implementation progress of the UNECE Strategy on ESD during 2006–2007. Also, CAREC in cooperation with the members of the CA Working Group (CAWG) has prepared a number of reviews on progress of new teaching programs for both EE and ESD in each of the CA countries based on these reports. To this end, individual country progress to promote both EE and ESD can be characterised in the following way:

(a) Kazakhstan:

Some positive changes have been implemented in terms of ESD at different educational levels in the Republic of Kazakhstan (RK). The Ministry of Education and Science of the RK is committed to the introduction of ESD within the frameworks of its cooperation with NGOs and higher education institutions. For example, ESD is one of the more important elements of the "Concept of Transfer of the Republic of Kazakhstan to Sustainable Development for 2007–2024" (at the strategic level), as well as being reflected in the EcoCode (at the legislative level). The Government Programme of the RK for 2007–2009 referenced a much desired Transfer to Sustainable Development at the national level, improvements in the quality of life and development of human capital. All of which is to be achieved through the integration of the principles of ESD into educational programs.

To help achieve the above stated goals, the main objectives of "The State program of Education Development" are:

- All levels of society to have access to quality education;
- Promotion of patriotism, tolerance, culture, respect for human rights and liberties;
- Renewal of content and structure of ESD based on local traditions (i.e. indigenous knowledge) and global experience; and,
- During the academic year 2008–2009 a bachelor study program, "Ecology and SD," was introduced into the higher education system to help integrate Kazakhstan into the global educational community.

NGO activity plays a significant role in the implementation of both EE and ESD. In fact, NGOs have been involved in this process for more than 15 years. Extra-curricular environmental activities are promoted and supported by "EcoObraz", Karaganda "EcoCentre", "Baiterek" and "Otrazhenie". All of which take an active part in the implementation of both national and regional projects on ESD in cooperation with the Ministry of Education and Science of RK. Also, CAREC makes a significant contribution towards the implementation of the UN Decade on ESD and the UNECE Strategy on ESD. To this end, in 2006, CAREC in cooperation with the Kazakh National Academy of Education named after Altynsarin, NGOs, and the OSCE Centre in Astana developed a draft National Action Plan for the integration of ESD into programs, state standards and courses to support implementation of the "Concept of Transfer of the Republic of Kazakhstan to Sustainable Development for 2007–2024".

(b) Kyrgyzstan:

Despite the emergence of a national dialogue between educators, experts and trainers in 2005 to date there has not been official endorsement of the Strategy and Action Plan for ESD. Although at the time of writing, some activities have been focused to provide the much needed further development and introduction of ESD for all levels of education. Despite these efforts Kyrgyzstan has no specialised regulatory framework which could serve as a base for implementation of the aim and objectives of the UNECE Strategy on ESD in accordance with the "Vilnius Frameworks of Implementation" (2005).

Nevertheless, 'The Concept of Environmental Safety of the Kyrgyz Republic' was approved by a Presidential Decree dated 23 November, 2007. This legislation highlights the need for the development of an Educational Concept for Sustainable Development. This law also provides a new impulse to sustain the process derived from the Second World Summit on Sustainable Development in 2002 and the Kiev Conference of Ministers of Environmental Protection in 2003. It is hoped that this piece of legislation will eventually assist Kyrgyzstan to contribute to the attainment of tangible goals for the UN Decade on Education for Sustainable Development.

Even though a legal base for environmental awareness has been created in the Kyrgyz Republic, the existing platform of legal documents on EE does not reflect these requirements. In addition, it is difficult for teachers and other professionals from the educational sector to access this information. The directions identified by this law include the following:

- Create a continuous environmental education system through introduction of environment and sustainability issues into curricula at all levels of education;
- Improve staff training and re-training in environmental expertise at all levels of obligatory and further education;
- Introduce new, as well improve existing curricula to provide a profound understanding of environmental problems and the requirement for the efficient use of natural resources; and,
- State support for EE.

(c) Tajikistan:

Despite the lack of development for measures and indicators for ESD in national strategies and programs, EE has been introduced in some form for all levels of education. However, ESD is still at the initial stage and is not yet a part of national development objectives. There are still insufficient training opportunities for teachers and very few priorities have been identified within the overall educational system. A significant amount of work is conducted by NGOs however a far greater effort is required to overcome the lack of interaction between ministries, as well as government agencies and NGOs.

The documents that underpin the Sustainable Development of Tajikistan are the "National Development Strategy until 2015" and the "Poverty Reduction Strategy for 2007–2009". These strategies identify the need to further develop EE through the creation of a network to increase awareness levels amongst managers and natural resources users. This will hopefully be achieved through training in environmental sustainability. Consequently, neither EE nor ESD principles were embedded within the framework of the Poverty Reduction Strategy list of priorities for education. Nor are EE or ESD principles included in the "Concept of Transfer of RT to Sustainable Development" adopted by the government in October, 2007.

Given all of the above, it is not surprising that a study conducted during 2006–2008 identified the following tends:

- Insufficient reflection of ESD issues in national strategic documents such as the National Development Strategy, the Poverty Reduction Strategy and the Concept of Transfer of the RT to Sustainable Development;
- Absence of a formal structure of inter-departmental co-operation on ESD;
- Key ESD topics are not profoundly reflected in formal education;
- ESD is not yet included in training programs for different levels of managers or for teacher training;
- Lack of capacity and manuals on EE, and absence of the exchange of information.

(d) Turkmenistan:

To date ESD has not been reflected as a priority in national programs and strategies. Moreover, ESD has not been incorporated into higher education curricula. On the other hand, ESD has been adopted in the "Strategy of Economic, Political and Cultural Development of Turkmenistan for 2020", the "National Programme "Bilim" (Education) and the NEAP (National Plan on Environmental Protection). However, the term ESD is substituted by the term "Environmental Education", which is also widely used in other national strategies, plans and programs".

There are mandatory standards for EE in secondary schools and higher education institutions that comply with the law of the President of Turkmenistan: "On improvement of the Education System in Turkmenistan". Some curricula changes have been introduced for certain disciplines connected with ESD. As a result of these changes students now study the basics of the environment, general approaches to conservation and environmental law, as well as international conventions on environment protection adopted by Turkmenistan. In addition, secondary schools and universities offer a range of extra-curricular activities, clubs, as well as discussions on EE issues and awareness.

(e) Uzbekistan:

The National Programme for staff training 1997–2010 identified the priorities for development in the educational sector over the short and long terms. There is to be an on-going re-consideration of curricula (which has happened within the framework of several TACIS Tempus projects funded by the European Commission), the development of new textbooks, the introduction of new teaching methods, greater accessibility to information technology, as well as the maintenance of the existing level of access and equality in the education system.

Uzbekistan is a member of the United Nations and cooperates with a number of its programs as well as UN bodies, i.e. UNEP, UNDP, UNESCO, WMO, WHO, UNIDO and two regional commissions, i.e. and the UNECE. To honour this co-operation, the Cabinet of Ministers Decree 19 September 2008 approved the "Action Programme on Environment Protection of the Republic of Uzbekistan for 2008-2012". The aim of the program is to direct the efficient use of natural resources and to introduce environmental methods for households as an additional measure with which to improve environmental protection. Additionally, there are some changes taking place in the development and introduction of ESD in this country at different levels such as the development of inter-departmental and inter-sectorial partnerships to incorporate sustainable development into all spheres of community activity. Nevertheless, there still remains quite a bit of work to be done to bring ESD in Uzbekistan into line with both sub-regional and global best practice (CAREC 2009.)

Based on the preceding information, it is apparent that EE as a component of ESD is still at vastly different stages across the sub-region. While steps are being taken by development aid organisations to encourage the CA countries to deal with other important areas of concern such as the health of populations, access to energy and food security, radical change hinges on the political will of the CA countries to address ESD as a priority because of the inter-relationships between environment, economy and society. Consequently, individual nations need to factor the concept of sustainable development into all significant, as well as relevant legislative and policy making processes, and to ensure that this change filters through to the educational sector. This change of course rests very much on the availability and constancy of development aid funding. To this end, the Tempus Program has been a driving force of change and has funded a number of undergraduate and post-graduate degree programs specifically aimed at INRM wherein Ministries of Education were project partners.

4 INRM in Central Asia

The term INRM was first coined in 1996 by the Consultative Group on International Agricultural Research (CGIAR)). INRM was first promoted at the 3rd CGIAR Systemwide External Review (CGIAR/TAC 1998) which recognised that a paradigm shift had occurred in 'best practice' of natural resource management. There had been a change from 'hard' reductionist science to more holistic approaches. Specifically, the 1998 review identified a move from classical agronomy to ecological sciences, from the static analysis of isolated issues to systems' dynamics, from top-down to participatory approaches, and from factor-oriented management to integrated management. The CGIAR subsequently set up a task force to coordinate work on INRM known as The Bilderberg Consensus (1999). In essence, the success of INRM relies on an ongoing dialogue, deliberation and negotiation between all of the involved stakeholders (Douthwaite et al. 2004).

As the term implies 'a holistic approach to natural resource management' new INRM degree programs funded under the Tempus instrument are the result of a well thought out synergistic process. Input is provided by local academics and their EU counterparts who are considered to be experts in their particular disciplines. These new degree programs aspire to give students the knowledge and skills with which to address both local and sub-regional environmental realities. From this bottom up approach, that is, with the integrated management of the environment taking the form of curricula designed around competence based education it is hoped that this change in the educational sector will eventually lead to the emergence of natural resource management legislation and policy that embodies the concept of sustainable development. Before this can occur however, national governments need to agree on sub-regional environmental goals and then initiate a dialogue between the educational sector and the managers, as ell as the users of natural resources.

More importantly, educators and managers of natural resources will have to adapt decision making to multi-disciplinary contexts. Also, listen to the collective voice of those who use natural resources. Nowadays, it is generally accepted that the use and management of natural resources is the subject of local social convention and negotiation, as well as the formal rules set down by government. It is not uncommon in countries with emerging economies to find that natural resource and other legislation still tends to override local norms. Yet, the relationships between natural resources, society and the economy are pressured in unpredictable ways by both local and distant economic change. Government is not always capable of reducing or eliminating this kind of uncertainty because of the range of exogenous and endogenous factors that affect these relationships. Consequently, the future managers of these complex relationships need to be trained so that they genuinely understand and concur with the perceptions, priorities, and actions of natural resource users. This type of "action research" which is central to the concept of INRM requires that the managers and users collaborate in sharing information and knowledge, as well as resources, also in undertaking joint activities including research. This kind of united front is really the only route for the realisation of the concept of sustainable development (Sterling 2001).

When the 'collective voice' is ignored, tension can arise between those who regulate and users. This is because government generally sees a geographically bigger and sometimes more complete picture, while the perspectives and interests of local people are often narrower and more self-centred. Notwithstanding, this is no reason for assuming that the broader view should automatically prevail. In order to manage these competing interests graduates of INRM degree programs need not only field specific expertise but also the transversal skills that assist them to arrive at some kind of consensus between the local and the geographically distant view of 'reality' (Frost et al. 2006). Whilst the INRM degree programs funded under the Tempus program are designed with local and trans-boundary sustainability goals in mind, there still exists the need for more equitable sharing of power and responsibility between government and the broader community.

5 Tempus Program and the Bologna Process in Central Asia

Tempus which ran from 1990 to 2013 was an external cooperation program funded by the European Union through its Education, Audio-visual and Culture Executive Agency (EACEA). This instrument has funded the modernisation of higher education systems in the Western Balkans, Central Asia, Eastern Europe, Northern Africa and the Middle East. Tempus was primarily designed to contribute to the reform and upgrade of higher education institutions, as well as systems in these partner countries. Some other benefits realised through cooperation were the cohesion of civil society and democratic governance, as well as an improvement in mutual understanding and intercultural dialogue between the European Union and its neighbours. The Tempus program provided a bottom-up approach to support positive change through national initiatives for higher education institutions. At the same time, the program provided the flexibility for the relevant ministries of partner countries to establish national and regional priorities in order to maximise the impact of the program on higher education reform processes.

The Tempus instrument contributed to the promotion of cooperation between higher education institutions of the European Union and the partner countries mostly through capacity building activities. Another sub-theme was the promotion of voluntary convergence of higher education systems in these countries with developments in higher education in the European Union, particularly alignment with the Bologna Process. Indeed, the Bologna Process has become a reference point for most Tempus partner countries by setting in motion a series of reforms to not only modernise higher education systems, also to make them more compatible across political sub-regions (Education, Audiovisual and Cultural Executive Agency 2011).

The Bologna Process was proposed at the University of Bologna in 1999 at which time the ministers of education from twenty nine European countries signed a declaration in the Italian city of Bologna. The process is constantly updated through a series of member nation Ministerial conferences. The Bologna Process aims to create the European Higher Education Area (EHEA) through the harmonisation of academic degree and quality assurance standards throughout Europe. The process supports the introduction of similar undergraduate and postgraduate levels in all countries that are signatories to the declaration. These reforms are set in a context of both sustainability and broader European policy issues. In other words, it is widely recognised that both education and educational cooperation play a role in the development of stable, peaceful and democratic societies.

Accordingly, the Bologna Declaration states that a Europe of Knowledge is an important factor for social and human growth. Consequently, the Bologna Process aims to incorporate and support the continuing development of ESD so that present needs be met without compromising the ability of future generations to meet their needs. In this regard, universities have been important partners in the building of transnational understanding and cooperation, thus also contributing to the European dimension of higher education. Nevertheless, universities have to continually rethink their position in society as educators. Some of the hallmarks of the Bologna Process are undergraduate degrees no shorter than 3 years, a European Credit and Accumulation Transfer System (ECTS) and increased opportunities for mobility for both students and teachers (European Communities 2009). An overarching framework for qualifications in the EHEA was adopted during the Bergen Conference of European Ministers Responsible for Higher Education 19–20 May, 2005. This framework comprises three cycles wherein the curricula of first two cycles are designed around learning outcomes, competences, as well as credit ranges (ECTS).

Competence based learning is more often than not an approach used for teaching students ‘concrete’ skills rather than abstract learning. Competences are used to guide the design of learning modules. While competences are not explicitly mentioned in the official documents on the Bologna Process they are closely related to an implicit objective of the reform efforts, that is, enhancing employability of higher education graduates. Hence, competences are central to the concept of employability of university graduates. Competences are classified as Key and Field Specific. The first group is considered to be comprised of social, self, systematic and domain related competences—or transversal abilities and skills—whilst field specific competences are related to labour market requirements (Schaeper 2008). Generally speaking, curricula for new degree programs funded under the Tempus instrument in CA are designed around generic descriptors—or competences—that equate with ministerial and industry requirements. Because perfect knowledge about the use and management of ecosystems does not exist, for the design of new degrees that provide the training for future natural resource managers, a question that needs to be considered is: which innovative approaches in teaching and learning are needed? (Copernicus Campus 2005).

To help answer this question a Needs Analysis is undertaken the results of which help to orient curricula design and development. The aim of these surveys is to determine what kind of competences students require to prepare them for labour market entry and to identify the transversal skills that provide the framework in which students can grasp the philosophy of INRM. Nevertheless, the national

Table 1 Differences in competences for INRM between Kazakhstan and Kyrgyzstan*

Kazakhstan—Competences	Kyrygzstan—Competences
Locate, understand and implement environmental legislation and policy	Locate, understand and implement environmental legislation and policy
Conduct Environmental Audits	Conduct Environmental Audits
Understand and apply methods for restoration and rehabilitation	Understand and apply methods of protection, rehabilitation and remediation of soil and water resources
Conduct fieldwork	Undertake laboratory and practical training
Use modern methods of studying ecosystems for the analysis of the processes	Acquire knowledge and skills for specialisation
Undertake laboratory research, analyse and present results	Understand the role of information technology in natural resource management
Model standard ecological situations and make predictions based on ecological models	Understand the local impact of trans-boundary environmental problems
Classify sources of pollution	Be able to communicate with peers and society
Understand the sustainable use of natural resources	Understand the role of and be able to work with international, sub-regional and national standards for the environment

* Based on results of Needs Analysis undertaken for Tempus Project CIBELES (2010).

obligatory component of ECTS leaves very little opportunity to include learning modules that introduce students to the multi-disciplinary context of INRM. An analysis of the competences that graduates from a new and recently accredited convergent degree program in soil and water management have to acquire helps to explain the prevailing situation which does not necessarily contribute to a bottom up approach to sustainable development that seeks to combine local knowledge with current global best practice (UNECE 2012). Table 1 provides a comparison between the Key and Field Specific Competences identified in Kazakhstan and Kyrgyzstan where the degree program has already been accredited by the relevant Ministries of Education.

This table reveals interesting facts about how sustainable development is understood at national levels which subsequently impacts at sub-regional level. The first is that in Kazakhstan students are required to understand the sustainable use of natural resources whilst in Kyrgyzstan this is neither a key or field specific competence. Yet, students in Kyrgyzstan are required to be able to work with international, sub-regional and national environmental standards. Implicit to this competence would be some understanding of the concept of sustainable development on the part of the student. This mismatch occurs in Kyrgyzstan because the sustainable use of natural resources is not reflected in national legislation that impacts on educational content. Whereas, in terms of key competences, students from Kyrgyzstan need to be able to communicate with their peers and this is not a priority requirement in Kazakhstan. Whilst in Kyrgyzstan there is an emphasis on understanding the local side of trans-boundary management of natural resources—

which would incorporate informal knowledge—there is not the same emphasis in Kazakhstan.

These differences in competences can be related to the political history of the sub-region and the 'culture of sustainability' in individual countries. In Kazakhstan for example, there is still a lack of external peer review for publications etc. which probably explains why there is no 'formal' requirement on students to be able to communicate with their peers. This oversight can be traced to the philosophy of the Higher Education Quality Assurance System that was established in 2008 which places emphasis on compliance rather than self-evaluation and improvement. At the moment in Kazakhstan there is not much emphasis on moving from the existing quality control mechanisms towards a stakeholder based quality assurance culture which is reflected in the relationships between universities and employers. Whereas at the global level it is accepted that employers really should be formally included in the quality assurance process, as well as assessment of study programs and research (Tempus 2011).

Meanwhile in Kyrgyzstan the labour market is not familiar with the idea of Bachelor and Master Programs. Consequently, a degree of scepticism exists regarding the equivalence of these programs with that of 'scientific specialist' which poses an obstacle for students who graduate from programs that align with the Bologna Process in terms of their entry to the labour market. As a result, there is an imbalance between the market demand and employable graduate supply. All of this is compounded by a lack of the systematic modernisation of structure and content of higher education or the removal of contradictions in the system. While there is still no adequate link between research and higher education some efforts have been made to merge scientific and educational institutions. However, many issues such as the development of 'scientific skills' have thus far been overlooked (National Tempus Office, Kyrgyzstan 2010). Although students from Kyrgyzstan will be skilled communicators they will have little 'real world experience' which will hinder the contribution this country can make to the integrated management of natural resources.

The situation outlined above is also evident in a convergent Master program which was developed and implemented within the framework of the SWAN Project. There are marked differences between competences required at national level for postgraduate water resource management professionals. In fact, in Kyrgyzstan one of the programs on offer requires that students undertake less than 20 ECTS in subjects associated with integrated water resource management. Again, the problem of National Qualification Frameworks that was mentioned earlier means that opportunities for professional specialisation are generally overtaken by Ministry of Education requirements that are applied across the board for postgraduate students. Moreover, the percentage of 'study hours' for these subjects sometimes outweighs 'specialised' research and practical orientation which are fundamental to notions of what constitutes postgraduate study programs (Brunner and Tillet 2007).

In Uzbekistan students are required to complete up to 756 h of study or one third of the course work for obligatory subjects that are enshrined in the National Qualification Framework for the same Master program. Of a total of 4,536 h, students spend 160 h in the laboratory which hopefully is augmented from the 1,240 h available for fieldwork/laboratory work for the research component of the dissertation. An obligatory teaching internship takes up 520 h of work which in the case of a field professional, is not really oriented towards field specific training. While in Kazakhstan where the system is more in tune with ECTS and Bologna, students still need to devote in excess of 12 ETCS to the study of pedagogy and a teaching internship. Again, this percentage of ECTS detracts from the highly specialised, as well as multi-disciplinary training that natural resource managers of the twenty first century require.

There will always be obstacles to overcome when competence based education and convergent degree programs are introduced in a political sub-region where ideas of sustainable development are still emerging. First, local teaching staff has to be able to design programs that comply with Ministry requirements on the one hand, meet national EE objectives on the other hand and at the same time contribute to a much broader understanding of what constitutes the integrated management of natural resources. The second but related situation is that INRM professionals from convergent degree programs graduate with different skills and competences which are probably quite relevant to deal with national environmental issues but not easily adapted to sustainable development at the trans-boundary level.

6 Conclusions

Highly complex political, socio-economic and cultural issues underpin the major environmental problems that exist in the CA sub-region. To help improve these circumstances the Tempus program has funded a number of initiatives that aim to contribute to the integrated management of the sub-region's natural resources. These initiatives took the form of new study programs that were intended to complement the Central Asian Education for Sustainable Development (ESD) and Environmental Education (EE) Programme developed during 2003–2006. Wherever possible, the aim of these actions has been to align educational sectors with the Bologna process in an effort to open up potential for exchange of research efforts between European INRM specialists and their Central Asian counterparts. In this vein, new curricula are framed around competence based education that is the result of stakeholder consultation. Two of these new study programs: CIBELES—Bachelor of Soil and Water Management and, SWAN—Master of Water Resource Management, have involved Ministries of Education as project partners in the hope that common methodologies for the design of curricula would be adopted across the sub-region. Certainly, some progress has been made in terms of harmonisation across national educational sectors. At the time of writing however, neither ESD nor INRM form of part of a larger environmental management picture.

ESD and INRM are relatively new and emergent philosophies across the sub-region. The integration of these new philosophies is hindered by the fact that governments and natural resource managers, as well as users have failed to grasp the complex inter-play of the environmental, economic and social aspects of life. The focus remains on national issues because of the push for economic growth in the region which comes at the expense of sustainable development for the sub-region. This is currently reflected in the fragmented management of some of the more important life support systems in the sub-region. Educational programs that fully embrace ESD will help to overcome this challenge (Blewett and Cullingford 2004). Some ways in which this can happen have already been identified as sub-regional objectives and have been funded by various development aid organisations. These include the creation of networks for the exchange of information, as well as training centres with similar missions in each country.

CIBELES and SWAN, amongst other projects, have paved the way for the kind of INRM to evolve that is required in Central Asia. National centres that support integrated water resource management were established within the framework of the SWAN project. One of the roles of these centres is to contribute to the mutual understanding of the limits of major shared waterways and a sub-regional dialogue that aims to diffuse conflict. The other role is to help promote the emergence of robust institutional and legal frameworks that can adapt to deal with the ever changing nature of water. Yet, professionals who graduate from this Master program have not had access to the kind of educational environment in which they can develop the creative thinking processes required to diffuse conflict over use of natural resources. Also, in terms of comparative and complementary field specific competences, the percentage of ECTS allotted to 'specialisation' and research skills are relatively low in terms of global best practice. Consequently, many of these postgraduates have very little understanding of the role adaptive and integrated management processes. Moreover, graduates from the convergent Bachelor of Soil and Water Management, which was developed within the framework of the CIBELES project, enter the Master program with very little 'specialist' undergraduate training.

There is very little synergy between the competences that natural resource professionals need to acquire because understandings of what equates with sustainability varies across the sub-region. These different understandings are reflected by what employers from both the public and private sectors consider to be critical natural resource problems and the preferred solutions with which to manage these environmental issues at the national level. At the moment in Kazakhstan there is no emphasis on development of communicative skills or self-evaluation and assessment which is reflected in the national approach to quality control for the higher education sector. Meanwhile in Kyrgyzstan students are expected to understand the concept of sustainable development which means that study programs need to be multi-disciplinary. However the neither the public or the private sector is disposed to employ professionals that have acquired this kind of holistic view of natural resource management. Consequently, at the national level the study of subjects directly related to INRM or any other life-long learning pursuit is not yet considered to be a priority.

These discrepancies between educational sectors and the related National Qualifications Frameworks can be traced to the varying rates of adoption of ESD at national level. While more progress has been made with the incorporation of EE study programs that integrate formal and informal knowledge—both of which underpin successful INRM—these programs only go some of the way to assisting students to identify and understand the relationships between environment, economy and society. The interaction of which, affect the concept of sustainable development as it is understood at local, national and sub-regional level. In fact, only Kazakhstan has a legislated objective to renew the content and structure of ESD based on indigenous knowledge and global best practice. The other four countries are lagging in terms of a similar commitment that is reflected in national legislation for the educational and natural resource sectors. Once national legislation supports a sub-regional sustainability perspective, the CA countries will be able to overcome the *piece meal* approach to training their natural resource managers. After all, the holistic management of the sub-regions life support systems relies on professionals who can identify, evaluate and act the local, national, as well as the global level.

The way forward requires INRM multi-lateral agreements, the adoption of a sustainable development charter for the sub-region and far more dialogue between the CA countries, the managers and the users of natural resources. At the same time these countries need to bear in mind that a sustainability dialogue that serves both the local view and the collective view should embrace the idea that development is not only driven by the push for national economic growth. Rather, it needs to be guided by the limits of natural resources when considered from both a national and sub-regional perspective. Once this happens then the introduction of textbooks, manuals and technology that promote ESD as an integral component of EE will help to speed up the process of change. All of which is predicated on the sub-region's ability to attract new investment, partnerships between the public and private sectors and much needed funding for the implementation of new programs, as well as action research projects for the educational sector.

Acknowledgments Tempus IV Project CIBELES Curriculum Invoking Bologna-Aligned Education Leading to Reform in Environmental Studies – TEMPUS JEP 511172-DE-2010 Tempus IV project SWAN Towards Sustainable Water Management in Central Asia - 158982-TEMPUS-ES-TEMPUS

References

Andersen A (1999) Cross cultural conflicts in fire management in North Australia: not so black and white. Conserv Ecol 3(1). http://www.consecol.org/vol3/iss1/art6/. Accessed 1 Sept 2013

Blewett J, Cullingford C (2004) Sustainability curriculum. Earthscan, London

Boldjurova IS (2007) Higher education in the Kyrgyz Republic: problems of modernization and internationalization'. http://libw01.kokushikan.ac.jp/. Accessed 21 Dec 2013

Brunner JJ, Tillett A (2007) Higher education in Central Asia: the challenges of modernization—an overview, World Bank. http://200.6.99.248/~bru487cl/files/RevAlldoc2c.pdf. Accessed 3 Oct 2013

CAREC (2007) Progress review on education for sustainable development in Central Asia: achievements, good practices and proposals for the future, Almaty. http://www.unece.org/fileadmin/DAM/env/esd/Implementation/ReportsStakehold/ESD_CAREC%20report_eng.pdf. Accessed 25 Sept 2013

CAREC (2009) The best practices on education for sustainable development in Central Asia, Review. Central Asian Regional Environmental Council, Almaty

Commonwealth of Australia (2004) Working with knowledge in natural resource management. (ISBN 064 255 0344)

COPERNICUS (2005) Guidelines for sustainable development in the European higher education area, how to incorporate the principles of sustainable development into the Bologna process. www.copernicus.campus.org. Accessed 9 Oct 2013

Douthwaite B, Ekboir J, Twomlow S, Keatinge JDH (2004) The concept of integrated natural resource management (INRM) and its implications for developing evaluation methods. In: Shiferaw B, Freeman HA, Swinton SM (eds), Natural resource management in agriculture: methods for assessing economic and environmental impacts, CAB International. http://boru.pbworks.com/f/Chapter_14+pg+321-340.pdf. Accessed 9 Oct 2013

Education, Audiovisual and Cultural Executive Agency (2011) Overview of the higher education system in the tempus partner countries. Central Asia, Asia, pp 29–46. (ISSN: 1831-97236)

European Communities (2009) ECTS Users' guide. http://ec-europa.eu.education/lifelong-learning-policy/doc/ects/guide-en.pdf

Fadeeva Z, Galkute L (2012) Looking for synergies: education for sustainable development and the bologna process. J Educ Sustain Dev 6:91–100

Frost P, Campbell B, Medina G, Usongo L (2006) Landscape-scale approaches for integrated natural resource management in tropical forest landscapes. Ecol Soc 11(2):30. http://www.ecologyandsociety.org/vol11/iss2/art30/. Accessed 1 Oct 2013

Galvin D, Aledo A, Escarré R, Sánchez JR (2006) Sustainable development as the basis for preservation of the environment. In: Conference Proceedings,Contemporary issues of environmental education, Udmurt State University, Ishvesk, Russia, 24–25 Oct 2006

Galvin D, Sánchez JR (2008) Cross Cultural Issues in Environmental Education. In: International education, noosphere and sustainable development, education conference held in the framework of Tempus TACIS Project JEP 25186/2004 Master in environmental law and policies in the Russian federation, Tambov State Technical University, Tambov, Russia, 5–6 June 2008

Guíllern Rodriguez FC (1996) Educación, medio ambiente y desarollo sostenible. Revista Iberioamericano de Educación Numero 11:103–110

Harris JM (2003) Sustainability and sustainable development. International society for ecological economics, international encyclopaedia of ecological economics. www.iescoeco.org.sus.dev.pdf. Accessed 3 Oct 2013

Higher education in Kyrgyzstan (2010) http://www.tempus.kg/index.htm. Accessed 25 Sept 2013 (the web-site of National Tempus Office in Kyrgyzstan)

Schaeper H (2008) The role of key competences in the bologna process: rhetoric and reality. XXIX International Congress of Psychology, Berlin

Sterling S (2001) Sustainable education: re-visioning, learning and change (Schumacher briefings). Green Books, Cambridge

The Bilderberg Consensus (1999) Bilderberg Meeting held, Sintra, Portugal, 3–6 June, www.bildergurg.org/1999. Accessedd 15 Sept 2013

The Environmental Association for Universities and Colleges (2009) Embedding sustainable development in the curriculum. www.eauc.org.uk. Accessed 23 Nov 2013

UNECE (2012) Learning for the future competences in education for sustainable development. Vormgeving Smidswater, Publicatienummer 1LVDO1202

UNECE (2011) Strengthening water management and transboundary water co-operation in Central Asia: the role of UNECE environmental conventions. http://www.unece.org/index.php?id=28204. Accessed 11 Nov 2013
UNESCO Asia-Pacific Regional consultation on an Education for ESD post 2014 programme framework. www.unesco.org/new/en/unescoevents. Accessed 1 Sept 2013
UNESCO, Education for Sustainable Development (2005–2014) http://www.unesco.org/new/en/education/themes/leading-the-international-agenda/education-for-sustainable-development/. Accessed 1 Sept 2013
UNESCO (2012) Institute for Statistics, 'Adult and Youth Literacy', September, 2012, No 20. www.uis.unesco.org/FactSheets.../fs20-literacy-day-2012-en-v-pdf. Accessed 15 Sept 2013
United Nations (2001) Secretary General Calls for Break in Political Stalemate over Environmental Issues, Press Release, Paris, 15 March2001
WCED (1987) Our common future. World commission on environment and development. Oxford University Press, Oxford
Zasada M, Serekpaev N, Orozumbekov A (2012) Forestry study programs at bachelor level in selected countries of the Commonwealth of Independent States. Studia i Materialy CEPL w Rogowie, R.14. Zeszyt 2 (31), p 155–162

Author Biographies

Denise Galvin Before coming to academia Denise Galvin (formerly Eisenhuth) was a consultant on environmental issues for some of the biggest privately funded public infrastructure projects constructed in Australia during the late 1990s. Since relocating to Spain in 2002 she has maintained the role of Adjunct Lecturer in the International Business Program, Faculty of Economics and Business Science, University of Alicante. The major part of her work in Europe has been in development aid projects funded by UNESCO, the European Union and the European Commission. She co-ordinated the research efforts of a major FP5 project, has provided capacity building and technical expertise in Tempus, ACP-Edulink and IPA instruments. Her PhD work was commenced in Australia with funding made available from the Murray-Darling Basin Commission, as well as the George Alexander Foundation and she is currently waiting award of the title from Oxford Brookes University. Denise's research interests lie with the development and implementation of policy for the participatory management of natural resources and ESD. She has published in both of these areas in English and Spanish.

Aibek Upenov Assoc. Prof. Aibek Upenov works in the Natural Sciences and Pedagogics Department, Talas State University, Kyrgyzstan. Until 2012 he held the role of Vice-Rector for International Relations and Science. Prior to taking up this role in 2004 he was the Dean of the Faculty of Economics and Ecology. He received his PhD in Soil Science in 2003 from Kyrgyz National Agrarian University named after K.I. Skryabin, Bishkek. Since 2005, Dr. Upenov has been involved in 10 Tempus Projects and 8 Erasmus Mundus. He was the regional co-ordinator for the Tempus Project EBAC which focused on building vocational capacity in higher education institutions in Southern Siberia and, CIBELES the focus of which was to align Bachelor and Master environmental programs in Central Asia with the Bologna Principals. Although he is more readily recognised for his work in Soil Science, Dr. Uponev also takes a keen interest in ESD and its implementation in Central Asia. He has published in Russian, German and English.

Institutional Practices Versus Student Needs and Its Implications for the Development of a Holistic Engineering Education for Sustainable Development (EESD) Framework

Subarna Sivapalan, Ganakumaran Subramaniam and Michael J. Clifford

Abstract

Sustainable development (SD) competences feature prominently in the 2012 Malaysian Engineering Accreditation Council (EAC) manual. The manual outlines 12 outcomes that undergraduate engineering students of Malaysian universities are expected to develop upon completion of their studies. These 12 outcomes, specifically those in relation to SD competences, are open for interpretation, in accordance to the vision, mission and educational philosophies of the respective universities and undergraduate engineering programmes. This paper highlights a Malaysian private engineering university's endeavours to include SD competences within its undergraduate engineering programmes. The paper first focuses on the extent to which SD is featured within the institution's programme educational outcomes and common modules. This was explored through qualitative means, namely manifest and latent content analysis. This is followed by a survey, to explore the present pedagogical practices within the undergraduate engineering programme to ascertain the extent to which it conforms

S. Sivapalan (✉)
Department of Management and Humanities, Universiti Teknologi PETRONAS, Bandar Seri Iskandar, 31750, Tronoh, Perak Darul Ridzuan, Malaysia
e-mail: subarna_s@petronas.com.my; subarna24@gmail.com

G. Subramaniam
Faculty of Arts and Social Sciences, The University of Nottingham Malaysia Campus, Jalan Broga, 43500 Semenyih, Selangor Darul Ehsan, Malaysia
e-mail: ganakumaran.subra@nottingham.edu.my

M.J. Clifford
Faculty of Engineering, The University of Nottingham, University Park, Nottingham, NG7 2RD, UK
e-mail: mike.clifford@nottingham.ac.uk

W. Leal Filho (ed.), *Transformative Approaches to Sustainable Development at Universities*, World Sustainability Series, DOI 10.1007/978-3-319-08837-2_28

to philosophies of education for sustainable development. Also highlighted are student stakeholders' views on approaches best suited to teach sustainable development within the undergraduate engineering programme. The paper then discusses findings of a thematic analysis of open-ended survey responses on students' needs that should be considered to help develop the desired sustainability learning experience in the university. A total of 12 categories were identified as a result of this thematic analysis, of which eight of these categories encompassed the common engineering modules i.e. (i) Practical versus Theoretical, (ii) Real sustainable development issues and situations, (iii) Sustainable development learning activities and assessment, (iv) The need for heightened exposure and awareness to sustainable development post-graduation, (v) Teaching and learning of sustainable development via knowledge of current technological trends, (vi) Sustainable development awareness through exposure within the engineering industry, (vii) Sustainable development content within current learning modules and (viii) Approach to teaching sustainable development. The remaining four categories identified were for the common non-engineering modules, namely (i) Communication and sustainable development, (ii) Approach to teaching sustainable development for non-engineering modules, (iii) Bringing real life sustainable development issues and situations into non-engineering modules and (iv) Relating engineering aspects with human and societal aspects. The paper ends with a discussion of the implications of these findings for the development of a holistic engineering education for sustainable development framework.

Keywords

Engineering education · Engineering education for sustainable development · Malaysia · Higher education · Whole institution approach · Institutional approach to sustainable development

1 Introduction

Malaysian undergraduate engineering programmes are developed in line with the criteria set by the Engineering Accreditation Council (EAC), in its Accreditation Manual. Interestingly, 66.7 %, or eight out of these 12 criteria are related to sustainable development competences (EAC Manual 2012). However, also apparent is that the EAC Manual has not been developed within the philosophy of engineering education for sustainable development (EESD). This conclusion was drawn based on the lack of evidence in the manual which suggests that sustainable development must be made a compulsory context within which all 12 undergraduate engineering programme outcomes must be developed.

In 2009, a study was carried out to look into the views of the Malaysian engineering industry employers on the present and expected competencies of the country's engineering graduates. Employers feedback were sought on 13 competencies, namely (a) 'ability to acquire and apply knowledge of engineering fundamentals, (b) theoretical and research engineering, (c) application and practice oriented

Table 1 Summary of main findings

Item	Competency	Current level of competency (%)	Expected level of competency (%)
1	Ability to acquire and apply knowledge of engineering fundamentals	54.3	83.6
2	Having the competency in theoretical and research engineering	47.4	73.2
3	Having competency in application and practice oriented engineering	52.4	85.5
4	Ability to communicate effectively, not only with engineers but also with the community at large	49.5	86.7
5	Having in-depth technical competence in a specific engineering discipline	48.8	82.5
6	Ability to undertake problem identification, formulation and solution	48.1	84.6
7	Ability to utilise a systems approach to design and evaluate operational performance	55.7	78.9
8	Ability to function effectively as an individual and in a group with the capacity to be a leader or manager as well as an effective team member	55.7	85.1
9	Having the understanding of the social, cultural, global and environmental responsibilities and ethics of a professional engineer and the need for sustainable development	51.2	80.3
10	Recognising the need to undertake lifelong learning, and possessing/acquiring the capacity to do so	49.3	80.1
11	Ability to design and conduct experiments, as well as to analyse and interpret data	42.4	74.6
12	Having the knowledge of contemporary issues	47.9	75.4
13	Having the basic entrepreneurial skills	24.4	57.6

(Azami Zaharim et al. 2009, pp. 411–414)

engineering, (d) communicate effectively, (e) in-depth technical competence in a specific engineering discipline (f) undertake problem identification, formulation and solution, (g) utilise a systems approach to design and evaluate operational performance, (h) function effectively as an individual and in a group with the capacity to be a leader or manager as well as an effective team member, (i) understanding of the social, cultural, global and environmental responsibilities and ethics of a professional engineer and the need for sustainable development, (j) recognising the need to undertake lifelong learning, and possessing/acquiring the capacity to do so, (k) design and conduct experiments, as well as to analyse and interpret data, (l) knowledge of contemporary issues, and (m) basic entrepreneurial skills' (Azami Zaharim et al. 2009, p. 411). The detailed findings of the study are illustrated in Table 1.

As presented in Table 1, findings on engineering graduates present level of competencies indicate that only five out of the 13 competencies listed a satisfaction level of 50 % and more. Competence 7 and 8 recorded the highest percentage at 55.7 % each, while the lowest was recorded for competence 13 at 24.4 %. These percentages are considered rather low, and suggest that engineers need to improve significantly in areas listed. As for the expected level of competencies, communicating effectively was listed as the competency most expected of Malaysian engineers, while least expected was entrepreneurial ability. In terms of sustainability competencies within the profession, only 51.2 % of the employers were satisfied with their engineers' present abilities in understanding 'social, cultural, global and environmental responsibilities and ethics of a professional engineer and the need for sustainable development' (Azami Zaharim et al. 2009, p. 411). This shows that almost half of the 422 employers surveyed thought their employees lacked this competence. Additionally, 80.3 % of the employers also indicated that they expected their engineers to be sustainability competent. These findings are significant, as it suggests that Malaysia's engineering education programmes are not adequately preparing its graduates to be sustainability competent. It also indicates a serious mismatch between the expectations of the industry of its engineers, and the quality of sustainability competent graduates produced by local universities.

In addition to the abovementioned 2009 study, there has also been much research conducted to better understand the Malaysian engineering education landscape. This is apparent in studies conducted by Ab Rahman et al. (2009), Omar et al. (2009), Azami Zaharim et al. (2010), Abd Hamid et al. (2005), Rohani et al. (2005), Johari et al. (2002), and Abdullah et al. (2005). Our review of these studies however indicates that undergraduate engineering education research in Malaysia has mainly focused on three specific areas. These are (i) engineering graduates' employability skills and other skills relevant for the engineering workplace, (ii) pedagogies for engineering education and (iii) studies on the development of a Malaysian engineering education model. It was also found that most research is not contextualized towards sustainable development or education for sustainable development (ESD). Research on the importance for engineering students to be sustainability aware, however, is evident through studies conducted by Azmahani et al. (2012), Sharipah et al. (2012), and Arsat et al. (2011).

To date, there is little evidence of EESD research conducted in Malaysia to address problems faced by universities and academicians to holistically integrate sustainable development within undergraduate engineering programmes. The focus areas of the studies highlighted in this paper, and the apparent lack of sustainable development integration guidelines provided by the EAC, is evidence of this limitation. The present study thus proposes to bridge this gap, through an investigation of the present educational practices and needs for sustainable development within the university. The implications of these findings, in light of the development of a holistic Malaysian undergraduate EESD framework, are also discussed. The paper first focuses on the extent to which sustainable development permeates within the institution's programme educational outcomes and common modules. This is followed by an investigation of the present pedagogical practices within the undergraduate engineering programme to ascertain the extent to which it conforms to philosophies of EESD. Also highlighted are student stakeholders' views on approaches best suited to teach sustainable development within the undergraduate engineering programme. The paper then discusses findings of a thematic analysis of open-ended survey responses on students' needs that should be considered to help develop the desired sustainability learning experience in the university.

2 Transformative Learning and EESD

Literature on pedagogies related to EESD within the higher education context have mostly highlighted the processes and strategies related to teaching and learning. In 2011, the United Nations produced the United Nations Decade of Education for Sustainable Development (DESD) Monitoring and Evaluation Report on currently accepted learning processes aligned with ESD. These processes, namely (i) collaboration and dialogue, (ii) engaging the whole system, (iii) innovation through transformative practice and (iv) active and participatory learning (Tilbury 2011) were also discussed from the perspective of processes and strategies of teaching and learning. Additionally, notions of learning such as experiential learning, deep learning, transformational learning, transdisciplinary and multidisciplinary learning, problem based learning, inquiry based learning, applied learning, active learning, participatory learning, critical emancipatory pedagogy and the use of environment and community as learning resources have also informed much of the research on pedagogies for ESD. Many of these notions of learning have also informed research on EESD. Cooperative learning, student-centred learning, deep learning and problem-based learning are some of the instances of these pedagogical notions apparent in EESD.

The pedagogical notions surrounding sustainable development are indicative of teaching and learning approaches associated with the theory of constructivism. The rationale behind this observation is the evident emphasis placed upon engagement and interaction between learners, as well as between learners and teachers, within a community of learning that promotes student centeredness, reflexivity and transformation. The need for educators to be facilitators and motivators of learning processes which advocate the need for learners to understand multiple perspectives, and be immersed in learning situated within the context in which it will be applied, is further evidence and indication of constructivism. Strongly linked to these philosophies of teaching and learning is the transformative pedagogy, an adult learning theory deeply rooted within constructivism.

Transformative learning has been described by Taylor as 'uniquely adult, abstract, idealized, and grounded in the nature of human communication' (2002, p. 5). According to McEwen et al. (2011), 'transformative learning is learning that takes the learner's knowledge and skills into a new domain, with a change or in cognitive and affective processes. It recognises that learning is not necessarily gradual, progressive and linear, but may have significant thresholds for change in understanding, and emotional intelligence' (p. 37). Learning from a transformational perspective is seen to manifest when learners are prompted to critically assess the very premises of their perception of the problem. In understanding transformative learning, Imel (1998) states the importance of considering the manner in which it can be cultivated within the learning context, the educator and the learner.

The theoretical orientation of the present study is thus informed by transformative pedagogies. This is due to several reasons. Transformative pedagogical notions are strongly linked to constructivist orientations. These orientations are seen as dominant in pedagogies related to ESD. The close association between the goals of ESD and transformative education which advocate the importance of being critical and reflective is another significant reason for this choice. Transformative pedagogies are also seen as the more significant preference, given its focus on adult education and teaching and learning processes within the context of higher education. As the present study is set within the context of higher education, the transformative paradigm thus serves as a fitting platform to better understand the pedagogical issues that surface from the findings of the study.

3 Exploring Institutional Practices and Student Needs Through Qualitative and Quantitative Approaches

Using an exemplifying case study approach, qualitative and quantitative means were used to explore the aims of the study. The study was conducted at a private engineering university located in Perak, Malaysia.

Qualitative analysis, namely manifest and latent content analysis proved useful in assessing the extent to which sustainable development features within the practices of the university. Content analysis can be pursued in two ways, namely through its manifest content or its latent content. However, note Fraenkel and Wallen (2000), the best method is to use both forms if possible. The manifest content of a communication refers to the 'obvious, surface content-the words, pictures, images, and so on that are directly accessible to the naked eye or ear' (Fraenkel and Wallen 2000, p. 475). An example would be to count the number of times a certain word appears in the particular type of content. Latent content on the other hand refers to the underlying meaning of the communication. An example would be to read through the whole communication and assess the extent to which the issue investigated is present in the communication. The interpretation of content analysis data is commonly conducted through the counting of 'frequencies and percentages or proportions of particular occurrences to the total occurrences' (Fraenkel and Wallen 2000, p. 475). They also note that a 'base or reference point for counting' (2000, p. 477) must also be recorded to enable comparisons to be made against the counted occurrences.

The quantitative approach used in the study was the survey approach. As the sole purpose of conducting the survey was to gauge perceptions, the data obtained from the survey was used for the purpose of understanding and describing the respondents' views. Likert scales of 1–5 (strongly disagree–strongly agree) were used as it aided with the determination of the students' opinions and attitudes on the questions posed in the questionnaire.

Respondents of the study were 388 final year undergraduate engineering students. The respondents comprised of 372 Malaysian students and 16 international students from Middle Eastern, African and Asian nations. Respondents ranged between the ages of 20–26, with 94.1 % of them within the age range of 21–23. The respondents were from five engineering programmes, namely Electric and Electronic Engineering, Chemical Engineering, Civil Engineering, Mechanical Engineering and Petroleum Engineering.

4 Findings on the Extent to Which Sustainable Development Is Featured Within the Educational Practices of the University

In assessing the extent to which sustainable development features within the university's educational practices, the following university documents were subjected to manifest and latent content analysis:

(i) vision and mission statements
(ii) research vision and mission statements
(iii) undergraduate engineering programme educational objectives
(iv) undergraduate engineering programme outcomes
(v) common undergraduate engineering and non-engineering module learning outcomes

The purpose of the analysis was to ascertain the extent to which the educational practices of the university were in accordance to sustainable development competences and EESD competences. The findings of the analysis are discussed below.

4.1 University's Vision and Mission Statements and Research Vision and Mission Statements

There is an absence of manifest content, but the presence of latent content related to sustainable development competences and EESD competences in the university's vision and mission. In relation to the university's research vision and mission, there is an absence of manifest content but the presence of latent content in relation to sustainable development competences and ESD competences.

4.2 Undergraduate Engineering Programme Educational Objectives of Chemical Engineering, Civil Engineering, Electrical and Electronic Engineering, Mechanical Engineering and Petroleum Engineering

The university's main programme educational objective was initially *To produce technically qualified well-rounded engineers and technologists with the potential to become leaders of industry and the nation*. This objective was later modified. The modified version presently consists of two objectives, namely *To produce technically qualified engineers with the potential to become leaders of engineering industries* and *To produce engineers who are committed to sustainable development of engineering industries for the betterment of society and nation*. The programme educational objectives and programme outcomes of all undergraduate engineering programmes offered in the university had been modified to include sustainable development outcomes. The former programme educational objectives of all undergraduate engineering programmes offered in the university did not contain any manifest or latent references to sustainability competences. However, the modified programme educational objectives of each programme contain one manifest and one latent reference. The phrase sustainable development indicates a manifest representation, while the phrase betterment of society and nation on the other hand denotes latent representation.

4.3 Undergraduate Engineering Programme Outcomes of Chemical Engineering, Civil Engineering, Electrical and Electronic Engineering, Mechanical Engineering and Petroleum Engineering

There are evidences of manifest and latent sustainability competences in all undergraduate engineering programme outcomes. However, the percentages of sustainability competences within all programmes differ in the former and modified outcomes. There has been an increase in sustainability competences in the programme outcomes of the Civil, Electrical and Electronic, Mechanical and Petroleum Engineering programmes. The Chemical Engineering programme however recorded a decrease. The results also suggest that the Mechanical Engineering programme has the highest difference in the percentage of sustainability competences in its former and modified programme outcomes, i.e. 29.5 %. This is followed by the Electrical and Electronics Engineering programme, Petroleum Engineering programme and finally the Civil Engineering programme with an 11.6, 10.1 and 9.1 % increase respectively.

In relation to the difference between the institution's current sustainability competences percentage and the Engineering Accreditation Council's sustainability competence percentage of 66.7 %, all engineering programmes recorded lower percentages, with the Chemical Engineering programme having the highest decrease at 22.3 %.

Table 2 Sustainability competences learning outcomes percentage in common undergraduate engineering modules

Name of module	Sustainability competences (%)
Introduction to oil and gas industry and sustainable development	44.4
Engineers in society	40
Engineering team project	40
Health, safety and environment	33
Engineering economics and entrepreneurship	0
Probability and statistics	0
Introduction to management	0
Professional communication skills	0
Academic writing	0
Malaysian studies	0

4.4 Common Engineering and Non-engineering Module Learning Outcomes

10 modules were identified for the analysis. Of the 10 modules, five non-engineering modules, namely *Engineering Economics and Entrepreneurship, Introduction to Management, Professional Communication Skills, Academic Writing* and *Malaysian Studies* are offered by the Department of Management and Humanities. The remaining five modules are offered by the Engineering departments. Table 2 depicts the summary of the results of the percentage of sustainability competences within the learning outcomes for all 10 modules.

As illustrated in Table 2, four of the 10 modules contain learning outcomes related to sustainability competences. The *Introduction to Oil and Gas Industry and Sustainable Development* module contains the highest percentage of sustainability competences with 44.4 %. This is followed by the *Engineers in Society* module and the *Engineering Team Project* module which each contain 40 % of learning outcomes related to sustainability competences. *Health, Safety and Environment* contains 33.3 % of learning outcomes related to sustainability competences. It can thus be summarized that 40 % of the total 10 common modules have learning outcomes related to sustainability competences.

5 Findings on Present Pedagogical Practices Within the Undergraduate Engineering Programme

This section describes findings from the survey which gauged respondents views on the pedagogical approaches currently practiced at the university. More specifically, the questions sought to determine if current pedagogical practices at the university reflected pedagogies related to the teaching of sustainable development. A total of 24 items were constructed. A five point Likert scale was used to obtain respondents' views. The scale used was an agreement scale. The five points of the scale denoted 1, for strongly disagree, 2 for disagree, 3 for undecided, 4 for agree and 5 for strongly agree. The mean score of each item indicates the level of agreement for the items.

The first item was *My engineering programme promotes the importance for all students to practice sustainability*. The results from the responses of the final year undergraduate engineering students seem to suggest that they agree that their respective engineering programmes do promote the importance for all students to practice sustainability. This is evident through the higher frequency of responses recorded under the agree (48.5 %) and strongly agree (22.2 %) categories. Nevertheless, there were 18.3 % of the students who were unsure if their respective engineering programmes did promote the importance for all students to practice sustainability, while 1.0 and 10.1 % of them indicated that they strongly disagree and disagree respectively with the statement. The mean score for this item was 3.81, indicating agreement.

Item 2, *My engineering lecturers discuss the importance for engineering students to practice sustainability through the courses they teach*, also indicated high responses for the agree and strongly agree categories with 46.6 and 16.5 % of responses recorded for these two categories. 24.2 % of the responses were undecided, 1.5 % strongly disagreed while 11.1 % disagreed with the statement. The mean score for this item was 3.65, indicating agreement.

The third item was *My language and communication lecturers discuss the importance for engineering students to practice sustainability through the courses they teach*. Interestingly, the highest number of responses for this statement was for the category undecided, instead of the agree or strongly agree categories. The frequency of responses for this category was 34.3 %, suggesting final year undergraduate engineering students were unsure if their English language and communication lecturers did discuss the importance for engineering students to practice sustainability in these modules. 33.0 % agreed while 12.6 % strongly agreed with the statement, while 14.7 % and 5.4 % responded that they disagreed and strongly disagreed respectively. The mean score for this item was 3.32, indicating disagreement with the statement.

My management lecturers discuss the importance for engineering students to practice sustainability through the courses they teach was the fourth item in the question. The results indicate that 42.8 and 19.8 % of the responses received were for the categories, agree and strongly agree respectively. 20.6 % of the responses received were for the undecided category, 3.9 % was for strongly disagree and the remaining 12.9 % was for the disagree category. The mean score for this item was 3.62, indicating agreement.

The next item was *My social science/humanities lecturers discuss the importance for engineering students to practice sustainability through the courses they teach*. 42.0 % of the responses received were for the category agree, 13.4 % for strongly agree, 30.4 % for undecided, 10.6 % for disagree and 3.6 % for strongly

problems, from the perspective of a future engineer, recorded the lowest mean score (3.47). This value is lower than the 3.50 average mean, indicating respondents disagreed that the language and communication modules taught them to reflect on issues and new ideas learnt from real environmental problems, from the perspective of a future engineer.

During lessons, students from different engineering programmes are given the opportunity to reflect on activities collaboratively (together) to share knowledge as a group was item 20. The highest percentage of frequencies recorded for this item were 41.2 % for agree, followed by 2.2 % for strongly agree. 18.3 % of the respondents were undecided, while 4.9 and 13.4 % of the respondents strongly disagreed and disagreed respectively. The mean score for this item was 3.62, indicating agreement.

The next item, item 21 was *Learning approaches in this university focus on experiences gained from my direct involvement in a particular learning situation involving environmental issues.* Although the highest frequency, 40.2 % was recorded for the category of agree, 31.2 %, the second highest frequency was made up of undecided responses. A combined disagreement of 17.8 % was recoded for this item. The mean score for this item was 3.40, indicating disagreement.

The subsequent item, item 22 sought to gauge respondents' agreement on real world learning opportunities. The statement was phrased as *Learning approaches in this university encourage students to apply ideas they have learnt and experienced through real world learning situations involving environmental issues.* Results for this statement indicate that 46.4 % (agree category) and 16.2 % (strongly agree category) of the respondents are in agreement that the university does encourage students to do so. However, 22.2 % were undecided, while 3.6 % and 11.6 % of the respondents strongly disagreed and disagreed to the statement. The mean score of 3.60 indicates that respondents agreed to the statement.

Item 23, *Learning activities in this institution require students to be actively involved in their own learning involving environmental issues* recorded most responses for the agree category, with 44.8 %, followed by 24.7 % for the undecided category. 2.1 and 11.1 % of the responses received were or the categories strongly disagree and disagree respectively. The mean score of 3.64 indicates agreement.

Item 24, the last item sought to gauge respondents' agreement to the statement *My university promotes the importance for all students to practice sustainability.* The results indicate that 39.9 % of the respondents agreed to the statement, while 23.7 % were undecided. These were the highest and second highest frequencies recorded for this statement. Nevertheless, the results also indicate that 3.6 and 11.9 % of the responses received were in disagreement that the university promoted the importance for students to practice sustainability. The 3.63 mean score recorded for this item indicates agreement.

6 Findings on Student Stakeholders' Views on Approaches Best Suited to Teach Sustainable Development Within the Undergraduate Engineering Programme

Undergraduate engineering students' preferences towards the teaching of sustainable development in the undergraduate engineering programme at the university were also gauged. A five point Likert scale was used to obtain respondents' attitudes and preferences. The scale used was an agreement scale. The five points of the scale denoted 1, for strongly disagree, 2 for disagree, 3 for undecided, 4 for agree and 5 for strongly agree. The mean score of each item indicates the level of agreement for the items. The summary of responses is as illustrated in Table 3.

Item A sought to determine respondents' preferences on whether sustainable development input should be taught as a separate engineering module on its own. The mean score of 3.32 recorded for this item however indicates that respondents were not in agreement with the teaching sustainable development in the undergraduate engineering programme as a separate engineering course on its own.

Item B focused on seeking respondents' preferences for sustainable development input to be taught as a separate non-engineering module on its own. The results suggest that respondents' preferred sustainable development input to be taught as two separate modules in the undergraduate engineering programme at the university. Once again, the lower than average mean score of 3.29 indicates that it should not be taught as a separate non-engineering course on its own.

Item C on the other hand sought to determine final year undergraduate engineering students' responses to whether sustainable development input should be provided through all engineering modules only. These include all courses, be it a common module, or otherwise. The mean score of 3.27 obtained for this item indicates that respondents were in disagreement of this approach.

Table 3 Approach to teaching sustainable development in the undergraduate engineering programme

Item	Statement	Mean	SD
A	As a separate engineering course on its own	3.32	1.24
B	As a separate non-engineering course on its own	3.29	1.17
C	Through all engineering courses only	3.27	1.19
D	Through all non-engineering courses only (language and communication, business/management and social science/humanities)	3.29	1.16
E	Within all engineering and non-engineering courses	4.06	0.92
F	The engineering lecturers should teach sustainability related content	4.34	0.72
G	The language/communication lecturers should teach sustainability related content	4.03	0.94
H	The management lecturers should teach sustainability related content	4.15	0.83
I	The social science/humanities lecturers should teach sustainability related content	4.18	0.85

Item D aimed to determine respondents' preferences to the teaching of sustainable development input through all non-engineering modules, i.e. language and communication, business/management and social science/humanities modules. The 3.29 mean score value obtained indicated disagreement in using this approach to the teaching of sustainable development in the undergraduate engineering programme.

Item E was on providing sustainable development input through all engineering and non-engineering modules, irrespective of if the module was a common module or otherwise. The 4.06 mean score value obtained indicates agreement to this approach.

The final four items sought to understand the 388 respondents' preferences on the lecturers who should teach sustainability related content to the undergraduate engineering students. Respondents were asked to indicate the extent to which the Engineering, English Language and Communication, Management and Social Sciences and Humanities should teach sustainability content. In the case of Engineering lecturers, 0.5 % of the responses indicated strong disagreement, 1.3 % indicated disagreement while 8 % denoted undecided responses. Higher percentage of the responses were recorded for the agree and strongly agree categories, with 45.4 and 44.8 % of responses accordingly. In relation to the English Language and Communication lecturers, a combined response of 77.6 % was recorded for the agree and strongly agree category. 14.4 % of the responses were undecided, while 1.5 and 6.4 % of the responses were in strong disagreement and disagreement respectively. For the statement The management lecturers should teach sustainability related content, only 1.3 and 2.8 % of the responses were recorded for the strongly disagree and disagree category respectively. 11.3 % of the responses were undecided, while the highest responses were for the agree and strongly agree category, with 48.2 and 36.3 % respectively. As for the Social Science and Humanities lecturers, most of the responses were for the agree and strongly agree categories, with 44.1 and 39.7 % of responses accordingly. 11.9 % were undecided responses, while the remaining 4.4 % were for the strongly disagree and disagree category. These results thus suggest that the respondents prefer all lecturers, regardless of their academic background to provide them with sustainability input. The mean scores obtained for these last four items were 4.34, 4.03, 4.15 and 4.18 respectively. These high scores indicate that respondents were in agreement that all lecturers regardless of their expertise should teach sustainability related content within the undergraduate engineering programme.

7 Students' Needs that Should Be Considered to Help Develop the Desired Sustainability Learning Experience in the University

In addition to Likert scale type items, the survey also consisted of an open-ended question. The purpose of the open-ended question was to elicit respondents' views on teaching and learning issues that should be considered to help develop the desired sustainability learning experience in the university.

Table 4 Open-ended responses categorized by type of module

Engineering modules		Communication/language/management/social science and humanities modules	
Categories	Number of references coded	Categories	Number of references coded
Practical versus theoretical (more practical exposure desired)	19	Developing communication skills for sustainable development	12
Inclusion of real sustainable development issues and situations	19		
Sustainable development learning activities and assessment for real world preparation	13	Approach to teaching sustainable development for non-engineering modules to be in accordance with the principles of education for sustainable development	50
The need for heightened exposure and awareness to sustainable development post-graduation	6		
Teaching and learning of sustainable development via knowledge of current technological trends	14	Bringing real life sustainable development issues and situations into non-engineering modules	13
Sustainable development awareness through exposure within the engineering industry	11		
Content within current learning modules should be sustainability driven	13	Relating engineering aspects with human and societal aspects	19
Approach to teaching sustainable development to be in accordance with the principles of education for sustainable development	30		

The section that follows describes the results of the NVivo analysis conducted to categorize the views provided by the respondents. A total of 219 open-ended responses were noted. The qualitative software NVivo version 10 was used to categorize these responses by type of module, i.e. engineering and non-engineering.

As seen in Table 4, a total of 12 categories were identified from the open-ended responses provided by the survey respondents. Eight of these categories were from the responses obtained for the Engineering modules, and the remaining four, for the non-engineering modules. Under the Engineering modules grouping, the category, *Approach to teaching sustainable development to be in accordance with the principles of education for sustainable development* had the highest number of responses, i.e. 30. The least number of responses were for the category *The need for heightened exposure and awareness to sustainable development post-graduation*, with six responses in total. As for the non-engineering modules grouping, the most number of responses were once again centred upon the category *Approach to teaching sustainable development for non-engineering modules*. A total of 50 responses made up this category. *Developing communication skills for sustainable development*, which had 12 responses, was the category which had the least number of responses under the non-engineering modules grouping.

8 Discussion on Implications of Findings for the Development of a Holistic EESD Framework

From the perspective of the undergraduate engineering curriculum, it appears that sustainable development and EESD do not feature prominently within the institution's academic and research vision and mission. Undergraduate programme outcomes indicate a moderate increase in the former and modified programme outcomes related to sustainable development competences. Findings on the extent of the inclusion of sustainable development and EESD learning outcomes of common undergraduate modules also suggest that there is inadequate emphasis of these outcomes, as merely 40 % of the common modules have learning outcomes related to sustainable development and EESD. The university is thus recommended to intensify its endeavours to make sustainable development and EESD more prominent within its undergraduate engineering curriculum. Having said that, these low percentages nevertheless indicate that the university is moving towards the goal of advancing sustainable attitudes through its undergraduate engineering curriculum.

In terms of sustainable development being made a learning context within the undergraduate modules, the findings of the study suggest that it does not feature within 90 % of the common modules. Additionally, none of the common non-engineering modules have learning outcomes related to sustainability. These findings are a cause for concern, as a narrow curriculum can be a hurdle to the implementation of sustainable development within the undergraduate engineering curriculum. In terms of pedagogy and practice, there seem to be attempts to use transformative teaching and learning approaches within the undergraduate engineering modules. The use of collaborative and active learning strategies, case studies, problem solving, individual activities and group based activities is an indication of its use. Open-ended findings however indicate that these activities have not been approached within a transformative and ESD pedagogical framework in mind, or with the aim of developing sustainability competences within the undergraduate engineering students.

EESD is also not seen to feature holistically within the university's educational practices. Sustainable development does pervade through the curriculum, but is not obviously emphasised. Pedagogically, attempts are made to approach the teaching and learning of sustainable development using methods aligned to the philosophies of transformative learning and EESD. However, as indicated in the findings of the study, these too are selectively practiced at the university. The university is therefore recommended to look into these gaps to enable effective development and implementation of EESD outcomes within its educational practices.

The findings clearly show that the university has taken some steps to include sustainable development in its undergraduate programme. However, sustainable development does not form part of the core curriculum across all engineering disciplines offered in the university's undergraduate engineering curriculum. In relating these findings to the context of higher education and EESD, it can thus be concluded that the university, has, at a rudimentary level, included sustainable development outcomes within the curriculum of the undergraduate engineering programme. This inclusion is however a result of the university's adherence to the accreditation guidelines stipulated by the Engineering Accreditation Council and not an initiative driven by the university's conscious effort to integrate sustainability in relation to the principles and practices of EESD. Findings of the present study suggest that the inclusion of sustainable development within the undergraduate curriculum has intensified as a result of adherence to accreditation policies set by the Engineering Accreditation Council. Yet, with merely 40 % its common undergraduate modules learning outcomes relating to sustainable development, the university has to formally institutionalize sustainable development and EESD within its undergraduate engineering curriculum.

There is also evidence of discipline bias, prescriptive content and cognitive learning approaches in the modules taught in the undergraduate engineering programme. These evidences point to a curriculum uncharacteristic of educational philosophies and practices that are transformative and sustainable in nature. It can thus be noted that the university is seen to espouse mechanistic and transmissive ideologies which are not in tandem with the philosophy of EESD. The university can therefore be categorized within the stage of accommodation, where the teaching and learning of sustainable development is conducted for the purpose of education about sustainability, and not for, or as sustainability. Although findings point to the use of EESD teaching and learning approaches such as case studies, problem solving activities and collaborative group discussions, there is nevertheless limited emphasis on the use of reflective and reflexive teaching and learning practices and real life problem solving approaches. This can be a continuous quality improvement (CQI) area for the university to look into, to promote transformative teaching and learning amongst its academicians and students.

As found through this study, the curriculum and pedagogical philosophies of a university hold the key to the development of a holistic EESD framework. The implications discussed in this paper highlight the significance of these criteria in establishing a successful EESD programme. It also provides vital pointers to universities, academicians and researchers working towards the development of a holistic EESD framework for their institutions. Hence, institutions of higher learning are recommended to look into these key areas in developing a holistic EESD framework for their institutions.

The findings of this study are also seen to be instrumental to the Ministry of Education. It will be particularly useful in formulating sustainability related higher education frameworks for the country's public and private engineering institutions of higher learning. The findings will also benefit institutions of higher learning in Malaysia that seek to introduce EESD within its undergraduate engineering programmes. Findings obtained from this study would allow university administrators and academicians make informed decisions on the curricula, pedagogical and institutional aspects that need to be revisited or expanded within their institution, so the incorporation of EESD could be carried out in a holistic manner.

9 Conclusion

This study looked into the educational practices and needs for sustainable development within a private engineering university in Malaysia. The paper first explored the extent to which sustainable development is featured within the institution's programme educational outcomes and common modules. Following this was an investigation of the pedagogical practices within the undergraduate engineering programme and students views on approaches best suited to teach sustainable development within the programme. Implications of these findings were also highlighted.

As with all research, the present study was also conducted within several limitations. The study was conducted as a single case study of a private Malaysian engineering university. Given the aims of the study and the stipulated duration to complete this research, the single case study approach was deemed to be the most suitable approach to adopt to obtain the data required within the permissible time frame of the study. Additionally, the study was limited to the Malaysian undergraduate engineering context. The postgraduate engineering context was not included in the study, given the time frame of the study. Data gathering through undergraduate module learning outcomes was limited to the common undergraduate modules as these were compulsory modules all engineering students had to complete to be able to graduate from the undergraduate engineering programme.

Several recommendations are suggested as future work for the present study. Researchers wanting to take up a similar study may also want to explore perspectives of first, second and third year undergraduate engineering student stakeholders, instead of those in the final year of their studies. A comparison between final year and non-final year undergraduate engineering students could be a possible angle to investigate using this approach. Researchers may also want to focus on postgraduate engineering students perspectives on the issues explored in this study. As the present study focused upon a single case study of a private Malaysian engineering university, future research can look into conducting the same study in other private engineering universities in the country, using a multiple case study approach. Alternatively, the study could also be conducted at Malaysian public universities that offer engineering programmes. Comparisons could then be made between findings from the public and private universities.

References

Abdullah A et al (2005) A Malaysian outcome-based engineering education model. Int J Eng Technol 2(1):14–21

Ab Rahman MN et al (2009) Assessment of engineering students perception after industrial training placement. Eur J Social Sci 8(3):420–431. http://www.eurojournals.com/ejss_8_3_06.pdf. Accessed 28 Mar 2010

Arsat M, Holgaard JE, de Graaff E (2011) Three dimensions of characterizing courses for sustainability in engineering education: models, approaches and orientations. In: Proceedings of the 3rd international congress on engineering education, Lisbon

Azami Zaharim M et al (2009) A gap study between employers' perception and expectation of engineering graduates in Malaysia. WSEAS Trans Adv Eng Educ 6(11):409–419

Azami Zaharim et al (2010) Practical framework of employability skills for engineering graduate in Malaysia. In: Proceedings of IEEE EDUCON education engineering 2010-The future of global learning engineering education, pp 921–927. April 14–16, Madrid, Spain

Azmahani AAA, Sharipah NSS, Khairiyah MY, Amirmudin U, Jamaludin MY (2012) Developing a structural model of assessing students' knowledge-attitudes towards sustainability. Procedia Social Behav Sci 56:513–522 (Proceedings of the International Conference on Teaching and Learning in Higher Education (ICTLHE))

Engineering Accreditation Council (EAC) (2012) Engineering programme accreditation manual 2012. http://www.eac.org.my/web/document/EAC%20Manual%202012.pdf. Accessed 14 Aug 2012

Fraenkel JR, Wallen NE (2000) How to design and evaluate research in education, 4th edn. McGraw Hill, Boston

Hamid MKA et al (2005) Crafting effective engineering problems for problem based learning: Universiti Teknologi Malaysia experiences. In: Proceedings of regional conference of engineering education, Johor, 12–13 Dec

Imel S (1998) Transformative learning in adulthood, ERIC Digest No.200. www.ericdigests.org/1999-2/adulthood.htm. Accessed 23 May 2009

Johari M et al (2002) A new engineering education model for Malaysia. Int J Eng Educ 18(1):8–16

McEwen L, Strachan G, Lynch K (2011) 'Shock and awe' or 'reflection and change': stakeholder perceptions of transformative learning in higher education. Learn Teach High Educ 5:34–55

Rohani JM (2005) Assessing the effectiveness of problem based learning (PBL) using quality function deployment (QFD): students perspective. In: Proceedings of regional conference of engineering education, Johor, 12–13 Dec

Omar MZ et al (2009) Measuring the outcomes from industrial training program. Eur J Social Sci 8 (4):581–588. http://www.eurojournals.com/ejss_8_4_06.pdf. Accessed 28 Mar 2010

Sharipah NSS, Azmahani AA, Khairiyah MY (2012) Perception on sustainable development among new first year engineering undergraduates. Procedia Social Behav Sci 56:530–536 (Proceedings of international conference on teaching and learning in higher education (ICTLHE))

Taylor EW (1998) The theory and practice of transformative learning: a critical review, information series no. 374. Columbus: Eric Clearinghouse on adult, career, and vocational education, Center on Education and Training for Employment, College of Education, Ohio State University

Tilbury D (2011) ESD: an expert reviewof processes and learning. Paris: UNESCO. http://unesdoc.unesco.org/images/0019/001914/191442e.pdf. Retrieved 30 Apr 2013

Author Biographies

Subarna Sivapalan is a Senior Lecturer in the Department of Management and Humanities at Universiti Teknologi PETRONAS, Malaysia, with research interests in engineering education for sustainable development and sustainability & higher education.

Ganakumaran Subramaniam is an Associate Professor in the Faculty of Arts and Social Sciences at the University of Nottingham Malaysia campus, with research interests in educational research.

Michael J Clifford is an Associate Professor in the Faculty of Engineering at the University of Nottingham UK, with research interests in appropriate and sustainable technologies.

The Role of Geography in Multi-inter-trans-disciplinary Study Programmes for Environmental Sustainability

Rudi Pretorius and Joan Fairhurst

Abstract

The sustainability challenge requires universities to create spaces in which the disciplinary mould can be substituted for transformational learning. "Transformative" implies emphasis on new ways of "being" and of thinking about the environment, facilitated by integration of a multitude of disciplines and perspectives. Due to its flexible, all-inclusive and integrative nature, Geography can offer much for sustainability education. The reality is that Geography lags in its contribution to study programmes in this field. This paper reflects on the role of Geography in facilitating transformative learning encounters for environmental sustainability. The focus is on the multi-inter-trans-disciplinary undergraduate programme in environmental management offered by the University of South Africa (Unisa) through open and distance learning. The view that environmental sustainability study programmes may benefit from an anchor discipline to provide structure and to facilitate the transformative learning experience, is debated. The value of Geography in this regard is highlighted, with reference to the experience gained from implementing this particular Unisa study programme. An outline is provided of how Geography can fulfill an anchor role in such study programmes, while maintaining its integrity as discipline. The paper will be useful to higher education practitioners involved in the implementation, offering and/or coordination of environmental sustainability study programmes, with specific reference to open and distance learning.

R. Pretorius (✉) · J. Fairhurst
Department of Geography, University of South Africa,
Private Bag X6, Florida, Johannesburg, 1710, South Africa
e-mail: pretorw@unisa.ac.za

W. Leal Filho (ed.), *Transformative Approaches to Sustainable Development at Universities*, World Sustainability Series, DOI 10.1007/978-3-319-08837-2_29

Keywords
Multi-inter-trans-disciplinarity • Sustainability study programmes • Transformative learning • Geography • Case study • Open and distance learning

1 Introduction

1.1 The Context of Sustainability Education and Environmental Sustainability Study Programmes

Nearing the end of the United Nations Decade of Education for Sustainable Development (UN DESD), a notable response of higher education institutions world-wide has been to implement a variety of study programmes focused on environmental sustainability (UNESCO 2012). Since these study programmes cover a wide spectrum of approaches, the need clearly exists for a flexible view of what Education for Sustainable Development (ESD) entails in different contexts, as expressed in the Bonn Declaration (UNESCO 2009). In sync with changing views of sustainability, such study programmes should reflect the move from technology and management towards integration between environmental, social and economic sustainability, with emphasis on cultural change (Liu 2011). Due to the persisting trend in academia of discipline based clusters guarding their turf, together with institutionalisation, the required transition from offering traditional science to offering integrated sustainability science, remains a challenge (Yarime et al. 2012).

1.2 Multi-inter-trans-disciplinarity as Setting for Environmental Sustainability Study Programmes

Dealing with the various sustainability-related issues the world faces, requires problem-solving skills regarding the integrated human-nature system, now recognized as a multi-inter-trans-disciplinary field of endeavour (Wiek et al. 2011). "Multidisciplinarity" implies purposeful combinations of disciplines, "interdisciplinarity" common ground between two (and more) disciplines and "transdisciplinarity" collective understanding of issues across disciplinary boundaries (Brown et al. 2010). Regarding problem-solving, multidisciplinarity involves collaboration using distinct disciplinary approaches, while inter- and transdisciplinarity involves dissolution of boundaries between disciplines, although not necessarily doing away with them (Cairns 2004). The move towards inter- and transdisciplinarity provides a frame of reference to prepare students to deal with sustainability issues (Brundiers et al. 2010). The success achieved however, still requires substantiation (Remington-Doucette et al. 2013), with the value of disciplinary grounding not to be underestimated (Mansilla and Duraising 2007).

1.3 Aim, Methodology and Value of Chapter

Firstly, this chapter links with Remington-Doucette et al. (2013) by exploring the value of disciplinary grounding in multi-inter-transdisciplinary environmental sustainability study programmes, with Geography as case in point. A critical assessment is provided of the view that such study programmes could benefit from an anchoring discipline (in this case Geography) to facilitate the transformative learning experience. Secondly this chapter acknowledges the growing need for locally relevant interpretations of ESD (UNESCO 2012), with focus on an environmental sustainability study programme tailored for the African context and utilizing a contextually relevant teaching and learning mode, namely open and distance learning (ODL).

Under scrutiny is the multi-inter-trans-disciplinary undergraduate study programme in Environmental Management (EM) offered by the University of South Africa (Unisa), an ODL institution. Geography is the major for this study programme, which is coordinated by the Department of Geography. This programme is reviewed in case study format, with focus on process rather than product. A case study approach is valuable as it allows detailed description and analysis of the study programme in the contextual setting it is offered, while allowing critical reflection. Following the case study, concluding thoughts are provided on the value of an anchor discipline in environmental sustainability study programmes.

This chapter contributes to the debate on incorporation of multi-inter-trans-disciplinarity into functional, locally relevant environmental sustainability study programmes, with reference to the value that linkage to specific disciplines can play. The case study being presented furthermore adds to the body of knowledge and experience on the unique challenges faced with implementation of ESD in varied contexts, and how these challenges can be addressed to serve the purpose of transformative learning towards more sustainable living most effectively.

2 Transformative Views of Sustainability Education

2.1 Supporting Sustainability Education with Transformative Learning

Transformative learning, with ultimate goal to guide individuals to realize the necessity of changing their worldviews, is more and more regarded to be central to ESD (Sipos et al. 2007). This has implications for ESD in terms of the required competences, the type of learning to take place and the design of associated learning experiences. Challenges in this regard include the difficulty to facilitate or design transformative encounters as a series of unconnected learning experiences, while a holistic, integrative approach is rather required. In addition, resistance may be encountered since students are challenged to leave their comfort zones. It is unfortunate that transformative learning, with its conception of a holistic role for higher education in terms of the need to address sustainability related issues, i.e. risk and uncertainty, is at odds with the market oriented new liberal notion of higher education focusing on producing productive capacity for the market economy (Blake et al. 2013).

2.2 A Transformative Pedagogy for Environmental Sustainability Study Programmes

A definite move can be distinguished towards adopting transformative sustainability learning (TSL) as the overarching pedagogy for environmental sustainability study programmes (Sipos et al. 2007). Such study programmes need to move beyond a mere focus on knowledge, skills and values relevant to sustainable development, to rather focus on learning geared towards asking relevant critical questions, clarifying values, envisioning more sustainable futures, thinking systemically, responding through applied learning and exploring interactions between tradition and innovation (Tilbury 2011). However, to implement the associated facilitating critical pedagogy, implies major shifts in order to accommodate the required critical reflexivity and level of multi-inter-trans-disciplinarity, while practicing experiential and place-based learning. Viewing ESD as simply an addition of sustainability related content to existing study programmes is no longer and in actual fact has never been, a viable approach (UNESCO 2012).

2.3 Realities and Implications in Terms of a Multi-inter-trans-disciplinary Context

In order to achieve the objectives set in terms of TSL, it is crucial for environmental sustainability study programmes to be developed, refined and offered within a multi-inter-trans-disciplinary context. The traditional unidirectional, hierarchical and reproductive approaches to teaching and learning within disciplinary silos are still found at far too many institutions of higher education. Failing to lead students to address sustainability problems in a holistic way tends to perpetuate these problems instead of contributing to solutions (Wals 2010). Rigid academic structures seem to work against more holistic ways of considering how the world functions, and have to be softened significantly. However, a balance should be struck, since some level of disciplinary knowledge always seems to be necessary as it has the potential to provide valuable input when addressing sustainability issues. The options for sharing and collaboration between disciplines through a multi-inter-trans-disciplinary approach, therefore seems to be the obvious route to follow (Stock and Burton 2011).

3 A Place for Geography in Environmental Sustainability Study Programmes

3.1 Nature of the Discipline of Geography

Geography can be considered in terms of two theoretical positions: "spatial-chorological" and "human-environment" (Turner 2002). Unification and an integrated identity seem difficult without favouring one of these positions above the other. Corresponding to increased awareness of environmental issues, and in sync with diversification of Geography into sub-fields, the human-environment position gained momentum since the latter part of the twentieth century, and is continuing to do so. The value of the human-environment position is that it presents a synthesis derived from integration of natural and human phenomena and processes. The generally accepted advantage of Geography so defined has to do with its power to supply relevant explanations and to be of value in terms of real-world problem-solving. However, the divergence between Physical Geography, with its natural science linkages and Human Geography, experimenting with various ways of knowing, continues to contest the ideal of unity in the discipline.

3.2 Geographical Perspectives and Sustainability Education

Recent advances in understanding sustainability and its achievement, with increased depth in studying its various dimensions, have led to rediscovery of the value of Geography. This specifically refers to the way environmental issues (physical/human) are dealt with by Geography within place-based contexts (Kates and Dasgupta 2007). Liu (2011) emphasises the relationship between Geography and Sustainability Science, both addressing human-environment interactions. Geography therefore supports ESD well, although there is variation in coverage (depth/scope) of sustainability concepts within various offerings by different higher education institutions (Higgitt et al. 2005). Ideally speaking geographers should be trained to step back from the detail and consider the bigger picture (Aplin and Batten 2004), which is valuable when dealing with sustainability issues. In addition, in terms of their integrative skills, their ability to adopt flexible approaches and to take location and scale into account, geographers have much to offer to this field.

3.3 Extent to Which Geography Features in Sustainability Education

Despite the logic of a prominent role for Geography in ESD, evidence concerning decisive action by the discipline in this regard is lacking (Higgitt et al. 2005; Liu 2011). Although the strong connection between Geography and ESD appears to offer an agenda for Geography, geographers seem to be under the impression that to provide only information on environmental processes is sufficient, whereas increasing awareness or insight regarding sustainability requires much more. The disciplinary boundaries of Geography are also regarded as potentially constraining achievement of multi-inter-trans-disciplinarity, because these boundaries are perceived as sustaining the discipline and not promoting ESD. According to Bednarz (2006) it appears as if Geography does not provide the amount and depth of input towards ESD that it is expected to do. The dichotomy between Physical Geography and Human Geography may be a contributing factor, with both regarding the linkage with ESD the responsibility of the other.

3.4 Possible Roles for Geography in Environmental Sustainability Study Programmes

Due to its nature, Geography is in a position to contribute to sustainability study programmes, especially through forging multi-inter-trans-disciplinary linkages (Holmes 2002). But this strength poses risks to the reproductive capacity of the discipline. In a multi-inter-trans-disciplinary context, Geography is pressured to provide tools and selections of geographical insights for vocational programmes, which may be at expense of attention to the requirements of the discipline. Following Aplin and Batten (2004), the value that Geography can add to environmental sustainability study programmes is not necessarily only interdisciplinarity, but also to address human-environment interactions and to incorporate spatio-temporality. As the human-environment position has a long standing in Geography, multi-inter-trans-disciplinarity need not necessarily be seen as threat but rather opportunity to link the human-environment approach, spatial technologies and measurements/observations enabled through it.

4 Case Study: The Undergraduate Study Programme in Environmental Management Offered by the University of South Africa (Unisa)

4.1 Contextual Setting: Unisa and the Context of Open and Distance Learning (ODL)

Although Unisa was regarded as one of the pioneers of classical distance education more than 100 years ago, the institution has since developed into a major role player in the ODL context (Tait 2008). In terms of student numbers, more than 300,000 students are enrolled across eight semi-autonomous colleges (Unisa 2012). With the vision "Towards the African University in the service of humanity", Unisa has joined the global call for greater sustainability by implementing several initiatives in support of the green campus drive. Since 2007 teaching and learning at Unisa has been transformed to optimise and streamline interaction between students and the university, with various innovations such as the increased use of appropriate information and communication technology (ICT) solutions, including a move towards e-learning (Unisa 2007). In this way student-peer and student-lecturer interaction is increasingly facilitated through the online learning platform (referred to as a virtual learning environment or VLE), known as myUnisa. Progress in this regard, however, has been constrained by amongst other things issues concerning access to ICT's by many students in South Africa (Oyedemi 2012) as well as general inadequate levels of computer literacy of students. However, in terms of not only its size, but also its pooled resources, experience and capacities, Unisa is in a position to provide crucial inputs in terms of ESD in Southern Africa and beyond (Unisa 2005).

4.2 Aim, Structure and Unique Elements of This Study Programme

The undergraduate EM Programme at Unisa was implemented in 2000. This was in response to an industry need for a B-degree preparing students to enter this field. The basic structural components of this programme are shown by the diagram in Fig. 1. Underlying this programme is the recognition of the value of multi-inter-trans-disciplinarity, the need for an integrated approach and a focus on the management of human decision-making about the environment. This programme differentiates between a BA and a BSc degree, and in this way recognises the value of specialisation in either the biophysical or the social sciences. Contributing to multi-inter-trans-disciplinarity in this programme is the fact that modules from six of the Colleges in Unisa feature in the curriculum. This spectrum of modules adds to the richness of the programme and is crucial in order to support the aims in terms of sustainability. In this regard the EM Programme at Unisa was one of the first, and remains one of the few multi-inter-trans-disciplinary study programmes at this institution. The most unique aspect of this programme is that it is offered through ODL, which increases its accessibility to potential students. Since remaining in their jobs and/or communities while studying, the opportunity exists for students to directly apply what they learn. Due to the urgency of the environmental dilemma, this synergy between learning and application, is invaluable.

Fig. 1 Structural elements of the Unisa EM Programme

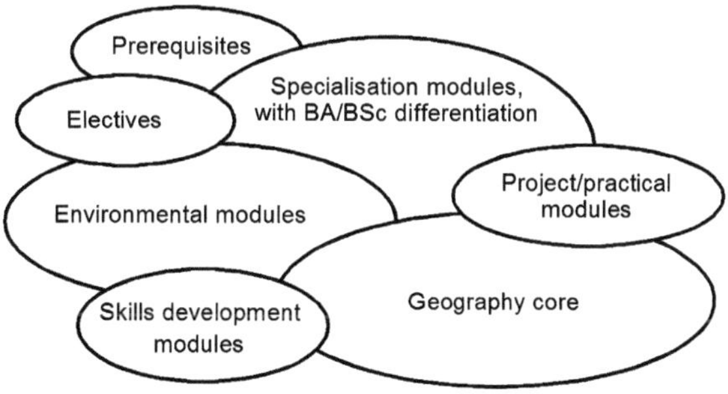

4.3 Reflection on the Role of Geography in This Study Programme

The Department of Geography was the driving force behind the implementation of the EM programme at Unisa, and has since its inception been responsible for managing, maintaining and coordinating it. Noteworthy is that the major subject for the EM Programme is Geography as well, with its inter- and multidisciplinary nature, integrative approach and sensitivity towards place and scale (Aplin and Batten 2004) complementing the overarching sustainability theme of this programme. Together with the implementation of the EM Programme, the Geography curriculum was redesigned in order to no longer provide separate options for BA and BSc students. This demanded integration of Human and Physical Geography and utilisation of environmental sustainability as the overall focus for undergraduate Geography. In addition, the curriculum was re-organised around themes or problems as foci for the various modules instead of the traditional arrangement in terms of the sub-disciplinary fields of specialisation in Geography. A global, continental and national focus was adopted for respectively the first, second and third levels of study, with skills integrated with the presentation of theory. Depending on the choices made by students, Geography can make up anything from 11 to 14 of the 30 required modules, and therefore acts as not only a binding force but also provides direction and a definite focus throughout all the levels of the EM programme.

4.4 Reflection on the Implementation Experience with This Study Programme

A pertinent issue during the development and implementation of the EM Programme concerns the existence of institutional silos, leading to turf protectionist and "territorial" behaviour and working against the cooperation and integration required by ESD. Referred to by Jucker (2002) as the shadow curriculum, this phenomenon can be a major constraint for successfully implementing ESD. Despite this constraint, the efforts of the Department of Geography to make cooperation possible in order to contribute meaningfully to ESD, kept the EM Programme on track. This determination also ensured survival of this programme during the restructuring of Unisa in the mid-2000s in terms of semi-autonomous colleges, together with the hiccups experienced while migrating from the "old" to the "new" structure. In this regard the EM Programme survived the test of time, being one of only a few long standing successful structured degrees at Unisa. In terms of student numbers, Zietsman and Pretorius (2006) observed a gradual increase over time, with students from all over South Africa and further afield. The fact that the EM Programme is perceived as successful has been assisting to raise the profile of the Department of Geography (as programme coordinators) and the discipline of Geography, thus contributing to safeguard the future existence of the department and the programme.

4.5 Reflection in Terms of Achievement of Multi-inter-trans-disciplinarity by This Study Programme

The aspect of multidisciplinarity is covered in the EM Programme in that environmental sustainability is not studied in only one, but in several of the participating disciplines (Geography, Economics, Law, Education, Development Studies, etc.), but without any real sharing of perspectives, methods and ways of knowing between the different disciplines. The controversy usually associated with inter- and transdisciplinarity because of the accompanying perceived dissolution of disciplinary boundaries, does not apply in the case of the Unisa EM Programme. This is because the discipline of Geography provides sufficient scope for coverage of inter- and transdisciplinary linkages through: (1) a focus on real-world issues (that are inherently interdisciplinary) in the curriculum, and (2) by moving assessment beyond the confined university boundaries to place-based applications that are part of the student's life-worlds, thus linking with the transdisciplinary agenda of integrating academic knowledge with that from other stakeholders. Problematic at Unisa is that because of departementalisation, modules from different disciplines in the EM Programme do not always speak to each other or to the overall aim of the programme in terms of fostering increased sustainability, which is a shortcoming to be addressed.

4.6 Reflection in Terms of Facilitation of Transformative Learning by This Study Programme

Transformative learning has been implemented as pedagogical approach feeding into changes required towards a more sustainable living in a number of the modules constituting the EM programme. This is specifically the case with many of the Geography modules, but not for the programme as a whole, because of the difficulties involved in prescribing to other disciplines *how* they should teach their modules and for that matter also *what* they should teach in them. Regarding the Geography modules, a definite move from first order to second and third order learning can be distinguished (Bateson 1972, as referred to by Sterling 2011), which links with and supports transformative learning for sustainability. The pedagogy in these modules is therefore no longer content driven and focussed on information transfer within accepted frameworks, but geared towards shifting the way of thinking and doing that frames student's perceptions of and their interaction with the world. This is achieved by using various forms of inquiry based teaching and learning within authentic contexts, as reported by Pretorius (2012). Allowing students to use their own living environments as authentic spaces for assessment tasks, Pretorius (2012) shows how this contributes to a shift in how they experience themselves and phenomena in the world around them, which in effect points towards transformative learning (Cranton 2009).

4.7 Concluding Reflection on the Role of Disciplinary Grounding in This Study Programme

The fact that the Department of Geography at Unisa is one of a few lone-standing, unmerged Geography Departments in South Africa and even further afield, for example Australia (Holmes 2002), puts it in a unique position in terms of its involvement with the EM Programme. This is in contrast to the trend for Geography Departments to become part of multidisciplinary schools or departments, but which usually comes at a cost to the intellectual core of Geography and which can eventually threaten disciplinary survival. The approach at Unisa instead capitalised on the interdisciplinary nature of Geography as a major strength and opportunity, rather than to consider other options (Fagan and Jacobs 1998). In this way the EM Programme could be structured with Geography as anchor discipline since it provides the multi-inter-trans-disciplinary linkages required for transformational sustainability learning. Apart from one or two modules (e.g. on impact assessment and environmental politics) which is not Geography, but had to be developed for the EM Programme, the integrity of the discipline of Geography at Unisa could be maintained. This in line with Abbott's (2001) viewpoint that interdisciplinary studies depend on specialised disciplines to feed into generation of new theories and methods, although this can be contested in terms of emerging ways of knowledge creation which are not discipline based, such as critical theory.

5 Discussion of Key Issues

This reflective chapter on the Unisa EM Programme and the role of Geography as anchor discipline, contributes to the pool of experience on ESD in different contexts, for which the need has been expressed during the UN DESD (UNESCO 2012). Situated in a relative rigidly structured mega ODL institution and in the technological challenged Global South, the experience obtained during the almost 15 years that this programme has been running, provides a unique perspective on ESD, with a number of important lessons that have been learnt.

5.1 Facilitation of Multi-Inter-Trans-Disciplinarity in Environmental Sustainability Study Programmes

The multi-inter-trans-disciplinary nature of environmental sustainability study programmes leads to several issues to be addressed, as illustrated by the EM Programme offered by Unisa. These include facilitation of collaboration between disciplines, transcendence of disciplinary boundaries, balance between "mainstream" and "service" teaching and the degree of maintenance versus dilution of disciplinary identity (Holmes 2002). At Unisa, implementation of the EM Programme was not accompanied by mergers between academic departments and also not really influenced by moves towards vocational training, therefore less initial antagonism was experienced, although some opportunistic buy-in did in fact occur. However, creation of a separate Department of Environmental Sciences later on, did in fact cause some issues. These could be settled by allocating responsibility for the field of environmental management to the Department of Geography for undergraduate qualifications, and to the Department of Environmental Sciences for postgraduate qualifications. Important is that mutual agreement led to a trade-off, and that ESD could continue unhindered at the institution.

5.2 The Need for a Transformative Pedagogy in Environmental Sustainability Study Programmes

"Sustainability education must therefore be prepared to deconstruct and reconstruct all aspects of teaching and learning" (Sipos et al. 2007). This process is to be accompanied by a move towards student-centered learning, through which students get responsibility in the active acquisition and processing of knowledge. Unisa is in a fortunate position that the ODL business model adopted after restructuring in 2004 is supportive of student-centeredness, with increased effort to optimise university-student, lecturer-student and student-student interaction (Unisa 2007). The type of changes that are required in order to be able to truly implement transformative sustainability learning (TSL) as pedagogy, are featuring in a number of the modules forming part of the Geography major of the Unisa EM programme. Examples of these strategies include inquiry based learning, analysis of case studies and project based work. The increased use of well-planned learning experiences through e-learning creates opportunities to improve on the collaborative aspects of leaning and the co-construction of knowledge, but presents challenges in terms of issues such as access to ICT's and relatively low levels of ICT literacy.

5.3 Pros Versus Cons of an Anchor Discipline in Environmental Sustainability Study Programmes

As shown by the Unisa example, several advantages may be associated with an anchor discipline in environmental sustainability study programmes. Chief among these is the structure thereby provided to such study programmes, assisting to organize and direct the learning experience in an integrated way over the total study period. For the EM Programme at Unisa this function is fulfilled by the Geography major. In the rigidly structured ODL institution with little room for implementation of new offerings, this was the obvious route to go. However, Geography at Unisa went through a major re-curriculation process prior to implementation of the EM Programme. This meant re-arranging the curriculum in terms of themes and within a problem-based context rather than according to traditional sub-disciplines. In this way Geography could assist in supporting the aims of ESD in terms of transforming to sustainability. A similar role can be fulfilled by other suitable disciplines, such as the emerging field of Sustainability Science (Benessia et al. 2012). A disadvantage of utilizing an anchor discipline, is that its presence may be perceived negatively as overpowering, but with careful facilitation this issue can be successfully down-managed.

6 Concluding Remarks

The Unisa undergraduate EM Programme provides a model of how an existing discipline (in this case Geography) can re-invent itself to serve both the reproductive capacity of the discipline and the agenda of transformation to sustainability in an innovative way. All in all, the EM Programme can be regarded as a success story, contributing towards not only institutionalisation of sustainability, but also providing Geography as discipline with an uncontested and acknowledged role in this regard at Unisa. Some sacrifices had to be made, such as substitution of core modules by the Department of Geography to provide service modules for the EM Programme not offered in any other department. In addition, coordination of the programme and related administrative tasks became the responsibility of the Department of Geography. However, in the long run it became a win-win situation from which the Department of Geography and ESD hugely benefitted. Geography at Unisa now serves the purpose of a major in the academic discipline as well as for the multi-inter-trans-disciplinary EM Programme.

For the Unisa programme, multi- and interdisciplinarity have been implemented with a fair amount of success, while inclusion of transdisciplinarity remains challenging. Due to the central coordination of this programme by the Department of Geography, it was possible to address some of the issues related to the multi-inter-transdiciplinary nature of the programme. To this end, the inter- and multidisciplinary linkages provided by Geography proved to be very useful. Although the model of using an anchor discipline for the EM Programme works well in terms of the ODL and Global South context of Unisa, there is no best way to implement sustainability in academic programmes (Remington-Doucette et al. 2013). Differing institutional contexts have to be considered and dealt with to design and successfully implement locally relevant environmental sustainability study programmes. The transformation toward e-learning provides the next challenge for the EM Programme, but includes the opportunity to improve on the transdisciplinary dimension of sustainability through collaboration and sharing within an online learning environment.

References

Abbott A (2001) Chaos of the disciplines. University of Chicago Press, Chicago, p 248

Aplin G, Batten P (2004) Open-minded geographers: their potential role in integrated adaptive environmental management. Aust Geogr 35(3):355–363

Bateson G (1972) Steps to an ecology of mind. Chandler, San Francisco, p 533

Bednarz R (2006) Environmental research and education in US Geography. J Geogr High Educ 30 (2):237–250

Benessia A, Funtowicz S, Bradshaw G, Ferri F, Ráez-Luna EF, Medina CP (2012) Hybridizing sustainability: towards a new praxis for the present human predicament. Sustain Sci 7 (Supplement 1):75–89

Blake J, Sterling S, Goodson I (2013) Transformative learning for a sustainable future: an exploration of pedagogies for change at an alternative college. Sustainability 5:5347–5372

Brown VA, Harris JA, Russell JY (2010) Tackling wicked problems through the transdisciplinary imagination. Earthscan, London, p 312

Brundiers K, Wiek A, Redman CL (2010) Real-world learning opportunities in sustainability: from classroom into the real world. Int J Sustain High Educ 11(4):308–324

Cairns J (2004) Sustainability and specialization. Ethics Sci Environ Polit 28:33–38

Cranton P (2009). Transformative Learning: Q & A with Patricia Cranton. Faculty Focus January 2009. http://www.facultyfocus.com/articles/instructional-design/transformative-learning-qa-with-patricia-cranton/ (last accessed: 6 April 2014)

Fagan R, Jacobs J (1998) Geography. In: Academy of Social Sciences in Australia (ed) Challenges for the Social Sciences and Australia, vol. 1. Australian Government, Canberra, pp 119–128

Higgitt D, Haigh M, Chalkley B (2005) Towards the UN Decade of education for sustainable development: introduction. J Geogr High Educ 29(1):13–17

Holmes JH (2002) Geography's emerging cross-disciplinary links: process, causes, outcomes and challenges. Aust Geogr Stud 40(1):2–20

Jucker R (2002) Sustainability? Never heard of it! Some basics we shouldn't ignore when engaging in education for sustainability. Int J Sustain High Educ 3(1):8–18

Kates R, Dasgupta P (2007) African poverty: a grand challenge for sustainability science. Proc Nat Acad Sci U.S.A 104(43):16747–16750

Liu L (2011) Where in the world of sustainability education is US geography? J Geogr High Educ 35(2):245–263

Mansilla VB, Duraising ED (2007) Targeted assessment of students' interdisciplinary work: an empirically grounded framework proposed. J High Educ 78(2):215–237

Oyedemi TD (2012) Digital inequalities and implications for social inequalities: a study of internet penetration amongst university students in South Africa. Telematics Inform 29:302–313

Pretorius RW (2012) The value of authentic learning spaces to facilitate education for sustainable development in open and distance learning. In: Filho WL (ed.) Sustainable development at universities: new horizons. Peter Lang, Frankfurt, pp 169–182

Remington-Doucette SM, Connell KYH, Armstrong CM, Musgrove SL (2013) Assessing sustainability education in a transdisciplinary undergraduate course focused on real-world problem solving: a case for disciplinary grounding. Int J Sustain High Educ 14(4):403–433

Sipos Y, Battisti B, Grimm K (2007) Achieving transformative sustainability learning: engaging head, hands and heart. Int J Sustain High Educ 9(1):68–86

Sterling S (2011) Transformative learning and sustainability: sketching the conceptual ground. Learn Teach High Educ 5:17–33

Stock P, Burton RJF (2011) Defining terms for integrated (multi-inter-trans-disciplinary) sustainability research. Sustainability 3:1090–1113

Tait A (2008) What are open universities for? Open Learn: J Open Distance e-Learn 23(2):85–93

Tilbury D (2011) Assessing ESD experiences during the DESD: an expert review on processes and learning for ESD. UNESCO, Paris. http://unesdoc.unesco.org/images/0019/001914/191442e.pdf (last accessed 31 March 2014)

Turner BL (2002) Contested identities: human-environment geography and disciplinary implications in a restructuring academy. Ann Assoc Am Geogr 92(1):52–74

UNESCO (2009) Bonn Declaration, UNESCO World Conference on Education for Sustainable Development, 31 March 2009 to 2 April 2009. UNESCO, Bonn. http://www.unesco.org/new/fileadmin/MULTIMEDIA/HQ/CLT/pdf/BonnDeclaration_EN.pdf (last accessed 30 March 2014)

UNESCO (2012) Shaping the education for tomorrow: 2012 report on the UN decade of education for sustainable development, abridged. UNESCO, Paris. http://unesdoc.unesco.org/images/0021/002166/216606e.pdf (last accessed 30 March 2014)
Unisa (2005) 2015 Strategic plan: an agenda for transformation. http://unesdoc.unesco.org/images/0014/001486/148654e.pdf (last accessed: 31 March 2014)
Unisa (2007) Self-evaluation portfolio for the commonwealth of learning trail audit. University of South Africa, Pretoria
Unisa (2012) An institutional profile of Unisa: Unisa facts & figures. http://heda.unisa.ac.za/filearchive/Facts%20&%20Figures/Briefing%20Report%20Unisa%20Facts%20&%20Figures%2020120215.pdf (last accessed: 7 February 2012)
Wals AEJ (2010) Mirroring, Gestaltswitching and transformative social learning—stepping stones for developing sustainability competence. Int J Sustain High Educ 11(4):380–390
Wiek A, Withycombe L, Redman CL (2011) Key competencies in sustainability: a reference framework for academic programme development. Sustain Sci 6:203–218
Yarime M, Trencher G, Mino T, Scholz RW, Olsson L, Ness B, Frantzeskaki N, Rotmans J (2012) Establishing sustainability science in higher education institutions: towards an integration of academic development, institutionalization, and stakeholder collaborations. Sustain Sci 7 (Supp. 1):101–113
Zietsman SS, Pretorius RW (2006) Learning programmes for environmental sustainability: a different approach to curriculum design. S Afr J High Educ 20(5):691–702

Author Biographies

Rudi Pretorius is senior lecturer in the Department of Geography, University of South Africa, and has lectured a variety of sustainability related courses over a number of years. He is the academic coordinator of the undergraduate programme in Environmental Management. He holds a Master of Science degree in Geography and a Master's in Business Leadership and is currently registered for a PhD in Geography focusing on the potential and role of Geography in education for sustainability in the South African context.

Joan Fairhurst has held a range of positions in her career as a geographer and an educator in primary and secondary schools, teacher training and academic institutions. She retired from the University of Pretoria and has a long-standing academic association with the Department of Geography at the University of South Africa. Her research interests cover aspects focusing on Environmental Education, Social Impact Assessment, Gender and Geography, Geography in Education and Geography and Development.

Education and Sustainable Development in Oman: The Contribution from Higher Education

Anthony C. Ijeh, Jeremy Brown and Ali Wad Haleeb

Abstract

This article presents a critical review of learning and teaching in a Higher Education Institute (HEI) in Oman using three key performance indicators (society, economy and environment). The performance indicators are used by The United Nations Educational, Scientific and Cultural Organisation (UNESCO) to monitor solution oriented approaches towards learning and teaching sustainable development in classrooms across global regions. The approach of this article towards identifying solution oriented practices was to qualitatively evaluate learning and teaching at a Higher Education Institute (HEI) in Oman over a sixteen week period. Data was collected using interview and observation of a targeted group of students studying the same course, asking about their opinions. The findings of this article show that assessment strategies play a key role in determining the effectiveness of solution oriented curriculum. The findings also showed that students were more satisfied with Solution Oriented Approaches to learning and teaching because they aligned with national problems with which they were familiar. The findings of this article are based on a single course and participation responses. The data collection method used in this article was restricted to observations and interviews. Future research could explore different institutions, and use pre/post studies of a group of courses.

A.C. Ijeh (✉) · J. Brown · A.W. Haleeb
Department of Management Information Systems, College of Business, University of Buraimi, Post Box 890, 512 Al Buraimi, Oman
e-mail: anthony.i@uob.edu.om

J. Brown
e-mail: jeremy.b@uob.edu.om

A.W. Haleeb
e-mail: ali.wad@uob.edu.om

W. Leal Filho (ed.), *Transformative Approaches to Sustainable Development at Universities*, World Sustainability Series, DOI 10.1007/978-3-319-08837-2_30

Future research could be undertaken over a longer period of time to allow for all HEIs within the Sultanate of Oman to partake. Results from such a study could be compared with those from the UNESCO monitoring station. This is currently being developed in association with the Research Council (TRC). This study found that the performance indicators used by UNESCO to monitor solution oriented approaches towards learning and teaching sustainable development in classrooms across global regions can accurately be used to explore the correlation of variables in different institutions using pre/post studies of a group of courses in the HEI. The research contributions from the selected student population used in this study shows correlation between the curriculums aim of preparing effective contributors to a sustainable society and assessment strategies used to obtain feedback on learning outcomes. This study found that the HEI to which the selected student groups belongs adopted the constructs developed by UNESCO to monitor solution oriented approaches towards learning and teaching sustainable development in classrooms across global regions. This studies value is that it provides insight into how one of Oman's HEIs is contributing towards Oman's commitment in meeting the 2014 target date set by the United Nations (UN) on Decade of Education for Sustainable Development (DESD).

Keywords
Higher education • Sustainable development • United nations • Solution oriented • Oman

1 Introduction

At its 57th session in December 2002, the United Nations General Assembly adopted Resolution 57/254 declaring the time period between 2005 and 2014 as the United Nations Decade of Education for Sustainable Development (DESD) in order to emphasize the critical role of education in moving towards a more sustainable world. The DESD offers an opportunity to promote the vision of a more sustainable and just global community through different forms of education, public awareness and training activities.

This relevance of this article to Resolution 57/254 is that it presents a critical review of learning and teaching in a Higher Education Institute (HEI) in Oman using three key performance indicators (society, economy and environment) used by the United Nations Educational, Scientific and Cultural Organisation (UNESCO) to monitor solution oriented approaches towards learning and teaching sustainable development in classrooms across global regions. The innovative approach used in this article set up a framework to detect if the four overarching goals of UNESCO stated in Chap. "Experiences of 'Reflective Action': Forging Links Between Student Informal Activity and Curriculum Learning for Sustainability" of Agenda 21 on education, training and public awareness, for which UNESCO was designated as task manager were present at a Higher Education Institute (HEI) in Oman were learning and teaching took place.

The novelty of the data collection process was to interview in language suited to their level of understanding a targeted group of students studying the same course, asking about their opinions of the four overarching goals: Promote and improve the quality of education, reorient the curricula, raise public awareness and understanding of the concept of Sustainable Development, train the workforce.

The relevance of the findings to Resolution 57/524 is that they suggest that the performance indicators used by UNESCO to monitor solution oriented approaches towards education and sustainable development in classrooms across global regions can accurately be used to explore the correlation of variables in different institutions using pre/post studies of a group of courses in the HEI. Furthermore the research contributions from the targeted student population used in this article shows correlation between the curriculums aim of preparing effective contributors to a sustainable society and assessment strategies used to obtain feedback on learning outcomes.

If as this study found, constructs developed by UNESCO to monitor solution oriented approaches towards learning and teaching sustainable development in classrooms across global regions have been adopted. Then this studies value is that it provides insight into how one of Oman's HEIs is contributing towards Oman's commitment in meeting the 2014 target date set by the United Nations (UN) on Decade of Education for Sustainable Development (DESD). The DESD vision provided the rationale and motivation for this article and it is hoped that the findings of this pilot research will in the future lead to research that includes all the Higher Education Institutes (HEI) in Oman.

The aim of this article was to establish if a selected Higher Education Institute (HEI) used as a case study has incorporated and is using the four goals recommended by the United Nations Educational, Scientific and Cultural Organisation (UNESCO) to govern the Decade for Education and Sustainable Development (DESD).

This aim was achieved by performing the following specific objectives:

- Objective 1: An extensive literature review was conducted, in order to investigate appropriateness of assessment strategies being used by the instructor (Literature Review section)
- Objective 2: The development of interview questions and observation criteria, in order to ascertain the appropriateness of the data being collected for analysis (Design /Methodology/Approach section)
- Objective 3: A comparative analysis between the curriculums goals of the Higher Education Institute (HEI) and the four over arching goals of United Nations Educational, Scientific and Cultural Organisation (UNESCO) to govern the Decade for Education and Sustainable Development (DESD).

By evaluating curriculum assessment strategies used in learning and teaching and their relevance towards achieving Chap. "Experiences of 'Reflective Action': Forging Links Between Student Informal Activity and Curriculum Learning for Sustainability" of agenda 21's four over arching goals created by the United

Nations Educational, Scientific and Cultural Organisation (UNESCO) this article contributes to the DESD debate on performance monitoring. It does this by outlining the four goals of UNESCO for DESD and comparing them against the learning outcomes of curriculum at a selected HEI in Oman.

2 Literature

A leading body reported that education was essential to sustainable development (UNESCO 2012). It also argued that curriculums should use a solution oriented approach towards education for sustainable development in the classroom (UNESCO 2012).

As curriculum and its assessment is key to how future contributors are equipped to work towards a sustainable society this article used a targeted population in order to evaluate different forms of curriculum assessment and their role in developing the competencies needed to equip future contributors to the Sultanate of Oman's sustainable society. The vision of Education and Sustainable Development is for education to be solution oriented so that contributors to a countries sustainable future are developed. Such opportunities for personal development can be through professional training (Ijeh et al. 2009a), capacity building schemes such as research work which addresses societal problems as shown in research work (Ijeh 2013). The benefits of solution oriented curriculums have been demonstrated through learning outcomes (Ijeh et al. 2009b; Ijeh 2010, 2011). These benefits enable contributors to use their education to transform the societies in which they live. It was with this in mind that the United Nations (UNESCO 2012) decided to launch the Decade of Education for Sustainable Development, so as to address the imbalance in global regions which were impoverished and underdeveloped.

In this article focus was placed on the strengths and weaknesses of assessment strategies used by an instructor at a HEI. This was so that the article could ascertain if assessment strategies could be used to equip future contributors to Oman's sustainable society? The article compares the strengths and weaknesses of all the methods used at the HEI to assess the targeted population against the DESD charter for learning and teaching solution oriented competences.

Existing literature show that assessment formats in use today include: essays, portfolios, patchwork text, negotiated learning and group projects, others are for practical assessments such as laboratory reports and computer based multiples choice tests (Glasner 1999). Leading writers report that essays can be used to assess knowledge and understanding of a topic (Brown et al. 1997). Others in support of this argument report that it also allows the instructor to showcase independence and academic writing skills and encourages student individuality (HEA Unknown). In contrast to this it is also argued that using essays for assessment does not necessarily mean that the essay will demonstrate deep learning by the student but that it can (Winter 2003). The study also reported that assessments had demonstrated that student essays frequently fail to demonstrate deep learning.

Other studies supported this argument which showed that conclusions were rarely reached in essays due to conceptual ambiguities (Brown and Knight 1994) and (Dignam 1999). In defence of using the assessment research work identified that essay questions were not made clear to the students, and recommended that further explanation of the structure of the essay be given by the assessor in order to clarify the expected feedback (Biggs and Tang 1999).

Research work identified weaknesses in the quality of dialogue with essay assessments as their not being recorded due to the limited time frame unlike portfolio assessment. The research work further argued that only portfolios allow for reflective developmental processes to take place (Brockbank and McGill 1998). Another reported weakness of essay assessment format was that the student may not know how to write essays and therefore could not write them (HEA Unknown). This view is supported by work which reported that a student's performance in writing an essay depends on how much prior practice that they have had (Brockbank and McGill 1998). Other assessment formats include portfolio, group project and negotiated learning. This section critical evaluates their strengths and weaknesses and why an instructor might choose one over the other.

Whilst portfolios reflect a student's academic skills the task given to the students might not be to test such skills. Again portfolios are gathered over time to allow student reflection on their work but can be bulky and restricted by the amount of words a student might be asked to write e.g. one thousand words. Group projects allow students access to creative thinking and peer tutoring which can benefit students who have difficulty in writing essays. These methods might however take away individuality and independence so that their work does not receive critique on their personal effort. It would appear that only the essay format of assessment allows personal critique, which might be its main strength. Negotiated learning allows for theory and practice to be assessed and whilst this choice of assessment format is determined by factors other than curriculum e.g. were an instructor feels that a student's view point cannot be made by arguments only without supporting evidence. This approach requires usually the production of an end product, which draws parallels with the solution oriented approach endorsed by the United Nations Educational, Scientific and Cultural Organisation (UNESCO 2012)

It was reported that assessment practices are used to communicate learning (Biggs 2003) to support this, another study defined learning experiences and the future character of students (Bloxham and Boyd 2008). The key advantage of using an essay for assessment is its ability to test the most important intellectual skills and demonstration of understanding (Winter 2003). Another advantage is its ability to allow for student individuality and writing style (Race and Brown 1998). The main weakness in using the essay for assessment are stated as its inequality (Race and Brown 1998) due to lack of experience in writing it and for teachers its subjective nature in addition to time consumed when being marked (Race and Brown 1998). The patchwork text is described as an innovators coursework assignment format developed because of its practical response to challenges to assessment problems in higher education (Winter 2003). It is also described as a collection of varied small sections of assessment covering learning outcomes which are stitched together over a period of time during the course of teaching (Winter 2003). The natural disadvantage of using patchwork text is its time value, which means it cannot be used for instant testing.

Negotiated task assessment is work based learning which allows future employers usually interested in students to assess them before employing them. This method bridges the gap between theory and practice and allows employers to partake in student assessment (Race and Brown 1998). It has the weakness of unreliability as placements tend to be individual and some students have better workplace opportunities to demonstrate learning outcomes than others (Race and Brown 1998). Student group work encourages the development of skills such as leadership, the ability to work in a team and creative thinking (ibid). For staff it has the ability to reduce assessment marking time, it however has the inability to assess individual members of the group and recognise their individual contributions (ibid).

Portfolios contain evidence that reflects student development and illustrates student behavioural progression (Race and Brown 1998). However they consume a lot of marking time and make it difficult to ensure reliability between staff marking in addition to encouraging cheating (ibid). All four assessment strategies described above were used to lecture students in the case study over a 16 week period. Leading writers suggest that the best form of assessment is to use all the above mentioned assessment instruments to test students, however there are restrictions in applying this school of thought due to the format of testing some of the learning outcomes in places like a computer laboratory. It is also argued that assessment practices in other non higher educational sectors are more established than those in higher education (Race and Brown 1998). This might be because a lot of the teachers in Higher Education are specialists in their subjects e.g. with terminal degrees and have not felt that taking a teaching qualification is necessary.

3 Method

The purpose has adopted a qualitative methodology to carry out the research. This is in order to explore and analyse general issues and to generate new knowledge. As qualitative research is subjective in nature it is suited to this study as the data collection methods of interview and observation involve examining and reflecting on the less tangible aspects of a research subject, e.g. values, attitudes, perceptions. However although this type of research can be easier to start, it can be often difficult to interpret and present the findings; the findings can also be challenged more easily.

Why is the research methodology used in this study qualitative? Sample size—small samples are often qualitative because of there is no need to conduct statistical analysis. Theories and Hypothesis—Part of the research will study the relevant literature to establish any gaps in the literature. Types of Data—quantitative data is highly precise and specific. Because measurement is an essential element of the research process. However under qualitative data the emphasis is quality and depth of the data.

Table 1 Data and information sources

Communication	*Secondary source*	*Data collected*
One way	Journals	Literature review data
One way	Conference Papers	Literature review data
Communication	*Primary source*	*Data collected*
Two way	Direct observation	Behaviour
Two way	Interview	Opinion

Location—qualitative research the study is usually carried out in the field which is where the survey data and interview data will come from. Reliability—under qualitative research it is not so important if qualitative measures are reliable. Validity—Is the extent to which the research findings actually represent what is happening in the situation. The research is qualitatively because the essence is capturing the richness of the data for explanation and analysis.

Other qualitative research methods used in this article include case study and participative enquiry, interview. Some of those not used include: Action research which is subjective and personal involvement rather than an objective stance, feminist perspective which is research which impacts on men and women and hermeneutics which is research which uses contemporary data rather than historical. The primary contribution of the research will be the design of the approach and the comparative analysis: This is because it would enable the research to find out about the design approaches suitability by critically analyzing and reading each literature source carefully. The research would also be able to develop and substantiate a credible opinion. This is necessary in order for the research to understand and create debate in an interesting way using up-to-date literature in order to enter discourse with other scholars in the field. This would be significant to other researchers if the study could provide an answer to the existing problem in the course of this review and research. This would be of tremendous interest to the public and private sectors alike as it would provide a solution that could mitigate the challenges caused. This is will be achieved by using the methods of how leading researchers in the field use primary and secondary resources to contribute to existing knowledge

In this article the overall approach to the research process, from the theoretical underpinning to the collection and analysis of data as shown in Table 2, the research work involved qualitative evaluation, and undertook a comprehensive review of all the literature sources used in the course of the article. The aim of the literature search was to identify as many items of primary and secondary data as shown in Table 1 as possible which were relevant to the article such as books, articles in journals, magazines and newspapers, conference papers, reports, archives, published statistics, united nation reports and white papers.

This enabled the article to obtain the required contemporary facts with which the article could investigate the comparisons authenticity. The article obtained its data from an interview and observation using a questionnaire and criteria data as shown in Tables 1 and 2 respectively. In order to obtain the relevant data sets the study undertook a pilot study. This was so the article could ascertain the confidence levels

Table 2 Methodology and method

Collection method	Methodology	Outcome of analysis
Interviews	Qualitative	Knowledge of participant
Observation	Qualitative	Know how in use of knowledge

of the interview questions and observation criteria. The outcome of the tests were analysed against already known values so as to confidently obtain the limiting factors that would affect the live interview

In designing the observation criteria and interview questions, the article considered UNESCOs Chap. "Experiences of 'Reflective Action': Forging Links Between Student Informal Activity and Curriculum Learning for Sustainability" of Agenda 21 on education, training and public awareness, for which UNESCO is designated as task manager, using four overarching goals. The rationale was to qualitatively evaluate learning and teaching at the Higher Education Institute (HEI) in Oman over a 16 week period and compare the findings with the four goals of DESD. The assessment strategies used at the HEI were compared against the four goals of DESD and an analysis undertaken to identify the similarities and differences

An initial investigation was undertaken to ascertain the following:

1. Is the assessment strategy inclusive, authentic and fair?
2. Does the assessment strategy evaluate feedback and feed-forward mechanisms?
3. Does the assessment strategy integrate teaching and assessment into one?
4. Does the assessment strategy promote learning?
5. Does the assessment strategy help students become active participants in learning?
6. Is the assessment strategy embedded within teaching practice and learning?

The following observations were made after participation of the students in class activity within the following scope: All the lectures were held in computer laboratories, so the students could use the internet and applications stored on the computers. Most students listened to the teacher but were not verbally active during lectures. Some students are challenged by the language barrier which makes it quite hard for the instructor to have effective feedback or feed forward. The instructor observed that students were able to work with the computers and search engines with ease. This means that students were able to learn a vocational course which is dependent on computers for instruction. Does this mean that the students would find essay assessments difficult? As essays are one of the main forms of assessment this is a serious challenge for learning and teaching. What about the other forms of assessment e.g. patchwork text, how do students fair when using them?

Making the assessment strategy inclusive, authentic and fair: To remove inequality in essay assessments students were given mock tests tasking them to write their answers in essay format. The tasks focused on realities that all students had experience of such as describing information systems used in the local shop.

In order to gain the benefit of using patchwork text assessment strategy, students were given tasks from week one which were collected in week eight during the mid-term break. Negotiation took place in the form of assessment strategies as students were only going to get a placement in their third or final year. Most students wanted to be tested by essay or report as it gave them a chance to express themselves. Students were placed in groups of four to six and asked to select an organisation which had the Information System they wanted to review. Students were then asked to keep their patchwork text in a portfolio along with their group project. This strategy helped the students reflect on what they had learned

Making the assessment strategy evaluate feedback and feed-forward mechanisms: As patchwork text is accumulated over time, before assessment feedback was given after every assessment. Students were also asked to maintain all accumulative assessments in folders for reflection. As negotiated learning is via choice of assessment all students in each class were asked to agree on an assessment strategy through which they would be tested. One key challenge faced in the region is the language barrier which can sometimes determine an assessment method. Students benefitted from group work as this allowed them to overcome the language barrier however it allowed lazy students to avoid doing any work. To overcome this barrier all students were retested on the tasks in the final examination. Portfolios were used although not exclusively, as each student had to maintain a folder of all project tasks which were to be bound and submitted in week eight. The exercise was rewarding for most students as it allowed them to understand why they were not awarded a mark (feedback) and how they could improve on their work in future tasks (feed-forward)

Making the assessment strategy integrate teaching and assessment into one: The instructor in this case study assessed all students and identified from existing records that some had challenges. Most students were female and culturally reserved as expected in the region. This meant that an integrated approach to learning was an obvious choice as it would overcome the challenges and barriers and ensure all students participated in class. Patchwork text was a useful tool in integrating the students as its accumulative format allowed all students to submit their projects on time. Negotiated learning was also a useful integration tool as its democratic philosophy allowed students of all backgrounds to express their opinions as a group as to what they wanted to learn and how. The group work was popular amongst the students as it brought down the language barrier which most students find a challenge. It however encouraged cheating which meant students were scored for work which they might not have contributed to. To balance this challenge the instructor re-tested learning outcomes for group work in the final examination. Portfolios were simply used to manage the projects assessments and patchwork accumulated over the semester. Integrated teaching and learning was maximised during this assessment because it allowed the student's time to reflect on their efforts.

Making the assessment strategy promote learning: In Patchwork Text assessment student perception was that they could work at their own pace (Surface learning approach). In Negotiated Learning assessment student perception was that they could work at their own pace (Surface learning approach). In essay assessments student perception was that they had to work hard (Deep learning approach). In portfolio assessment student perception was that they could work at their own pace (Surface learning approach). The perception of students for each assessment determined their attitude, willingness to work and final performance in the module

Making the assessment strategy help students become active participants in learning: Deep learning approach should be used in Patchwork Text assessment. Deep learning approach should be used in Negotiated Learning assessment. Surface learning approach should be used in Essay assessments. Deep learning approach should be used in portfolio assessment. Students should be encouraged also to combine portfolios with essay assessments which according to research evidence bring out the best in students when combined as an assessment strategy

Embedding assessment strategies within teaching practice and learning: The lectures in question at the higher education institution were for students majoring in Management Information Systems. The students are on average one hundred and ten and they come from varying backgrounds, mostly Bedouin. In Oman education is free and so people from all walks of life walk through the gates of Oman's Universities to collect their educational entitlement. The level of achievement brings with it challenges and teachers face those challenges in class. From experience students are hands on with computers and can use the internet to search for information. This is in contrast to their ability to write an essay or report. Having tested the abilities of the students, they seem to perform well when using diagrams. This encourages a line of thought which is to test the students on a variety of assessments so that they grasp the concept and the idea being taught. The semester in which the students were assessed saw the module using the following strategies: essay, group work, project, examination, tests and multiple choices. There could and should have been more.

The assessment methods and class activity has explained this research works understanding of how education and can be used to:

- Promote and improve the quality of education
- Reorient the curricula
- Raise public awareness and understanding of the concept of SD
- Train the workforce

It also shows how procedures can be used to help students become active participants in the process of their own assessment, how assessment strategies can make assessment practices inclusive, authentic and fair, the importance of design and the feedback and feed-forward mechanisms, how assessment for learning can become integrated into teaching and assessment practices in an educational setting and how the author has embedded within his own teaching practices key aspects of assessment for learning. This article has just scratched the surface of learning to teach effectively using assessment strategies.

After each class the instructor reflected on class activity. At the start of the lecture the three learning outcomes were clearly explained and the aim of the lecture was clearly spelt out. During the lecture examples using knowledge of local business in Oman to reinforce points and concepts were given. Examples were repeated several times so that the students were made to hear the key concepts over and over again. Students were asked to use their mobile phones to explain what an Information System was to reinforce learning. To support this, the instructor paced the class in order that they acquired understanding of the concepts and learning outcomes at their reasonable pace. A case study was used during the lecture to focus the student's attention on the learning outcomes and during the lecture students were asked to state, describe and explain various words and concepts.

For better understanding the instructor could have asked more questions of the students after each slide to reinforce their understanding of the topic. Technologies used in the classroom include an overhead projector which was used to show PowerPoint slides with diagrams and text. Some slides were linked to previous lectures to enable the students understand the topic being taught and for them to build on them. Students used learning from other classes and modules to explain their understanding of topics and were allowed to change their understanding to suit the area under discussion. These encouraged formative and summative assessments were students responded to questions in order to understand what was being said. At the end of the lecture a recap of the learning outcomes was done. Areas for development include etiquette in the classroom. Academically students should be given as many examples as possible and asked to give examples of their own to show their understanding. In addition group tasks should be used in class to encourage peer tutoring were students don't understand terminology or concepts being described. Finally, tutorials should be undertaken in class as students are not bound by the same discipline when a teacher is not in class.

The interview questions asked were developed from the four goals of the DESD shown in Table 3. Behavioural Questions Related to Problem Solving (questions 1–4). Roles and portfolio Items (question 5–15)

1. Can you describe how what you have learnt on your degree program has enabled you solve a problem?
2. Can you describe how your degree program has improved your time keeping and multi-tasking skills?
3. Can you describe your recent role in a group and what challenges you faced if any?
4. Can you describe what social activities you took part in recently with you classmates?
5. Can you describe any research project you are involved in?
6. What was your role in the research project?
7. What leadership roles have you held recently?
8. Can you describe what Green IT is?
9. Does your degree program have a course on Green IT or sustainability?
10. Have you attended any conference of seminars recently?
11. What was the topic of the seminar or conference?
12. Does your college have a sustainability club?

Table 3 Development of interview questionnaire

UNESCO Agenda 21—DESD four goals	Interview questionnaire approach
Promote and improve the quality of education: The aim is to refocus lifelong education on the acquisition of knowledge, skills and values needed by citizens to improve their quality of life;	1. Can you describe how what you have learnt on your degree program has enabled you solve a problem? 2. Can you describe how your degree program has improved your time keeping and multi-tasking skills? 3. Can you describe your recent role in a group and what challenges you faced if any? 4. Can you describe what social activities you took part in recently with you classmates?
Reorient the curricula: From pre-school to university, education must be rethought and reformed to be a vehicle of knowledge, thought patterns and values needed to build a sustainable world;	5. Can you describe any research project you are involved in? 6. What was your role in the research project? 7. What leadership roles have you held recently? 8. Can you describe what Green IT is? 9. Does your degree program have a course on Green IT or sustainability?
Raise public awareness and understanding of the concept of SD: This will make it possible to develop enlightened, active and responsible citizenship locally, nationally and internationally;	10. Have you attended any conference of seminars recently? 11. What was the topic of the seminar or conference? 12. Does your college have a sustainability club?
Train the workforce: Continuing technical and vocational education of directors and workers, particularly those in trade and industry, will be enriched to enable them to adopt sustainable modes of production and consumption	13. Are you ICDL certified? 14. What vocational certifications do you have? 15. Did you start and not finish any certification?

13. Are you ICDL certified?
14. What vocational certifications do you have?
15. Did you start and not finish any certification?

4 Results

The results presented a comparative analysis between the DESD goals of UNESCO and curriculum of the HEI

Behavioural Questions Related to Problem Solving

1. Can you describe how what you have learnt on your degree program has enabled you solve a problem? Respondents to question 1 of the interview questionnaire were asked to describe how what they had learnt on their degree

program had enabled them solve a problem? From the responses received it became apparent that most respondents had acquired problem solving skill which aligns with Goal 1 of DESD. However the study did receive responses that said they had not received any, these number of respondents were less than 5 % of the total respondents.

2. Can you describe how your degree program has improved your time keeping and multi-tasking skills? Respondents to question 2 of the interview questionnaire were asked to describe how their degree program had improved their time keeping and multi tasking skills? From the responses received it became apparent that most respondents had acquired time keeping skills and multi tasking skills which align with Goal 1 of DESD. However the study did receive responses that said they had not acquired any, these number of respondents were less than 2 % of the total respondents.
3. Can you describe your recent role in a group and what challenges you faced if any? Respondents to question 3 of the interview questionnaire were asked to describe their recent role in a group and the challenges if any that they faced? From the responses received it became apparent all respondents had taken part in a group project. The common challenge faced was that some group members were not contributing to the group or that the time the meeting was arranged for was convenient for them or some other member to attend. The skills acquired from group work align with Goal 1 of DESD.
4. Can you describe what social activities you took part in recently with you classmates? Respondents to question 4 of the interview questionnaire were asked to describe how what social activities they had taken part in recently with their classmates? From the responses received it became apparent that all respondents had taken part in social activities with their classmates recently. The networking and communication skills acquired from socialising align with Goal 1 of DESD.

Questions Related to Portfolio Items (Curricula, Awareness, Vocation)

5. *Can you describe any research project you are involved in?* Respondents to question 5 of the interview questionnaire were asked to describe any research project they were involved in? From the responses received it became apparent that only 9 % of the respondents had any involvement with a research project which aligns with Goal 2 of DESD. The remaining 91 % of respondents were not involved in research projects. It is worth mentioning that research projects are usually funded by the Research Council.
6. *What was your role in the research project?* Respondents to question 6 of the interview questionnaire were asked to describe their role in a research project they were involved in? From the responses received it became apparent that only 9 % of the respondents had any involvement with a research project which aligns with Goal 2 of DESD. The remaining 91 % of respondents were not involved in research projects. It is worth mentioning that research projects are usually funded by the Research Council.

7. *What leadership roles have you held recently?* Respondents to question 7 of the interview questionnaire were asked to describe how what leadership roles they had held recently? From the responses received it became apparent that 20 % of the respondents had acquired leadership skills which align with Goal 2 of DESD. The remaining 80 % of respondents were not involved in leadership roles. It is worth mentioning that those in leadership roles were group captains of class teams or research projects.
8. *Can you describe what Green IT is?* Respondents to question 8 of the interview questionnaire were asked to describe what they knew about Green Information Technology (IT)? From the responses received it became apparent 88 % knew what Green IT was which aligns with Goal 2 of DESD. However the remaining 12 % did not know what Green IT was.
9. *Does your degree program have a course on Green IT or sustainability?* Respondents to question 9 of the interview questionnaire were asked to explain what part of their degree program contained Green IT or sustainability? From the responses received 88 % could explain what part of their program contained an area on sustainability which aligned with Goal 2 of DESD. However the remaining 12 % felt their degree program did not contained Green IT or sustainability.
10. *Have you attended any conference of seminars recently?* Respondents to question 10 of the interview questionnaire were asked to describe what seminars or conferences they had attended recently? From the responses received 80 % had attended a conference or seminar which aligns with Goal 3 of DESD. However 20 % of the respondents had not attended a conference or seminar.
11. *What was the topic of the seminar or conference?* Respondents to question 11 of the interview questionnaire were asked to describe the topic which was presented at the seminar or conferences they had attended recently? From the responses received 80 % had attended an ethical seminar on behaviour which aligns with Goal 3 of DESD. However 20 % of the respondents had not attended any conference or seminar.
12. *Does your college have a sustainability club?* Respondents to question 12 of the interview questionnaire were asked to describe the type of sustainability club their college of University had? From the responses received none of the respondents were aware of any sustainability club which aligns with Goal 3 of DESD.
13. *Are you ICDL certified?* Respondents to question 13 of the interview questionnaire were asked if they were certified in ICDL. From the responses received none of the respondents were certified in ICDL which aligns with Goal 4 of DESD.
14. *What vocational certifications do you have?* Respondents to question 14 of the interview questionnaire were asked the vocational certificate that they had. From the responses received it became apparent that none of the respondents had any vocational certificate which aligns with Goal 4 of DESD.
15. *Did you start and not finish any certification?* Respondents to question 15 of the interview questionnaire were asked if they had started a course and taken

examinations for a vocational certificate. From the responses received all the respondents had started a course and taken exams for a vocational certificate which aligns with Goal 4 of DESD.

5 Discussion

After reviewing the assessment strategies used within the HEI in Oman the challenges faced by the respondents within the HEI were highlighted. These had to be placed in the context of learning and teaching in order for the goals of UNESCO as shown in Table 3.

Interviews and observations were used to obtain data from respondents, to enable the article identify which of UNESCOs goals had been successfully implemented at the HEI as shown in Table 4.

Curriculum used at the HEI identified some of the ways in which UNESCOs goals were implemented as shown in Appendix 1–5.

In Fig. 1 a Spider Chart showing the HEI Curriculum areas which are aligned with UNESCOs goals and those that are not

Figure 1 clearly shows that DESD Goal 2 is not being implemented by the HEI. Respondents answered negative when asked about their research projects, leadership roles, Green IT and sustainability.

Figure 1 clearly shows that part of DESD Goal 3 is not being implemented by the HEI. Respondents answered negative when asked if their college had a sustainability club.

Figure 1 clearly shows that part of DESD Goal 4 is not being implemented by the HEI. Respondents answered negative when asked about vocational certification. Whilst respondents had started vocational education none of them seemed to finish and earn a certificate for it

Practicality of articles findings to Higher Education Institutes

(a) Doesn't require infrastructure,
(b) Aligns Higher Education Institutions (HEI) with the strategy for DESD
(c) Cost effective
(d) Easy to adopt model/system

The HEI in Oman would need to align its curricular in the following areas in order to meet UNESCOs goals. These three areas each represent a goal for DESD and need to be met.

- Research work
- Raising awareness of Sustainability
- Vocational Courses and certification

Table 4 Respondent Statistics

Interview Questionnaire Approach	Number of respondents able to explain	Number of respondents unable to explain
Can you describe how what you have learnt on your degree program has enabled you solve a problem?	120	6
Can you describe how your degree program has improved your time keeping and multi-tasking skills?	123	3
Can you describe your recent role in a group and what challenges you faced if any?	126	0
Can you describe what social activities you took part in recently with you classmates?	126	0
Can you describe any research project you are involved in?	7	119
What was your role in the research project?	7	119
What leadership roles have you held recently?	16	110
Can you describe what Green IT is?	70	56
Does your degree program have a course on Green IT or sustainability?	70	56
Have you attended any conference of seminars recently?	101	25
What was the topic of the seminar or conference?	101	25
Does your college have a sustainability club?	0	126
Are you ICDL certified?	0	126
What vocational certifications do you have?	0	126
Did you start and not finish any certification?	126	0

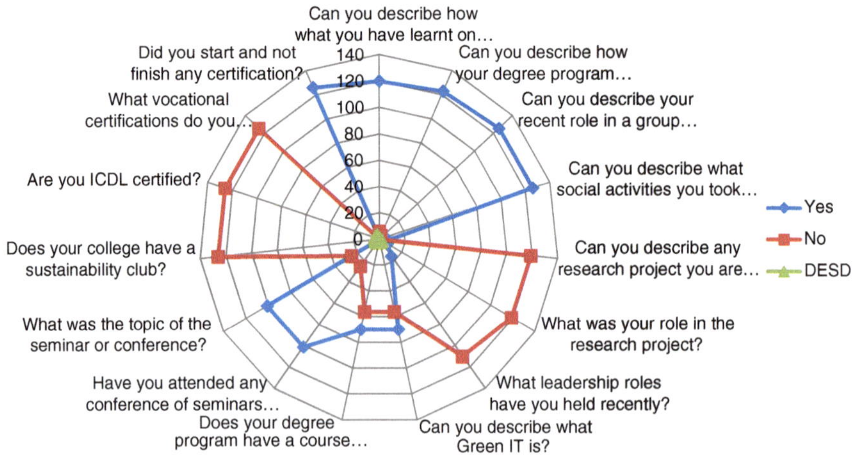

Fig. 1 Spider chart showing HEI curriculum areas aligning with DESD

6 Conclusion

The purpose of this section is the evaluation of the article which was covered in the last six sections. This section will therefore consider whether the aims and objectives set out in section one were achieved. The aim was to present a critical review of learning and teaching in a Higher Education Institute (HEI) in Oman using three key performance indicators (society, economy and environment) as set out by the DESD.

This aim was achieved by performing the following specific objectives:

1. **Objective 1**: An extensive literature review was conducted, in order to investigate appropriateness of assessment strategies being used by the instructor (Literature Review section)
2. **Objective 2**: The development of interview questions and observation criteria, in order to ascertain the appropriateness of the data being collected for analysis (Design /Methodology/Approach section)
3. **Objective 3**: A comparative analysis between the curriculums goals of the Higher Education Institute (HEI) and the four over arching goals of United Nations Educational, Scientific and Cultural Organisation (UNESCO) to govern the Decade for Education and Sustainable Development (DESD).

7 Future Research

The findings of this article are based on a single course and participation responses. The data collection methods used in this article was restricted to observations and interviews. Future research could explore different institutions, and use pre/post studies of a group of courses. Future research could be undertaken over a longer period of time to allow for all HEIs within the Sultanate of Oman to partake. Results from such a study could be compared with those from the UNESCO monitoring station. This is currently being developed for The Research Council (TRC) Open Research Grant.

References

Biggs J, Tang C (1999) Teaching for quality learning at university: what the student does. The Society for Research into Higher Education and Open University Press, vol 3. McGraw Hill, pp 1–357

Biggs JB (2003) Teaching for quality learning at university: what the student does. The Society for Research into Higher Education and Open University Press, 2nd edn. Buckingham, Chap 7

Bloxham S, Boyd P (2008) Developing effective assessment in higher education: a practical guide. Open University Press, Buckingham, p 2008

Brockbank A, McGill I (1998) Facilitating reflective learning in higher education. The Society for Research into Higher Education and Open University Press. Buckingham, pp 1–283

Brown S, Knight P (1994) Assessing learners in higher education. Teaching and Learning in Higher Education. Kogan, London, pp 1–161

Brown G, Bull J, Pendlebury M (1997) What is assessment? Assessing student learning. Routledge, London, pp 1–317

Dignam A (1999) An Astaire and Kelly model of argument. The Times Higher Educational Supplement, Feb 26th, p 36

Glasner A (1999) Innovations in student assessment: a system-wide perspective. Assessment matters in higher education, choosing and using diverse approaches (Eds. Brown, S and Glasner, A), Press, 1, pp 14-27

HEA (2013) Different Forms of Assessment In Higher Education Academy. Social Policy and Social Work (SWAP). Available online at. www.differentformsofassessment.pdf Accessed 21 Apr 2013

Ijeh AC, Brimicombe AJ, Preston DS, Imafidon CO (2009a) Geofencing in a security strategy model. In: Jahankhani H, Hassami A, Hsu F (eds) Global security, safety and sustainability, vol 45, 1. Springer, New York, pp 104–111

Ijeh AC, Preston DS, Imafidon CO (2009b) Opportunities for personal development through Professional Doctoral training. Int Conf Prof Doctorates 1(1):70–71

Ijeh AC (2010) Wireless security model. Lodged with the intellectual property office UK: Date Lodged: 27 Oct 2010. Applicant: Ijeh AC (Journal 6342) Patent GB1018091.7, Journal No. 6342, 1–13

Ijeh AC (2011) Geofencing as a security strategy model. Professional Doctorate Thesis, University of East London, UK. DInfoSec (Doctor of Information Security), Research Open Access Repository (ROAR) 1, 1, 1–126

Ijeh AC (2013) Impact of road safety campaigns on human behaviour in vehicles on Oman roads. Faculty mentored undergraduate research award program. Research Council, Oman. Agreement No FRP/ASU/13/001, pp 1–45

Winter R (2003) Contextualising the patchwork text: addressing problems of coursework in higher education. The patchwork text, a radical re-assessment of coursework assignments (Innovations in education and teaching international, Special Issue) vol 40, 2, pp 1–22

Race P, Brown S (1998) The lecturer's toolkit, 2nd edn. Kogan Page, London

Wals AEJ (2012) Shaping the education of tomorrow. Full length report on the decade of education for sustainable development, commissioned by UNESCO. Monit Eval Rep 1(1):8–27

Author Biographies

Dr Anthony Ijeh has spent the last 20 years teaching, researching and consulting in sustainable Information Systems and related areas. He received his Doctorate in Information Security from the University of East London, in (2011) where he researched sustainable Wireless Security Models (Patent Journal 6342, GB1018091.7) then consulted internationally on sustainable Information Systems (IS). He first taught at University of East London in England, where he received and managed several grant awards from: Haberdashers New Entrepreneurs Start up Grant (2009), Muslim Association Fund (2010), Emerald Fund (2010), IP21 (2010), National Physical Laboratory (2010), Royal Academy of Engineering (2010) and various University of East London grants to support his work and upkeep. He joined the School of Architecture, Computing and Engineering (ACE) in 2007 and soon after developed models for sustainable Wireless Security. His work saw him earn awards from IET for Security Innovation (2009), BCS for Research Development (2010), and IEEE for Best Doctoral Thesis in UK & RI (2011). He also taught classes on Business Information Systems and has consulted nationally and internationally on different aspects of sustainable Information Systems, more recently on In-Vehicle Information Systems, in the Sultanate of Oman as part of a team funded by the Research Council (2013).

Presently he is involved in various projects covering steering alert systems, environmental alert systems, tacit innovation management and solution oriented curriculums as part of teams funded by the Research Council in the Sultanate of Oman. He would like to thank all his colleagues and students who have worked with him on any project over the years for their creative visioning and for providing inspiration for this paper.

Dr Jeremy Brown has spent over 20 years teaching and researching in Information Systems, eLearning and Community Informatics. He received his doctorate in Information Systems from the University of Sunderland (UK) in 2013, where he investigated societal and environmental factors which act as barriers and/or motivators to the use of telecentre sites located in rural communities in the Philippines. He first taught at Jubail University College, in Saudi Arabia in 1992 where he worked closely with local industries on training needs analysis projects. Since then he has worked at higher education institutions in the UK, Japan, Singapore, New Zealand and Ireland. In 2005 he joined University Brunei Darussalam where he played a central role in a joint project with the Ministry of Education to implement eLearning systems in educational institutions throughout Brunei. He joined University of Buraimi in Oman in September 2013, where he is currently developing research projects investigating the use of Telecentres in Oman, and eLearning capacity in higher education in Oman.

Ali Wad Haleeb has over 20 years teaching, researching and consulting experience in Information Systems. He received his Master Degree from Strathclyde University in 1982 where he researched the North Sea oil rigs, designed its communication systems and then consulted internationally on it. He holds membership of the IET, IEEE and has been a registered Chartered Engineer by Royal Charter since 1984. His research work contributed five solutions to the Y2 K computer problem and inventing three systems to protect mobile phone users from radiation. At an Omani University he led a research team which in developed K-Map simplification techniques and another to discover a new artificial intelligence searching technique. Recently he designed a new mobile digital logic design laboratory which he presented at an international conference for research and innovation. Ali is currently working on a mobile analogue communications laboratory as part of curriculum development for Higher Education in the field of science within the Sultanate of Oman.

Part III
Problem-Solving and Integrative Practices

Environment and Social Sustainability Approaches in Policy and Practice: A Case Study of Symbiosis International University, Pune, India

Prakash Rao, Yogesh Patil, Viraja Bhat and Shilpa Kulkarni

Abstract

Environmental issues, development concerns and sustainability have become a major interest area for society as critical issues for economic growth and societal development. The recent Rio+20 summit in 2012 has strengthened the need for building sustainable development as an integral part of educational institutions through integrated learning and use of innovative pedagogical methods including the internationalization of educational systems. The present chapter seeks to focus on sustainability as a core vision and its concepts in policy and practices by implementing a unique standardized curriculum integrating various levels of sustainable development by building curriculum in key disciplines like energy, environment, infrastructure, climate change, global governance and corporate sustainability at the Symbiosis International University. The case study also discusses a preliminary assessment of the sustainability operations of the University through quantification of its energy, water consumption patterns and other initiatives. The study uses pilot student driven projects to describe sustainable development in practice. The study addresses some of the key issues and challenges in building environmental aspects as a part higher education policy and as a sustainable practice across constituent institutions of the University.

P. Rao (✉) · V. Bhat · S. Kulkarni
Department of Energy and Environment, Symbiosis Institute of International Business, Symbiosis International University, Hinjewadi, Pune 411057, India
e-mail: prakash.rao@siib.ac.in

Y. Patil
Symbiosis Institute of Research and Innovation, Symbiosis International University, Lavale, Pune 412115, India
e-mail: head_respub@siu.edu.in

W. Leal Filho (ed.), *Transformative Approaches to Sustainable Development at Universities*, World Sustainability Series, DOI 10.1007/978-3-319-08837-2_31

Keywords
Sustainability · Symbiosis · Infrastructure · Curriculum · Policy · Practice

1 Introduction: Concept of Sustainability in Higher Education

Over the past few years, there has been a growing realisation on the need to see sustainability as the centre piece of all anthropogenic activities leading to rational use of environmental resources and their conservation. The concept of sustainability emerged from the Brundtland Commission Report as product of the World Commission on Environment and Development (WCED) in 1987 and Agenda 21 from the Earth Summit of the United Nations Conference on Environment and Development (UNCED) in 1992 (NCSE 2003; Scheunpflug and Asbrand 2006; Springett 2005). The central question has been "to what extent is the use of natural resources and environment possible, if our economy is to exist on at least the present level in the very long run"? (Malovics et al. 2008). In recent times, apart from incorporating sustainability related curriculum in educational institutions, it has also been suggested that the institutions of higher learning should also measure their sustainability quotient in terms of tracking various energy and ecological foot prints and impacts (Leal Filho 2012). A sustainable business management practice has to acknowledge the "embeddedness in social, environmental and economic systems, and focuses on management and relationships to meet the environmental, social, and economic requirements of many different stakeholders in its networks" (Van Kleef and Roome 2007).

Sustainability itself, however, is a complex term that has been open to a variety of interpretations (Bonnett 2002). In the past, although the concept of sustainability tended to be centred squarely on environmental issues, as in the framing of a balance between economic growth and ecological carrying capacity, more recent understanding of the term have come to include socio-cultural aspects as well (Dempsey et al. 2009). Approaches to sustainability-related issues are diverse and situations are uniquely different between nations and cultures (Enderle 1997; Matten and Moon 2004). The most frequently seen factors used in performance measurement are: economic, environmental and social.

The university courses are the economic front where the courses duration, faculty and the fees structure plays a greater role to define economic viability of the course (Velazquez et al. 2009). The social aspect of the university not only encompasses the faculty and staff but it also involve the student and community in large. India being an emerging economy, in the post liberalization era, most of Indian Business Schools and Universities have started programmes focusing on sustainable development, energy and environment aspects in their curriculum; and waste management systems is one of the integral parts of such programmes (Rao et al. 2013).

From an educational perspective there have been several advances in the field of sustainability through the development of policy and operational aspects. University systems around the world have developed several approaches and models to promote sustainability policies as well as carry out implementation of activities leading to sustainable development (Leal Filho 2011). The UNCED in 1972 sowed the seeds for creating importance around key issues related to environmental conservation, biodiversity protection and environmental awareness. This was further reiterated through the Tbilisi Declaration of 1977 where a framework for environmental education activities was developed to be considered at local, regional, national and international levels through a set of fundamental principles. In 1987, sustainable development was formally adopted as a term by the Brundtland Commission (WCED 1987). Subsequently, several other agreements and declarations have been set in motion for integrating sustainable development across institutions of higher learning. Some of these include the Copernicus Universities Charter for Sustainable Development (1994), the Luneburg Declaration on Higher Education for Sustainable Development (2001) and more recently the 2010 G8 University summit—statement of action (Leal Filho 2011).

One of the major decisions of the Rio summit of 1992 was to promote Agenda 21—as a series of recommendations aimed at promoting sustainable development across various dimensions. These included education, training and awareness generation amongst institutions, civil society, business and industry and other relevant bodies and agencies. As such the key focus of this activity was to help develop capacity in educational institutions through explaining the various processes related to physical and biological environment along with inter linkages to socio-economic environment and human development. In doing so, it was envisaged that university programs in individual countries would create a system of training young graduates to inculcate sustainable aspects for society and development as in the case of Fluminense Federal University (UFF), Brazil (Romero 1995). Campus related initiatives have further been strengthened by the development of an integrated decision making process wherein all aspect of sustainability ranging from policy, teaching, research and practice have been created through tools using pressure state exposure effect framework and a multi bottom line approach (Waheed et al. 2011). Haigh (2010) in a treatise aptly points that curriculum greening is the only solution to infuse sustainable development activities across institutions of higher education and suggests that our educational enterprises must be fully geared to meet the needs of the future instead of relying on destructive and disruptive characteristics of the current economic era.

Academic institutions have in recent years apart from focusing on sustainable development based curriculum are also considering the prospect of incorporating some of the emerging disciplines into their own sustainability models. Sterling (2001) points out that development of sustainable education framework requires in depth visioning, design, and implementation at levels for achieving environmentally sustainable lifestyles. Several other authors (Madeira et al. 2011; Mitchell 2011; Speller 1992; Stubbs and Cockling 2008) have also highlighted the need for incorporating sustainable development into the framework of University institutions.

The desire to achieve sustainable development often has its fair share of challenges. Often educational institutions are seen to have certain misconceptions of sustainability and its implications for society. While most would agree that environment and conservation activities are seen as a positive sign of societal development, serious challenges remain in implementation of such actions. Some of the barriers to such actions include resource availability, lack of relevant human resource capacity, motivation among institutional staff about environment issues, management support, etc. (Leal Filho 2000, 2011).

2 Educational Institutions and Sustainability Institutions in India

Sustainable development at Universities and academic institutions in India has been mainly seen from the point of introducing curriculum which is related to various aspects of sustainability and environmental conservation (Rao 2011). Since the introduction of environment as a subject across post graduate education institutions in India in 1985, many Indian universities have introduced environment related curriculum as compared to niche based Universities which have emerged only recently. Traditional science based colleges and engineering institutions have at some stage incorporated energy efficiency, power sector related courses as a part of their electrical engineering discipline.

Since the initiation of economic reforms process in India from 1991, several institutions have started to consider the importance of environment as a compulsory element of education curriculum at primary and secondary level as well as at higher education institutions.

From the operational perspective there are very few examples of how sustainable development practices are implemented at Universities and other educational institutions (Roy et al. 2008; Spranger 2011). While some use fulltime residential courses and curricula on energy and environment (Rao 2011) as an attempt to promote sustainability, others follow a paradigm of undertaking short courses aimed at niche based target groups.

In India, practices are either restricted to niche based strategies or through institutions like the Indian Institute of Science, Bangalore, which has been collaborating on reducing its carbon emissions through various in-house related research and development activities. On the other hand there has been extensive research carried out at national and city levels to determine the environmental (Living Planet 2012) impacts and carbon footprint of cities and regions (ICLEI 2009). The key aspect of environmental sustainability of the institution itself has not been studied in detail barring a few studies at the International level (Leal Filho 2012).

As Higher Education Institutions (HEI) embark on adopting sustainability led curriculum and operational practices in their overall institutional framework, the use of several tools like GASU (Lozano 2006). Auditing Instrument for sustainable higher education (Roorda 2001) and STAUNCH (Lozano 2010) become very relevant. Sustainability planning and action has led to other forms of policy related initiatives e.g. the Association for the Advancement of Sustainability in Higher Education (AASHE) which helps promote and coordinate the sustainability activities across Universities (AASHE 2012).

Also, in the past two decades, especially after post liberalisation era in 1991, curriculum development in India in the broad area of environmental science and management got restricted or confined to a particular niche area because of lack of suitable human resource and unavailability of necessary teaching resource. Depending upon the available expertise the curriculum over the period of time got conversant with either biology or chemistry or geology or geography. Focus on the management aspects was completely lacking. Moreover, the entire curriculum completely lacked the integration with other disciplines like social sciences, economics, etc. This practice in the past generated human resource that lacked the holistic (sustainable) approach to resolve environmental problems. Secondly, although the energy and environmental related programmes and courses are being taught in large number of universities across India, there are very few instances in which the acquired knowledge is being put into practiced for developing sustainable/ green campuses due to one or other reasons, mainly being the financial constraints. These have been a matter of concern and the issues of resource use and conservation needs to be addressed immediately as the situation gets more vulnerable.

In the light of above, the authors narrate the case study of Symbiosis International University, Pune, India concerning the development of curriculum and initiatives (policy) related to sustainability and pilot assessment of operational aspects (practice) of sustainability.

3 Methodology

The present chapter deals with the initiatives taken at the Symbiosis International University, Pune, India, which is spread across six different campuses in Pune. In adopting a methodology, the study focused on the curriculum (policy) related to sustainable development coupled with the preliminary assessment of the operational aspects (practices) of sustainability at the University. In 2013, a curriculum development exercise was initiated at Symbiosis International University to evolve and standardize a uniform curriculum across its constituent colleges in the field of sustainability studies and infrastructure management. In order to study the policy perspective the methodology took into account the process of developing an integrated approach to curriculum development at the Symbiosis International University as well as considered the various sustainability related courses being imparted at undergraduate and post graduate levels.

For the operational aspects of sustainability the study focused on undertaking preliminary assessments of environmental sustainability through indicators like water, energy, landscape, waste, resource utilisation, etc. from May to October 2013 across six different campuses of Symbiosis International University viz. Senapati Bapat Road, Atur Centre, Vimannagar, Khadki, Lavale, and Hinjewadi.

4 Sustainability Education in Symbiosis International University: A Policy View

The genesis and origin of the Symbiosis International University (SIU) established in 2002 lies in the foundation of the 'Symbiosis Society' by Padma Bhushan Dr. S.B. Mujumdar, a renowned educationist in the year 1971 with a motto *'Vasudhaiva Kutumbakam'* which effectively means "world as one family". As a deemed University SIU is recognised by the University Grants Commission (UGC) and accredited by NAAC with A grade in 2009. Symbiosis scientifically means living together for mutual benefit and the University is today a true representation with 43 institutes imparting education in diverse disciplines at under graduate, post graduate and doctoral levels. SIU is known to be a multicultural, multi lingual and multinational institution with academic democracy and nurturing innovation, passion and compassion along with Internationalisation (www.siu.edu.in). At SIU, sustainability lies at the core and the entire team works together through many initiatives at all levels of organizations. The education at SIU has pedagogy which has both policy and practices in its curriculum.

4.1 Sustainability Curriculum Visioning at SIU

As a part of the growing need for convergence and integration, the SIU conducted a curriculum development and review exercise aimed at standardising inter disciplinary courses in sustainability and other relevant disciplines into a uniform set of courses across its constituent institutions and colleges. This also meant avoiding repetitive courses and overlaps across Symbiosis Institutions (Cayuela et al. 2013).

The exercise was initiated at the university level where an expert committee was involved in review, design and modification of course offerings and evaluation criteria as per the level of understanding of the participants (Bloom's Taxonomy). The exercise involved subject matter experts, faculties and practitioners. The curriculum development exercise was held in two phases with phase one essentially a gap analysis and brain storming session, while phase two being the actual designing of the courses with syllabi. As a part of the exercise the authors reviewed existing courses on sustainability with faculty members and through a process of deliberations and brainstorming, courses were standardized including developing module specific syllabi. Initially at the start of the exercise 90 courses were available for under graduate and post graduate students as part of sustainability studies. Based on the various reviews of existing courses, a total of 77 courses were finalised which meant that 13 courses were not included for reasons like, being an overlap course, too theoretical, not industry relevant, etc. From a University point of view, courses in sustainability were heavily biased toward post graduate studies (65) as compared to under graduate studies (12). The exercise also revealed a few courses (11) which were common to various symbiosis constituent Institutions as a critical component of their curriculum on sustainability (Table 1).

Table 1 Common courses in sustainability and infrastructure at Symbiosis International University

Sr. No.	Common courses on sustainability across Symbiosis International University	UG/ PG
1.	Concepts and applications in sustainability	PG
2.	Environment impact assessment	UG/ PG
3.	Oil and gas economics	PG
4.	Project management software	PG
5.	Public private partnerships	PG
6.	Field/Research project	UG/ PG
7.	Governance and corporate sustainability	PG
8.	Project management	PG
9.	Transmission and distribution management	PG
10.	Urban and industrial waste management	PG
11.	Renewable energy sources	PG

4.2 Sustainability Discipline as Stand Alone Courses at Symbiosis International University

In 2009, the Symbiosis International University conceptualised a unique post graduate course which was aimed at integrating energy development and environment concerns and equipping aspiring global managers with managerial, economic, and technical competence. Core focus areas include sectors like sustainable energy development, conventional energy (coal, oil/gas), renewable energy, energy economics, carbon markets and trading, corporate environmental sustainability, energy management, environmental assessments, natural resources management etc. Subsequently post graduate programmes were also introduced related to Infrastructure management in the context of increasing focus on infrastructure development and economic growth in India.

While most educational institutions develop niche based courses with specific focus on creating a particular domain knowledge for students, these stand alone courses at SIU were conceived with a view to integrate some of the critical issues linking the energy, environment and infrastructure sectors with issues of social development and equity were therefore in a way trend setting courses. Some of these integrated approaches are detailed below:

Power sector: The recent power sector reforms in the country following the Electricity Act, 2003 mandated the development of a transparent, fair and equitable process in the generation, transmission and distribution of electricity across the country. This has led to several private sector entities trying to capture a market share of the power sector generation. As a consequence, several environmental concerns like mining, and rights acquisition, loss of ecologically rich ecosystems, wetland reclamation, forest cover loss, etc. have been often neglected in the desire to set up new and ultra-modern power plants (Areendran et al. 2013). The post graduate programmes have tried to include these issues as part of a rigorous curriculum which will also sensitise future managers through curriculum focused on project management, environmental impact assessments rules and regulations, wastewater management, sustainable energy development, infrastructure project development, etc.

Corporate sustainability: The current initiatives in Indian industry to build responsible business as a sustainability activity as an integral part of core business is limited and seems all the more a difficult choice to make given the current economic growth conditions around the world. However with the adoption of the recent Companies Bill of 2013, it is hoped that businesses will develop frameworks to create an integrated approach to addressing responsible business activities through the adoption of the triple bottom concept of environmental, social and economic performance. The post graduate programmes at the University have incorporated various sections in the curriculum to inculcate the business case of sustainability for clarity and thinking on the balance between infrastructure growth and sustainable development through use of methodologies and tools for developing sustainability standards and guidelines.

Water energy nexus: In today's market and growth centred economy, resource consumption and utilisation are often seen as key to the sustainable development of society and its stakeholders. In this context the linkages between water and energy use has been studied by many scientists and is now being seen as a crucial aspect of economic growth. Moreover, organisations like the World Business Council on Sustainable Development (WBCSD), UNEP have emphasised the importance of water and energy as directly relevant to sustainable business operations. The curriculum at Symbiosis has attempted to integrate these two disciplines in an effort to provide a holistic approach to the water energy nexus. The new focus of the curriculum seeks to study the linkages between water—energy use through adoption of demand side management options, optimizing efficiency of existing systems e.g. steam engineering and water savings, recycling of waste water, alternate energy use including their application across different infrastructure sectors.

Climate change and development: According to the Fourth Assessment Report of the Intergovernmental Panel on Climate Change (IPCC), climate change is likely have tremendous impacts on biodiversity, natural resources, abundant freshwater resources, local livelihoods and many other sectors. Serious societal challenges are seen as many industries rely on natural resources for production and operations. The role of business schools in creating leadership to tackle sustainability challenges has been a key factor (Adams et al. 2011) in driving the development of curriculum aimed to creation of sustainable or low carbon universities. At Symbiosis, the under graduate and post graduate programmes have developed targeted courses which address the impact of climate change not only for business risks but also its implications to society and for policy makers. The curriculum focuses on the direct linkages between climate change impacts and energy development and also on recent market mechanisms (Renewable energy certificates, Perform Achieve and Trade) and emerging policy mandates like NAMAs. Such mechanisms offer vast scope for various business and industry players including small and medium enterprises (SMEs) (Saini et al. 2012). The courses related to renewable energy have been conceived from undergraduate level itself in the light of the growing focus by policy makers in increasing the overall share of renewable energy in the energy mix of India.

Public private partnership: The Symbiosis Integrated programme curriculum has also introduced innovative course modules which seek to conceptualise the governance model of public private partnerships as a solution to rapid economic growth. Ever since the PPP process has evolved globally for the past two decades, there has been mixed results on its efficiency, performance monitoring and viability across different growth sectors. At the same time, it is of interest to note that several PPP led initiatives have been taken up in the water and sanitation, transport, power sector to promote sustainable infrastructure growth. At Symbiosis, courses at the post graduate level have been developed on the PPP led governance model with a focus on sustainable development for management education.

Industrial ecology: Since industrial development has rapidly grown in developing countries, particularly Asian countries and given the constraints as mentioned previously, it is a pressing need to accelerate the need for economic development with simultaneous protection of environmental resources. The present programme curriculum at Symbiosis decided to introduce a new integrated industrial planning and management mechanism in its curriculum on the discipline of Industrial Ecology and springs from interests in integrating notions of sustainability into environmental and economic systems (Ehrenfeld and Gertler 1997). Thus, model of industrial symbiosis in developing and underdeveloped countries would have an opportunity to manage their waste to become resource without many investments in technology of waste management (Bhat et al. 2012; Itankar et al. 2013; Patil 2012). It will also (a) help industries to improve their environmental performance, strategic planning and will become more competitive; (b) help local communities develop and maintain a sound industrial base and infrastructure without sacrificing the quality of their environments; and (c) help local, regional and national government to formulate policies and regulations in order to improve environmental protection with simultaneous building of business competitiveness.

Courses have been designed related to Life Cycle Assessment (LCA) as a method to assess environmental impacts associated with all the stages of a product's life from cradle-to-grave (Finnveden et al. 2009). LCA completely avoids the restricted viewpoint towards environmental concerns by way of: (a) accounting an inventory of relevant energy and material inputs and environmental releases; (b) evaluates the potential impacts associated with identified inputs and releases; and (c) interpreting the results to help make a more informed decision (Scott 2010).

5 Sustainability at Symbiosis International University: A Practical View

The Symbiosis International University has six campuses in Pune city, India catering to the education requirements of several thousand students in India and overseas. Many students stay on campus 24/7. The sustainability practices presented here by the authors are an initial assessment of the practices in the areas of education (policy) and energy, water and food waste (operations) which was primarily driven by students. The various initiatives were started as a pilot project at the Symbiosis Infotech Campus (SIC) at Hinjewadi initially and were then taken up at the other campuses.

5.1 Promoting Campus Wide Energy, Water and Food Conservation Through Assessments

The students of Symbiosis Institute of International Business held an observatory survey in the SIC campus hostel which is a residence to more than 1,000 students. This initiative aimed to understand the use of various electrical appliances, tube lights, fans and plug points during the period when the students were not in the rooms for a period of 11 (eleven) days. This was followed by posters which were put to spread awareness among students towards energy conservation due to disregard towards energy saving. The data was collected and the analysis in terms of the cost incurred is shown in Fig. 1.

Food waste management: Food wastage has always been an important concern in India, and a lot of food gets wasted on a day-to-day basis. The maximum amount of food wastage comes from hotels and restaurants. However, lately it has been noticed that there is a considerable amount of food wastage generated in residential colleges as well because of food served in the hostels. The food wastage was noted down for a period of one month for breakfast, lunch and dinner on all the days and the data collected is presented in Fig. 2. In order to manage and reduce the quantum of waste generated from food SIU has initiated the process of setting up bio gas plants at two of its locations in Pune.

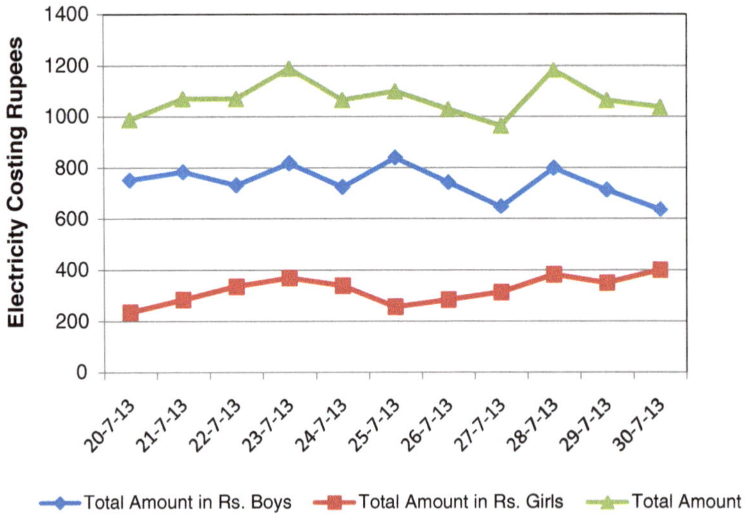

Fig. 1 Total cost incurred for different unused electrical appliances on different dates

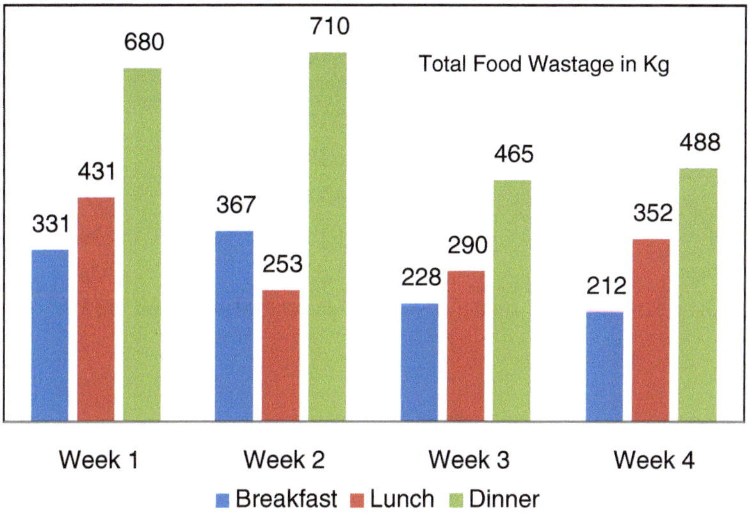

Fig. 2 Food wastage (in kg) at SIC Campus (August 2013)

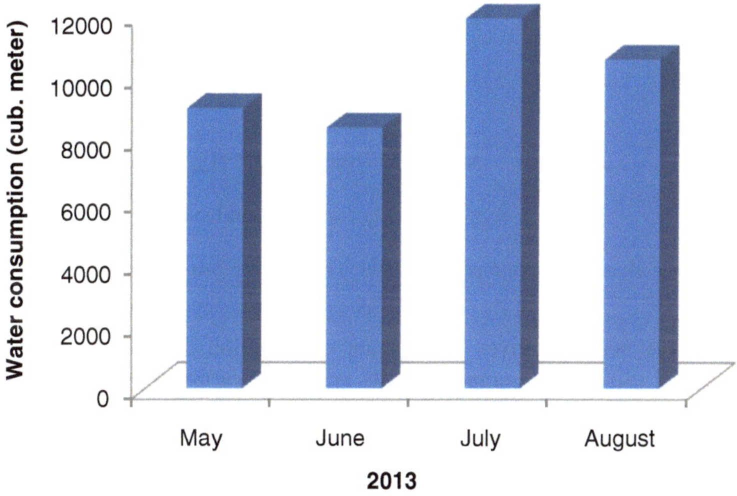

Fig. 3 Month-wise total water consumption at SIC Campus

Water consumption: Water is one of the most precious commodities and its demand, availability, consumption and use, wastewater generation and its treatment, recycling and disposal are crucial aspects for a developing and emerging economy like India. The data related to water consumption was collected for the period of 4 months to understand the usage of water in SIC campus (Fig. 3). It was

Table 2 A comparative data of resource consumption and food wastage at various campuses

Campus location	Avg. electricity consumption (Units)	Avg. water consumption (l)	Avg. food wastage (Kg)
SIC	104,103	2,150	45
SIU	290,116	6,504	50
SIMS	51,202	3,776	28
S.B. Road	33,491	6,133	30
Viman Nagar	67,270	2,055	376

Note All the values mentioned are the average of 6 months (May–October 2013)

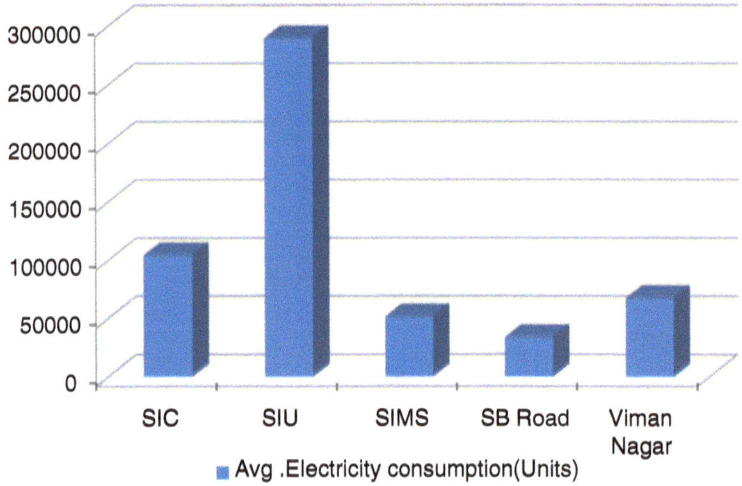

Fig. 4 Average electricity consumption (in units) across Symbiosis Campuses in Pune

found that the new initiatives include the construction of rain water harvesting system at hill top Lavale campus of SIU and three earthen dams at the hill base having storage capacity of 50 million litres of water. Sewage water generated at Lavale campus is being treated using biological treatment method like Reed Bed Technology. Further, the treated wastewater is used for the gardening purpose.

Assessment of food, water, electricity in the SIU Campuses in Pune city: Data related to food, water and electricity were collected for a period of 6 months between May and October 2013 and a comparative analysis is depicted in Table 2 and Figs. 4, 5 and 6. Based on the comparison of data at individual campuses various measures are being initiated to reduce and rationalise consumption practices across the campuses.

Environmental observatory display program: This initiative was to spread awareness about significant parameters which affect environment. The local parameters displayed included maximum and minimum temperature and humidity,

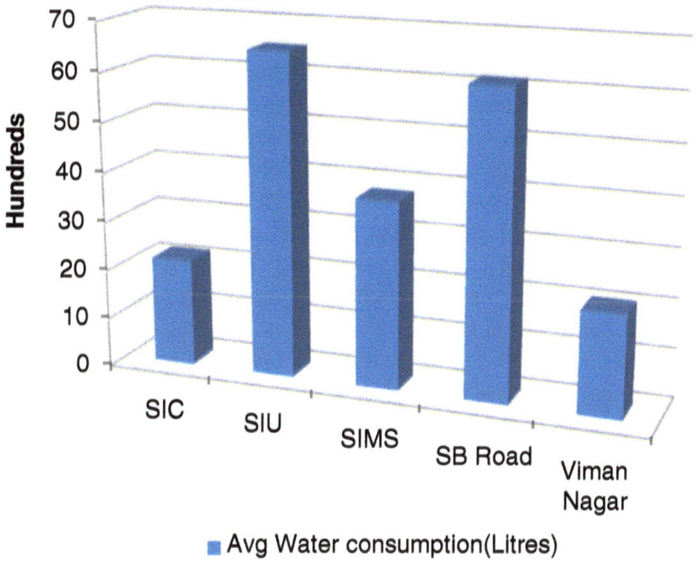

Fig. 5 Average water consumption (in litres) across Symbiosis Campuses in Pune

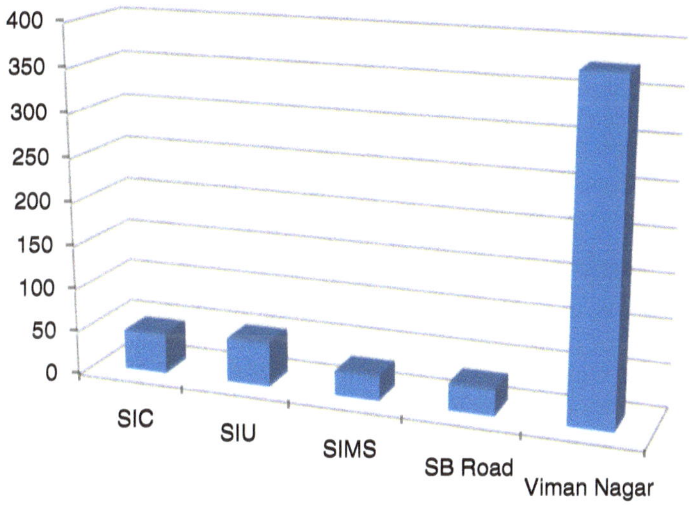

Fig. 6 Average food wastage (in Kg) across Symbiosis Campuses in Pune

while the global parameters on display monitored emissions of CO_2, CO, NO_2, SO_2, O_3, CH_4, N_2O, CFC-12. A database of these parameters is being maintained for the last 4 months since November 2013 and will be used to create awareness about climate change in the long run.

Computer energy savings program: The individual institutes under SIU have various policies to conserve ambient energy during computer usage. At the SIIB campus, asleep mode policy is implemented which switches off the monitor when not in use. In a class of 2 h duration if the faculty gives an offline assignment or quiz with a break of 10 min, the students generally leave the computer switched on. Due to the sleep mode policy the monitors go to the sleep mode and saves almost 2 paise in their electricity bill per computer. The SIIB laboratory houses 75 computers which results in a saving of 75 × 2 paise = 150 paise (1 INR = 100 paise) of electricity bill for duration of 2 h. There are totally 100 computers in each campus and if the same policy is applied everywhere on an average Rs. 2 is saved in the electricity bill for the duration of 2 h.

Transportation Policy: In an effort to reduce greenhouse gas emissions from students, faculty, and university vehicles, all the SIU campuses are declared as Green zones where no student vehicles are allowed. There are shuttle buses for the students to commute from the hostel and also to the city. The faculty and staff use the shuttle buses to travel from city to the campuses. There are two electric vehicles in the university premises used for commuting within the university campus. Pool of bicycles is kept at the university premises for student usage. Faculty who stay nearby use carpools to reduce vehicle fuel usage and help in reduction of greenhouse gas emissions.

Sustainability Practices through CSR activities: The World Business Council for Sustainable Development defines CSR as '*the continuing commitment by business to behave ethically and contribute to economic development while improving the quality of life of the workforce and their families as well as of the local community and society at large*'. Sharing these views and working in the lines of Symbiosis's vision the university students put in consistent efforts for holistic development of Society. Since the year 2004, SIIB students contribute to the curriculum development, study plan for the students of VIIIth class depending and implement the study plan by visiting twice a week to delivering lectures in Computer, English and Mathematics. The contribution is just not limited to teaching; students celebrate Teacher's Day, Christmas, New Year, Children's Day, thus sharing a symbiotic relationship.

In the year 2008, an innovative way of tutoring 11 children of SIC Staff was started. The children ranging from Nursery to XIIth Std were tutored by a group of 23 students in the subjects the parents claim they are weak in. The performance of 11 students with whom the initiative took was monitored and has shown considerable improvement in their learning capacities.

6 Conclusion: Sustainability as a Foundation for Change at SIU

The importance of environmental sustainability as the future direction for societal action is perhaps the need of the hour. This is particularly relevant in the context of some of the most challenging and complex global environmental issues the world is facing. Our current understanding of some of the key issues of impact of climate change (Pachauri and Reisinger 2007) only seems to suggest that there is very little time for the world at taking action in mitigating the effects of greenhouse gas emissions which are like to increase in an exponential way in a business as usual scenario (Meinhausen et al. 2009). The rising demand for energy and its consumption in order to achieve higher economic growth is cited as a key driver of higher GHG emission rates. The increasing urbanisation rate and local environmental stresses from over population, industrial development, migration, etc. (Mukhopadhyay and Revi 2009) are an indicator of some of the imbalances humans are likely to face. This could mean strategic involvement of not only country governments but also business and industry, academic institutions and civil society in alliance that will help to build a sustainable and low carbon economy.

The example of Symbiosis International University recounts the initiatives focussed on development of UG and PG curriculum (policy) related to sustainable development and pilot scale assessment of the operational aspects (practices) of sustainability in the various campuses. Taking forward the motto of '*Vasudhaiva Kutumbakam*' (world as one family), the University as its first step through the exercise of curriculum development tried to integrate all the constituent institutes that otherwise are imparting education in diverse courses and disciplines. The faculty members along with other academicians and subject experts, practitioners after a long process of deliberations and brainstorming reviewed, modified, standardized and optimized the sustainability related courses. This exercise helped in removing the redundant elements thereby enhancing the quality and need based substance. The sustainability related courses introduced at SIU are stand alone courses and were envisaged to integrate critical issues linking energy, environment and infrastructure sectors with social development and equity. However, the question remains whether such stand alone courses can make a difference to sustainability education across a large University like Symbiosis. At the present moment only a handful of student community take these sustainability course offerings each year as this is not seen as a main stream in the larger scheme of things. According to Hegarty et al. (2011) such stand-alone courses can lead the way for developing sustainability action model across campuses. Future directions must include as series of faculty development programmes which can target non mainstream disciplines towards the realisation of sustainability as a key and integral aspect of education.

On the operations front (practices), pilot assessments of the resources used and wasted were examined on various campuses of the university. For this project, students were involved in large number which sensitized and provided them with the requisite first hand practical experience. In the era of climate change, it is imperative for all the organisations, companies, educational institutes and universities to measure the efficacy of the resource use in terms of water use and recycling, food wastage and energy use and conservation. While there are no shortages of these resources on any of the campuses of the University at this moment, the authors believe that with the increase in population, urbanisation, industrial development, and other human activities, it is certain that there will be huge resource crunch of these basic resources in an emerging economy like India in the years to come. In order to tackle these problems and prepare for the future, the authors suggest that HEI like Symbiosis International University will certainly need to take a "Leadership Role" by imparting and practicing sustainability education that will integrate the three dimensions viz. environment, economic and social in a true sense. The example of Symbiosis International University of visioning campus greening curriculum as well as implementing operational aspects of sustainability at various campuses requires in depth imagination and an ability to imbibe market based changes since sustainability policies are seen as very dynamic and linked clearly to societal needs and pressures.

The importance of environment as a compulsory element of primary and secondary school curriculum has already been implemented by the University Grants Commission, Government of India. It is imperative to build and strengthen interdisciplinary sustainability based curriculum across HEI. The authors suggest that this can be achieved through a clear vision and developing education frameworks and policy guidelines across the academic institutions and inclusion of new governance approaches.

References

AASHE (2012) Sustainability tracking, assessment and rating system (STARS) version 1.2 technical manual. Association for the advancement of sustainability in higher education, p 300, http://www.stars.aashe.org. Accessed 24 Mar 2014

Adams C, Heijltjes MG, Jack G, Marjoribanks T, Powell M (2011) The development of leaders able to respond to climate change and sustainability challenges: the role of business schools. Sustain Acc, Manag Policy J 2(1):165–171

Areendran G, Rao P, Raj K, Mazumdar S, Puri K (2013) Land-use/Land-cover change dynamic analysis in mining areas of Singrauli district in Madhya Pradesh, India. Trop Ecol 54(2):239–250

Bhat V, Rao P, Patil Y (2012) Development of an integrated model to recover precious metals from electronic scrap: a novel strategy for e-waste management. Procedia: Soc Behav Sci 37:397–406

Bonnett M (2002) Education for sustainability as a frame of mind. Environ Educ Res 8:9–20

Cayuela A, Robinson JB, Campbell A, Coops N, Munro A (2013) Integration of operational and academic efforts in sustainability at the University of British Columbia. In: Caeiro S, Leal Filho W, Jabbour C, Azeiteiro UM (eds) Sustainability assessment tools in higher education institutions—mapping trends and good practices around the world. Springer International Publishing, Switzerland, Part IV, Switzerland pp 223–236

Dempsey N, Bramley G, Power S, Brown C (2009) The social dimension of sustainable development: defining urban social sustainability. Sustainable development. http://www3.interscience.wiley.com/journal/122406493/abstract. Accessed 24 March 2014

Ehrenfeld J, Gertler N (1997) Industrial ecology in practice—the evolution of interdependence at Kalundborg. J Ind Ecol 1(1):67–79

Enderle G (1997) A worldwide survey of business ethics in the 1990s. J Bus Ethics 16:1475–1483

Finnveden G, Hauschild M, Ekvall T, Guinee J, Heijungs R, Hellweg S, Koehler A, Pennington D, Suh S (2009) Recent developments in life cycle assessments. J Environ Manag 91(1):1–21

Haigh M (2010) Greening the university curriculum: appraising an international movement. In: Chalkley B, Haigh M, Higgitt D (eds) Education for sustainable development. Routledge, London, p 320

Hegarty K, Thomas I, Kriewaldt C, Holdsworth S, Bekessy S (2011) Insights into the value of a 'stand-alone' course for sustainability education. Environ Educ Res 17(4):451–469

ICLEI, South Asia (2009) Energy and carbon emissions profiles of 54 South Asian cities. Report for the British High Commission, p 82

Itankar N, Bhat V, Chourey J, Barve K, Kulkarni S, Rao P, Patil Y (2013) Resource recovery from wastes containing valuable metal species using low-cost biomaterials—an approach of passive bioremediation and its newer applications. In: Patil YB, Rao P (eds) Applied Bioremediation—active and passive approaches. Intech Open Science Publishers, Croatia, pp 255–283

Leal Filho W (2000) Dealing with misconceptions on the concept of sustainability. Int J Sustain High Educ 191:9–19

Leal Filho W (2011) About the role of universities and their contribution to sustainable development. High Educ Policy 24:427–438

Leal Filho W (2012) (ed) Sustainable development at universities: new horizons series: Umweltbildung, Umweltkommunikation und Nachhaltigkeit/Environmental Education, Communication and Sustainability, vol 34. Peter Lang Scientific Publishers, Frankfurt, pp 994

Living Planet Report (2012) WWF International Publication

Lozano R (2010) Diffusion of sustainable development in universities curricula: an empirical example from Cardiff University. J Clean Prod 14(9–11):963–972

Lozano R (2006) A tool for graphical assessment of sustainability in universities (GASU). J Clean Prod 14(9–11):963–972

Madeira AC, Carravilla MA, Oliveira JF, Costa CAV (2011) A methodology for sustainability evaluation and reporting in higher education institutions. High Educ Policy 24:459–479

Malovics G, Csignene NN, Kraus S (2008) The role of corporate social responsibility in strong sustainability. J Socio-Econ 37:907–918

Matten D, Moon J (2004) Corporate social responsibility education in Europe. J Bus Ethics 54:323–337

Meinhausen M, Meinhausen N, Hare W, Raper SCB, Frieler K, Knutti R, Frame DJ, Allen MR (2009) Green house gas emission targets for limiting global warming to 2 °C. Nature 458:1158–1163

Mitchell R (2011) Sustaining change on a Canadian campus: preparing Brock University for a sustainability audit. Int J Sustain High Educ 12(1):7–21

Mukhopadhyay P, Revi A (2009) Keeping India's economic engine going: climate change and the urbanisation question. Econ Polit Wkly 44(31):59–70

NCSE (2003) Recommendations for education for a sustainable and secure future. In: Blockstein DE, Greene J (eds) National council for science and the environment. Washington, DC

Pachauri RK, Reisinger A (2007) Contribution of working groups I, II and III to the fourth assessment report of the intergovernmental panel on climate change. Core Writing Team. IPCC, Geneva, pp 104

Patil Yogesh B (2012) Development of an innovative low-cost industrial waste treatment technology for resource conservation—a case study with gold-cyanide emanated from SMEs. Procedia: Soc Behav Sci 37:379–388

Rao P, Patil Y, Gupte R (2013) Education for sustainable development: trends in Indian business schools and universities in a post liberalization era. In: Caeiro S, Leal Filho W, Jabbour C, Azeiteiro UM (eds) Sustainability assessment tools in higher education institutions—mapping trends and good practices around the world. Springer International Publishing, Germany, Part IV, pp 417–432

Rao P (2011) Integrating sustainability into global business practices—an emerging tool for management education. In: Gupte R, Venkataramani B, Gupta D (eds) Internationalisation of higher education. Excel Book Publishers, New Delhi, pp 60–73

Romero MJR (1995) The role of the university in sustainable development: challenges and opportunities. High Educ Policy 8(4):26–29

Roorda, N (2001) AISHE: Auditing instrument for sustainable higher education. Dutch Committee for Sustainable Higher Education. www.eauc.org.uk/file_uploads/aishe-book1_5.pdf. Accessed 2 Apr 2014

Roy R, Potter S, Yarrow K (2008) Designing low carbon higher education systems: environmental impacts of campus and distance learning systems. Int J Sustain High Educ 9(2):116–130
Saini S, Rao P, Patil Y (2012) City based analysis of MSW to energy generation in India, calculation of state-wise potential and tariff comparison with EU. Procedia: Soc Behav Sci 37:407–416
Scheunpflug A, Asbrand B (2006) Global education and education for sustainability. Environ Educ Res 12:33–46
Scott B (2010) A gate-to-gate life-cycle inventory of solid hardwood flooring in the EasternUS. Wood and Fiber Science, Society of Wood Science and Technology. The Sheridan press, Hanover, Pennsylvania, USA
Speller, P (ed) (1992) Building partnerships for sustainable development: Federal University of Mato Grosso. High Educ Policy 5(1):31–32
Spranger S (2011) Calculating the carbon footprint of universities. Unpublished Master's thesis in economics and informatics; Erasmus School of Economics, p 107
Springett D (2005) Education for sustainability in the business studies curriculum: a call for acritical agenda. Bus Strategy Environ 14:146–159
Sterling S (2001) Sustainable education: re-visioning learning and change. The Schumacher Society, Schumacher Briefing No. 6. Green Books, Dartington, p 94
Stubbs W, Cocklin C (2008) Teaching sustainability to business students: shifting mindsets. Int J Sustain High Educ 9(3):206–221
Van Kleef JAG, Roome NJ (2007) Development capabilities and competence for sustainable business management as innovation: a research agenda. J Clean Prod 15:38–51
Velazquez L, Nora M, Margarita S (2009) Deterring sustainability in higher education institutions —an appraisal of the factors which influence sustainability in higher education institutions. Int J Sustain High Educ 6(4):383–391
Waheed B, Khan F, Veitch B, Hawboldt K (2011) An integrated decision making framework for sustainability assessment: a case study of Memorial University. High Educ Policy 24:481–498
World Commission on Environment and Development (WCED) (1987) Our common future. Oxford University Press, Oxford

Author Biographies

Dr. Prakash Rao is an Associate Professor and Head—Energy and Environment at Symbiosis Institute of International Business, Symbiosis International University, Pune, India.

Dr. Yogesh Patil is an Associate Professor and Head—Research Publications at Symbiosis Institute of Research and Innovation, Symbiosis International University, Pune, India.

Ms. Viraja Bhat is an Assistant Professor at Symbiosis Institute of International Business, Symbiosis International University, Pune, India.

Ms. Shilpa Kulkarni is an Assistant Professor at Symbiosis Institute of International Business, Symbiosis International University, Pune, India.

Problem-Based Learning in Developing a MSc Curriculum on Sustainable Development

N. Tikhomirova, S. Zenchanka and S. Malchenka

Abstract

The transition to Sustainable Development demands the training of specialists knowledgeable in different areas of sustainability. Education for Sustainable Development is one of the driving forces for change. That is why the United Nations General Assembly in its Resolution 57/254 declared a Decade of Education for Sustainable Development (2005–2014). There are two approaches to curricular development. The first approach is a subject-based learning (SBL) connected with a traditional type of curriculum as a set of different disciplines combined by interdisciplinary links. The second approach is problem-based learning (PBL), each problem being discussed from different points of view. The paper describes the approach of MESI to the development of the second type of programme. It entails a problem-based approach and includes global issues of sustainability. PBL approach implies the analysis of the main problems of Sustainable Development such as "Resource Efficiency", "Disasters and Conflicts", "Climate Change", "Poverty", "Economics of Sustainability" and some others from the interdisciplinary and knowledge management points of

N. Tikhomirova
Moscow State University of Economics, Statistics and Informatics,
7, Nezhinskaya Str., Moscow 119501, Russia
e-mail: office@rector.mesi.ru

S. Zenchanka (✉) · S. Malchenka
Minsk Branch of Moscow State University of Economics,
Statistics and Informatics, 127, Build. 2, Mayakovskogo Str.,
220028 Minsk, Belarus
e-mail: szenchenko@mesi.ru

S. Malchenka
e-mail: smalchenko@mesi.ru

W. Leal Filho (ed.), *Transformative Approaches to Sustainable Development at Universities*, World Sustainability Series, DOI 10.1007/978-3-319-08837-2_32

view. This approach is in line with practice-oriented learning. All courses include environmental, economic and social aspects which enable a wide assessment of different problems of sustainability.

Keywords
Sustainable development • Education • Master programme • Problem based learning

1 Introduction

Sustainable Development is a rather new area of knowledge. Typically new areas of science and their learning are developed as subject areas, but sustainable development is an integrated area of knowledge, which includes economic, environmental and social aspects of development, demands a different approach.

The education process at classic universities and applied universities was usually constructed on subject principle, with a division into more narrow areas as specializations (e.g.: Physics: optics, thermodynamics; Biology: microbiology, zoology, ecology, etc.).

For teaching matters related to sustainable development it is necessary to use different approaches based on the synthesis of knowledge from different areas giving synergetic effect. Since Master programmes are usually directed towards specialists training in some specific area of knowledge for sustainable development, it is necessary to develop completely new programmes permitting to train specialists having synergetic skills in all areas of sustainable development.

The aims of this paper are:

- To compare different approaches to the learning process;
- To consider the current state of working-out MS programmes for Sustainable Development in Russia;
- To present a new approach in the development of MS programme based on PBL and interdisciplinary approaches.

2 Education for Sustainable Development

The definition of sustainable development was suggested in the 1987 in Brundtland Commission Report as "the development that meets the needs of the present without compromising the ability of future generations to meet their own needs" (WCED 1987).

There are other definitions of Sustainable Development depending on the ways and places where it used (Leal Filho and Manolas, 2012; Van Dam-Mieras et al. 2008).

The concept of Sustainable Development was adopted at the Conference in Rio (Agenda 21 Agenda 1992). Agenda 21 considered all aspects of Sustainable Development and paid special attention at Education for Sustainable Development (ESD). In particular Article 36.3 defines:

> Education, including formal education, public awareness and training should be recognized as a process by which human beings and societies can reach their fullest potential. Education is critical for promoting sustainable development and improving the capacity of the people to address environment and development issues.

World Conference on Education for Sustainable Development recognized Education for Sustainable Development "as one of the key drivers for moving society in the direction of sustainable development. ESD connects cultures and, through that, countries. It has the capacity to bring a rights-based and global justice perspective to development issues for inclusive societies" (UNESCO 2009).

Understanding the importance of ESD, the United Nations declared the Decade of Education for Sustainable Development (2005–2014) (UN 2002a, b).

Many authors consider ESD as essential step for leaders' training for serious transformations (Arima 2009; Shriberg 2012; GPSS-GLI 2013). The aim of Joint Master's programme in Sustainable Development (Course Catalog 2012–2013) was defined as education of scientists who will be able to make a substantial contribution to the transition to a sustainable society through their scientific research and their skills in the area of societal interventions.

To promote the ideas of ESD the Regional Centers of Expertise (RCE) on Education for Sustainable Development were created over the world. In preface to Special issues of International Journal of Sustainability in Higher Education devoted to Regional Centers of Expertise (RCE) on Education for Sustainable Development Van Dam-Mieras et al. (2008) wrote that "RCE develop their activity at regional level, the level of human daily life in a social and cultural context, but also take into account pressing global problems such as climate change, renewable energy, and all those most important issues contained in the Millenium Development Goals". Starting from 7 RCE on 2005 there are 117 acknowledged RCEs worldwide now (http://www.ias.unu.edu/sub_page.aspx?catID=108&ddlID=1847).

To reach a success the ESD should use different teaching approach to learning such as subject-based learning, interdisciplinary approach and problem-based learning.

3 Subject-Based Learning Approach

Traditional teaching approach to the specialists training is disciplinary or subject-based learning. As mentioned in "A policy response from the IoI Education Forum" (Forum 2012), subjects give stability to a curriculum and provide the boundaries within which teachers establish their professional identities and pupils develop their identities as learners. Hence this approach supposes clear and stable boundaries between different subjects and is based on leading role of teacher who tells what students need to know. The role of students is to memorize lectures and study the subject according to the discipline curricular.

Subject-oriented learning is mainly used now to train specialists in natural sciences such as physics or chemistry and in mathematics. This approach closely interacts with interdisciplinary approach, that is, during a lecture on some areas of mathematics the examples from physics are given and this corresponds to the statement of Spence (2002) about lecture illustration: "the problem assigned to illustrate how to use it".

4 Interdisciplinary Approach

The interdisciplinary (transdisciplinary) approach to education combines more than one discipline and uses links between different subjects or disciplines to improve learning. As our world is interdisciplinary this approach enables students to see all subjects as part of their lives, rather than as separate disciplines.

Considering the advantages and disadvantages of interdisciplinary teaching Jones (2009) wrote that it allows students to see different perspectives, work in groups, and make the synthesizing of disciplines the ultimate goal. As a result this approach helps students and their teachers to develop critical thinking, communication, creativity, pedagogy, and other competences.

Clark and Button (2011) suggested "sustainability transdisciplinary education model (STEM)", a contemporary approach linking art, science, and community that provide university students and society at large shared learning opportunity. As a result the convergence of science, art, and aesthetics enabled the participants to develop a deeper spiritual awareness and understanding of eco-justice for the promotion of a sustainable society.

Considering different aspects of interdisciplinary learning (IDL) and its link with ESD McEwen (2011) found that "students recognized some learning outcomes associated with IDL in ESD more strongly than others. For example: 60.7 % students perceived that their worldview had changed through their study on the course; 70.5 % students perceived that they were developing skills in systems thinking through their course; 90.2 % perceived that they had developed a capacity to engage in both subjective and objective thinking".

These examples show that the training of specialists in sustainable development demands implementation of interdisciplinary approach.

5 Problem-Based Learning Approach

According to Kates et al. (2001) the sustainability field differs considerably in term of structure, methods and content from science as commonly understood by the scientific community therefore sustainable development requires changes in the approach to learning.

In the middle of the last century Problem-based learning (PBL) was developed for medical education (Barrows and Tamblyn 1980) and is now applied in different areas (Aydoğdu 2012; Du et al. 2009; Planet 2001; Grace 2000).

As can be seen from various publications, the main difference of PBL from SBL is that students develop many new skills of learning in teams and solving some professional problems. PBL is a technique that permits students to use an active, task-oriented, and self-directed approach to their own learning.

Savin-Baden (2000, 2007) proposed different models of PBL covering the objective of PBL and involving the perception of knowledge, learning, problems, students, the teacher roles, and the assessment. These models are: (1) attainment of knowledge, (2) PBL for professional work, (3) PBL for interdisciplinary understanding, (4) PBL for cross-discipline learning and (5) PBL for critical competencies.

Savery (2006) defined PBL as an instructional (and curricula) learner-centered approach that empowers learners to conduct research, to integrate theory and practice, and to apply knowledge and skills for development a viable solution to a defined problem.

There are four basic steps in problem solving:

1. Defining the problem;
2. Generating alternatives;
3. Evaluating and selecting alternatives;
4. Implementing solutions.

Students are largely responsible for the implementation of the second and third steps. As a result the PBL method requires students to become more responsible for their own learning.

Besides that the role a PBL teacher is changed. He/she becomes a facilitator of students learning. Facilitators make key aspects of expertise visible through questions that students are learning through modeling, coaching, and eventually feeding back some of their support. In PBL the facilitator is an expert teacher, able to model good strategies for learning and thinking, rather than providing expertise in specific content (Hmelo-Silver and Barrows 2006). At the same time "a majority of faculty members in sustainability programs experience the paradoxical situation of being responsible for training students in areas in which they themselves have not trained" (Brundiers et al. 2010).

Sustainable development is connected with solving of such problems as Climate Change, Sustainable Governance, Resource Depletion, etc. Vilches et al. (2012) wrote that "the obstacle for incorporating the state of the world into the curricula, teaching practice and research lies in the lack of tradition in education when it comes to approaching these types of global problems, which demand systemic treatment. This is a serious obstacle, because, although each problem has a particular importance and deserves individual attention, none of them can be understood or treated without taking into account the others". Hence they wrote that "an explicit demand to analyze the Earth's situation on a global scale allows us to understand the close connection between the different problems, as well as their local and global repercussions. Consequently, we need to keep in mind the necessity of holistic approach if we want to avoid the "natural" tendency towards local and isolated treatment as well as causal reductionism".

6 Review of Master's Programs in Sustainable Development

The best 100 Master programmes in Environmental Management and Sustainable Development (Ranking 2012–2013) are presented at site http://www.best-masters.com/ranking-master-sustainable-development-and-environmental-management.html. These programmes refer to environmental management, resource management, energy, corporate responsibility, law, etc. Approximately 20 % of Master courses are directed to training in sustainable development.

MSc programme in Sustainability Science and Policy in Maastricht University uses PBL approach (www.icis.unimaas.info).

Japanese universities are not presented in this ranking. At the same time the University of Tokyo presented Graduate Programme in Sustainability Science "Global Leadership Initiative" providing the necessary training that enables future global leaders to make a profound impact on sustainability science and sustainable development (GPSS-GLI 2013; Onuki and Mino 2009). Research Institute for Sustainability Science in Osaka University introduced the educational programme where students learn Sustainability Science by interacting with different academic and cultural backgrounds (Uwasu et al. 2009).

The special issue of International Journal of Sustainability in Higher Education was devoted to different aspects of ESD. Considering the state of ESD in the Asia-Pacific region Ryan et al. (2010) noted that this region "offers many creative initiatives and shows considerable progress in ESD and in understanding the learning dimensions of sustainability. At the same time, it mirrors global trends in that further work is needed to promote systemic change in educational arenas, particularly in terms of strategic integration within HE institutions. The Asia-Pacific contributions to this collection demonstrate the need to harness national policy, to develop local and regional initiatives and to work effectively towards more profound change in HE curricula and through collaboration with external communities and stakeholders."

In the South Pacific Islands the concept of ESD is new as was pointed by Corcoran and Koshy (2010), but "the national educational curricula have attempted to address sustainability issues under natural systems, life system, and global system themes. In order to move forward from these, a seamless transition involving the reorientation of the curriculum of formal education and the practices of non-formal and informal educational sectors to promote the lifelong learning will be required".

In India some institutions introduced the courses on Technology and Sustainable Development in their curricula and certain universities have created special programmes such as Master in Sustainable Development (Judavpur University, University of Madras), Master in Public Policy and Sustainable Development and MBA Business Sustainability (TERI University), Leadership Programme on Nutrition Security and Sustainable Development (India Gandhi Open National University) (Chhokar 2010).

Master programme in Environmental Management and Sustainable Development in the UNEP-Tongji Institute of Environment for Sustainable Development (Shanghai) was organized in 2005. One of its aims was to prepare students for leadership role in SD (Niu et al. 2010).

The analysis shows that primarily all these programmes are interdisciplinary but curricula are subject oriented.

7 Development of MSc Curriculum in Russia

Considering the Education for Sustainable Development in Russia Gromoff et al. (2012) claimed the existence of the following problems and trends:

- the average period for innovations formed after getting the degree is 10 years, the average number of students being capable of innovation activity is approximately 10 % from the whole number of students, but after 10 years not more than 1 % of graduates are involved into the innovation practice, others are lost;
- the education system is not aimed at forming the innovation approaches to the problem-solving because of the absence of structured understanding of knowledge creation process;
- the lack of interaction among innovation actors—firms, the public sector, academia, and society—in modern innovation ecosystems.

These problems are the reasons for creation of the master courses directed to introduction of innovation curricula in education process and to training the leaders.

In leader Russian universities such as Moscow State University (http://edu.msu.ru/catalog/department.php?id=8), Sankt Petersburg State University (http://spbu.ru/obrazovatelnye-programmy-spbgu), Siberian Federal University (http://edu.sfu-kras.ru/oop/search/66/14/0/K/0), Baltic Federal University (http://www.kantiana.ru/apprentice/) and many others the Master's programmes in ecology and nature using are developed in accordance with Federal Educational Standards. These programmes include Sustainable Development as essential part. These programmes are subject oriented and use interdisciplinary approach.

During 2008–2013 about 20 Master's programmes in the field of Environmental Management were realized in different Russian universities with supporting of the TEMPUS Programme. These programmes were subject oriented too and consider some areas of Sustainable Development.

This year none of the Russian universities have programmes on Environmental Management and Sustainable Development ranked among the best 4000 Master Programs (Ranking 2012–2013).

As was mentioned above ESD should be oriented on training specialists for solving problems of SD and most suitable learning method for ESD is Problem-based learning.

8 Problem-Based MSc Programmes

The analyses of best Masters programmes in Sustainable Development and Environmental Management (Ranking 2012–2013) shows that in common all these Masters Programmes are subject (professionally) oriented and include sustainability as additional part.

The key question to be raised is the competence for sustainable development. The answer will depend on the goal of Masters programme. In subject base learning the competence should be oriented to some specific area of sustainability; in problem based learning the competences should be wider.

The models described above suppose more and more narrow specialization at every new level of education. But in the conditions of Sustainable Development it is necessary to train specialists being capable of thinking globally and taking decision sat highest level of governance.

Let's consider a model of Master programme in Sustainable Development based on the principles of Agenda 21 and UNEP approaches. One aspects of such approach were considered by Tikhomirova et al. (2012).

The main areas of Sustainable Development were considered at conference in Rio (Agenda 21):

- social and economic dimension;
- conservation and management of resources for development;
- strengthening the role of major groups.

World Summit in Johannesburg (Report 2002) confirmed "that poverty eradication, changing consumption and production patterns and protecting and managing the natural resource base for economic and social development are overarching objectives of essential requirements for sustainable development" and adopted Plan of Implementation of the World Summit on Sustainable Development.

The outcome of the Conference Rio+20 named "The future we want" underlined (Outcome 2012):

> We also reaffirm the need to achieve sustainable development by promoting sustained, inclusive and equitable economic growth, creating greater opportunities for all, reducing inequalities, raising basic standards of living, fostering equitable social development and inclusion, and promoting integrated and sustainable management of natural resources and ecosystems that supports, inter alia, economic, social and human development while facilitating ecosystem conservation, regeneration and restoration and resilience in the face of new and emerging challenges.

UNEP concentrated its efforts on six programmes: “Climate Change”, “Disasters and Conflict”, “Ecosystem Management”, “Environmental governance”, “Harmful substances and Hazardous Waste”, “Resource Efficiency” (http://www.unep.org/) which cover all main areas of Sustainable Development given above.

The suggested model of Master Science education uses UNEP approach.

Each programme is considered from three scientific aspects i.e. economy, environment and society. The first step is to state a problem. This is a task for teacher (facilitator) who must define a problem and present its specific features. Then students work in small teams and define the reasons of problem, interrelations between different scientific aspects and generate alternative solutions (the second step). After work in small teams students discuss their results and evaluate and select the most suitable solutions (the third step). The last stage involves implementation of the developed proposals. Possible results for the problem “Climate Change” are presented in Table 1.

Results presented in Table 1 reflect only visible reasons and interrelations. Actually they are more complex. The reasons for economy component depend on macro- and microeconomic development. Economical interrelations depend on competition, market conditions, national and international legislation and WTO regulations. In environmental component the pollutions can be natural and

Table 1 Steps for problem-based learning climate change

Problem	Climate change		
Components	Economy	Environment	Social aspects
Reasons	Industrial development in the own country, transfer of industry abroad	Natural and anthropogenic pollution, greenhouse gases (GHG)	Industrial development in own country, transfer of industry abroad, state of the environment, poverty
Interrelations	Competition, common market, WTO, legislation	Transboundary and transsphere (atmosphere-hydrosphere, hydrosphere-soil) pollutions, legislation	Population migration, sickness rate of population, legislation
Alternative solutions	Clean production, social responsibility, “green” economy, smart economy	Pollution decrease,“pollutant pays”, environmental management,“green” economy	Social responsibility, “green economy”
Common decision	Smart society, Sustainable development		

anthropogenic generated, it is necessary to consider all their types and their physical condition—physical, chemical and biological. Knowledge of these features and physical, chemical and biological processes in atmosphere, hydrosphere and lithosphere permits to estimate the transboundary interaction of pollutants. Environmental legislation permits to regulate some of transboundary processes. The reasons of social component depend on economic and environment aspects and on poverty that results in power health of population and its migration. The possible decisions for all components of Climate Change problem are approximately the same. The common decisions are smart society and sustainable development. This example shows the complexity of realization of PBL and interdisciplinary approaches to education for sustainable development which supposes involving the teachers from different areas of knowledge cooperating in problem-based learning.

Such model supposes high competence of a teacher (a facilitator). J. Parker (2010) noted: "Although much remains to be done to document un-sustainability, our assessment of positive ways forward and strategies for adaptation, mitigation and restoration of life systems (including human social systems) must be based on more joined-up forms of knowledge". Taking into account the inter- and multidisciplinary approach used in our model several teachers should be involved in problem development.

9 Conclusion

The analysis of different approaches to education such as interdisciplinary learning, subject-based learning and problem-based learning shows that the latter should be used in Education for Sustainable Development.

Suggested model of MSc programme in Sustainable Development aimed attraining of governance specialists and top-managers of companies is suggested for the discussion. The model is based on problem-based learning approach and more close using of interdisciplinary approach when each global problem is discussed. The model includes two sessions under the supervision of a facilitator and the students' sessions. The facilitator is to pose a problem during the first session and to help students to find the common decision among the alternatives. During their sessions the students discuss the problem, find the reasons of the problem and suggest possible decisions.

The main difference of the suggested model from known the models is the analysis of the problems of sustainable development which are defined by UNEP as inter- and cross-disciplinary problems. This model corresponds to the fourth and the fifth models of Savin-Baden (2000, 2007).

The presented example of the analysis of the Climate Change problem shows the complexity of the educational process and the need for new competences of the facilitators.

The suggested model of MSc education is aimed at training the top managers for municipal and regional governance.

References

Agenda 21 (1992) http://sustainabledevelopment.un.org/content/documents/Agenda21.pdf. Access 20 Nov 2013

Arima A (2009) A plea for more education for sustainable development. Sustain Sci 4:3–5

Aydoğdu C (2012) The effect of problem based learning strategy in electrolysis and battery subject teaching Hacettepe Üniversitesi Eğitim Fakültesi Dergisi. H U J Educ 42:48–59

Barrows HS, Tamblyn RM (1980) Problem-based learning, an approach to medical education. Springer, New York

Brundiers K, Wiek A, Redman CL (2010) Real-world learning opportunities in sustainability: from classroom into the real world. Int J Sustain High Educ 11(4):308–324

Chhokar KB (2010) Higher education and curriculum innovation for sustainable development in India. Int J Sustain High Educ 11(2):141–152

Clark B, Button CH (2011) Sustainability transdisciplinary education model: interface of arts, science, and community (STEM). Int J Sustain High Educ 12(1):41–54

Corcoran PB, Koshy KC (2010) The Pacific way: sustainability in higher education and the South Pacific Island nations. Int J Sustain High Educ 11(2):130–140

Course Catalog (2012–2013) International Joint Master's Programme in Sustainable Development. www.jointdegree.eu. Access 20 Nov 2013

Du X, Erik de Graaff E, Kolmos A (eds) (2009) Research on PBL practice in engineering. Education Sense Publisher Rotterdam, Boston, Taipei

Forum (2012) Towards a subject-based curriculum. A policy response from the IoI Education Forum. http://www.instituteofideas.com/transcripts/SubjectCurriculum.pdf. Access 07 Feb 2014

GPSS-GLI (2013) The University of Tokyo graduate program in sustainability science global leadership initiative. http://www.sustainability.k.u-tokyo.ac.jph. Access 20 Nov 2013

Grace O (2000) Is PBL suitable only for the health sciences curricula? CDTL Brief Centre Dev Teach Learn 3(3):4–6

Gromoff A, Stavenko J, Evina K, Kazantsev N (2012) Subject oriented approach to embed an innovation in education process. Far East J Psychol Bus 8(2):53–65

Hmelo-Silver CE, Barrows HS (2006) Goals and strategies of a problem-based learning facilitator. Interdisc J Prob-Based Learn 1(1):21–39

Jones C (2009) Interdisciplinary approach—advantages, disadvantages, and the future benefits of interdisciplinary studies. ESSAI: Vol 7, Article 26. http://dc.cod.edu/essai/vol7/iss1/26/

Kates RW, Clark WC, Corell R, Hall JM, Jaeger CC, Lowe I, McCarthy JJ, Schellnhuber JH, Bolin B, Dickson NM, Faucheux S, Gallopin GC, Grübler A, Huntley B, Jäger J, Jodha NS, Kasperson RE, Mabogunje A, Matson P, Mooney H, Moore B III, O'Riordan T, Svedin U (2001) Sustainability science. Science 292(5517):641–642

Leal Filho W, Manolas E (2012) Implementation sustainable development in higher education. In: Fernando G, Ruth P, Walter LF, Ulisses MA (eds) Contribution to the UN decade of education for sustainable development, Peter Lang GmbH, Frankfurt am Main, Berlin, Bern, Brussels, New York, Oxford, Vienna

McEwen L (2011) Interdisciplinary learning in ESD at taught postgraduate level: research-informed capacity building for curriculum development. http://www.heacademy.ac.uk/assets/documents/sustainability/glos_interdisc_final.pdf. Access 20 Nov 2013

Niu D, Jiang D, Li F (2010) Higher education for sustainable development in China. Int J Sustain High Educ 11(2):153–162

Onuki M, Mino T (2009) Sustainability education and a new master's degree, the master of sustainability science: the graduate program in sustainability science (GPSS) at the University of Tokyo. Sustain Sci 4:55–59

Outcome of the Conference Rio+20 (2012) The future we want. Access 20 Nov 2013

Parker J (2010) Competencies for interdisciplinary in higher education. Int J Sustain High Educ 11 (4):325–338

Planet (2001) Special edition two Case studies in problem based learning (pbl) from geography, earth and environmental sciences. http://www.gees.ac.uk. Access 20 Nov 2013

Ranking (2012–2013) The best 100 Master programmes in environmental management and sustainable development. http://www.best-masters.com/ranking-master-sustainable-development-and-environmental-management.html. Access 20 Nov 2013

Ryan A, Tilbury D, Corcoran PB, Abe O, Nomura K (2010) Sustainability in higher education in the Asia-Pacific: developments, challenges, and prospects. Int J Sustain High Educ 11(2):106–119

Savery JR (2006) Overview of problem-based learning: definitions and distinction. Interdisc J Prob-Based Learn 1(1):9–20

Savin-Baden M (2000) Problem-based learning in higher education: untold stories. Open University Press/SRHE, Buckingham

Savin-Baden M (2007) Second life PBL: liminality, liquidity and lurking keynote speech. In: Reinventing Problem-based learning conference republic polytechnic, Singapore 7–9th March 2007

Shriberg M (2012) Building sustainability leaders: a framework to prepare students to thrive on complexity and lead transformative changes. In: Leal Filho W (ed) Sustainable development at universities: new horizons, Peter Lang GmbH, Frankfurt am Main, Berlin, Bern, Brussels, New York, Oxford, Vienna

Spence LD (2002) Problem based learning: lead to learn, learn to lead. School of Information Science and Technology. http://n-imagecache.aldenhosting.com/~n1studyg//pblhandbook.pdf. Access 20 Nov 2013

Tikhomirova N, Malchenka S, Zenchanka S (2012) Sustainable development of Universities and the University of Sustainable development: experience from Russia. In: Leal Filho W (ed) Sustainable development at Universities: new horizons. Peter Lang GmbH, Frankfurt am Main, Berlin, Bern, Brussels, New York, Oxford, Vienna

UN (2002a) Report of the world summit on sustainable development. http://www.un.org/jsummit/html/documents/summit_docs/131302_wssd_report_reissued.pdf. Access 20 Nov 2013

UN (2002b) UN Resolution 57/254. United Nations decade of education for sustainable development. http://www.un-documents.net/a57r254.htm. Access 20 Nov 2013

UNESCO (2009) World Conference on education for sustainable development 31 March–2 April 2009 Bonn, Germany. Proceeding. UNESCO/BMBF/German Commission for UNESCO

Uwasu M, Yabar H, Hara K, Shimoda Y, Saijo T (2009) Educational initiative of Osaka University in sustainability science: mobilizing science and technology towards sustainability. Sustain Sci 4:45–53

Van Dam-Mieras MCE (Rietje), Fadeeva Z, Mochizuki Y (2008) Preface. Int J Sustain High Educ 9(4):367–368

Vilches A, Marques L, Gil-Perez D, Praia J (2012) The need for contributions to the decade of education for a sustainable future: an ethical commitment. In: Fernando G, Ruth P, Walter Leal F, Ulisses Miranda A (eds) Contribution to the UN decade of education for sustainable development, Peter Lang GmbH, Frankfurt am Main, Berlin, Bern, Brussels, New York, Oxford, Vienna

World Commission on Environment and Development (1987) Our Common Future. http://www.un-documents.net/our-common-future.pdf. Access 20 Nov 2013

Author Biographies

Natalia Tikhomirova (Dr. Sc. and Professor in Economic Sciences) is Rector of Moscow State University of Economics, Statistics and Informatics (MESI). Her scientific interests lie in the areas of knowledge management, quality management and innovation management. She is an expert in e-learning. As a rector, Natalia Tihomirova conducts a large amount of administrative work.

Siarhei Zenchanka (Professor) is Head of the Department of International Relations and Marketing at the Minsk Branch of MESI, and has a PhD in Applied Science. His research focus is in the areas of sustainable development, quality management and environmental management, where he is an auditor.

Siarhei Malchenka (Professor) is Director at the Minsk Branch of MESI, and holds a PhD in Chemistry. His research areas are knowledge management, innovation management, environmental protection and life safety. As a director, he undertakes a large amount of administrative work.

ECOPUMA, the Strategy for a Sustainable University at UNAM

Mireya Imaz, Dalia Elizabeth Ayala, Luis Gutiérrez and Marjory González

Abstract

The National Autonomous University of Mexico (UNAM) is an important cultural project of Mexico, being the main educator of Mexican professionals, and a national referral for public policies on sustainability. The university has 330,382 students, and produces 35.2 % of the scientific research in Mexico, as well as maintaining installations in Spain, US, Canada and China. According to Webometrics UNAM is ranked 36 globally, and is ranked at 72. In order to reduce the environmental impact of the university's fundamental activities of teaching, research, and cultural diffusion, as well as to analyze and update curricula, UNAM developed the Strategy for a Sustainable University (Eco-PUMA). The strategy includes eight axes: energy, water, waste management, responsible consumption, green areas, mobility, sustainable construction, and electronic administration. Furthermore, as a measure of the progress in sustainability implementation in all campus, UNAM developed a certificate on environmental performance, which is based on international standards and is becoming a national reference. This paper presents results from the first years of EcoPUMA.

Keywords

UNAM · Ecopuma · Sustainability

M. Imaz (✉) · D.E. Ayala · L. Gutiérrez · M. González
Programa Universitario de Medio Ambiente, Universidad Nacional Autónoma de México, Edificio de Programas Universitarios, Circuito de La Investigación Científica, Ciudad Universitaria, CP 04510, DF, Mexico, Mexico
e-mail: puma@unam.mx

W. Leal Filho (ed.), *Transformative Approaches to Sustainable Development at Universities*, World Sustainability Series, DOI 10.1007/978-3-319-08837-2_33

1 Introduction

As mentioned by Burnside et al. (2012), humanity has a dual nature: we depend on the same laws and forces of nature as the rest of the species on the planet, but we take two thirds of the gross primary production of the world, as well as having large variations in energy consumption, huge cultural diversity, and an apparent social organization. In this context, how can we face the challenges of linking development with the natural limits of the planet?

Current patterns of consumption and resource exploitation create severe environmental impacts that affect us all, and threaten the existence of human beings as well as the rest of species of the planet. Among these we could mention biodiversity and soil loss, contamination of water, soil and air, generation of great amounts of waste, and increasing concentration of greenhouse gases with the resulting disruption of global climatic patterns.

We are on an unsustainable path that imposes complex challenges and requires mobilization of resources from government, academics, and civil society. Sustainable development represents an alternative for building integral solutions to environmental problems, and a way for nations and persons to join together in the achievement of a dignified life for all.

Higher Education Institutions must assume leadership in forming professionals capable of developing innovative and interdisciplinary strategies to attend the needs of humanity within the limits of the ecological systems of the Earth. Universities provide opportunities for professional development, research, and cultural diffusion, so they have an enormous responsibility in the construction of new paradigms.

In this context, the National Autonomous University of Mexico (UNAM), the most important university in Latin America, has a social responsibility as the main educator of professionals in Mexico and is a national referral body for sustainability policies. Hence, the Strategy for a Sustainable University (EcoPUMA) was developed by UNAM to reduce its environmental impact, given the complexity of operation of the university, which includes a large number of stake holders who actively participate in the development of proposals to achieve a more sustainable university.

To date the EcoPUMA strategy is the largest and most important effort of a public university to move towards a more sustainable operation, and to construct new and different ways of education and research that help to reduce environmental and social problems in Mexico. Another innovation is that this strategy puts sustainability at the core of operations and not just as an addendum.

2 Strategy for a Sustainable University of the UNAM

The National Autonomous University of Mexico (UNAM) is an important cultural project in Mexico, providing education for 330,382 students from high school to postgraduate levels, and producing 35.2 % of the scientific research in Mexico.

The university has installations in 24 states in the country, as well as in Spain, USA, Canada, and China (UNAM 2014). UNAM is ranked 36 among global universities by Webometrics, and in UNAM is ranked 17 among Specialized Higher Education Institutions, 35 among urban campus universities, and 72 in the general ranking (Greenmetrics 2014).

UNAM maintains the National Seismological Service, National Astronomy Observatory, National Botanical Garden, the National Library, National Mareographic Network, National Herbarium, the monitoring program of the Popocatepetl Volcano, and three ecological reserves, one of them within the main campus, something unique in the world. The university's budget for 2014 was around $3.5 billion US dollars (UNAM 2014).

UNAM has incorporated the environmental dimension in education, research and cultural diffusion through an important number of publications, new courses, workshops, conferences and technical certificates. The university is also making a great effort to create new degree courses with a main axis of environmental responsibility and sustainability, which respond to the need for interdisciplinary studies to deal with the complexity of socio-environmental systems.

UNAM, as a public institution, has a social responsibility as the main educator of Mexican professionals, and has an important national role in reviewing public policies for sustainability. The Strategy for a Sustainable University (EcoPUMA) was developed within this framework, and aims to create a new culture of sustainability in teaching and research activities, while reducing its environmental impact.

EcoPUMA promotes the confluence and multiplication of efforts undertaken by the UNAM community, with the conviction that the future urges us to become active agents in the construction of a new development paradigm that is socially equitable, economically viable and environmentally responsible.

The strategy is articulated around everyday operation of the UNAM and has eight action and evaluation axis: green space, mobility, energy, water, waste, responsible consumption, sustainable construction, and electronic administration.

Over the last decades, UNAM has made efforts to reduce its environmental impact and to use resources more efficiently, although an integral strategy for the coordination, promotion and monitoring of these actions was lacking. The EcoPUMA strategy seeks to coordinate efforts, and guarantee that these are undertaken with adequate and permanent evaluation tools. Here we present the advances of the EcoPUMA Strategy for a Sustainable University.

3 Advances in the Eight Axis of EcoPUMA

The *green spaces axis* of EcoPUMA aims to design and compose green spaces within the different university campiin accordance with local water availability, and type of soil and terrain, as well as to promote the use of native plant species of the distinct ecosystems where university infrastructure is located.

Among the actions for green spaces is the installation of green roofs that work as bioclimatic systems, and serve as spaces for research, teaching and environmental education. A green roof was installed on the Scientific Research Coordination building, where research projects are being conducted on the role of green roofs in carbon capture, carbon dioxide-oxygen production, mitigation of urban heat, and local climate regulation, as well as studies on dry, freezing and diseases resistance, capacity of plants to capture heavy metals including lead, and the absorption of atmospheric toxic compounds by edible plants.

The main UNAM campus is located within a unique ecosystem known as the "Pedregal", which originated after the eruption of the Xitle volcano more than 1500 years ago. The vegetation of this ecosystem is comprised mainly of shrubs with a high tolerance to drought, and has a great diversity of arthropods, birds, reptiles, and small mammals. In the mid-80's, the Pedregal de San Ángel Reserve was created, representing an important step in actions for sustainability by UNAM as a third of the main university campus protects native biodiversity. This makes UNAM the only university in the world that has a reserve within a campus in an urban area.

The *mobility axis* of EcoPUMA seeks to improve mobility of people and vehicles within and away from the university, favouring and promoting the use of non-motorized vehicles, and providing disincentives for the use of cars. To cope with the challenges of mobility in the main campus UNAM developed two programs: Bicipuma and Pumabus.

Bicipuma is a program based on the loan of bicycles for travel between different facilities in the main campus. The system includes a cycle-path of almost 6 km long, and the program has 950 bicycles, making 4,000 loans per day, and 582,000 trips each year. This system has promoted the use of alternative ways of transport in UNAM, and become a model for non-motorized mobility in other cities of the country.

The other program is Pumabus, which is a free bus transport system for safe and efficient transportation within the main UNAM campus that aims to reduce the use of private cars. This system includes 12 routes and 60 buses that travel along an exclusive bus lane within the campus. Pumabus also connects to two subway and two metro bus stations near to the campus enabling the displacement of members of the university community. There are also several routes of low or no-charge buses to take members of the university community from the subway stations, metro bus and other bus stations, to the UNAM main campus, and to other academic and administrative facilities.

These systems represent alternative mobility options that promote civic culture around pedestrians, the use of bicycles and public transport, and reinforce the sense of community and belonging among university members. At the same time, there is a reduction in the amount of atmospheric pollutants emitted by UNAM through its daily operation.

Another project in this axis is the Science Walk, which seeks to modify the infrastructure of the second most important access for pedestrian and cyclists in the main campus. This project is based on a participative design for making this path a safe and inclusive space for community integration, incorporating educational and cultural elements with a low environmental impact.

The objective of the *energy axis* is to reduce and optimize energy consumption in UNAM and promote research and implementation of technologies based on renewable sources of energy, that contribute to the reduction of greenhouse gas emissions.

Based on the International Panel of Climate Change (IPCC) protocol, an inventory was made of greenhouse gas emissions associated with energy consumption in the main UNAM campus. The results showed that in 2011, the campus produced 49,580 tons of CO_2 derived from electricity, diesel and gas consumption, which corresponds to 0.1 % of all emission in the Federal District (Escobedo et al. 2014). The study also presented a scenario of increased emissions to 2020, and proposed several mitigation measures, which would also reduce the costs of electricity consumption in the university. This inventory presents a base-line from which the institution can evaluate and quantify the success of actions implemented to reduce the carbon footprint.

Another action was the installation of a new system for heating the university Olympic swimming pool, one of the biggest in Latin America (6 million L and 4,000 m^2), based on solar energy and heat pumps, together with a cover to prevent heat loss at night. The aim of this system is to reduce operating costs, allowing investment in more efficient technologies, especially those based in the use of renewable energy resources. The use of the solar system will substitute up to 46 % of the annual gas consumption, and reduce by 42 % the greenhouse gases emissions each year. The system will be monitored to evaluate its effectiveness, and the possibility of further development and expansion.

Also, the lighting system of the Olympic Stadium was replaced by more efficient systems to save energy by up to 86 %. Energy efficient lighting is also being used for almost 3,000 street lights in the main campus, and other university installations.

Since 2007, UNAM has implemented actions for a more efficient use of water resources. This includes the substitution of WC and washroom facilities for water efficient models, along with the installation of 170 consumption meters in several university buildings. A permanent monitoring program is maintained to detect and repair leaks, and to analyze water quality for human consumption. In coordination with the green spaces axis, exotic plant species have been replaced with native drought-resistant species in gardens and other green areas to reduce water consumption for irrigation. Most of the main campus buildings collect rain water to replenish wells.

In addition, an alternative system for collecting rain water was installed. This system feeds a water dispenser, and serves as a demonstration, teaching and research device, raising awareness of water conservation systems among the university community. The installed system could collect around 104,000 liters of water per year, and could represent a viable option for places with limited access to drinkable water.

The goal of the *waste management axis* is to reduce waste generation and make waste management in UNAM more efficient. This contributes to save energy, prime materials, water and other resources employed in the production and distribution process of items we consume daily, as well as to reduce pollution associated with the fabrication of new objects and final waste deposit.

The operation of UNAM generates 15 tons of waste daily. To cope with the problem of waste in all university campus and facilities, diverse actions have been implemented over the last decades. One of the first was the creation of a compost plant in 1993. Years later organic and inorganic waste separation was implemented, and later this included separation of PET and paper waste. However, in this system less than 20 % of waste is recovered for recycling, which besides the environmental impact, represents great economic costs for waste disposal in landfills.

As part the activities of this axis, a diagnosis was carried out of the generation and management of solid urban waste in the university. The study showed that 50 % of all waste could be sent to recycling chains, thereby reducing from 84 to 32 % the amount of waste sent to landfills. This study was the basis for the design of a new program for the comprehensive management of waste that includes separation and recycling, as well as strategies to reduce waste generation.

The new system of waste management considers five separation categories: (1) organic, (2) PET and PEAD, (3) aluminium, tetra pak and glass, (4) paper and cardboard, and (5) other wastes that do not correspond to any of the previous categories. Waste separation containers have been installed in each university department where members of the university can deposit waste items in the appropriate container for later collection and transport to a storage center. From the storage center each type of residue follows a specific route. Organic wastes go to the compost plant, which is in process of renovation to fulfill the requirements of the new waste-management system. The recyclable materials are sent to recycling plants, and remaining waste is sent to landfills.

The entire infrastructure necessary for this waste management system, including containers and the construction of storage centers, was designed and produced by UNAM also as a result of the diagnostic study. The system was tested over a pilot period in the Veterinary Science Faculty, Sciences Faculty, and in the postgraduate studies building. During this period, the operation of the system was monitored constantly to make any adjustments necessary.

The objective of the *responsible consumption axis* is to promote in every UNAM department the acquisition of supplies with least environmental impact, and facilitate recycling and reutilization of materials. One of the actions in this axis was the elaboration of guidelines for the acquisition of supplies that use fewer resources in their production, transport and deposition, and are easy to recycle or that are long-lasting.

The criteria for the acquisition of supplies for stationary, disposable cups and other materials, as well as lights and vehicles with lower environmental impact, were published in the institutional magazine 'Gaceta UNAM' on 22nd August 2011, representing an important step for the transformation of institutional consumption patterns in the short-term (Gaceta UNAM 2011). The guidelines are compulsory and all academic and administrative entities must follow them. In addition to these criteria, there is a catalog of appropriate products and distributors elaborated after an exhaustive investigation of the available products in the market.

Evaluation of the acquisitions reported in 2013 for every department in the UNAM main campus determined a 40 % compliance with the criteria by university entities. These criteria have also been incorporated in cafeterias and small food stores within the university to reduce and eventually eliminate the use of styrene and non-recyclable plastics.

UNAM has education and research centers in 24 states, and promotes the creation of new campus to encourage education, research and cultural diffusion of benefit to the entire country. In order that the growth in infrastructure at UNAM is consistent with the objectives of the EcoPUMA strategy, the *sustainable construction axis* aims to use materials and systems with least environmental impact, as well as incorporating efficient use of energy and water, and adequate waste management, while guaranteeing the wellbeing of users and employees.

To accomplish the objective of this axis, guidelines for sustainable construction were developed with the participation of several university entities involved in design, construction and maintenance of buildings. The guidelines include criteria for: selection of sites for new constructions, measures to minimize environmental impacts during construction or remodeling, incorporation of green spaces with native plant species, actions to ensure secure and efficient mobility of users, systems for efficient use of energy and water, as well as inclusion of long-lasting, recyclable materials that are free of CFCs. The guidelines also consider the inclusion of residual water systems, and materials that allow water infiltration, as well as the implementation of UNAM's new system of waste management. The first building constructed following these guidelines was the Laboratory of Sustainability Sciences in the main campus.

The objective of the *electronic administration axis* is to develop online and digital platform alternatives to reduce the use of paper in administration at UNAM. Advances in this axis include the implementation of advanced electronic signature systems that allow teachers to upload notes, information, and course materials, as well as online systems for inscription and other administrative procedures, and the use of email to send internal notifications and memos.

4 UNAM Environmental Certificate

The *UNAM Environmental Certificate* was designed to evaluate the environmental performance of university entities, and assess four components of the EcoPUMA Strategy for a Sustainable University: energy conservation, water use, generation and management of waste, and compliance with the responsible consumption criteria.

This assessment instrument is a system of credits based on four standards: Sustainability Tracking, Assessment and Rating System (STARS); Leadership in Energy and Environmental Design (LEED); the certification system for sustainable construction in the Federal District of Mexico City; and the criteria for acquisition of supplies with lower environmental impact of UNAM. EcoPUMA awards the

UNAM Environmental Certificate at three levels: Basic, Blue, and Gold. Along with the certificate there is a detailed document with results of the evaluation and recommendations to improve environmental performance and achieve the next level of certification. The achievement of a certification level requires that scores in all four components correspond to the desired level. The certificate is valid for 3 years, though entities can request a new evaluation at any time.

Groups of trained youngsters associated with PUMA visit the university entity to compile information on sustainable use for later digitalization and analysis. To date evaluations have been conducted of 65 academic and administrative university entities, as well as at 19 high schools incorporated with UNAM. In addition to these university entities, evaluations have been conducted for nine schools of the National Institute of Arts, and five federal government departments, including the Department of Environment and Natural Resources. In this way, the *UNAM Environmental Certificate* is becoming a national instrument for assessment of environmental performance.

5 Conclusion

The EcoPUMA Strategy for a Sustainable University is the most important effort of the National Autonomous University of Mexico aimed at reducing the environmental impact of the institution. The strategy also contributes to the formation of professionals and citizens aware of the ecological heritage of Mexico, and committed to its conservation, and the responsible use of resources.

In the first years of EcoPUMA, there have been relevant advances with great impact not only in reducing the environmental cost in daily operation of such a large and complex university as UNAM, but also by transforming the way the UNAM community carry out daily activities, think of their university, conduct their citizenship, and participate in the construction of sustainable societies.

Over the next years, the EcoPUMA Strategy for a Sustainable University will develop further programs, guidelines, evaluation tools, environmental education activities, and communication materials, to attain the goals of each axis, making UNAM the most sustainable Higher Education Institution of Mexico, and maintaining its position as the most important referral body for society and sustainability policies in the country.

It is important to highlight that the willingness, enthusiasm and participation of all sectors of the university community is what makes possible the commitment of UNAM to confront and attend to the current environmental challenges. However, this is just the beginning, as we look ahead to the construction of a better future for all, with social equity, fair economic development, and healthly and functional ecosystems.

Acknowledgments We thank Katherine Renton for revision of the manuscript and her helpful comments and suggestions.

References

Burnside W, Brown J, Burger O, Hamilton M, Moses M, Bettencourt LM (2012) Human macroecology: linking pattern and process in big-picture human ecology. Biol Rev 87(1):194–208

Escobedo A, Briceño S, Juárez H, Castillo D, Imaz M, Sheinbaum C (2014) Energy consumption and GHG emission scenarios of a university campus in Mexico. Energy Sustain Dev 18:49–57

Gaceta UNAM (2011) Universidad Nacional Autónoma de México. Secretaría Administrativa. Criterios para la adquisición de bienes con menor impacto ambiental. EcoPUMA. Consumo-responsable. 22th August. http://www.dgcs.unam.mx/gacetaweb/historico.html. Accessed 18 Mar 2014

UI GreenMetric World University Ranking (2014) Result of green metric ranking 2013. http://greenmetric.ui.ac.id/id/page/ranking-2013. Accessed 29 Mar 2014

Universidad Nacional Autómoma de México (2014) Portal de Estadística Universitaria. Numeralia. http://www.estadistica.unam.mx/numeralia/. Accessed 15 Mar 2014

Author Biographies

Mireya Imaz is a Biologist form UNAM. She has a Masters Degree in Ecology and Environmental Sciences. Since 2008 she is the Coordinator of the Environmental University Program (PUMA), where she leads environmental projects for public and private institutions, as well as projects with civil society organizations. She also coordinates the Strategy for Sustainable University EcoPUMA and she is the general coordinator of the specialized course "The Environmental Dimension in the Design and Implementation of Public Policies". She also had collaborated with several private institutions, including *McKinsey* & Company.

Dalia Ayala is a Biologist from UNAM with a Masters Degree in Systematics. At present she coordinates Environmental Education projects in the University Environmental Program (PUMA) at UNAM. She teaches courses on environmental education, biodiversity, sustainability, avian conservation and other topics related to Sustainability Science for high school and graduate students. She also gives talks on environmental issues to elementary, secondary, high School and undergraduate students. She previously worked on the Biodiversity and Environment Information System project of the Institute of Biology at UNAM, and has participated in several bird conservations projects in Mexico.

Reducing Carbon Emissions in a Third Level Educational Institution in Sub-Sahara Africa

Izael Da Silva, Geoffrey Ronoh, Clint Ouma and Caren Jerono

Abstract

The effort to reduce carbon emissions as the arguably most prevalent cause of global warming has been a positive trend in most African countries. One of the most successful strategies towards reaching that goal is the shift from fossil fuel power generation to renewable sources of energy such as wind, hydro, geothermal and solar. As Kenya sits on the equator it enjoys an all year round insolation between 5 and 6 kW/m^2/day which is more than double of the average insulation in Germany, a country where solar energy is widely used. Taking advantage of a green line of financial support created by the French Government, Strathmore University embarked in a project to install a 600 kW roof-top, grid connected solar PV system to cater for its electricity needs. Having as a background of the newly instituted Feed-in-Tariff regulation, the system is designed to produce more than the required self-consumption such that the extra power can be sold to the utility via a PPA (power purchase agreement) and the revenue used to pay for the electricity used by the university at night. This paper describes the whole process from the technical, regulatory, educational and financial aspect highlighting the positive and negative events along the path such that it can be useful for other private sector institutions interested in greening

I. Da Silva (✉) · G. Ronoh · C. Ouma · C. Jerono
Strathmore Energy Research Centre, Madaraka Estate, Ole Sangale Road,
PO Box 5985700200, City Square, Nairobi, Kenya
e-mail: idasilva@strathmore.edu

W. Leal Filho (ed.), *Transformative Approaches to Sustainable Development at Universities*, World Sustainability Series, DOI 10.1007/978-3-319-08837-2_34

their sources of energy, invest in renewable energy and thus reduce their operation costs. The authors have written this work having in mind not only countries in Africa but all other countries which sit in the so called "solar belt".

Keywords
Sustainability · Carbon emissions · Solar PV technology

1 Introduction

The concentration of GHG emissions has been on the increase over the years in line with the temperature of the earth which was determined to have risen by 0.8° between the years 1900 and 2005. Despite a rising awareness on greenhouse gas (GHG) emissions, Diabat et al. (2010) show that, over the last 150 years, the last decade was recorded as the hottest with the year 2005 being the hottest year. As a result, concerns over global warming have penetrated all parts of society such as the corporate and industrial world and recently institutions of higher learning.

In 2008, CO_2 emissions in Sub Saharan Africa (SSA) was reported at 0.84 metric tons per capita (Trading Economics 2013). These carbon emissions were generated by burning of fossil fuels to generate electricity, drive the transport industry, and to run industrial processes such as cement manufacturing. The current trend is that higher institutions of learning are taking a leadership role in driving sustainability initiatives. Thus, they are well placed in playing the role of fighting climate change since their focus is to educate future generations of leaders. This responsibility covers the institutions' carbon emission reduction, conservation of energy and water and other sustainability initiatives. Velazquez et al. (2006) define a sustainable university as a higher institution of learning that involves, addresses and promotes on a global and regional basis, the reduction of negative environmental, societal, economic, and health effects that result from the use of their given resources to meet their functions in outreach and partnership, teaching, research and stewardship to help society to achieve a transition to sustainable lifestyles.

This paper outlines sustainability initiatives that Strathmore University has put into place to reduce carbon emissions and ensure a green environment around campus. Through these initiatives, Strathmore University aims innovatively create awareness amongst its 5,000 student plus population. The paper focuses on a solar photovoltaic (PV) project that the University invested in to replace electricity provided by the national utility. A description is made of the process followed from the technical, regulatory, and financial aspects highlighting the positive and negative events along the path such that it can be useful for other tertiary educational institutions interested in greening their sources of energy, invest in renewable energy and thus reduce their operation costs. The paper also aims to serve as a case study to other institutions of higher learning on the processes and challenges in undertaking a sustainability initiative in a commercial manner.

Fig. 1 Management science building at Strathmore University

2 Sustainability Initiatives at Strathmore University

Strathmore University (SU) is a higher learning institution which was established in 1961 as a college and has over the years grown into a renowned private university. The University has an established policy of integrating business with environmental conservation which has been embraced by all employees towards environmental sustainability. With regards to its built environment, the University has adopted green buildings as a way of improving the benefits to students, staff, workers, the community, its bottom line.

The Student Centre (SC), Management Science Building (MSB) and the Strathmore Business School (SBS) which add up to 22,000 square meter of space were constructed using the LEED (Leadership in Energy and Environmental Design) standards. These green buildings consist of mainly offices, lecture halls, conference halls, recreational facilities and a cafeteria. Compared to conventional buildings, the energy consumption has been reduced by 40 % (Da Silva and Ssekulima 2011). A Building Management System (BMS) is integrated into the buildings to control the resource utilization. The BMS used is based on SNAP PAC System Architecture with OPTO-SNAP controllers. User defined control-programming is used to define the functioning of the various components such as motion detectors, power cards and lighting control. The BMS uses room orientation and time-of-day to disable lighting fixtures that are close to the windows when sufficient natural lighting is available. It also disables all lighting in individual rooms when the BMS Motion Detectors indicate that the area has been vacated (Fig. 1).

The buildings have in place a full-building voltage stabilizer to help protect all electronics, including the light ballasts from the recurrent voltage fluctuation on the National Grid. In addition, the buildings have incorporated water evaporation cooling system in addition to natural ventilation. Other sustainability initiatives in the University include:

- Landscaped flower beds with ornamental shrubs and trees to increase the aesthetic value of its landscape.
- On-site waste separation, processing and location of bins in strategic places around the University.
- Common printing system aimed at reducing paper use in the university. A user password is utilized within the system to ensure wastage during printing is controlled.
- Water conservation through rain water harvesting from roof catchments through incorporation of pervious paving systems. Harvested rain water is utilized in irrigating flower gardens. Part of the rain water is pre-treated to be used for non-drinking purposes.
- Green roofs help reducing temperature in the buildings and a 10 kW solar PV roof-mounted, grid-connected is installed in the Student Centre to help offsetting the base load.

3 University Electrical Load Profile

The University spends approximately US $300,000 per annum on electricity costs. Major electrical loads mainly comprise of lighting, cooking, air ventilation, hot water heating, lifts, escalators, computers and other appliances. During the week the demand increases from around 04:30 h and peaks at around 11:15 h and then drops at around 20:45 h. On weekends and holidays, the demand is low due to reduced number of activities (Fig. 2).

Month-by-month consumption varies from lows of 80 MWh to highs of 140 MWh as shown in below:

In order to reduce its operational costs, the University is constructing a grid-connected roof top solar-PV system designed to supply Strathmore University's entire campus with electrical energy, for a period of no less than 20 years. In this regard, Strathmore University is set to be the first carbon neutral University in Sub Saharan Africa (Fig. 3).

4 The Strathmore University Solar PV Project

In December 2012, the Strathmore University Council approved the development of a roof-top grid-tied solar PV project as part of its sustainability initiative. A number of factors were important in the decision to invest in a solar PV power plant at the University:

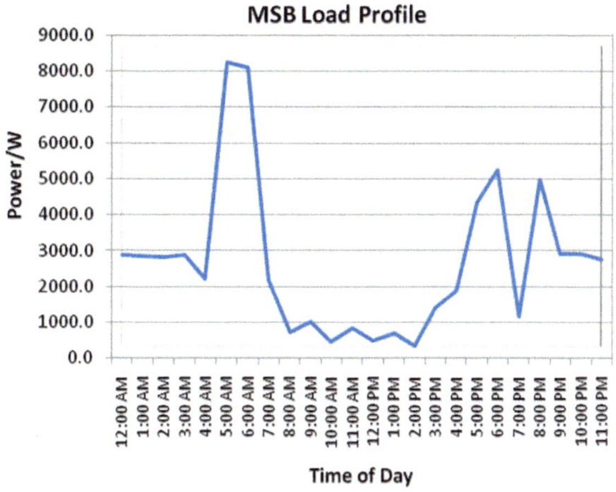

Fig. 2 Daily load profile for the MSB building

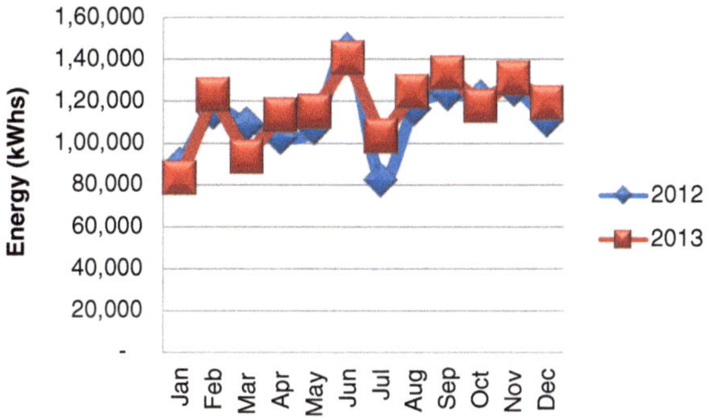

Fig. 3 Monthly consumption for the entire university for 2012 and 2013

- Awareness and knowledge of the University's senior management about solar PV technology, and its pioneering spirit.
- The University's location close to the equator giving it an advantage in tapping solar power due to the average insolation of 4–6 kWh/m^2/day without the need for sun tracking.
- The existence of a green energy facility, set up by the French Development Agency, offering project financing at concessional rates.
- Additionally, the existing feed-in-tariff regime allowed the possibility of a grid-connected system without the need for storage which is always expensive and availing the possibility of selling to the utility excess energy.

4.1 Regulatory Processes

The initial concept behind the project was that the national utility, Kenya Power (KPLC), would accept to partner with the University in piloting an Electricity Banking Arrangement (EBA) which would have allowed the University to store excess energy in the grid. In the end, the discussions with KPLC on the EBA were not successful. Instead, KPLC referred Strathmore to the existing possibility of signing a Power Purchasing Agreement (PPA), under the feed-in-tariff regime, that would commit KPLC to purchase surplus solar power from Strathmore at \$0.12/kWh (Fig. 4).

The FiT route presented a number of challenges, namely:

- The PPA process is lengthy and costly as is best suited for large-scale projects aimed primarily at selling power to the grid. The processes consist of three major steps, all of which require substantial investment in funds and time.
- A number of procedures still need to be streamlined, using process tools to maximize efficiency of the FiT approval process. For example, guidelines on how project developers can access the required data to be used for grid connection from the utility within a reasonable period of time.
- While a standardized PPA exists for projects below 10 MW, it is not suited for projects in which a large percentage of the energy generated is utilized for captive use on site. A number of clauses need to be amended and negotiated, which leads to delays.

4.2 Impact of the Value Added Tax (VAT) Bill, 2013

On September 2, 2013, the Kenya Government gazetted the new VAT bill which introduced 16 % VAT on hitherto exempted goods. Included in this category is specialized solar PV equipment and related products e.g. deep cycle batteries.

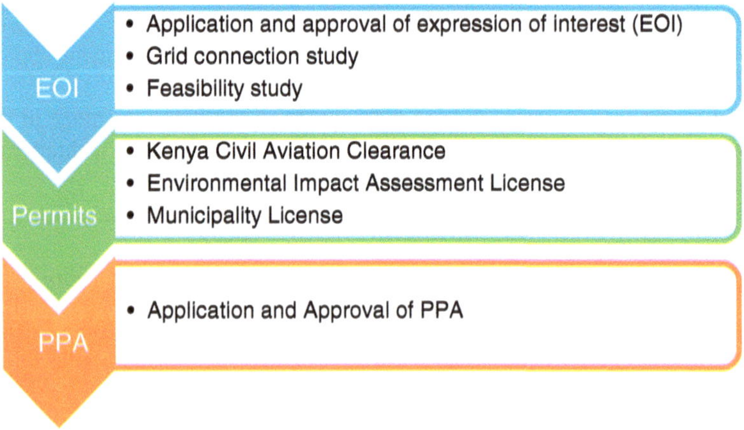

Fig. 4 The PPA process in Kenya

The impact of this new regulation was to introduce an additional cost of 16 % into the project (or approximately US $220,000). This development jeopardized the financial viability of the project. At the time of writing this paper, special negotiations are on-going to obtain a VAT-exemption on the project.

4.3 Project Sizing

The high costs of electricity in Kenya, averaging $0.225/kWh presents an opportunity for replacing electricity from the grid with "cheaper" solar PV electricity thus reducing operating costs. An optimal system size will maximize on-site consumption and minimize excess sales to the grid due to relatively low FiT of $0.12 per kWh.

The minimum plant size eligible for the FiT regime in Kenya in 500 kW, thus, this represented the smallest size that the University could consider. To arrive at an optimal plant size, modeling tools were applied to various plant sizes. The sizing analysis was undertaken through two models, namely;

- SAM (a renewable energy application designed by the National Renewable Energy Laboratory, an agency of the US Government); and
- A financial model in Excel that takes some of the results from the R model to calculate Net Present Value (NPV) of systems of various sizes.

Assuming a basic system cost of $2.40 per watt plus VAT, as well as current energy prices of $0.225 per kWh, and a feed-in tariff of $0.12 per kWh, the optimal size is 600 kW as shown below:

Figure 5 above shows how the NPV varies as system size changes. Beyond 600 kW the losses from energy exported to KPLC at a lower price than the Levelized Cost of Electricity (LCOE) begin eating into the savings achieved by Self Consumption (SC) of PV electricity. A system that is too small will not reap all possible savings offered by the $0.225 tariff.

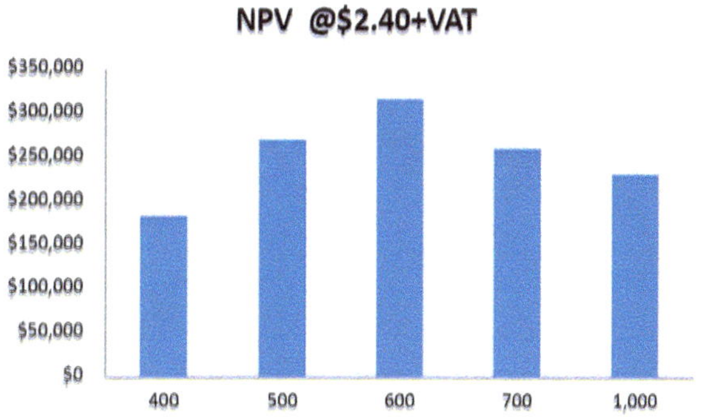

Fig. 5 NPV of various solar PV project sizes considered

The difference between the power produced by the solar system and that purchased from KPLC is the power saved and valued at $0.225 per kWh. At certain times, there would be excess power from the solar and it would be exported to KPLC. Exported power is valued at a maximum of $0.12 per kWh.

4.4 Project Financing and Economics

To finance the project, the University secured credit availed by a local bank, Co-operative Bank of Kenya, under a credit line provided by the AFD Green Line of Credit at 4.1 % yearly interest rate over a 10 year period, with one year moratorium on the principal. Prior to the project, the University monthly spending on electricity varied from KES 1.5 million (US $17,000) to KES 2.5 million (US $28,700) per month as shown below, and averaging an annual amount of KES 24 million (US $280,000) (Fig. 6).

Assuming a basic system cost of $2.40 per watt plus VAT, as well as current energy prices of $0.225 per kWh, and a feed-in tariff of $0.12 per kWh, the ratio of captive use to export is 3:1, the expected financial returns from the project are as shown below (Fig. 7):

The project is expected to turn to turn cash positive within the second year as savings from purchases from the grid and exports outweigh the cost of imports. Payback period is estimated at 7 years (Fig. 8).

5 Conclusions

5.1 Green Initiatives are Good for the Bottom Line

The 600 kW project at Strathmore University demonstrates that the sustainable initiatives in institutions of higher learning can result in reduced operational costs. However, critical factors such as cost of capital need to be evaluated (Fig. 9).

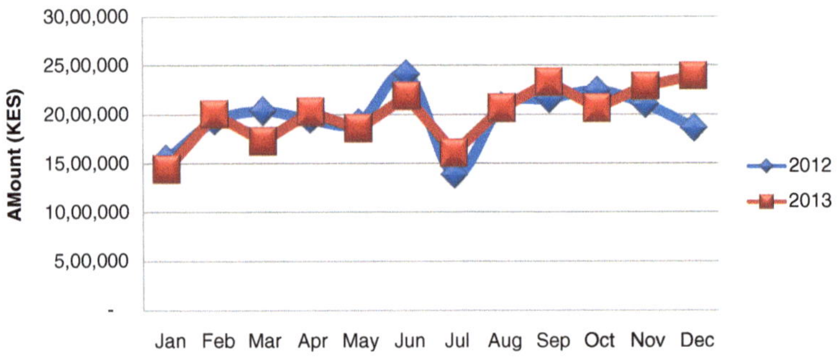

Fig. 6 Monthly expenditure on electricity at Strathmore University for 2012 and 2013

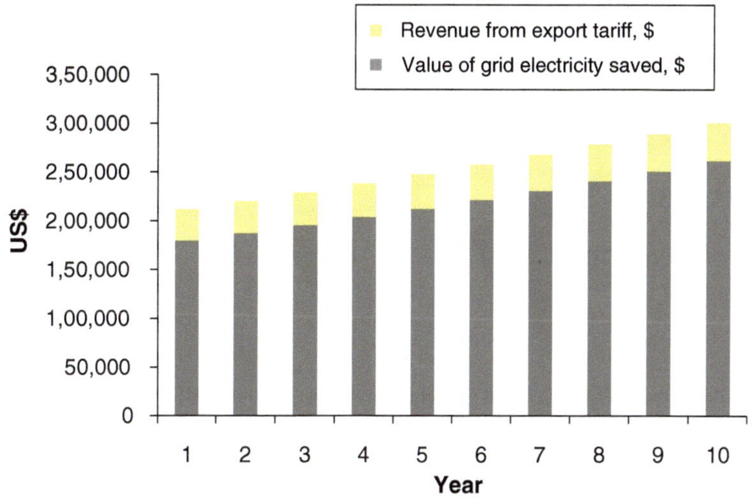

Fig. 7 Projected combined revenue and savings (first 10 years of the Solar PV project)

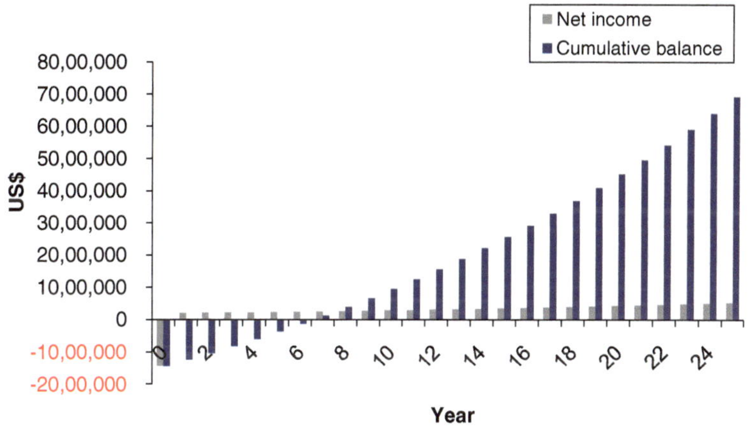

Fig. 8 Cumulative project balance and net income for 25 years

The Weighted Average Cost of Capital (WACC) is as important a factor in determining the Levelized Cost of Electricity (LCOE) from Solar PV as solar radiation (Ondrazcek et al. 2013).

The fact that Strathmore University could access relatively cheap financing at 4.1 % yearly interest rate compared to a market rates in Kenya in excess of 10 %, made the project commercially viable without any subsidies. As a result, the sustainability initiative not only results in reduced carbon emission, but also lowers the cost of energy to the University (lower solar PV LCOE compared to costs of electricity).

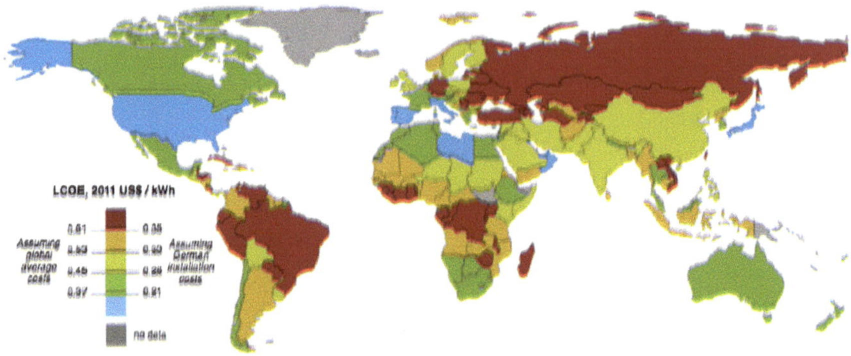

Fig. 9 Global map of LCOE, based on weighted average national GHI nationally-specific WACC

5.2 Regulatory Environment

Undertaking sustainability initiatives at tertiary level institutions requires an enabling policy environment. In the case of Strathmore University, the solar PV project has encountered numerous challenges that have delayed the implementation of the project. The cyclic annual nature of solar production, combined with University operating hours requires that the supply of excess electricity production be sent to the National Grid, for later use when the load exceeds the available generation capacity or the sun energy is not available. However, a net-metering policy is non-existent in Kenya at the moment while the existing FiT regime was not suitable to the University needs of using captive power on site. Further, changes to the VAT bill resulted in additional costs to the project which severely impacted its financial viability.

5.3 Measuring Progress

Lack of data concerning energy consumption and carbon emissions patterns results in lack of clarity on priority areas to focus on. For example, Davis (2014) highlights that keeping close tabs on energy usage is a valuable practice that leads to significant energy savings. Analysis of data from many areas of operations for instance, refrigeration and air conditioning, commuting and waste provides an understanding into the factors influencing emissions from higher institutions of learning. In this regard, Strathmore University intends to develop a metric to measure the impact of a number of sustainable initiatives at operational level in the institution and utilize the metric as an indicator to gauge their progress towards achieving a more sustainable system.

5.4 Sustainability Initiative as a Change Agent

Strathmore University has over 5,000 students on its campus. Through the project, the University aims at creating awareness amongst its student population by explaining to them the incorporated sustainability initiatives brought into the built environment. Through the Strathmore Energy Research Centre, the University intends to conduct seminars to educate the students on the application of technologies such as solar PV. It is envisaged that these students will then become agents of change in their homes or work places in the topic of sustainability. Further, senior management from other tertiary institutions will be invited to open days to tour our green facilities and get familiar with details behind the 600 kW project.

The project will also serve as a hands-on training for solar technicians from various vocational training institutions in the region. Grid-connected solar PV is a relatively new area in Kenya with a shortage of qualified technicians.

References

Da Silva IP, Ssekulima EB (2011) Energy efficient building envelope designs for institutional buildings in East Africa. In: International conference, towards sustainable energy solutions for the developing world. Domestic use of energy conference. Cape Peninsula University of Technology, Cape Town, South Africa, pp 79–84

Davis C (2014) Big data can drive energy savings. Control engineering—industrial energy management, February 2014 edition. http://bt.e-ditionsbyfry.com/publication/?i=196923&p=88. Accessed 14 Apr 2014

Diabat A, Simchi-Levi D (2010) A carbon-capped supply chain network problem. In: IEEE international conference on industrial engineering and engineering management, 2009. IEEE, Piscataway, NJ, 523–527

Ondrazcek J, Komendantova N, Patt A (2013) WACC the dog: the effect of financing costs on the levelized cost of solar PV power. FNU Working Paper No. 201, May 2013

Paron P, Olago D, Omuto C (2013) Kenya geo-environmental resources and hazards. Elsevier Science, Burlington. http://public.eblib.com/EBLPublic/PublicView.do?ptiID=1495674. Accessed 14 Apr 2014

Schäfer M (2011) Micro perspectives for decentralized energy supply proceedings of the international conference, Technische Universität Berlin, Univ.-Verl. Der TU, Univ.-Bibliothek, Berlin, 7–8 Apr 2011

Trading Economics (2013) CO_2 Emissions (metric tons per capita) in sub Saharan Africa.http://www.tradingeconomics.com/sub-saharan-africa/co2-emissions-metric-tons-per-capita-wb-data.html. Accessed 14 Apr 2014

Velazquez L, Munguia N, Platt A, Taddei J (2006) Sustainable university: what can be the matter? J Cleaner Prod. http://www.esd.leeds.ac.uk/fileadmin/documents/esd/10._Sustainable_University__What_Can_be_the_Matter.Velazquez-2.pdf. Accessed 14 Apr 2014

White SS (2009) Early participation in the American College and University Presidents' climate commitment. Int J Sustain High Educ 10:215e227

Author Biography

Professor Izael Pereira Da Silva is a Renewable Energy specialist with over 15 years of experience in research and academic leadership. Professor Da Silva is an Associate Professor at Strathmore University and the Deputy Vice Chancellor Academic Affairs. Under his mandate lays the creation of the Strathmore Energy Research Centre—SERC in partnership with GIZ in which he is the director. SERC offers various activities including training, research and consultancy in energy related topics. Previously he founded and directed CREEC—Centre for Research in Energy and Energy Conservation in Makerere University, Kampala, Uganda. He left CREEC in 2010 and joined Strathmore. CREEC is very much relevant for Uganda and currently runs a good number of renewable energy projects sponsored by the World Bank, GIZ, ADA (Austria Development Agency), UNIDO, Ministry of Energy and Mineral Development, Uganda National Council for Science and Technology, Sida/SAREC, NORAD, Global Alliance for Clean Cook stoves, etc. Professor Da Silva has a PhD in Power Systems Engineering from the University of Sao Paulo (Brazil). Professor Da Silva has been working with Government Ministries, Development Agencies such as the Deutsche Gesellschaft für Internationale Zusammenarbeit (GIZ), the Swedish International Development Cooperation Agency (SIDA), the World Bank, etc. His research interests are: Renewable Sources of Energy and Efficient Use of Energy, Renewable Energy Policy, Rural Electrification and Sustainable Development. He is also a Certified Energy Manager with the Association of Energy Engineers in the United States of America.

Students' Views of Environmental Issues: The Experience of a Federal Brazilian University

Marisa Sartori Vieira, Patricia Cristina Silva Leme and Vânia Gomes Zuin

Abstract

Many tools have been used to measure and disseminate sustainable actions in universities, as shown by the Association of University Leaders for a Sustainable Future (ULSF 2008). In 2011, the University of São Paulo developed a virtual platform called *Information, Awareness and Assessment of Sustainability in Higher Education Institutions* to promote the participation of the university community (students, lecturers and administrative staff) in the assessment of sustainability in the university. The initiative later attracted the attention of other Brazilian universities. This work presents the results of the "sustainability test" applied to a group of students of the Federal University of São Carlos. In their opinion, the university should invest more in environmental education and improve management systems related to waste reuse/recycling and clean technologies. The views concerning these and other environmental issues have contributed to improve the actions of university managers and teachers towards a sustainable future.

Keywords

Sustainability · Students · Assessment

M.S. Vieira · V.G. Zuin
Postgraduate Program in Education, Federal University of São Carlos,
Rodovia Washington Luiz, Km 235, São Carlos, SP, Brazil

P.C.S. Leme (✉)
Environmental Management Office, University of São Paulo,
Av. Trabalhador Saocarlense, 400, São Carlos, SP 13566-590, Brazil
e-mail: pazu@sc.usp.br

W. Leal Filho (ed.), *Transformative Approaches to Sustainable Development at Universities*, World Sustainability Series, DOI 10.1007/978-3-319-08837-2_35

1 Introduction

In the last decades Environmental Education (EE) in Latin America has become stronger, especially by identifying potentials and facing socio-environmental problems. Higher education institutes (HEI) have great responsibility for educating individuals with autonomy, critical thinking and potential to transform socio-environmental relations (Rupea[1] 2007).

The inclusion of environmental topics in universities, known as *greening universities,* is happening gradually and independently in teaching, in the curriculum, in research and extension, and in campus management and institutional policies.

The National Environmental Education Policy (PNEA) of Brazil (law 9795/1999) states that EE should be present in all levels and forms of teaching. "It should not be taught as a specific discipline in the curriculum", but developed in an integrated and permanent form.

During the Rio+20, in 2012, the National Council of Education approved the National Curricular Guidelines for Environmental Education (Brasil 2012), which, among other principles and objectives, determines that HEI should:

> approach sustainability, in its various aspects, by means of curricular activities and mandatory interdisciplinary projects promoting the study of environmental legislation and management, according to the professional profile of the various undergraduate courses, teaching degrees, technological graduation and their respective post-graduation courses (Brasil 2012).

Besides including specific content in curricular disciplines, the university has tried to incorporate sustainability in administrative routines and has dialogued with the community. According to Leff (2001), to address sustainability is to "open up to the domain of ethical values, practical knowledge and traditional wisdom". Research with the population can grasp socio-environment conflicts from their origin and provide the necessary knowledge for its application.

In the last years, this movement became more intensive and new strategies for evaluating environmental practices were created. Some internationally known tools, such as the Assessment Instrument for Sustainability in Higher Education (AISHE),[2] Sustainability Tracking, Assessment and Rating System (STARS),[3] Learning in Future Environments (LIFE),[4] have contributed to this analysis by characterizing the universities' progress in approaching sustainability.

---

[1] The universities' network of programs of environmental education for sustainable societies (RUPEA), created in 2001, today consists of 15 groups and 11 Brazilian universities.

[2] It was developed in the Netherlands (2000–2001) and has been intensely applied in at least 11 countries.

[3] Initially launched in US and Canada (2010) by the Association for the Advancement of Sustainability in Higher Education, it is now used by more than 350 institutions in 10 countries.

[4] It was developed in UK, Australia and New Zealand in 2011.

This article tries to identify and analyze how sustainable a higher education institution in Brazil is from the point of view of its students. Managers and professors participation in sustainability assessment is common. But the students' views of environmental issues can help the university to have more efficient actions.

Researches involving tools for sustainability assessment are still incipient and this field is yet to be explored (EUAC 2014). So, besides appreciating the students' views, this work aimed at using a new *online* tool: the "Sustainability Test" (Leme and Pavesi 2012). Unlike other instruments, developed in Europe and North America, this test considers the Latin American context and allows the academic community to express its views and knowledge related to the activities that involve the environment and education in the university *campus*.

This diagnosis, from the students' point of view, could provide further information for the development of strategies and the improvement of EE practices in this and other HEI.

2 Greening Universities and the Challenge of Evaluating Sustainability

Addressing sustainability in universities is strongly connected to the intense process of institutionalizing EE, originating from social and environmental movements and historical events worldwide: the Stockholm Conference on Human Environment (1972), the Tbilisi Declaration (1977) and the United Nations Conference on Environment and Development (1992). With advancing industrialization and increasing problems, education is an important element to face socio-environmental crises.

The discourse around sustainability gained visibility with the Brundtland Report (1987), organized by the United Nations, in which governmental representatives, educators/teachers and entrepreneurs assumed a new understanding of development: "sustainable development is that which responds to the needs of present generations without compromising the capacity of future generations to fulfill their own needs" (Brundtland 1991, p.46).

Although this document brings advances aligning economic growth with environmental preservation and social equity, the government priorities of the capitalist consumption system have defined other arrangements (Lima 2002). Still, despite the limits imposed by the market, the many political events mentioned above are important for the discussion around sustainability, as well as for the elaboration of proposals for the HEI, concerning *campus* management, formation of employees, reorientation of the curriculum, research and public participation. Leal Filho (2011) surveys various significant declarations that have contributed to explain the role of universities in the *greening* process, such as the *The Magna Charta of European Universities* (1988), *The Talloires Declaration of University Presidents for a Sustainable Future* (1990), *The COPERNICUS Universities Charter or Sustainable Development* (1994), *The Lüneburg Declaration on Higher Education for*

Sustainable Development (2001) and *G8 University Summit: Statement of Action* (2010).

Brazilian universities have participated in international and interdisciplinary programs and projects, for instance the *Red de Ambientalización Curricular de los Estúdios Superiores* (Red ACES), gathering eleven universities from Europe and Latin America (Guerra and Figueiredo 2014). Recently, interchange of experience was promoted by four seminars, named *Sustainability in the University* (2010–2013), organized by public and private institutions, also involving the governments of Brazil, Colombia, Chile, Argentina, Spain, Portugal and England (Leme et al. 2012).

The effort to stimulate dialog and to put sustainability in the agenda aims at not simplifying EE, or merely reducing it to "good environmental behaviour", or, still, disseminating the naive idea that there is a consensus among educators and a common understanding of the environment (Carvalho 2011). According to Foladori (2001), the contradictions of human society are not biological, but social, based on economic history, "which does not have roots in ecological contradictions in general, but in those related to class differences and social sectors in particular" (p. 45). Poverty, hunger, illiteracy, violence and negligence of other basic needs of developing countries, like in Latin America, are part of the environmental illnesses and must be considered in the pursuit of development for sustainability (Gaudiano 2001). In this work we defend a critical EE, beyond behavioral activism, an inquisitive education in favor of autonomy and emancipation, according to Adorno (2000) urgent elements for human survival against barbarism.

This is a challenge for the HEI, because, as in any institution, a dominant culture legitimates and reproduces itself, stressing the inequalities and the disputes of different scientific fields (Bourdieu 2003). The lack of dialog between the curricular disciplines and of collaborative experience, a competitive environment, unequally distributed financial resources, non-transparent decision-making structures, among others, are still difficulties to be overcome in the HEI (Oliveira 2012).

Most scientific and technological activities in the universities study concrete environmental problems. Thus, investigations in the field of natural sciences outweigh those in human and social sciences (Sáenz 2014). Almost all interventions are merely technical, used to remedy or minimize visible impacts, such as the reduction of energy consumption or waste generation.

The report about sustainability in universities—*Higher Education's Commitment to Sustainability: From Understanding to Action* (Tilbury 2011)—emphasizes notable experiences in Australia, Africa, Japan and China, supported by the United Nations Environment Programme (UNEP). This research demonstrates the significant investment in the areas of technology for sustainability, but also the growth, in the last 10 years, of educational practices stimulating people to critically reflect and act politically, mainly in the academic extension field. According to the author, the increase in extension activities and partnerships related to environmental education indicates the acknowledgement of the university's social role in taking knowledge to the community's practices.

Various systems have been created for evaluating HEI's progress towards sustainability (Tilbury 2011). In order to group the organizations that developed evaluation systems and contributed to improve environmental actions in universities, the UNEP subsidized the implementation of the Platform for Sustainability Performance in Education. The virtual space allows access to instruments of support for the HEI, such as the AISHE, Alternative University Appraisal (AUA),[5] Conference of Rectors of Universities in Spain (CRUE),[6] LiFE, STARS e The Green Plan.[7] PegadaEcológica[8] and Global Reporting Initiative[9] also deserve mention. All of them have sustainability indicators, such as the production and use of materials, pollutant emissions, environmental aspects in the curriculum, urbanization, relationships within the academic community, institutional plans etc. Some were applied in Brazilian universities, such as the AISHE, to evaluate the sustainability of undergraduate courses at the University of Passo Fundo. However, according to Brandli et al. (2012), this tool has abstract criteria, difficult to understand.

It is fundamental that sustainability indicators are aligned to the social, political and cultural context of each evaluated institution. Therefore, certain systems of indicators developed in European countries may be inadequate in Latin-American institutions.

3 The Platform of Sustainability and the Greening of the Federal University of São Carlos

To contribute to the evaluation of sustainability in Latin-American HEI, the platform of *Information, sensitization and evaluation of sustainability in the university* was created.[10] The focus is the *Sustainability Test*, to be used by the academic community (management, students, teachers, lecturers and other employees). According to Leme and Pavesi (2012), the aspect that most distinguishes it from previous tools "is the firm intention to promote the participation of the entire community in the evaluation, as well as to privilege sensitization concerning sustainability" (p. 198).

[5] Developed in Japan in 2009.

[6] Since 2007 a work group evaluates environmental and sustainable strategies in Spanish universities.

[7] Created in 2010 by a French partnership between CGE (French Conference of Grandes Écoles) and the Ministry of Ecology.

[8] This methodology of environmental accounting evaluates the pressure of human consumption on natural resources. Expressed in hectares, it allows different patterns of consumption and verifies if they are within the ecological capacity of the planet.(http://www.wwf.org.br/natureza_brasileira/especiais/pegada_ecologica/o_que_e_pegada_ecologica/)

[9] Non-governmental organization, founded in 1997, which supports companies and other institutions in the elaboration of sustainability reports, technical protocols and indicators.

[10] The platform was developed within a cooperation international project between the University of São Paulo (USP) and the Universidad Autónoma de Madrid (UAM), with the support of the Spanish Agency for International Development Cooperation (AECID)—2009–2011.

Besides evaluation, this digital space intends to be a communication channel for the HEI, to disseminate and stimulate the development of sustainable actions. In 2011, the highlight of the project was the joining of new partners: Pontifícia Universidade Católica do Rio Grande do Sul (PUC-RS) and UFSCar (Federal University of São Carlos).

The University of São Carloshas has about 16 thousand students and the environmental discussion is well incorporated in many areas. It was one of the pioneer institutions in Brazil in establishing, in 1993, an organ responsible for environmental issues: the Special Coordination for the Environment (Coordenadoria Especial para o MeioAmbiente—CEMA). This organ is involved in three big programs: o Programa de Educação Ambiental-PEAm (The Program of Environmental Education); o Programa Agro-ecológico (the Agroecological Program) and the Programa de Conservação de Energia e Controle de Resíduos (Program of Energy Conservation and Waste Control). In 2013 this Coordination was renamed the *General Secretary of Environment and Sustainability Management* (Secretaria Geral de GestãoAmbiental e Sustentabilidade (SGAS).

Other research and extension initiatives in this field come from study groups, such as the Study and Research in Environmental Education group, and the Study and Research in Green Chemistry, Sustainability and Education group. Both target projects involving curriculum, teaching, teacher preparation, technologies and sustainability. Regarding institutional politics, conceived in 2004 by the academic community in a process coordinated by the rectory, the first Institutional Plan for Development of UFSCar (UFSCar PDI 2012) tried to contemplate the environmental issue in various aspects. However, in 2012, its guidelines were reviewed and the objectives, detailed:

> **(2004) Original guideline**–3.1.21 To include concepts and practices in the curricula focusing on the environment.
> **(2012) Suggestion of revised guideline**–To include in the pedagogical projects of undergraduate and postgraduate courses and in daily routines concepts and practices focusing on relevant socio-environmental issues of each area of knowledge, with special attention to teacher preparation courses, adopting the inter and transdisciplinary approach according to the National Policy of Environmental Education (p. 21).

In the area of teaching, various lecturers in the Red ACES project participated, from 2002 to 2004, developing research and plans for greening the curriculum, mainly in teacher preparation courses related to natural sciences and their technologies. In the Chemistry course for teachers at UFSCar, for instance, *Green Chemistry* is part of the curriculum. According to its founders (Anastas and Warner 1998), Green Chemistry can be defined as "the creation, development and application of products and chemical processes to reduce or eliminate the use and generation of substances harmful to humans and to the environment" (p. 135). A research carried out by students taking this course pointed out that most of them perceived environmental problems focused on preservation and pure science, whose main solution was to "adopt a behavior of conservation, obtain knowledge in environmental sciences and develop capacities related to environmental management and scientific experience" (Zuin et al. 2009, p. 564). The insertion of new disciplines and the collaborative work of some lecturers contributed to disseminate the production of material, processes and technologies considered more environmentally adequate, expanding the environmental view of the students. It is our premise that the appropriation of the subject by the academic community should be gradual and continuous, so as to not lose its complexity.

Considering the relevance of sustainability in HEI, the curricular initiatives, the research and the implementation of institutional politics at UFSCar, a question arises: how do the students evaluate the greening of the campus in relation to the curriculum, waste management, green technologies and the role of the university towards environmental problems? This work aims at identifying and analyzing the view of a group of students (of Chemistry, Environmental Management and Analysis and Education) concerning the level of greening of UFSCar—São Carlos campus, in the areas of teaching/curriculum and environmental management.

4 Methodology

The research deals with a case study with a qualitative approach, highlighting aspects for the understanding of the context to be analyzed (Lüdke and André 1986).

The initial idea was to involve a maximum number of students in teacher formation courses in Chemistry at UFSCar (São Carlos) and also the research group GPQV, composed of postgraduate students in the area of Education and Chemistry. In the beginning of 2014, these students were invited by social networks and emails to take the *Sustainability Test*, available for 15 days in the link http://www.projetosustentabilidade.sc.usp.br/index.php/eng. During this period, students of the Environmental Management and Analysis course showed interest and were included as participants.

In its original format, the *Sustainability Test* consists of 44 closed multiple choice questions, divided in two parts. The first—*Is your campus sustainable?*—asks about actions in the teaching, research, extension and management areas. The second part—"*What is your engagement with the socio-environmental sustainability actions of your campus*?"—presents questions that invite the participant to reflect on his/her own actions in the university.

One of the advantages of this tool is its easy adaptation to local realities. Thus, the test was adapted by the authors to the context of UFSCar, and a third part, with open questions, was added. The objective was to analyze the students' views about environmental education but, in the end, the test had 68 questions dealing with environmental management and education, environmental commitment, participation, wastes, water, sewage, energy, mobility/accessibility, green areas, sustainable purchasing and atmospheric contamination.

This research analyzed the answers to ten questions about the insertion of environmental education in the curriculum, the university's role towards environmental problems, waste management, green/clean technologies and the involvement

Fig. 1 Quantity of participants, by course

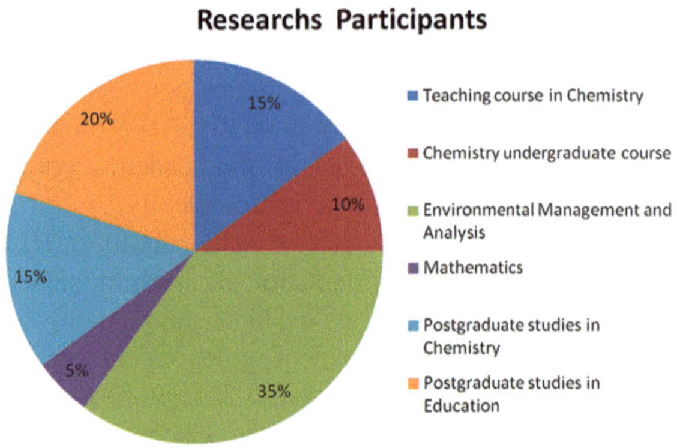

of students in this matter. The texts of the open questions were separated into meaningful units and then divided into categories (defined a posteriori) to provide better understanding of the investigated topics (Moraes and Galiazzi 2006).

Twenty participants were chosen for the analysis: 3 from the teaching preparation course in Chemistry, 2 from the Chemistry undergraduate course, 7 from the Environmental Management and Analysis, 1 from the teacher preparation course in Mathematics, 3 from postgraduate studies in Chemistry and 4 from postgraduate studies in Education. See Fig. 1.

5 Results

5.1 The Role of the University Towards Environmental Problems

As already mentioned, the university has double responsibility, both in the formation of protagonists, prepared for citizenship, and as an example of how to apply sustainability in its space (Rupea 2007). Both aspects were contemplated in the open questions:

(a) *In your opinion, what should be the role of the university towards environmental problems?*
(b) *Comment on the role of the university towards the development and application of green technologies.*

In the first question (a) three categories for analysis were established: (1) environmental education, (2) research and (3) environmental management. The role of the university according to 65 % of the participants fits in category (1), or rather, the institution should invest in the environmental formation of the students, creating a specific discipline and developing more projects and programs to involve students and promote critical education.

The university should assume the responsibility to contribute to the environmental formation of its students (…) University should use its resources to disseminate more projects related to this area. (Student from the teacher training course in Chemistry)

The role of the university is to contribute to expand knowledge in this field through teaching (Postgraduate student in Chemistry)

Two answers (10 %) mentioned the importance of research (category 2):

The role of the university is to find ways of solving environmental problems by encouraging research and disclosure. (Student from the teacher training course in Chemistry)

Since the university produces basic technological, warfare and industrial knowledge, among others, it is the main responsible for the generation of toxic and contaminating material. It is worth considering who the university works for. (Undergraduate Chemistry student)

Management elements (category 3) were present in 15 % of the answers, all of which emphasized more sustainable management, recycling and reuse of campus waste:

In my opinion, the university should create mechanisms, considering how much is wasted and what could be avoided, from food in the university restaurant to wastes in lab classes; (it should) promote the understanding of green chemistry. (Undergraduate Chemistry student)

In the second open question (b), about clean or green technologies, 90 % of the students mentioned the relevance of the topic and how little the university was involved in research, politics, teaching projects, extension and other practices in this field.

I believe the university should develop, apply and divulge available green technologies more. And somehow encourage undergraduate students to help develop and implement new green technology. For instance, I live in the students' housing in the campus. I believe they could install solar panels to capture energy during the day, so that electricity bills could become much cheaper. (Student from the teacher training course in Chemistry)

According to Brazilian law, educational institutions have to promote EE through disclosure and use of technologies for sustainable management of natural resources. Thus, educational efforts must be accompanied by management, structural and operational actions, concerning wastes, clean technologies or other environmental issues.

5.2 Greening the Curriculum and Environmental Management

Considering the importance of teaching and environmental formation, three questions were analyzed:

(a) *Is Environmental Education treated in any discipline or activity in your undergraduate course? If so, how is it dealt with and in which disciplines?* (Open question)

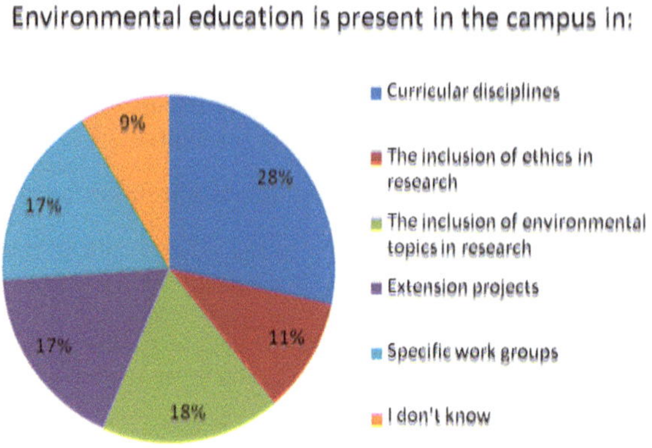

Fig. 2 Percentage of areas where EE is present

(b) *Has your undergraduate course contributed to your environmental formation? Comment.* (Open question)
(c) *Where and how is EE present in your university?* (Closed question)

The answers to question (c) were categorized by course. The participants of the teacher preparation course in Chemistry (15 %) said that only some lecturers had put EE into action, mainly in lab practice and chemical waste management. All the students of Environmental Management and Analysis (35 %) confirmed that the subject is well addressed in every discipline and in interdisciplinary projects. On the other hand, students of mathematics and undergraduate chemistry (15 %) did not identify any of these actions.

In relation to question (d), 75 % of the students answered "yes": the courses have contributed to their environmental formation by the effort of lecturers, disciplines or projects.

The closed question (e) indicated that students perceive EE in curricular disciplines (28 %), the insertion of socio-environmental topics in research (18 %), extension projects (17 %), specific EE groups (17 %) and the inclusion of ethics in research (11 %). Only 4 (9 %) mentioned they did not know such practices, as shown in Fig. 2.

Curricular greening still needs improvement, since the rigid structure of the disciplines hardly permits the dialog between the fields of knowledge, mainly with the exact sciences. The structure of the curriculum is part of a scientific field which, like any other, is a place for dispute, with concepts legitimized by whom deters prestige and power, therefore making it difficult to insert other forms of thinking (Zuin 2011). Apart from purely technical education, EE is seeking formation, in Adorno's (2000) proposition, towards reflection, autonomy, and understanding of the relations of production and reproduction of human life in society and in relation to nature.

Besides investing in the formation of its community, as pointed out by the students, the university should adopt sustainable practices in campus management. In the question "How do you evaluate waste management in your campus?", most

Fig. 3 Evaluation of campus waste management

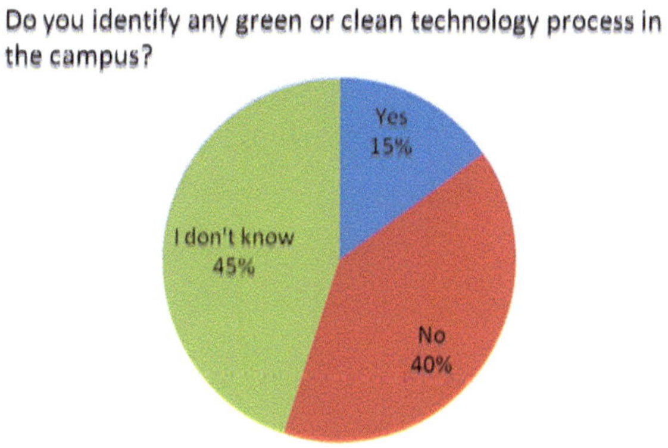

Fig. 4 Identification of the use of green technologies in the campus

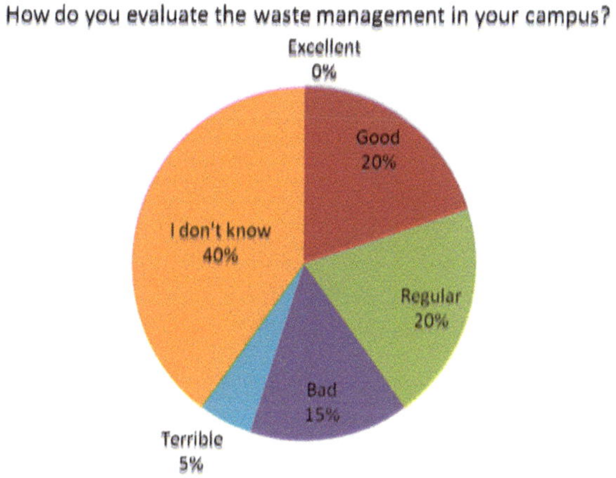

of the participants (40 %) declared not to have any knowledge, followed by "regular" (20 %), "good" (20 %), "bad" (15 %) and "terrible" (5 %)—Fig. 3. Some qualified their answers in the item *Comment*, indicating that wastes constantly accumulate in garbage bins and are disposed of in other inadequate places.

Another highlight was the lack of knowledge about clean or green technologies. In the question "Do you identify any kind of green process or technology in your campus?" 45 % answered "I don't know", followed by "no" (40 %) and "yes" (15 %)—Fig. 4.

One positive answer was that half of the students recognized that there are organs in charge of EE and management in the campus. To the question "Is there any organ in your campus dedicated to environmental issues and problems?" 50 % answered "yes" and even identified some groups: CEMA or SGAS, UGR, Grupo GIRe[3] (Grupo de Incentivo a Redução, Reutilização e Reciclagem) e Grupo GAIA (Grupo Ambiental Ipê Amarelo). Good part of the students (45 %) declared they did not know, and only 5 % answered "no"—Fig. 5.

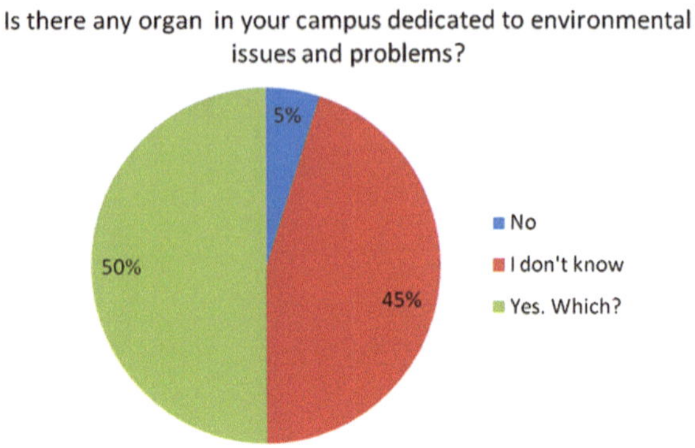

Fig. 5 Knowledge of organs responsible for environmental issues in the campus

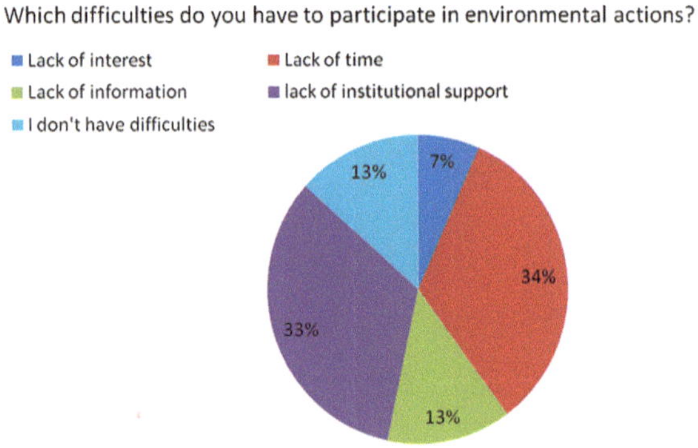

Fig. 6 Students' difficulties to get involved in environmental practices

The last answers analyzed were based on the students' involvement with environmental issues: "Do you feel motivated to participate in actions related to the environment in your campus?". 70 % answered "yes" and 30 %, "no". For the question "Which difficulties do you have to participate in environmental actions?", only 13 % mentioned *none*. Most students answered "Lack of time" (34 %), followed by "lack of institutional support" (33 %), "lack of information" (13 %) and "lack of interest (7 %)—Fig. 6.

With this same tool, Borges (2013), in his studies at Pontificial Catholic University of Rio Grande do Sul, also pointed out that most students (53.1 %) do not have time to get more involved. Characteristic of modern western society are the amount and speed of the flow of information. Activities need to be increasingly faster, more specific, more fragmented, unfortunately compromising memorization and significant connections between what goes on and what touches us (Bondía 2002). Time is a fundamental element to enhance experience and consequently contribute to change values.

Beyond sustainability assessment, the platform tends to be a place for reflection, containing texts and videos. As every tool, it has limits and deficiencies, at this moment not worth discussing.

The last open question for evaluating the test itself—*Did you like the test? Did it help you in any way?*—raised positive elements: all those who concluded the test liked its approach, despite considering it long and tiring. Receptive to criticism and suggestions, the test is constantly improving. At the moment it is being adapted so new uses can be made concerning new direct contacts and greater involvement of the academic community.

6 Conclusions

In synthesis, most participants believe that the role of the university is to actually strengthen EE. Few, or none, perceive it inserted in disciplines and exact science courses such as graduation in Chemistry or teaching Mathematics. Waste management and the use of green technology were not well evaluated, the last topic being even less known to the students. On the other hand, organs and institutional groups responsible for this issue have certain visibility, at least for half of the participants. In other words, UFSCar (São Carlos) has a lot to improve in addressing sustainability, in the curriculum and campus management, including the adoption of clean technologies (Zuin 2011).

The pilot project carried out for this article provided a first environmental diagnosis, which resulted political, ethical and aesthetical interpretations, concepts and judgments. It also permitted a review of the *Sustainability Test*, defining its limits and potentials, and adapting it to local conditions in the campus. The data presented brings interesting elements to the field of research on sustainability assessment. In order to obtain subsidies to improve the dialog between managers and students, the next step is to involve more students and to strengthen links with the professors.

We hope this movement helps the academic community to think about environmental issues and teachers and managers to perform their daily activities in a way that treats sustainability as part of the institution, as a source for ideas and not as a marginal issue. The search for an education based on dialog and criticism is a long and challenging process that, in this institutional context, is limited by material and human resources, besides centralized power relationships. Fomenting research about university greening and promoting dialog are strategies to face these challenges.

References

Adorno TW (2000) Educação e emancipação, 2nd edn. Tradução de Wolfgang Leo Maar. Paz e Terra, Rio de Janeiro, p 190

Bondía JL (2002) Notas sobre a experiência e o saber da experiência. Revista Brasileira de Educação, 19:20–28. http://www.scielo.br/scielo.php?script=sci_arttext&pid=S1413-24782002000100003&lng=en&tlng=.%2010.1590/S1413-24782002000100003 (last accessed 4/09/2014)

Bourdieu P (2003) O campo científico. In: Ortiz RA (ed) Sociologia de Pierre Bourdieu. Olho D'água, São Paulo, pp 123–155

Brasil. Lei n. 9795 de 27 de abril de 1999. Dispõe sobre a educação ambiental, institui a "Política Nacional de Educação Ambiental" e dá outras providências. Brasília: Diário Oficial da União

Brasil. Parecer CNE/CP nº. 14/2012 que estabelece as "Diretrizes Curriculares Nacionais para a Educação Ambiental". Brasília: Diário Oficial da União

Brandli LL, Frandoloso MAL, Fraga KT, Vieira LC, Pereira LA (2012) Avaliação da presença da sustentabilidade ambiental no ensino dos cursos de graduação da UPF". Revista da Avaliação da Educação Superior (Campinas) 17(2):433–454. http://www.scielo.br/scielo.php?script=sci_arttext&pid=S1414407720120002000008&lng=en&tlng=pt.%2010.1590/S1414-40772012000200008 (last accessed 4/09/2014)

Brundtland (1991) Nosso Futuro Comum—Comissão Mundial sobre Meio Ambiente e Desenvolvimento. 2nd edn. da Fundação Getúlio Vargas, Rio de janeiro, p 430

Carvalho ICM (2011) Educação Ambiental: a formação do sujeito ecológico, 5th edn. Cortez, São Paulo, p 256

Foladori G (2001) Limites do Desenvolvimento Sustentável. da UNICAMP, Campinas, São Paulo, p 696

Gaudiano EG (2001) Una nueva lectura a la historia de la educación ambiental en América Latina y el Caribe. Revista do Desenvolvimento e Meio Ambiente 3:141–158. http://www.ecologiasocial.com/biblioteca/GonzalezGhisotiraEducAmbALat.pdf. da UFPR (last accessed 4/09/2014)

Borges JAS (2013) Sustentabilidade e acessibilidade no ensino superior: contribuições para um diagnóstico socioambiental da PUCRS. Dissertação (Mestrado em Educação), Porto Alegre, p 144

EUAC (2014) Environmental association for universities and colleges, EAUC—United Kingdom. http://www.eauc.org.uk/theplatform/why_sustainability_in_education (last accessed 4/09/2014)

Leal Filho W (2011) Applied sustainable development: a way forward in promoting sustainable development in higher education institutions. In: Leal Filho W (ed) World trends in education for sustainable development. Peter Lang Scientific, Frankfurt, p 462

Leff H (2001) Epistemologia ambiental. Cortez, São Paulo, p 240

Lima GFC (2002) Educação e Sustentabilidade: possibilidades e Falácias de um Discurso. Anais do I Encontro da Associação de Pós-Graduação e Pesquisa em Ambiente e Sociedade, Indaiatuba-SP, p 15, http://www.anppas.org.br/encontro_anual/encontro1/gt/sociedade_do_conhecimento/Gustavo%20F.%20Costa%20Lima.pdf(last accessed 4/09/2014)

Lüdke M, André MEDA (1986) Pesquisa em Educação: abordagens qualitativas. Pedagógica e Universitária, São Paulo, p 99

Leme PCS, Pavesi A (2012) A plataforma da sustentabilidade como base para a construção coletiva de comunidades universitárias sustentáveis e solidárias". In: Leme PCS, Pavesi A, Alba DH, Díaz MJG (Org.) Visões e Experiências Ibero-Americanas de Sustentabilidade nas Universidades. Gráfica Alhambra, Madrid-ES, pp 197–203

Leme PCS, Pavesi A, Alba DH, Díaz MJG (Org.) (2012) Visões e Experiências Ibero-Americanas de Sustentabilidade nas Universidades". Gráfica Alhambra, Madrid-ES, p 357

Moraes R, Galiazzi MC (2006) Análise textual discursiva: processo reconstrutivo de múltiplas faces. Revista Ciência Educação, Bauru 12(1):117–128

Oliveira HT (2012) Contextos e desafios na produção de sentidos sobre sustentabilidade e ambientalização da educação superior. In: Leme PCS, Pavesi A, Alba DH, Díaz MJG (Org.) Visões e Experiências Ibero-Americanas de Sustentabilidade nas Universidades. Gráfica Alhambra, Madrid-ES, pp 37–42

Sáenz O (2014) Panorama de la sustentabilidad en las universidades de América Latina y el Caribe. In: Ruscheinsky A, Guerra AFS, Figueiredo ML, Leme PCS, Ranieri VEL, Delitti WBC (Org). Ambientalização nas instituições de educação superior no Brasil: caminhos trilhados, desafios e possibilidades. EESC/USP, pp 25–38

Rupea (2007) Rede Universitária de Programas de Educação Ambiental. Mapeamento da Educação Ambiental em instituições brasileiras de Educação Superior: elementos para

discussão sobre políticas públicas". Documento Técnico nº 12. Brasília: Órgão Gestor da Política Nacional de Educação Ambiental
Tilbury D (2011) Higher education for sustainability: a global overview of commitment and progress. In: GUNI (ed) Higher Education in the world 4. Higher education's commitment to sustainability: from understanding to action. palgrave. Macmillan, Barcelona
ULSF (2008) Association of University Leaders for a Sustainable Future. Secretariat for signatories of the Talloires Declaration. http://www.ulsf.org/ (last accessed 4/09/2014)
UFSCar, PDI (2012) Plano de desenvolvimento institucional. Versão preliminar para apreciação do Conselho Universitário. Universidade Federal de São Carlos (UFSCar), São Carlos-SP
Zuin VG, Farias CR, Freitas DA (2009) Ambientalização curricular na Formação Inicial de Professores de Química: considerações sobre uma experiência brasileira". Revista Eletrónica de Ensenãnza de las Ciencias 8(2):552–570
Zuin VG (2011) A inserção da dimensão ambiental na formação de professores de Química, 1st edn. Átomo, Campinas, p 179

Author Biographies

Marisa Sartori Vieira has a degree in Pedagogy and is a postgraduate student of the Program in Education at the Federal University of São Carlos. She researches Environmental Education and Sustainability in Higher Education. She was a technical assistant (2011) of the International Cooperation project in a partnership between the University of São Paulo (USP) and Universidad Autónoma de Madrid, Spain (UAM), developing a Virtual Platform of Sustainability in Universities (http://www.projetosustentabilidade.sc.usp.br/index.php/eng).

Patricia C. Silva Leme is an environmental educator at the Environmental Management Office of the University of São Paulo (USP, http://www.sga.usp.br). She has a degree in Biology at the Federal University of São Carlos, Brazil (UFSCar), a Master Degree and a PhD both in Education at UFSCar (2008), with a part-time scholarship at the Universidad Autónoma de Madrid, Spain (UAM). She was a research fellow at the University of South Africa (April–July 2013), coordinator of the International Cooperation project which aims at addressing sustainability in universities (2009–2011) and has been coordinating the Virtual Platform of Sustainability in Universities (http://www.projetosustentabilidade.sc.usp.br/index.php/eng) since 2010 In the last years she has been researching and publishing in the areas of training of environmental educators and addressing environmental education and sustainability in universities.

Vânia G. Zuin is a professor at the Federal University of São Carlos (UFSCar, Brazil). Her background is in Science (Analytical Chemistry—São Paulo University, Brazil, and Università degli Studi di Torino, Italy, 2001) and Education (Environmental Education—São Paulo University, Brazil, 2010), with major interests in developing green methodologies to determine bioactive high-value organic substances extracted from brazilian agroindustrial wastes. Her research also focuses on the development and dissemination of Green Chemistry and Environmental Education material for universities. She wrote several publications of high impact in her areas and coordinated international projects, such as *Green Chemistry in Higher Education: toward a green chemistry curriculum for Latin America and Africa universities*, sponsored by IUPAC. She is a member of the IUPAC Subcommittee on Green Chemistry and, since 2012, the coordinator of the Green Chemistry Section of the Brazilian Chemistry Society (SBQ). As a member of the Alexander von Humboldt Foundation research network (AvH, Germany), she was elected vice-president of the Humboldt Kolleg in Brazil in 2013. Recent distinctions include the 52 National Literary Jabuti Bronze Award (Brazilian Chamber of Book) and the title of Honorary Visiting Fellow at the University of York (UK).

Experiences of 'Reflective Action': Forging Links Between Student Informal Activity and Curriculum Learning for Sustainability

Christine Willmore and Hannah Tweddell

Abstract

This paper uses research carried out at Bristol to explore 'reflective action' as a tool for building relationships between formal curricular structures, informal activity by students and the community to enable sustainability education to engage both theoretical understanding and practical experience. It explores the synergies available when activity can move between the formal curriculum and student activity outside of their formal studies and looks at the benefits and challenges for institutions who have sought to bring engaged student activity focussed upon real world problems into formal learning to provide active learning for sustainability. It reflects on the impacts students perceive this to have on the community in which the University is situated and the university: community relationship.

Keywords

Education · Reflective action · Action learning · Curriculum · Engagement

C. Willmore (✉)
Bristol Green Academy, Reader in Sustainability and Law, University of Bristol, Wills Memorial Building, Queens Road, Bristol BS8 1RJ, UK
e-mail: chris.willmore@bris.ac.uk

H. Tweddell
University of Bristol Students' Union, Queens Road, Bristol, UK
e-mail: hannah.tweddell@bristol.ac.uk

W. Leal Filho (ed.), *Transformative Approaches to Sustainable Development at Universities*, World Sustainability Series, DOI 10.1007/978-3-319-08837-2_36

1 Introduction

Student action for sustainability has gathered pace, despite the waxing and waning of governmental commitment. At least a third of students at Bristol University are engaged in informal curriculum sustainability opportunities, but this is just the easily documented tip of a larger iceberg. Given that 12 % of the population of the city of Bristol are university students, and nationally there are 7 million students and 600 student unions, this offers a large, but largely unseen, potential impact.

University is often the first period away from home and initiatives are predicated on the premise that practices developed at this stage become habits for life. This process of habit building operates through the informal and subliminal curriculum, but this paper suggests that whilst habits of this kind can contribute to sustainability education, (Lipscombe 2008), internalisation is strengthened when linked to sustainability learning in the formal curriculum (Ryan and Cotton 2013). Theories about influences on positive sustainability behaviours are not always reconcilable, so it is not possible to articulate precise pathways through which actions may become embedded and normative (Ajzen 1991; Jackson 2005; Kolimuss and Agyeman 2002; Stern 2000). The emphasis in this paper is upon developing reflective awareness to achieve internalised change. It is easy to assume that this internalisation is happening—but internalisation requires reflection. Reflection is a learnt skill, which deepens with practice. To promote transformative education and reflective practice requires opportunities to practice those skills. This article examines one way of doing this 'reflective action' projects.

2 The Bristol Approach

The authors often find academics will be recycling, walking where possible, actively supporting local charities and even be passionate self identifiers as committed to sustainability—but will somehow separate that part of their lives from their academic work. This ability to split the self into different aspects presents a challenge to sustainability education.

Bristol University maps the opportunities for influence and domains of student experience using five circles. Student experience results from the interplay of these domains.

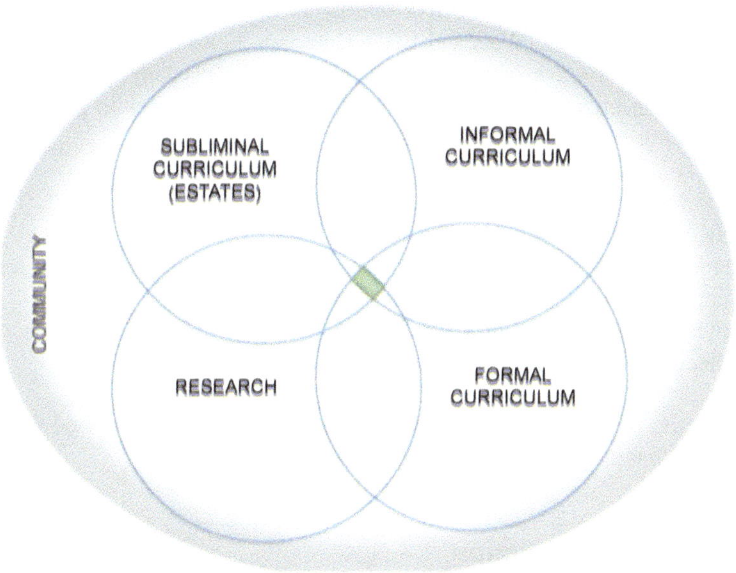

The formal curriculum is only one part of the student experience. The informal curriculum includes activities in societies and informal social networks. The subliminal curriculum reflects the structuring of practice resulting from the physical estate and cultural ethos of the institution, the provision of access to some resources and the relative inaccessibility of others. And finally, living in a research environment shapes student experience through explicit organising norms. Although the diagram shows the potential for areas of learning to overlap, it is possible for them to operate in isolation, splitting theory from practice, their lived lives. Sustainability education offers opportunities for holistic insight into how the facets of their lived experience come together: to experience the overlap point. If students have an opportunity to experience this point of unity, they may experience the transformative potential of sustainability education. The claim is that 'reflective action', as a variant of action learning offers a rich tool for a transformative sustainability education.

Work carried out by Tierney and Tweddell (2012) at Bristol University with student focus groups identified a sense of disconnect between the formal curriculum offer and the student's practical experiences and a request for more explicit links between the formal, informal and subliminal curriculum.

This article explores one particular approach to strengthening the theory: practice relationship, through the embedding of 'reflective action' projects as part of the summative assessment of a student. Whilst the focus is upon developing a student's own alignment of theory and practice, the creation of space for those opportunities requires institutional and organisational alignment.

Almost all students must carry out a final year project, offering a broad potential for students to carry out reflective action research rooted in the theoretical and methodological paradigms of their studies, but providing opportunities to embed sustainability in practical action.

Central to the aim of 'reflective action' is the provision of opportunities for transformative learning (Wals and Blaze Corcoran 2006) focussed on the ability to integrate multiple ways of understanding issues, the capacity to cope with complexity and uncertainty, and to reconcile theoretical insights with practical action.

Our premises for this research are that:

- Action without reflection risks activism
- Reflection without action is theory
- There is no action without learning, but that reflection enriches that learning
- It cannot be assumed that reflection occurs
- Hence, the model of action + reflection = learning was developed as a form of 'reflective action'.

Reflective action is a process through which students are supported to bring into play the different domains of their experience as a student.

3 'Reflective Action'

'Reflective action' is a response to Kolb's (1984) experiential learning cycle, aiming to add a structured opportunity to increase the likelihood normative embedding through reflection.

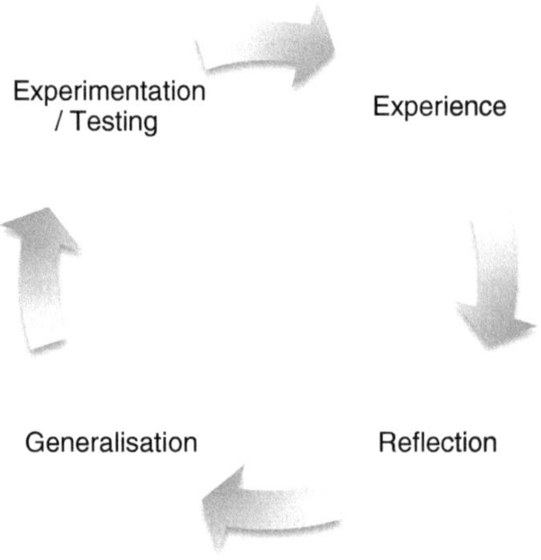

3.1 Kolb's Learning Cycle

'Reflective action' requires action in a community external to the classroom. It is located within the family of action learning approaches (Brockbank and McGill 2004; Beaty and McGill 1992; Rooke et al. 2007; Raelin 2011; Pedler and Abbott 2013), but has a dual emphasis upon reflection and engagement with external communities.

'Action learning' is generally attributed to Revans in the 1960s deriving from his comment 'there can be no learning without action and no sober and deliberate action without learning' (Pedler 2011; Revans 1982). Raelin (2011) identified common features including their dialectic basis, the development of contextualised insight, with learners as active participants in the process, emphasising reflection-in-action rather than reflection-on–action, meta-competence, facilitated learning, practice based learning outcomes, comfort with tentative outcomes/uncertainty. These characteristics make the family of action learning approaches fertile ground for new sustainability pedagogies. Beaty (2001) observed of action learning that "members engage in learning from experience in order to change rather than simply repeating previous patterns." 'Reflective action' uses that, but stressed the importance of support for reflection to strengthen opportunities for transformational learning.

'Reflective action' also draws upon action research approaches (Bradbury Huang 2010; Reason and Bradbury-Huang 2001), but the emphasis is upon the development of the student researcher as opposed to the research itself.

Writing on experiential learning tends to emphasise the process of problem solving, as opposed to the opportunities for normative reflection, although Schön's work (1983) around reflection-in-action does focus more upon surfacing intuitive understanding. Amulya (2004), in the context of practitioner education, uses the concept of 'reflective practice' to articulate more reflective focus: "Reflection is an active process of witnessing one's own experience in order to take a closer look at it, sometimes to direct attention to it briefly, but often to explore it in greater depth. This can be done in the midst of an activity or as an activity in itself. The key to reflection is learning how to take perspective on one's own actions and experience… By developing the ability to explore and be curious about our own experience and actions, we suddenly open up the possibilities of purposeful learning—learning derived not from books or experts, but from our work and our lives."

Reflective action is used in this article to describe a process of giving students an opportunity to develop their own approach to the theory: practice relationship; to develop reflective practice and enables students to see reflexivity in action. That explicit process is more likely to produce normative transferrable life practices (Cantor 1995; Cranton 1989; Knowles 1977).

4 Movement Between Formal and Informal: Case Studies

Reflective action in the formal curriculum can exist at a several levels simultaneously:

- Individually brokered projects
- Embedded elective opportunities
- Embedded programme provision
- Whole institution approaches.

As the number of students involved increases there is a growing need for institutional alignment, but also a growth in the potential to influence institutional culture, and to strengthen community and formal/informal curriculum relationships. However the relationship between formal and informal curriculum activity depends as much on what might be called 'organisational fit' as scale.

4.1 Individual Initiatives

Examples of individuals who have 'knitted their own' links are common. Individually brokered opportunities occur when there is an alignment of an identified community aspiration and a student's ideas. Such opportunities are individual, opportunistic, and do not give rise to sustained opportunities. Methodologically and pedagogically they span a wide range of processes. Central to successful projects was a prior relationship, between a community group and an academic, student or a student society. Community partners were well motivated and active. The outcome sought by the community was a single piece of work, whose timing mapped onto institutional timelines. There was not necessarily a full exploration of reflective action approaches and whilst individuals in those studies reported positively on opportunities for reflection and developing reflexivity awareness, these were not always explicitly part of the methodology or assessment and there were no peer support networks. Whilst these can be rich, transformative experiences for the individual, they are resource intensive and cannot form the basis for whole institutional transformations.

4.2 Embedded Elective Opportunities

Embedded elective opportunities offer these opportunities to more students in a supported manner and can form part of an institutional approach, but still do not require institution alignment. The Bristol University Farmers Market illustrates this approach and the importance of formal, informal and subliminal curriculum communication. A module, Global Environment and Human Health, required students to develop and deliver a community health project. One cohort identified a lack of outlets to purchase locally sourced healthy foods close to the University precinct and developed a precinct public farmer's market to address this.

The pedagogies of this unit derive from action research so the methodology and purpose was about learning through doing, exploring reflexivity and reflecting upon the process. For the original students, learning outcomes were associated directly with their role as change agents when healthcare professionals.

At the module's conclusion, the initiative was adopted within the informal curriculum by the Student Union then normalised in institutional practice, (the subliminal curriculum), supported by estates. The way in which the project moved seamlessly through the formal, informal and then subliminal curriculum illustrates an important issue: the ethics of designing engagement projects which, if they have an impact, should continue beyond the life of the student project. For this movement to occur institutional and organizational alignment is necessary, or at least, communication space must exist. Such opportunities, integrated into a module, are moving towards more sustainable models of theory: practice alignment.

4.3 Whole Programme Examples

When moving from individual modules to whole programme embedding of reflective action a different set of challenges emerge. It is possible to develop a whole cohort project, but the Bristol work indicates that whole programme reflective action projects require a process of brokerage, where a plurality of potential projects are identified, which facilitated mapping of student interests onto those projects. Two examples have informed the authors' thinking on this.

A Bristol University led STEM Project (Miller 2011) explored the incorporation of community-based placements in the undergraduate Civil Engineering curriculum. It was not specifically a sustainability project, but sends light on the processes required. This was conducted as an action learning project, designed to deliver engaged project opportunities for a cohort of students, but also to learn about the issues associated with whole programme approaches. Third year student interns who were undertaking the course worked with community partners to scope projects for the third year project i.e. their own cohort. That study identified a number of important issues reflected upon later.

In a second example, for the Environmental Policy and Management M.Sc., a recent graduate from the course was employed to scope projects with community partners suitable for an M.Sc., dissertation, but using a different pedagogic positioning, in which the community idea formed a backdrop to a project, which did not necessarily have to use reflective action methodologies.

Building upon this work, Bristol University carried out research, designed to identify how it might adopt a whole institution approach to the provision of opportunities for embedded community research, utilised semi-formal interviews with staff, students and community representatives who had already been involved in community partnerships or similar initiatives (Wakefield and Leggett 2012).

5 Challenges of Embedded 'Reflective Action' Projects

Experiences from Bristol demonstrate that there are challenges of community based learning projects. Competing priorities, multiple expectations, conflicting agendas and power relations were found to exist between different partners.

Universities traditionally have a top down approach to community interaction through institutional partnerships and engagement strategies and risk being inflexible to this type of work. One way to provide sufficient institutional flexibility, within a whole institution strategy is through the scholarship of engagement language of 'critical engagement' (Van de Ven 2007). Whilst Fear et al. (2006), have warned against what they perceive as a tendency in the critical engagement literature to see engagement as an institutional and organisational question which "misleads us into believing that enhancing and advancing engagement is an organisational development matter", the Bristol experience suggests there are institutional and organisational challenges to be addressed to create the space within which 'reflective action' can flourish. The adoption of a holistic view of student experience is itself an institutional decision. The impact of the Farmers Market example was only possible in a context of an institutional approach facilitating diverse networks of interaction.

Academic resistance, noted by Wakefield and Leggett (2012), may stem from a number of causes. Embedding reflective action projects in the curriculum requires considerable planning, before even exploring the appetite of potential partners, and in essence needs to both emerge from and in turn shape the relationships between institutions, students and the communities to which they relate. Whilst institutional policy can create a climate of encouragement and support, the process is about relationships and a commitment to co-creation of insight, must be fostered not imposed. All projects explored in the Bristol study were driven by academics, informed by student aspirations expressed through feedback, and whilst students were heavily engaged in designing the particular project, they did not originate from a co-creation approach to curriculum development. The evidence suggests that the best projects result from students and academics working as partners to identify an appropriate approach from the outset. Inevitably such inquiry will be shaped by a disciplinary focus, but cannot be focussed exclusively upon it at least as the starting point for exploring complexity. Reflective action involves partnership, recognition of multiple sources of expertise, decentring the expertise of the academic (Pilling-Cormick 1994), using new methodologies to accommodate research for real world problems. Non-traditional approaches tend to receive less recognition from senior managers (Boyer 1990), however there is some evidence in Wakefield and Leggett (2012) that these initiatives could actually be positive in terms of income and publications. Some welcome the challenge to academic thinking, as a supervisor responded to Miller (2011): "We had to come up with different ways of approaching the problem so that we could get hold of useful information, and that was certainly a challenge for me".

The process of reflective learning through community projects places students in contexts of uncertainty and complexity: on the edge of chaos. Academics themselves may be uncomfortable with such uncertainty, however rich a learning place, particularly for the sustainability change agent skills (Kleiman 2011).

5.1 Learning Framework

Articulation of the learning sought needs to reflect some fundamental thinking about the nature of any particular reflective action project. Reflective action projects focus upon self reflection, understandings of reflexivity and the methodologies of working in a community. The central aim for the student may be a grade, but an ulterior aim is to develop their own mediation between theory and practice: to provide the opportunity for what can be holistic and transformative insights of their own. This contrasts with placements, where students demonstrate that they have successfully applied their employer's appreciation of the theory/practice relationship to sustainability.

Academics and students need to identify their rationale for embarking upon this —what are the real drivers? Are they turning communities into a living laboratory, or are they seeing the process as one of co-creation where the student and community partner jointly create fresh insights and potentially act as catalysts for change? Is it a spin off from an academic's own research or a student's informal curriculum activity or a potential source of fresh research insights and challenges? Is it outcome or process the focal point?

Emphasis within the skills abilities and knowledge specified in sustainability learning outcomes varies (Rowe and Johnston 2013; Scott 2007). The particular feature of learning outcomes that reflective action engages is a set of skills described by the American College Personnel Association (ACPA 2008a) led learning outcomes framework as 'change agent skills', a detailed specification including partnership, participation, interdisciplinarity, reflexivity, and the ability to link theory and practice (ACPA 2008b).

Appropriate definition of learning outcomes is essential to the successful embedding of reflective action and to avoid mismatches of perception. Intended Learning Outcomes (ILOs) need to be articulated, for individual modules and, in the case of programmes requiring reflective action, for the programme. Articulations will vary, it may be defined by reference to students articulating a theory: practice relationship, to be able to deliver partnership projects, or may extend to referencing the holistic, transformational experience, although the inherent unpredictability of transformation points makes them generally unsuitable for inclusion in ILOs. A key question to ask in phrasing the ILO is whether this particular approach is for all, or simply one way of doing a project, alongside laboratory or theoretical ones.

This leads into the task of articulating the mode of assessment and assessment criteria. The focus of reflective action is upon co-creation of new insights into subjects or relationships in a 'learning community'. This requires a re-negotiation of concepts of summative working. External to the academy, co-creation is the norm. Within the University it is circumscribed by conventions of individual assessment. The particular pedagogic issues associated with assessment of group work with external partners need explicit consideration.

What is the thing that is to be assessed? Journal? Project? Portfolio? The choice depends upon the precise focus of the ILO. If the focus is upon inward reflection reflective journals might be used. Is the measure of success the articulation of a student's own theory/practice relationship? Is the process sought one of reflection-in-action or reflection-on-action? Traditional academic criteria? Translational skills? In the authors' experience, consideration of the nature of the 'thing' to be assessed is a good vehicle for exploration of the real nature of the module and its hidden assumptions. Is the expectation that the student will deliver one document, for use by the partners or two documents, one, the community report designed for use by the partner, and another for the purposes of academic assessment? The 'two masters' solution is less challenging as a starting point, but it reinforces theory/practice divisions and the power of the academy rather than dismantling them. Bringing these two together into one product is more challenging for academics and students.

5.2 Methodological Framework

The second big set of challenges is around methodology. What is the position of the community partner in the research? Who is setting the question and methodology? Are community partners the subject or object of the research?

Complex real world questions require epistemological pluralism, (Miller et al. 2008). An action-learning project would expect to be using Revan's methodology (Coghlan and Pedler 2006). However the proposal here is not a full use of action learning, but rather a real world engagement as a catalyst to learning using a variety of methodologies.

Students are trained in the methodologies most commonly deployed in research in their own discipline. This may include action research and co production, but these are seldom taught as research methodologies early in a disciplinary career, so students may approach their project without prior experience of such methodologies. It cannot be assumed that students will naturally and inherently be able develop and apply appropriate methodologies. So, including reflective action projects in a curriculum requires reflection upon the methodological training provided through the curriculum to ensure it is appropriate. At a fundamental level this makes the notion of reflective action projects a driver for a wider reflection upon the curricular experience.

5.3 Skills Framework

Working with partners requires particular skills, notably the change agent skills considered earlier. Reflective action provides both an opportunity to practice many of those skills in an integrated manner, but also requires prior experience of them as scaffolding to the work. To maximise the prospects of success, students need practice in interdisciplinary practice and group skills as well as appropriate methodological training. Some of the training needs to be specific to the project e.g. the 'two master' issue, but much is generic.

There is also a need for support throughout the project itself. This can come in the form of peer-to-peer support; mentoring by prior participants in reflective action or the provision of a resource centre. This most normally takes the form of a resource centre with examples of past projects. However, there are opportunities for richer community constructed multi media resources to illustrate the stories developed so far in reflective action projects, and open up questions for future projects to build a visible weave of relationships, so that those involved in individual reflection action projects can appreciate their connection to a bigger whole.

5.4 Partnership Framework

Beyond internal requirements of curricular consistency and pedagogic structures, there is a key concern, of ensuring the process does not lose sight of the community partner, recognising this process has at least three loci of partnership: student, community and academic.

It is vital to ensure the project is of benefit to all partners: student, community and academic. If it requires the academic or community partner to do more, or do different (which at least in the early stages means doing more), takes them out of their comfort zone or away from their own chosen approaches they are less likely to engage, unless there is a benefit in excess of the perceived burden.

This leads into the central topic of relationship building. Long-term relationships between students, academics, and community maximize the effectiveness of 'reflective action' approaches. They foster shared understanding, lead to reduced transaction costs, and reduce the turnover in partners. As shown in the Farmers Market example, involving student societies within the partnership framework can offer both sustainability and enhanced opportunities for work to move between the informal and formal curriculum.

The strongest relationships exist where the University has decentred itself from the process, and there is:

- Articulation of drivers/benefits for all parties
- A culture of co-creation
- Equal partners respecting the expertise that each brings

However committed individuals are to such partnership working, there are embedded power problems (Down and Nurse 2007). It is very easy for HEIs to be seen or experienced as centres of privilege and expertise, with a strong sense of self, to claim control through assertions of expertise and to disempower communities as a result. The institutional ethos socializes academics, and students to the idea of the other as lacking. Reflective action requires decentring that expertise, recognising the community as possessing a different expertise. This decentring poses a significant challenge to Universities. One respondent to Wakefield and Leggett (2012) warned that students should not be seen as a "delegator" but as a "collaborator". Being seen as a delegator can lead to research being done about, done to or done for the community group rather than done with leading to disempowerment. Even if individuals feel in a position to transcend such power relationships, subliminal or historic feelings of disempowerment may require a lengthy process of relationship building in relation to those who have been most marginalized and disempowered. Reflective action projects require explicit work to recognize and share what others bring to the table, and provide space for shared conceptualization, what Down and Nurse (2007) have called "rules for engagement to help us de-centre ourselves".

Power issues also operate within community partners, themselves complex organizations with multiple power loci, making it important to reflect upon power structures within the partner organization (Brown and Gaventa 2010).

5.5 The Two Masters Challenge

A number of differences in practical expectations between the institution, student, academic and partner can bedevil the process unless carefully mapped.

Ethics approval is one such difference. Methods for securing ethics approval within HEIs are not always conducive to timely development of student projects. The community partner, however, may operate in a very different ethical context, and may therefore be frustrated by the requirements of HEI ethics approval. Making the navigation of such tensions explicit can itself be seen as a valuable sustainability learning opportunity.

The two masters issue extends to questions of timescale and outcome. Community groups are not short of ideas that would benefit from research by students. Defining the questions in appropriate timescales is harder. Universities operate on an academic year basis—students sign up for projects typically just over a year before they submit the project with only one window per year. Postgraduate projects often work to an even tighter timeline. This may not fit with the community partner's timescales. Mismatches of expectation need careful management. Long term relationships and careful brokerage, can address some of these, but prior informal curriculum relationships are particularly useful, enabling formal curriculum projects to be seen as part of a continuum of relationship with the University: community partners do not generally see the formal/informal curriculum distinction to be as important as the institution does.

5.6 Student Brokerage

Beyond projects established by individuals, reflective action projects require the development of a brokerage scheme. Whilst Eulogy operates a national on line brokerage, and the National Union of Students in the UK is experimenting with a very light touch on line brokerage in 2014/15, most current projects have a substantial brokerage demand.

Brokerage requires the development of a relationship between students, academics and community partners to ensure that projects. Finalisation of the proposal will lie with the student and their partner, but the initial framing of a question needs sensitive nurturing.

The Bristol work highlights the success of using paid student interns who are undertaking, or had recently undertaken, the relevant course as brokers. Whilst it can present some continuity challenges, it provides valuable experience of project brokerage for the individual interns, and is particularly beneficial in scoping projects. Interns have the subject and technological knowledge to understand what projects would appeal to students and can inspire their peers leading to an improved uptake of projects (Miller 2011). They can help shape the projects to appeal and help students to understand what could be drawn from a particular project.

5.7 Sustainability

There is a need to embed reflective action project processes in the normal structures of teaching and assessment to avoid the perception of the work involved being additional and therefore marginal. To move beyond being a pet project, consideration needs to be given to the longer term sustainability of the project or relationship. Student, and community expectations are built up and need to be managed.

Whilst establishing the first set of relationships requires work, the relationships need sustaining, and refreshing. There is a constant turn over in partners, as their needs and capacity to support such a project changes. Is there going to be a sufficient flow of partnerships to meet student demand? If not, how can this be managed?

For the partner, the nature of the project itself can lead to sustainability issues, which need to be addressed in project design by the brokers working with the partner. If the project develops a new area of work for the partner, will this be sustainable? What is the impact of the project on the partner?

If a useful practice emerges, there are ethical issues associated with not continuing this practice into normative processes. As we saw with the Farmers Market, student societies, as part of the informal curriculum have a useful role as a bridge from one off projects, sustaining relationships and indeed sustaining the project itself beyond the timeframes of particular curricular initiatives.

6 Voices of Partners

Students taking part in reflective action projects report positively on the experience (Wakefield and Leggett 2012). In most cases, however, students had a choice between reflective action and theoretical/laboratory projects, so to some extent the students are a self-selected group. Students offered reflective insights into their experience:

Satisfaction of tangible results

Opening up life possibilities not previously considered

Deeper understanding from contextualisation

Working with communities was far less solitary than research can be

Compatibility of traditional single-disciplinary research and real-world community outcome requirement: new methodologies and logics

Benefits of and need for reciprocity

Increased awareness of the need for reflexivity in the real world.

There was reflection upon the distinction between placements/internships and reflective action projects: "More autonomy than an internship but under the University wing".

Whilst there are the inevitable concerns around practicalities and timelines, community partners who responded to Wakefield and Leggett (2012) were positive. Whilst initially there were fears about the process, these did not materialize:

> Very often expectations were exceeded, in fact far exceeded. There is often a multiplier effect. There have been very few times when nothing has been delivered. Delivery is *always* at the very least interesting and useful for discussion or provocation. Instances where insights contributed which would not have materialised otherwise.

> It's dangerous to expect too much of them, best to take the opportunity to go with the flow… Having said that, any expectations I did have were superseded—my every day expectations were mundane, it has developed into something more, something much bigger.

Limited capacity and time constraints can affect the ability of community partners to support projects. Miller (2011) and Wakefield and Leggett (2012) highlighted that this could be a 'killer' issue for smaller group. Problems can lead to additional calls on academic time (Miller 2011).

There is also an inherent risk of disappointment in any model with more potential projects than students: not all community partners will be matched to a student partner. Sustaining a relationship through such a failure affects future engagement.

Research has so far focused upon those who participated in a project. The reasons for not participating may be surmised, including the capacity of the organization and objective, personality or timeline misalignments; but there is a need for more work into the decision to participate and the impact of investing time in developing an opportunity.

7 Conclusion

This work started from a concern that student action for sustainability may not lead to embedded normative practice. It considered the opportunities for 'reflective action' within the curriculum to deepen student reflection, develop change agent skills and foster transformative opportunities, in particular in relation to their own navigation of theory: practice relationships. It provides evidence to support the case that beyond a certain point, institutional alignment is important to this process. Where there is a mapping of the formal curriculum project onto informal curriculum activities in either a planned or incidental way, this provides more holistic learning opportunities.

The research has identified a number of challenges to the successful provision of 'reflective action', but indicates they can be successfully overcome at individual, elective, programme or institutional level if appropriately planned.

Student feedback, gathered by students, indicated the richness of the opportunities afforded through the provision. Two quotes from students interviewed by Wakefield and Leggett (2012) identified the seeds of transformation within the institution itself arising from 'reflective action' becoming a more widely adopted approach:

> It breaks down in many ways the biggest problem of the ivory tower scenario: students/academics can be totally isolated from societies, losing an understanding of how society works and how the rest of the world understands things.
> By sensitising the University to it you are also making the University learn better what it is in society—this is the most valuable bit: research partnerships encourage reflexivity both within the university and the community partners.

The decentring of expertise associated with reflective action, if successful, may contribute to broader.

Such insights suggest a potential for institutional transformation and co-creation of the construction of the educational institution itself arising from wider use of 'reflective action' approaches. Students themselves perceived reflection action projects as having the capacity to change their own and the institution's relationship to the city itself: "Builds substantial relationships with the community: reduces student transience". "It is part of what a University is, to contribute to the community".

This article has mapped the challenges and opportunities for others considering reflective action in the curriculum as a tool for assisting in normative embedding of sustainability learning. The last word lies with a student interviewee, whose words indicate he had done just that:

> What's the point in being in Bristol if you're not going to contribute to it?

References

ACPA (2008a) ACPA. Towards a sustainable future. http://www.myacpa.org/sites/default/files/ACPA_Sustainability_Monograph.pdf. Accessed 31 March 2014

ACPA (2008b) ACPA. Presidential taskforce on sustainability. http://www.myacpa.org/sites/default/files/Change_Agent_Skills.pdf. Accessed 31 March 2014

Ajzen I (1991) The theory of planned behaviour. Organ Behav Hum Decis Process 50:179–211

Amulya J (2004) What is reflective practice. Centre for reflective community practice, MIT. http://www.itslifejimbutnotasweknowit.org.uk/files/whatisreflectivepractice.pdf. Accessed 31 March 2014

Beaty L (2001) Action learning: A Practitioner's guide. Routledge, London

Beaty L, McGill I (1992) Action learning: a practitioner's guide, republished 2013. Routledge, London

Boyer E (1990) Scholarship reconsidered: priorities of the professoriate. Princeton University Press, Princeton

Bradbury Huang H (2010) What is good action research? Action Res 8(1):93–109
Brockbank A, McGill I (2004) The action learning handbook: powerful techniques for education, professional development and training. Routledge, London
Brown LB, Gaventa J (2010) Constructing transnational action research networks: reflections on the citizenship development research centre. Action Res 8:5–28
Cantor J (1995) Experiential learning in higher education, linking classroom and community (Report no. 7). Graduate School of Education and Human Development, The George Washington University, Washington DC. http://files.eric.ed.gov/fulltext/ED404949.pdf. Accessed 31 March 2014
Coghlan D, Pedler M (2006) Action learning dissertations: structure, supervision and examination. Action Learn Res Practice 3(2):127–140
Cranton P (1989) Planning instruction for adult leaners. Wall and Thompson, Toronto
Down L, Nurse H (2007) Education for sustainable development networks, potential and challenge: a critical reflection on the formation of the Caribbean regional network. J Educ Teach 33(2):177–190
Fear F, Rosaen C, Bawden R, Foster-Fishman P (2006) Coming to critical engagement: an authoethnographic exploration. University Press of America, Lanham
Jackson T (2005) Motivating sustainable consumption: a review of evidence on consumer behaviour and behavioural change. Sustainable development research network, UK, p 154
Kleiman P (2011) Learning at the Edge of Chaos. AISHE-J 3(2):62.1–11
Knowles M (1977) The modern practice of adult education: Andragogy versus Pedagogy. New York Association Press, New York
Kolb DA (1984) Experiential learning: experience as the source of learning and development. Prentice Hall, Englewood Cliffs
Kolimuss A, Agyeman J (2002) Mind the gap: why do people act environmentally and what are the barriers to pro-environmental behaviour? Environ Educ Res 8(3):21
Lipscombe B (2008) Exploring the role of the extra-curricular sphere in higher education for sustainable development in the United Kingdom. Environ Educ Res 14(4):455–468
Miller K (2011) The challenges and opportunities of incorporating community-based learning into STEM subjects. http://www.hestem-sw.org.uk/project?id=38&pp=469. Accessed 31 March 2014
Miller TR, Baird TD, Littlefield CM, Kofinas G, Chaplin F III, Redman C (2008) Epistemological pluralism: reorganizing interdisciplinary research. Ecol Soc 13(2):46
Pedler M, Abbott C (2013) Facilitating action learning: a practitioner's guide. Open University Press, Maidenhead
Pedler M (2011) Action learning in practice, 4th edn. Gower, Farnham
Pilling-Cormick K (1994) Resistance by educators to using a self-directed learning perception scale. In: Hiemstra R, Brockett RG (eds) New directions for adult and continuing education, chapter 8. Jossey-Bass, San Francisco
Raelin J (2011) The action modalities. Action learning's good company. In: Pedler M, Action learning in practice, 4th edn. Gower, Farnham, pp 369–379
Reason P, Bradbury-Huang H (2001) Handbook of action research: participative inquiry and practice. Sage, London
Revans RW (1982) The origin and growth of action learning. Chartwell-Bratt, Brickley
Rooke J, Altouhan C, Young A, Young S (2007) Doers of the word, an inquiry into the nature of action in action learning. Action Learn Res Practice 4(2):119–135
Rowe D, Johnston L (2013) Learning outcomes. In: Johnston L (ed) Higher education for sustainability. Routledge, Abingdon, pp 45–60
Ryan A, Cotton D (2013) Times of change. In: Sterling S, Maxey L, Luna H (eds) The sustainable university progress and prospects. Earthscan, Abingdon, pp 151–168
Schön D (1983) The reflective practitioner: how professionals think in action. Basic Books, New York
Scott C (2007) How well do we engage our students? J Law Med and Ethics 35

Stern PC (2000) Toward a coherent theory of environmentally significant behaviour. J Soc Issues 56(3):407–424

Tierney A, Tweddell H, Willmore C (2016) Measuring education for sustainable development: experiences from the university of Bristol. Int J Sustain High Educ 17:1

Van de Ven A (2007) Engaged scholarship: A guide to organizational and social research. Oxford University Press, Oxford

Wakefield M, Leggett M (2012) To create a culture where students are enabled and supported to research in partnership with communities. University of Bristol, unpublished

Wals A, Blaze Corcoran P (2006) Sustainability as an outcome of transformative learning. In: Holmberg J, Samulesson E (eds) Drivers and barriers for implementing sustainable development in higher education. UNESCO education for sustainable development in action, Technical paper no. 3

Author Biographies

Christine Willmore is Reader in Sustainability and Law, combining community and professional practice with over 30 years teaching, researching and in practice as a barrister exploring sustainability, based throughout at the University of Bristol. A barrister by qualification she practiced in environmental law and been active in a number of sustainability organisations, but her primary focus has been upon teaching sustainability in the law school and latterly in leading institutions initiatives for sustainability education. She led the University's work to develop environmental law and sustainability in the law curriculum, established an interdisciplinary sustainable development module, available to all students that won a Times Higher and Green Gown Award in 2007, and in 2013 jointly won a Green Gown Award for whole institution approaches. Together with Hannah Tweddell and Ash Tierney she has developed an ESD resource wiki that receives over 50,000 hits a year. She currently leads Bristol Green Academy work, organizes and speaks at ESD conferences and has responsibility for all undergraduate programmes as University Academic Director of Undergraduate Studies.

Hannah Tweddell obtained a B.A. in Peace Studies and a PGC in Regeneration and Project Management from Bradford University, and an M.Sc., in Environmental Policy and Management from Bristol University. She was a sabbatical officer at Bradford University Student Union for 2 years, and then spent time as development worker for a national campaign organizer for CND, before coming to Bristol. At Bristol she was appointed the University's first sustainability intern, has co-organised conferences, curriculum development work and student action. She now works for the University Student Union as ESD Coordinator with the Student Green Fund Project. A focus of her work is the relationship between the informal and formal curriculum.

School Gardens and Ecovillages: Innovative Civic Ecology Educational Approaches at Schools and Universities

Constantina Skanavis and Evangelos Manolas

Abstract

To prevent further damage to the natural ecosystem, it would be necessary to produce environmental stewards capable of making knowledgeable and conscientious decisions regarding the environment. Due to renewed environmental awareness, new civic ecology educational trends, like school gardening and ecovillage projects are enjoying an intense interest. Environmental knowledge alone is not sufficient to solve conservation problems, and the role of civic ecology in solving these problems has become increasingly important. Conventional classroom learning from books, wall charts and memorization often results in youth's negative attitudes towards environmental sciences. Environmental education researchers therefore have suggested that classroom interventions or combinations with field experiences that actively involve youth may promote pro-environmental behavior, knowledge and positive attitudes towards the environment. Research has shown that school gardens and ecovillages enhance learning, promote experiential learning, and teach environmental education and environmental dispositions. School gardens in primary and secondary education as well as undergraduate and graduate university programs connected with ecovillages cover a continuum of efforts to increase the benefits of environmental education in real world conditions backed up with hands on

C. Skanavis (✉)
Department of Environment, University of the Aegean, University Hill,
81100 Mytilene, Greece
e-mail: cskanav@aegean.gr

E. Manolas
Department of Forestry and Management of the Environment and Natural Resources,
School of Agricultural and Forestry Sciences, Democritus University in Thrace,
193 Pantazidou Street, 68200 Orestiada, Greece
e-mail: emanolas@fmenr.duth.gr

W. Leal Filho (ed.), *Transformative Approaches to Sustainable Development at Universities*, World Sustainability Series, DOI 10.1007/978-3-319-08837-2_37

experience in miniature environments, where youth works in sympathy with nature. The purpose of this paper is to describe the infrastructure of school gardens and ecovillages and assess their style of learning. Furthermore the focus would be in determining, through successful case stories, what are the academic, behavioral, recreational, social, political, and environmental remediation benefits through these civic ecology experiences.

Keywords
School gardens · Ecovillages · Innovative educational approaches · Schools · Universities

1 Introduction

According to the Environmental Protection Agency (EPA 2014a, b)

> Environmental education is a process that allows individuals to explore environmental issues, engage in problem solving, and take action to improve the environment. As a result, individuals develop a deeper understanding of environmental issues and have the skills to make informed and responsible decisions.

EPA also lists the components of environmental education which are:

- *Awareness and sensitivity* to the environment and environmental challenges
- *Knowledge and understanding* of the environment and environmental challenges
- *Attitudes* of concern for the environment and motivation to improve or maintain environmental quality
- *Skills* to identify and help resolve environmental challenges
- *Participation* in activities that lead to the resolution of environmental challenges.

It is also important to note that:

> Environmental education does not advocate a particular viewpoint or course of action. Rather, environmental education teaches individuals how to weigh various sides of an issue through critical thinking and it enhances their own problem-solving and decision-making skills (Environmental Protection Agency 2014a).

It is crucial to support students into identifying their values, so they can envision how these values might play out in action. Thus, we advocate not for any specific responsible environmental behavior, but rather for a lifestyle where actions are consciously and reflectively driven by values and a vision for the world. Some environmental thinkers, like John Muir, believe that if people have more opportunities to see or enjoy natural landscapes this will make them value and preserve these places more.

John Muir, and the educational models he inspired, believes experiencing and knowing the natural environment is sufficient in providing us with incentives to act on behalf of this world. A summary of Muir's philosophy of action might be presented as follows:

- Experiences with the environment will increase our familiarity and knowledge of the natural world.
- We become emotionally attached to and posit value in those things we know about.
- We act to preserve things we are emotionally attached to and in which we posit value.
- We will act to preserve the environment if through our experiences with it we become more knowledgeable about it (Goralnik and Nelson 2011).

This philosophy of environmental action is also the implicit philosophy of several environmental organizations and outdoor education programs. For example, as Goralnik and Nelson (2011) point out The Sierra Club has not stopped supporting Muir's central point of taking people into the natural world in order to encourage them to take action in favor of the places they see, and links these experiences to action. A phrase stated by John Muir is even cited in The Sierra Club's Web site: "If people in general could be got into the woods, even for once, to hear the trees speak for themselves, all difficulties in the way of forest preservation would vanish".

Environmental concern can range from fear and worry about environmental problems to a general desire to care for the environment (Rickinson 2001). Although environmental fear, known as ecophobia, is a critically important concept, few studies provide information on how ecophobia impacts children's immediate and long-term participation in pro-environmental activities. The studies that have started to research this question reveal that environmental concerns, especially fear and anxiety, relate to feelings of helplessness, which may impact children's willingness to participate in environmentally friendly activities. Strife (2012) provides an excellent review of research conducted on this question. She points out that it is of paramount importance to understand children's perspectives since children both now and in the future will influence and be influenced by environmental issues in many ways. The increase in ecophobia is partly rooted in society's emphasis on distant ecological problems such as rainforest destruction, or effects of climate change as symbolized by the lone polar bear floating on an isolated iceberg. Focusing on such distant and vague issues may contribute to creating and perpetuating ecophobic feelings in children (Strife 2012).

Goralnik and Nelson (2011) believe that people act on behalf of the things to which they feel emotionally attached. Consequently, if people become emotionally attached—possibly via a physical attachment—to the environment, then they will also act in environmentally friendly ways. Outdoor education is a successful part of environmental education when we aim on emotional attachment through physical experiences. The term outdoor education is a general term that is frequently applied to programs or activities that can be, and usually are, conducted in the out-of-doors (Parkin 1998). In this paper we will present school gardens and ecovillages, as part of outdoor education, which promote environmental sensitivity. Furthermore, we attempt to evaluate how these affect environmental behavior. Finally, we will try to reveal, through successful case stories, what are the academic, behavioral, recreational, social, political, and environmental remediation benefits through these civic ecology experiences.

2 School Gardens

School gardens are not a new idea. In all levels of education, gardens have offered students places of experiencing aesthetic beauty, of developing feelings of attachment with the natural world, opportunities for investigating a variety of life sciences and learning outdoors. The first official school garden in the U.S. was created in Roxbury, MA at the end of the 19th century. Until 1950, schools in many big U.S. cities had gardens and WWII saw the advent of school victory gardens growing food to support the war effort. In the 1950s as a result of the national focus on technology, interest in school gardening became less important (Genkins 2014).

School gardening can have many purposes: academic, behavioral, recreational, social, political, and environmental. Gardens may increase children's respect for nature and support their physical, mental or moral development. The benefits of student involvement in school gardening can be classified in three broad development areas: cognitive, physical, and social/emotional (Genkins 2014). School gardens offer students personal observation of and complex experiences in nature which teachers use to support essential curriculum. School yard gardens are also useful for outdoor science, environmental education or hands-on learning.

Up until the late 1960s and early 1970s school gardens remained common at many schools. In the period from the 1970s to the early 1980s school gardening programs were not a priority (Gigliotti 1992). However, due to renewed interest on the environment, new educational trends and insights into how the environment of a school affects how young people behave and develop, school gardening and beautification projects are witnessing a revival (Heffernan 1997).

Using school gardens for teaching improves student learning (Sheffield 1992) and environmental dispositions (Alexander et al. 1995; Barker 1992; Waliczek 1997; Wotowiec 1975). A garden is a most interesting and exciting place for children to play, work, and learn (Herd 1997). The school garden can provide the right experience for children to acquire knowledge about their natural environment (Pale et al. 2001; Waliczek and Zajicek 1999). A garden and gardening activities make it possible for pupils to learn about the environment and also experience the natural world firsthand. In addition, a gardening-based curriculum can cover several subjects while giving the students the chance to experience nature firsthand (Eames-Sheavly 1994; Klemmer et al. 2005; Pigg et al. 2006). Outdoor educational programs when compared to traditional biology teaching are generally believed to improve environmental attitudes and knowledge (Fančovičová and Prokop 2011) and at the same time they combat the problem 'plant blindness' and give pupils more incentives to study the subject.

Children who grow their own food are more likely to develop healthy eating habits such as eating fruits and vegetables (Bell and Dyment 2008; Libman 2007; Lineberger and Zajicek 2000; Morris et al. 2001; Pothukuchi 2004) and that they greatly improved their knowledge about nutrition (Canaris 1995; Koch et al. 2006; Pothukuchi 2004). They are also more likely to keep healthy eating habits for the rest of their lives (Morris and Zidenberg-Cherr 2002), which can contribute to preventing or delaying chronic disease conditions (Heimendinger and Van Duyn 1995).

Depending on their level of engagement in greening projects, they can also see that they have a right to take part in decisions that affect their quality of life (Dyment 2004; Hart 1997). Gardening in particular can provide a chance to deal with losses and failures, and to see firsthand the responsiveness of plants to care and nurturing. When fully engaged in the greening process, young people can build skills related to democracy, participation and citizenship which they can potentially use in their lives in the future (Dyment 2004; Hart 1997).

Research on children with learning disabilities who engaged in gardening discovered that they improved their nonverbal communication skills, developed awareness of the advantages of order, learned how to work cooperatively with others, and formed positive relationships with adults (Dyment and Bell 2006). Juvenile offenders who enjoy gardening show improved self-esteem, interpersonal relationships and attitudes towards school (Cammack et al. 2002; Flagler 1995; Waliczek et al. 2001). Furthermore, gardening has long been recognized as having a healing power which means having a positive influence on mental health and well-being (Bell and Dyment 2008; Ulrich 1999).

3 Ecovillages

Robert Gillman (1991) described the greatest challenge of our time this way:

> For humankind at the end of the 20th century there is hardly anything more appealing—yet apparently more elusive—than the prospect of living in harmony with nature and with each other.

On a global level there is a need for positive models which demonstrate a sustainable future for people and nature. Ecovillages are such a model.

Sustainability is based on this principle: Our natural environment provides, directly or indirectly, everything we need for our survival and health. Sustainability is about the conditions under which humans and nature can exist in productive harmony, that permit fulfilling all the requirements necessary for the survival and well-being of present and future generations (Environmental Protection Agency 2014b). In this day and age, we have become an urban species where our lives are full of stress, materialism, and pollution. We are tied to technology in all aspects of our life, including business, education, and entertainment.

As economic policies grew more harmful to the environment and more environmental problems began to spring up in the late 20th century, groups of people felt fed up with these aspects of modern life and began forming communities that heavily incorporate ecological principles (Sevier et al. 2008; Van Schyndel Kasper 2008). Although the first ecovillages appeared in the early 1930s (Ecovillage Jarma, Sweden and Ecovillage Solheimar, Iceland) the modern ecovillage movement started to develop in the 1970s (The ecovillage movement 2013). The term ecovillage first appeared on the scene in 1991 in a sustainability report commissioned by the Gaia Trust (Sevier et al. 2008). Ecovillages are intentional communities united by common ecological, economic, social and spiritual values. They aim to reduce their ecological footprint. Ecovillage principles can be useful to both urban and rural areas.

In 1995, the Global Ecovillage Network (GEN) was founded to link several diverse ecovillage communities and related projects together and to help spread information about them across the world.

According to Jonathan Dawson former president of the Global Ecovillage Network, ecovillages have five key principles:

(1) They are not government-sponsored projects, but grassroots initiatives.
(2) Their residents value and practice community living.
(3) Their residents are not overly dependent on government, corporate or other centralized sources for water, food, shelter, power and other basic necessities. Rather, they attempt to provide these resources themselves.
(4) Their residents have a strong sense of shared values, often characterized in spiritual terms.
(5) They often serve as research and demonstration sites, offering educational experiences to others (Christian 2007).

An ecovillage is therefore defined as "a human-scale, full-featured settlement, in which human activities are harmlessly integrated into the natural world in a way that is supportive of healthy human development, and can be successfully continued into the indefinite future" (Gilman 1991). Ecovillages do not have to be fully self-sufficient or isolated from the surrounding community.

Ecovillages can be formed in any location; they need not be only in rural areas. The Global Ecovillage Network explains how ecovillages encompass aspects of ecological design, permaculture, ecological building, green production, alternative energy, community building practices and much more (Taggart 2009).

The purpose of these communities is to create an environment where people can live and innovate in peace with the Earth. The residents of ecovillages achieve this purpose by growing crops to provide their own food, composting to reduce waste and create fertilizer, and build from local and green materials (Sevier et al. 2008). To reduce their consumption of fossil fuels and natural resources ecovillages often use alternative technologies for heating, electrical and water systems. Using systems common to the whole community along with sharing cars and many tools reduces the ecovillage's impact on the environment and makes it more self-reliant. It may be said that ecovillages practice and promote a way of life which "... embraces the conscious decision to live more simply, thereby consuming less" (Sevier et al. 2008).

Findhorn's ecological footprint, in Scotland, is half the UK national average. Some examples of Findhorn footprints are: for home and heating 21.5 %, food 37 %, travel 43 % and consumables 46 % of the national average (Findhorn Foundation 2014). With regard to Sieben Linden ecovillage in Germany a 2003 study by the University of Kassel showed that residents of the village have a three times smaller footprint than the average individual in Germany. Some examples of Sieben Linden footprints: the energy requirement for heating is only 1/3 of the German average and water consumption is reduced to 2/3 of the national average (Sieben Linden Ecovillage 2014a). Also, at Sieben Linden per capita carbon dioxide emissions are just 28 percent the national average (Worldwatch Institute 2014). At Ithaca Ecovillage, in the USA, the houses consume 40 % less energy and resources than the country average (Dawson 2006).

One major way ecovillages disseminate their values and achievements is through education. As Dawson (2010) points out

> The various educational packages developed within ecovillages reflect the core ethics of the communities themselves in that they are holistic—exploring interdependence and the relationships between issues and subjects that are generally considered independently in more conventional settings—and experiential, in that they engage all of the learner's faculties—head, heart, and hands.

In this sense ecovillage education should be understood as part of the wider trend toward environmental education based on systems thinking. Learning happens in the context of live experiments and engagement "in such living laboratories can be a profound transformation for students as they experience in a very tangible way the dynamic relationship between values, lifestyle, and community structures" (Dawson 2010).

Some examples of ecovillage educational initiatives as they appeared in the last 15 years:

North Carolina State University in the U.S. has created a program called Eco-Village, located in Bragaw Hall, in which students go beyond the classroom to lead, create and solve complex energy, environmental and sustainability issues at local and global level. Their work is supported by experienced mentors (North Carolina State University 2014).

In Kentucky, in the U.S., Berea College has created Berea College Ecovillage, an ecologically-designed residential and learning complex for student families. In addition to 50 apartments, a child development and daycare center, Berea College Ecovillage includes a commons house, a Sustainability and Environmental Studies (SENS) demonstration house but also vegetable gardens, fruit trees and a permaculture food forest.

The ecovillage uses 50 % less energy than conventional residences in the area and almost 50 % of the waste stream is diverted to recycling, reuse or composting. The ecovillage also incorporates various 'green design' elements such as passive solar heating and photovoltaic panels while roof-top capture of rainwater contributes to landscape irrigation and production of fruits and vegetables (Berea College 2014).

The Ecovillage Training Center (ETC) at The Farm, in Tennessee, U.S.A., offers courses in fields such as permaculture, herb logy, installing solar electricity and water heating, cob, earth bag, round pole, and straw bale construction, midwifery, constructed wetlands and sustainable farming (The Farm 2014).

Crystal Waters Ecovillage in Australia offers a permaculture design course. Topics include dam construction and methods towards drought proofing, improving soils, seed saving, seed raising, pest management, making liquid fertilizers, maintenance of fruit trees, establishing and maintaining forest systems, windbreaks and their design, energy efficient design in buildings, myths and facts about energy systems and materials, school gardening schemes, school farms (Crystal Waters 2014).

Sieben Linden Ecovillage in Germany offers courses for school and university students, groups of people doing a volunteer service, etc. Courses include Permaculture Design Courses, the Ecovillage Design Education (how to set up ecovillages), straw bale building, plastering, etc. (Sieben Linden Ecovillage 2014b).

At Findhorn Ecovillage, Scotland, Findhorn Foundation College in partnership with the School for the Built Environment, Heriot-Watt University, Edinburgh, offers two 3-week intensive residential courses which are part of an M.Sc. degree program in Sustainable Community Design (Findhorn Foundation College 2014).

4 Conclusion

Education such as that offered in school gardens and ecovillages is holistic education. Time spent alone or in small groups in natural areas motivates. An essential difference between outdoor learning and indoor learning is that "indoor learning" can promote cognitive development but the nature of the immediate classroom environment seriously limits the range and depth of sensory experience and affective learning. Outdoor learning "offers a holistic mode of learning through direct, sensory, affective and cognitive engagement with ecological systems and processes, such that the consequences of individual and collective actions may have immediate and real outcomes for a person" (Lugg 2007).

Outdoor experience can also offer alternative world-views and practical approaches to more sustainable living. School gardens and ecovillages are experiments, living models of sustainability, and examples of how action can be taken immediately. Can such experiments ever hope to be effective as a transformative force of global capitalism and consumerism? This is a difficult question to be answered at this present time (Litfin 2012) because there is still limited generation of data and research in this area of concern. This aspect of sustainability movement has not yet been satisfactorily developed; it has not yet established itself. And questions still abound: How can such experiments work out if implemented on a larger scale? How can we find more ways to overcome the obstacles that are keeping them from fully realizing their ideals? Are there more worthy ideals they could be pursuing? How can we encourage replication of their successes? More research is needed on all these questions (Ashlock 2010).

Yet, such efforts do constitute seeds of hope. As Litfin (2012) points out they are *not* the answer to the crisis we face. They are just *one* answer. A key contribution of such experiments is the power of their example. They teach practices which need to be applied. If our way of life is not sustainable, it will cease. Whether the end of the current order is precipitous or gradual, what is certain is that any successful experiments will become enormously salient or at least more important. Although still relatively young, such efforts have the potential to help reshape human communities in a sustainable and holistic manner.

References

Alexander J, North MW, Hendren DK (1995) Master Gardener classroom garden project: an evaluation of the benefits to children. Child Environ 12(2):256–263

Ashlock CR (2010) Lessons learned from the ecovillage movement: a global analysis focusing on Francophone countries. Annandale-on-Hudson, NY. http://ashlockcharlotte.files.wordpress.com/2013/10/the-ecovillage-movement.pdf. Last accessed 26 March 2014

Barker SL (1992) The meaning of a youth gardening program: a naturalistic inquiry. Unpublished doctoral dissertation, Indiana University, Bloomington

Bell AC, Dyment JE (2008) Grounds for health: the intersection of green school grounds and health-promoting schools. Environ Educ Res 14(1):77–90

Berea College (2014) Ecovillage. http://www.berea.edu/SENS/ecovillage/. Last accessed 12 March 2014

Cammack C, Waliczek TM, Zajicek JM (2002) The Green Brigade: the effects of a community-based horticultural program on the self-development characteristics of juvenile offenders. HortTechnology 12(1):82–86

Canaris I (1995) Growing foods for growing minds: integrating gardening and nutrition education into the total curriculum. Child Environ 12(2):134–142

Christian DL (2007) Finding community: how to join an ecovillage or intentional community. New Society Publishers, Gabriola Island, 264 p

Crystal Waters (2014) Permaculture design course. http://crystalwaters.org.au/permaculture-design-course-april-2014/. Last accessed 12 March 2014

Dawson J (2006) Ecovillages: new frontiers for sustainability. Green Books, Devon, 96 p

Dawson J (2010) Ecovillages and the transformation of values. State of the world 2010. The Worldwatch Institute. http://www.gaiaeducation.org/docs/Ecovillages%20Transformation%20Values.pdf. Last accessed 18 March 2014

Dyment JE (2004) 'At that age, you just accept what you have … you never question things': a case study of student participation in school ground greening projects. Child Youth Environ 14 (1):130–152

Dyment JE, Bell AC (2006) Our garden is colour blind, inclusive and warm: reflections on green school grounds and social inclusion. Int J Inclusive Educ 12(2):169–183

Eames-Sheavly M (1994) Exploring horticulture in human culture: an interdisciplinary approach to youth education. HortTechnology 4(1):77–80

Environmental Protection Agency (2014a) What is environmental education? http://www2.epa.gov/education/what-environmental-education. Last accessed 2 Feb 2014
Environmental Protection Agency (2014b) What is sustainability? http://www.epa.gov/sustainability/basicinfo.htm. Last accessed 14 March 2014
Fančovičová J, Prokop P (2011) Children's ability to recognize toxic and non-toxic fruits: a preliminary study. Energy Educ Sci Technol Part B: Soc Educ Stud 3(3):497–504
Findhorn Foundation (2014) Ecological footprint is the lowest! https://www.findhorn.org/2007/04/ecological-footprint-is-the-lowest/. Last accessed 11 April 2014
Findhorn Foundation College (2014) Findhorn residential within MSc sustainable community design. http://www.findhorncollege.com/postgrad/msc_modules.php. Last accessed 14 March 2014
Flagler J (1995) The role of horticulture in training correctional youth. HortTechnology 5 (2):185–187
Genkins M (2014) The ABCs of school gardens. http://www.mastergardenersmecklenburg.org/uploads/6/0/7/6/6076512/the_abcs_of_school_gardens_genkins_0412.pdf. Last accessed 18 March 2014
Gilman R (1991) The eco-village challenge. http://www.context.org/iclib/ic29/gilman1/. Last accessed 12 March 2014
Goralnik L, Nelson MP (2011) Framing a philosophy of environmental action: Aldo Leopold, John Muir, and the importance of community. J Environ Educ 42(3):181–192
Gigliotti LM (1992) Environmental attitudes: 20 years of change? J Environ Educ 24(1):15–25
Heffernan M (1997) Horticulture rediscovered: the flowering of American Schoolyards. ERIC Document Reproduction Service No. ED446965
Heimendinger J, Van Duyn MA (1995) Dietary behavior change: the challenge of recasting the role of fruit and vegetables in the American diet. Am J Clin Nutr 61(6):1397–1401
Hart R (1997) Children's participation: the theory and practice of involving young citizens in community development and environmental care. Earthscan, London, 220 p
Herd M (1997) Learn and play in the garden. Barron's educational series. Hauppauge, New York, 128 p
Klemmer CD, Waliczek TM, Zajicek JM (2005) Growing minds: the effect of a school gardening program on the science achievement of elementary students. HortTechnology 15(3):448–452
Koch S, Waliczek TM, Zajicek JM (2006) The effect of a summer garden program on the nutritional knowledge, attitudes, and behaviors of children. HortTechnology 16(4):620–625
Libman K (2007) Growing youth growing food: how vegetable gardening influences young people's food consciousness and eating habits. Appl Environ Educ Commun 6(1):87–95
Lineberger SE, Zajicek JM (2000) School gardens: can a hands-on teaching tool affect students' attitudes and behaviors regarding fruit and vegetables? HortTechnology 10(3):593–597
Litfin K (2012) A whole new way of life: ecovillages and the revitalization of deep community. In: De Young R, Princen T (eds) The localization reader: adapting to the coming downshift. The MIT Press, Cambridge, 376 p
Lugg A (2007) Developing sustainability-literate citizens through outdoor learning: possibilities for outdoor education in higher education. J Adventure Educ Outdoor Learn 7(2):97–112
Morris JL, Neustadter A, Zidenberg-Cherr S (2001) First-grade gardeners more likely to taste vegetables. Calif Agric 55(1):43–46
Morris JL, Zidenberg-Cherr S (2002) Garden-enhanced nutrition curriculum improves fourth-grade school children's knowledge of nutrition and preferences for some vegetables. J Am Diet Assoc 102(1):91–93
North Carolina State University (2014) College of sciences, 2013–2014 undergraduate catalog. http://catalog.ncsu.edu/undergraduate/collegeofsciences/. Last accessed 11 March 2014
Pale PE, Thompson KF, Keyes M (2001) Students, standards, and exploration: a responsive, relevant, and engaging curriculum. Middle Sch J 33(1):35–40
Parkin D (1998) Is outdoor education environmental education? Int J Environ Educ Inf 17 (3):275–286

Pigg AE, Waliczek TM, Zajicek JM (2006) Effects of a gardening program on the academic progress of third, fourth, and fifth grade math and science students. HortTechnology 16(2):262–264

Pothukuchi K (2004) Hortaliza: a youth 'nutrition garden' in southwest Detroit. Child Youth Environ 14(2):124–155

Rickinson M (2001) Learners and learning in environmental education: a critical review of the evidence. Environ Educ Res 7(3):207–320

Sevier L, Henderson M, Naidu N (2008) Ecovillages: a model life? The ecologist. http://www.theecologist.org/investigations/society/268714/ecovillages_a_model_life.html. Last accessed 24 March 2014

Sheffield BK (1992) The affective and cognitive effects of an interdisciplinary garden-based curriculum on underachieving elementary students. Unpublished doctoral dissertation, University of South Carolina, Columbia

Sieben Linden Ecovillage (2014a) Ecological footprint. http://www.siebenlinden.de/index.php?id=52&L=2. Last accessed 14 March 2014

Sieben Linden Ecovillage (2014b) http://www.permacultureglobal.com/projects/801-sieben-linden-ecovillage. Last accessed 14 March 2014

Strife SJ (2012) Children's environmental concerns: expressing ecophobia. J Environ Educ 43 (1):37–54

Taggart J (2009) Inside an ecovillage: born of aligned ecological values and design, ecovillages are found in over 70 countries around the world. Altern J 35(5):20–21

The ecovillage movement (2013) http://www.selba.org/EngTaster/Worldview/TheEmergingHolisticWVEcovillageMovement.html. Last accessed 1 Aug 2013

The Farm (2014) Ecovillage training center. http://www.thefarm.org/etc/etcabout.html. Last accessed 12 March 2014

Ulrich RS (1999) Effects of gardens on health outcomes. In: Marcus CC, Barnes M (eds) Healing gardens: therapeutic benefits and design recommendations. Wiley, New York, 624 p

Van Schyndel Kasper D (2008) Redefining community in the ecovillage. Hum Ecol Rev 15 (1):12–24

Waliczek TM (1997) The effect of school gardens on self-esteem, interpersonal relationships, attitude toward school, and environmental attitude in populations of children. Doctoral dissertation, Texas A&M University, College Station

Waliczek TM, Zajicek JM (1999) School gardening: improving environmental attitudes of children through hands-on learning. J Environ Hortic 17(4):180–184

Waliczek TM, Bradley JC, Zajicek JM (2001) The effect of school gardens on children's interpersonal relationships and attitudes toward school. HortTechnology 11(3):466–468

Worldwatch Institute (2014) Green acres: communities reduce ecological footprints. http://www.worldwatch.org/green-acres-communities-reduce-ecological-footprints. Last accessed 13 April 2014

Wotowiec PJ (1975) Students' and parents' ratings of the tract garden program in the Cleveland Public Schools. Dissertation Abstr Int 36:3395-A

Author Biographies

Constantina Skanavis is a Professor in Environmental Communication, Education and Didactics at the Department of Environment at the University of the Aegean (Mytilene, Greece). She is also the Head of the Research Centre of Environmental Education and Communication. She joined the University of the Aegean 10 years ago. Before that she was a Professor at California State University, Los Angeles. She has developed several courses on issues of environmental health and education. She currently teaches environmental education, environmental communication and environmental interpretation courses in undergraduate and postgraduate levels at the University of the Aegean. Professor Skanavis has numerous publications on an international basis and has given presentations all over the world.

Evangelos Manolas is an Associate Professor in Sociology and Environmental-Forest Education at the Department of Forestry and Management of the Environment and Natural Resources, School of Agricultural and Forestry Sciences, Democritus University of Thrace, Orestiada, Greece. He has written, edited and co-edited nine books and has published more than 100 papers and book chapters.

Overcoming Obstacles to Classroom Based Cross-Border Environmental Education in Universities

Eric Pallant, Beth Choate and Derek Reno

Abstract

Solving issues of global environment importance depends upon cross-boundary and cross-cultural cooperation. The Global Liberal Arts Alliance is an organization of 27 liberal arts colleges and universities, located in 15 nations, whose purpose is to strengthen education in the tradition of the liberal arts through the exchange of experience and the development of mutually beneficial programs. The Alliance encourages and facilitates classroom partnerships between professors in different countries around the globe. Professors submit potential courses to an international coordinator who then finds suitable matching courses in other countries. Faculty then meet for planning in the summer prior to teaching to coordinate activities, syllabi, and student-to-student interactions. The authors have engaged Allegheny College students (Pennsylvania, USA) in four different environmental courses with students in parallel courses in Bulgaria, Pakistan, and Morocco. Many students expressed frustration with difficulties in communication, while faculty complained of issues pertaining to technology. Issues of cross-cultural understanding were still achieved,

E. Pallant (✉) · B. Choate · D. Reno
Department of Environmental Science, Allegheny College,
520 N. Main St, Meadville, PA 16335, USA
e-mail: epallant@allegheny.edu

B. Choate
e-mail: bchoate@allegheny.edu

D. Reno
e-mail: renod@allegheny.edu

W. Leal Filho (ed.), *Transformative Approaches to Sustainable Development at Universities*, World Sustainability Series, DOI 10.1007/978-3-319-08837-2_38

although sharing environmental perspectives was less successful. This paper will describe the many challenges and benefits of having undergraduate environmental students cooperate across time zones and radically distinct cultures.

Keywords
Sustainability education · International collaboration · Cross-cultural communication

1 Introduction

Teaching sustainability is increasingly being incorporated into curricula of higher education institutions, especially in multidisciplinary contexts (Dale and Newman 2005). Like cultural, economic, and other societal issues, many environmental issues are global matters that transcend national boundaries (for example: Bennet et al. 2012; Pinar 2003). Topics covered in environmental education courses often reach beyond national borders (Pinar 2003). Therefore, it is important for institutions of higher education to participate in cross-cultural and transnational cooperative programs that encourage the placement of environmental issues in a global context which will improve their chances of being successfully resolved (Buchan et al. 2007; Pinar 2003; Stephens et al. 2008). Literature exists on localized environmental problem solving and sustainability initiatives (Pallant et al. 2013; Eatmon et al. 2013) as well as international comparison between sustainability initiatives at various institutions of higher education (Beringer 2007; Ferrer-Balas 2008; Lu and Zhang 2014). Furthermore, much has been published on international and cross-cultural conflict resolution approaches (Marsella 2005; Abu-Nimer 1996; Vayrynen 2001; Ramsbotham et al. 2011; Avruch et al. 1991). However, little has been written on cooperative, educational, cross-cultural programs at undergraduate institutions aimed at improving knowledge in the realm of sustainability, which often goes hand-in-hand with the resolution of human conflict (Lundegård and Wickmann 2007).

Cross-cultural communication is crucial in order to solve international issues, including environmental problems (Bennet et al. 2012; Ciurana and Filho 2006; Pinar 2003; Stephens et al. 2008). To promote international communication concerning environmental issues in institutions of higher education, professors of environmental science at Allegheny College and Denison University of the Great Lakes College Association worked through the Visiting Fellowships program of the Global Liberal Arts Alliance to coordinate communication and cooperation between cross-cultural, parallel courses in Bulgaria, Morocco, Pakistan, and Greece. This paper details the efforts and results of these transnational, sustainable education connections.

The Great Lakes College Association (GLCA) is a coalition of colleges from Ohio, Pennsylvania, Michigan, and Indiana intended to strengthen and advance liberal arts and sciences (Great Lakes College Association). The Global Liberal Arts Alliance (GLAA) is a worldwide partnership of colleges and universities founded by the GLCA aimed at strengthening liberal education while helping institutions to overcome institutional issues, improve educative programs, and support international discussion and understanding of societal challenges (Global Liberal Arts Alliance 2011). As a way to encourage international cooperation in May of 2009, the GLAA began to facilitate and fund international visits of faculty and administrators as part of their Visiting Fellowships initiative (Global Liberal Arts Alliance 2011).

2 Methods

Beginning in Fall 2012 the Global Liberal Arts Alliance coordinated pairings of college and university classes from different countries. Five of these paired courses were environmentally themed. Each pairing consisted of one undergraduate, American-style, liberal arts college in the United States with one similar institution in another country (See Fig. 1.)

All of the partnered colleges shared similar methodologies. Faculty preparing classes for partnership met face-to-face in the summer prior to the year their shared courses were delivered. They planned a range activities for garnering integration

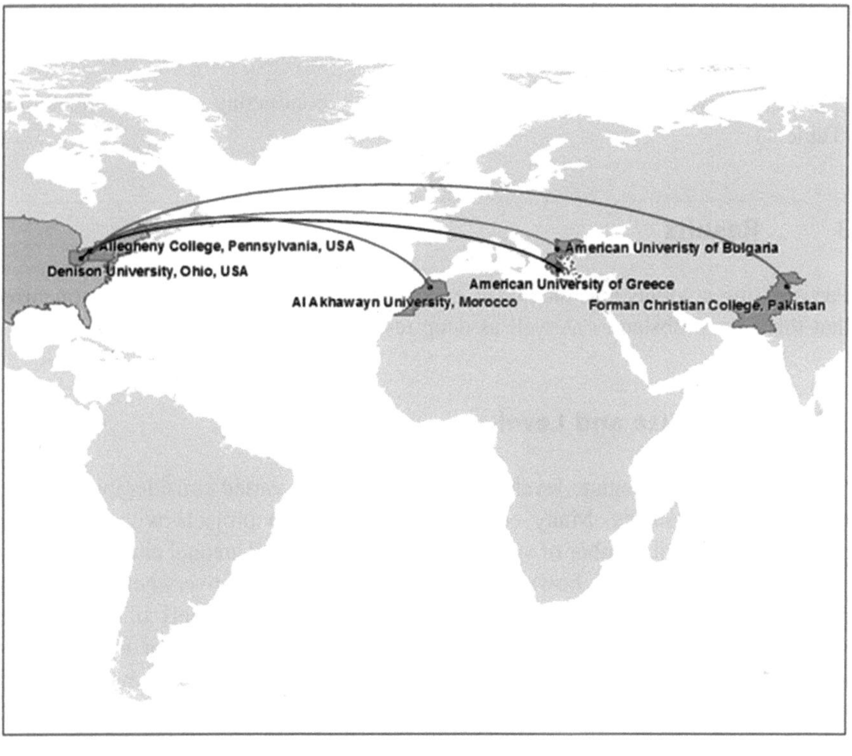

Fig. 1 Map of environmentally themed course pairings and their institutions

and cooperation among students and faculty. Using videoconferencing technology (VCT) professors provided lectures and PowerPoint presentations to students in foreign classrooms. VCT was also used to facilitate discussions between full classes. Some classes required assignments by their students be presented to students in foreign classrooms and in addition some classes assigned joint term papers and research projects for groups of students combined from the different countries. Individual interactions between students were accomplished with social media such as Facebook, Skype, SMS, Instant Messaging, and Email. It should be noted that most paired classes used a combination of techniques for facilitating international cooperation on learning about the environment, but each pairing used a unique balance of shared lectures, projects, videoconferencing, readings, and assignments.

Feedback from students was obtained through discussions with instructors, as well as examining student evaluations from Allegheny College students. Participating faculty were sent a questionnaire comprised of the following open-ended questions:

1. What were the benefits for you and your students to participating in a joint class venture through the GLAA?
2. What were the costs for you and your students to participating in a joint class venture through the GLAA?
3. Would you do this again and if so what would you do differently? If not, why not?

Feedback was obtained from all instructors conducting course connections. (Table 1).

3 Results

Participating students and faculty in environmentally themed classes all concurred that there were obstacles as well as deep rewards.

3.1 Class Size and Level

The number and collegiate level of the students often varied considerably between the connected courses. Many courses assigned group projects with the aim of partnering an equal number of students from each course. Unequal class size did not allow for even pairings, however, and groups were often overwhelmingly comprised of individuals from one institution. The collegiate levels of the students varied as well. When introductory courses were paired with upper level courses, students in the upper level courses seemed to feel as though they were more concerned with the quality of projects than their younger partners.

Table 1 Details of courses partnered through the global liberal arts alliance global course connections program

Term	Class	Level	Colleges and Location
Fall 2013	ENVSC 340: World regional geography	3,4	Allegheny college, Meadville, Pennsylvania, USA
	GEOG 133: Geographical profile of Pakistan	1,2	Forman christian college, Lahore, Punjab, Pakistan
Fall 2013	ENVSC 370: Insect ecology and the environment	2,3,4	Allegheny college, Meadville, Pennsylvania, USA
	SCI 160: Introduction to environmental science	1,2,3,4	American university in Bulgaria, Blagoevgrad, Bulgaria
Fall 2013	Econ 202/ENVS 290: Economic growth and environmental sustainability	3,4	Denison university, Granville, Ohio, USA
	ES 3117: Principles of environmental management	3	American college of Greece, Athens, Attica, Greece
Spring 2013	ENVSC 230: Soil to plate	1,2	Allegheny college, Meadville, Pennsylvania, USA
	GEO 130101: Introduction to geography	1,2,3,4	Al Akhawayn university, Ifrane, Morocco
Spring 2014	ENVSC 230: Soil to plate	1,2	Allegheny college, Meadville, Pennsylvania, USA
	GEO 130101: Introduction to geography	1,2,3,4	Al Akhawayn university, Ifrane, Morocco

3.2 Time Zones, Sense of Time, and Accents

Many students around the world lead busy lives. They take multiple classes and have several out-of-class responsibilities and appointments. Finding a suitable time for students in the same class to meet to work on a group project is often problematic. This difficulty is multiplied when trying to find a single time to schedule, for example, a joint videoconference of two classes in widely separated time zones or isolating a time for a guest lecture by the foreign professor. Instructors viewed having to come to class at a time different from their designated class time as a major "cost" for the students. In addition, many students found it difficult to find an appropriate time of day or night to communicate with one another via Skype, SMS, or Instant Messaging. Additionally, partnered colleges and universities had varying academic calendars which in several cases led to difficulties in communication between students for joint project work.

A second problem turned out to be more insidious than the question of time zones. Different cultures, as well as different professors within the same culture, maintain distinct senses of punctuality and deadlines. A due date agreed upon during the summer for a joint project may have meant different things to different people. For example, a deadline for a joint term paper on agriculture and the environment was supposed to be completed by students in Morocco and

Pennsylvania by the date that classes ended at the school in the United States. This date was 1 week before classes ended in U.S. and several weeks before the end of the semester officially ended in Morocco. While the American students had a strict deadline for completing their paper, many were frustrated by slow responses from their Moroccan counterparts who appeared to be working with a less determinate deadline. Other paired classes had similar conflicts.

A third issue was one of accents. Every participant, by definition, was a foreigner, and consequently when communicating with peers in another country had to overcome issues of dialect, pronunciation, speed, and accent. As anyone who has tried to communicate in a foreign language over the telephone can attest, technology often makes communication more difficult than when it occurs in person. Facial features that offer clues to meaning are often indistinct and out-of-focus when videoconferencing full classes. Also, random syllables were frequently lost by technology which reinforced the difficulty of fostering complete communication across multiple time zones.

The resolution to at least some communication issues was to move more toward asynchronous forms of communication such as email, Facebook, and online postings of lectures and PowerPoints. This was less satisfying, at least to the professors, than the personal interactions between students they had hoped to foster around environmental issues.

3.3 Political and Cultural Differences

Cooperation for the solution to global environmental problems can be marred by differences in political goals and perspectives. This difference played an important role in at least one partnership between students at Allegheny College in Meadville, Pennsylvania and Forman Christian College in Lahore, Pakistan. Foreign relations between the governments of the United States and Pakistan have been strained for many years. Compounding matters of difference in perception of world politics was the subject matter of the classes. In addition to covering the environmental geography of their respective regions, the classes also discussed the politics and environmental issues inherent in the conflict between Israel and its neighbors. American students arrived to the conversations largely ignorant of politics and history in the Middle East and Pakistan. Pakistani students generally came to conversations with preconceived antipathy toward Americans and Israelis. Conversations between students during videoconferences and on Facebook were sometimes quite tense and had little to do with issues of the environment.

Conversely, an assignment designed to point out cultural distinctions between American and Moroccan college students found fewer differences than expected. Students in each country prepared a journal of all the foods they consumed in a day and then shared Power Points of their findings with one another on a class website (https://sites.google.com/a/allegheny.edu/soiltoplate/food-log-powerpoints). The surprising finding was that the dominant feature was globalization. Students in Morocco and the

United States to a surprising degree ate similar processed foods prepared by school cafeterias, ate out at fast food restaurants like McDonalds, purchased quick, packaged foods at local dispensaries, and drank caffeinated softdrinks like RedBull.

3.4 Technology

The success of a connected course seems to hinge on the technological capabilities of both educational institutions, as well as student knowledge of programs that allow face-to-face interaction. A connected course between Allegheny College in Pennsylvania and Al Akhawayn University in Morocco was able to videoconference once with no difficulties. The compatibility of the videoconferencing systems made this an easy and effortless meeting on the part of the professors.

The connected course between the American University in Bulgaria and Allegheny College planned two videoconferences during the summer planning session. Bill Clark from Bulgaria planned to discuss medical pests of the Roma population in Bulgaria, while Beth Choate from Pennsylvania planned to discuss issues of invasive insects in the US and Bulgaria. When consulting with information technologies support staff it was indicated that the technology was in place for PowerPoint presentations to be projected and shown through the videoconferencing technology. Students would be able to see and hear presentations, as well as interact with the "visiting" professor. Due to incompatibility of the videoconferencing systems, as well as limited time availability from the IT support staff, the professors were forced to find other methods of connecting. Skype was employed in both instances; however, PowerPoints could not be utilized and in one instance, the bandwidth was too slow to allow for actual video. Both of these situations diminished interactions between professor and students, and created a less inviting atmosphere for remote lecturing.

Communication between students when conducting group projects was often limited to e-mail and Facebook. Allegheny College students were required to complete group projects with students in Morocco and Bulgaria. Students were encouraged to set up Skype appointments with their partners. In an effort to facilitate these interactions, professors collected Skype names and exchanged them between groups. Many Allegheny College students were not familiar with Skype and set up Skype accounts for these interactions. However, due to limited knowledge of the program many did not feel comfortable or make the effort to communicate face-to-face. Instead, many students used Facebook, googledocs or e-mail to communicate with group members. The final product, as well as the excitement of students, seemed to increase when students put forth the effort to speak directly with their group members.

Faculty at Denison established a discussion forum through Blackboard; however, this proved not very inviting for students at the American University of Greece. While all faculty saw the inherent value in course connections for teaching issues of cross-cultural understanding, all agreed that modifications were necessary for the program to be successful.

3.5 Group Projects

All connected courses that we investigated contained a group project component. As is often the case with group work, certain members of the group were responsible for completing the majority of the required assignments. In some cases, portions of the group from both courses did connect; however, in other situations, students from one course were not able to connect with their group members from abroad. Students involved in the Pennsylvania-Bulgaria connection had difficulty sharing a google doc with their group members. The students believed that they had made every effort to reach out to their American University of Bulgaria (AUBG) group members by sending them a google doc invite; however, the AUBG students did not use gmail and therefore did not have access to this document. These are not uncommon issues with group projects; however, many students complained that adding cultural and political differences, as well as the fact that their partners were across the globe made these assignments difficult to complete.

One of the twenty students involved in the Pennsylvania-Morocco classes commented that group papers are difficult even when your group members are in the same country. Two Allegheny College students working with students in Bulgaria felt as though "no additional learning was received from the group project" and that it was difficult "because of lack of preparation." The project was described as "confusing" and "stressful." Seven of the sixteen students enrolled in World Regional Geography at Allegheny College that paired with Geographical Profile of Pakistan at Forman Christian College in Pakistan commented negatively about the group work involved with the connection. Comments focused on the difficulty of collaborating with students in the connected course. One individual commented that "it was next to impossible to establish a dialogue between myself and my Pakistani partners to fully complete the project." Two students noted that the videoconferencing with their connected course was "interesting," but then went on to comment that the group projects did not add anything to the experience. The following comment does a thorough job of summing up the concerns of many others in the class. "It's too hard to coordinate with an Allegheny student who is just as busy as I am plus two Pakistani students who are just as busy and in a different time zone."

Faculty echoed the students in their opinions, stating that students saw the group work as a "burden" more than an opportunity. This was especially true in instances where the assignments associated with the course connection did not contribute to their final course grade. Faculty also expressed frustration at the hesitation of students to contact their group project partners and begin working on their assignments.

3.6 Planning

The cost for faculty brought forward by most of the course instructors was an increase in the amount of preparation time. Initial planning of courses was facilitated by a meeting of the GLCA Global Liberal Arts Alliance members participating in these connected courses. Summer planning was productive and ambitious. During the semester, planning the connections between courses and facilitating student projects required additional preparations and communications to already busy schedules; difficulties with cultural or political differences and/or technology required significant effort. Coordination of technology seemed to be the major time expenditure, followed by designing extra assignments and staying on campus late due to evening videoconferences. Although all courses, connected or not, require time for planning, being aware of the potential time commitment necessary to complete a connected course is important when considering the cost/benefit analysis.

3.7 Rewards

For students Despite frustration with group work, students expressed appreciation for specific interactions that resulted from the connections. Students participating in the course connection between Allegheny College and the American University of Bulgaria enjoyed the guest lectures that were given by the respective professors. Both groups of students appreciated the time and energy that the professors put into speaking with them and preparing for each other's courses. Despite issues with technology, a positive review was provided by many students. This seems to be one course connection activity that students felt as though helped them learn more about a specialized topic. Two student evaluations from the course connection between Pennsylvania and Pakistan commented that videoconferencing with students in Pakistan "helped [me] to think outside the Allegheny (College) box" and "gave perspective". One student noted that the discussion with the class in Pakistan was "interesting". An Allegheny College student involved in the course connection with Morocco commented that "I did enjoy the time spent with them (the Moroccan students)".

For faculty Faculty benefited from course connections in a variety of ways. When lectures were exchanged between classes, faculty felt as though they learned through observation of a different style of teaching. Faculty from Denison University and American University of Greece continued their collaboration and are currently working on a collaborative research project on climate change perceptions on the two campuses.

For the educational process The primary reward of partnering two classes from different countries was how accurately the partnerships modeled real-world scenarios that are part of trying to solve international environmental problems. In sum, nearly all the partnerships encountered unforeseen difficulties. Communication between students from different cultures was difficult because of challenges involved in understanding one another's accents, ability to select a common language in which both partners were equally proficient, or because there were real or perceived differences in political perspectives. Technology was not always reliable. Widely disparate time zones and pressing commitments to complete unrelated projects interfered with the ability of students to collaborate. Projects or research papers designed for completion by students in partnered classes in separate countries were particularly hard to bring to fruition.

Yet, the very obstacles that hindered student success are precise matches for what professionals encounter whenever they try to work across international borders. Government and NGO professionals working in foreign countries to solve environmental problems also must overcome time zones, accents, politics, technology, and disparities in expectations. In this regard, creating partnerships between students in foreign countries was a successful modeling of environmental problem solving in the real world.

3.8 Friendship Building Exercises

The literature on peace education and conflict resolution offers some tools for overcoming obstacles emerging from groups in conflict (Bar Tal and Rosen 2009; Salomon 2004.) In fact, while participants in different countries may concur that progress on environmental issues is necessary, their perceptions of the sources of environmental degradation and their perceived roles in ameliorating them may be widely disparate (Chaitin et al. 2004). Political animosities that interfered with communication between American and Pakistani students, for example, finally began to erode when the two groups shared Power Points with one another solely dedicated to describing the personal lives and native villages of the participants. This exercise, unfortunately, came at the end of the semester, when it really should have been used at the outset. Learning from this mistake, a combined class on Food, Agriculture, and the Environment (United States and Morocco) began with shared Power Points describing the foods that students loved to eat and included photographs of students preparing or eating their favorite foods.

4 Conclusion

There are certain lessons that emerged from these partnerships that enhanced the success of the partnerships. Probably most important was for the faculty involved to engage in meta-teaching, e.g., explaining at the outset that not only will obstacles emerge over the course of the upcoming semester, but that the emerging obstacles can be expected from any international interaction. Faculty must use the partnership to teach students that the solutions to working across borders to solve environmental problems requires a lot more than simply good science and engineering. In addition, professional practitioners must learn the skills of patience, persistence, planning, and cultural sensitivity. In short, one main object of the international partnership is to go beyond learning about environmental issues in a foreign country. A significant goal is to learn *how* to work with people who are different from ourselves.

References

Abu-Nimer M (1996) Conflict resolution approaches. Am J Econ Sociol 55(1):35–52

Avruch K, Black PW, Scimecca JA (1991) Conflict resolution: cross-cultural perspectives. Greenwood Press, New York

Bar Tal D, Rosen Y (2009) Peer education in societies involved in intractable conflicts: direct and indirect models. Rev Edu Res 79(2):557–575

Bennet DC, Cornwell GH, Al-Lail HJ, Schneck C (2012) An education for the twenty-first century: Stewardship of the global commons. Liberal Education 98(4). Retrieved from http://www.aacu.org/liberaleducation/le-fa12/bennett_cornwell_al-lail_schenck.cfm

Beringer A (2007) The Lüneburg sustainable university project in international comparison: an assessment against North American peers. Int J Sustain High Educ 8(4):446–461

Buchan GD, Spellerberg IF, Blum WE (2007) Education for sustainability: developing a postgraduate-level subject with an international perspective. Int J Sustain High Educ. doi:10.1108/14676370710717553

Chaitin J, Obeidi F, Adwan S, Bar-On D (2004) Palestinian and Israeli cooperation in environmental work during the "Peace Era". Int J Polit Cult Soc 17(3):523–542

Ciurana AM, Filho WL (2006) Education for sustainability in university studies: experiences from a project involving European and Latin American universities. Int J Sustain High Educ. doi:10.1108/14676370610639263

Dale A, Newman L (2005) Sustainable development, education and literacy. Int J Sustain High Educ 6(4):351–362

Eatmon T, Pallant E, Lawrence S (2013) Food production as an integrating context for campus sustainability

Ferrer-Balas D, Adachi J, Banas S, Davidson CI, Hoshikoshi A, Mishra A, Motodoa Y, Onga M, Ostwald M (2008) An international comparative analysis of sustainability transformation across seven universities. Int J Sustain High Educ 9(3):295–316

Global Liberal Arts Alliance (2011) About us. http://www.liberalartsalliance.org/about-us. Accessed 13 Feb 2014

Great Lakes College Association (n.d.) About us. http://glca.org/about-glca. Accessed 13 Feb 2014

Lu S, Zhang H (2014) A comparative study of education for sustainable development in one British university and one Chinese university. Int J Sustain High Educ 15(1):48–62. doi:10.1108/IJSHE-11-2012-0098

Lundegård I, Wickmann P (2007) Conflicts of interest: an indispensable element of education for sustainable development. Environ Educ Res 13(1):1–15. doi:10.1080/13504620601122566

Marsella AJ (2005) Culture and conflict: understanding, negotiating, and reconciling conflicting constructions of reality. Int J Interc Relat 29(6):651–673

Pallant E, Boulton K, McInally D (2013) Greening the campus: the economic advantages of research and dialogue. In: Leal WF (ed) Sustainable development at universities: new horizons. Peter Lang Scientific Publishers, Frankfurt, pp 373–382

Pinar W (2003) International handbook of curriculum research. L. Erlbaum Associates, Mahwah

Ramsbotham O, Miall H, Woodhouse T (2011) Contemporary conflict resolution. Polity, Cambridge

Salomon G (2004) A narrative based view of coexistence education. J Soc Issues 60:272–287

Stephens JC, Hernandez ME, Román M, Graham AC, Scholz RW (2008) Higher education as a change agent for sustainability in different cultures and contexts. Int J Sustain High Educ. doi:10.1108/14676370810885916

Vayrynen T (2001) Culture and international conflict resolution: a critical analysis of the work of John Burton. Manchester University Press, Manchester

Author Biography

Eric Pallant is Professor of Environmental Science, Beth Choate is Assistant Professor of Environmental Science, and Derek Reno is an undergraduate major of environmental science all at Allegheny College. Pallant's research incorporates international sustainable development in small communities including college campuses; Choate's research focuses on insect ecology and sustainable pest management in managed ecosystems; Reno is studying environmental science and environmental problem solving.

The Digital Studio: A Case Study of an Online Sustainable Design Studio

Robert Fryer

Abstract

Universities see online education as a means to address decreasing enrollment and growing tuition costs, and many are offering online courses as a result. Architecture and design programs are hesitant to take their courses online, however. Replacing the hands-on and direct interaction central to studio-based education with virtual interactions is worrisome to some educators. This worry is often caused by the assumption that essential design skills, techniques and studio culture will be lost. This paper presents the results of an online sustainable design studio. The results have implications for studio culture, conceptual and skill based learning outcomes, and site specific design development. Results seem to indicate less sophisticated site specific design interventions and a significant loss of direct student to faculty interaction. However, conceptual and skill based learning outcomes are consistent with the traditional on campus version of the same course. The paper is meant to offer insight into how studios could be taught virtually, and to critically evaluate the challenges and opportunities of a digital future for sustainable design education. It also discusses the implications that online programs may hold for sustainability.

Keywords

Sustainable design · Online education · Virtual studio · Sustainable development

R. Fryer (✉)
MS Sustainable Design Program, College of Architecture and the Built Environment, Philadelphia University, 4201 Henry Avenue, Philadelphia, PA 19144, USA
e-mail: FryerR@philau.edu

W. Leal Filho (ed.), *Transformative Approaches to Sustainable Development at Universities*, World Sustainability Series, DOI 10.1007/978-3-319-08837-2_39

1 Introduction

1.1 Why Courses Are Going Online

Higher education in the US has undergone changes since the financial crisis in 2008. At that time, enrollment in college began to decline due to household finances. Although the trend has changed since 2008, the higher education sector has not yet recovered completely. As reported in US News, "Overall college enrollment fell for the second year in a row in 2013, from about 20.2 million students in the fall of 2012 to about 19.9 million this year." The decline in enrollment, however, is not only caused by economics, but also is attributable to the decreased number of graduating high school seniors (Bidwell 2013). At the same time, the cost of attending college has grown, but the increase in tuition rates has slowed even though it's still increasing. The Chronicle writes, "Average published tuition and fees at private nonprofit four-year colleges for 2013–14 grew to $30,094, up from $28,989 the year before." But, contributing to the financial pressure students are experiencing is the lack of federal funding that was more abundant and easily accessible before the financial crisis (Supiano 2013). In response, universities are offering online courses, since they are seen as a way to offer an education for less cost to students and universities. Online courses don't require the same cost-intensive resources, such as facilities, furniture, or fixtures. These savings may be passed on to the student in lower tuition and room and board costs, while there are fewer expenses for the university per student as enrollment in online courses increases. However, architecture programs are reluctant to offer their courses online, especially studios. The hesitation is caused mostly from the legacy of teaching studio courses in an intense setting lasting hours at a time with lots of personal interaction between student and faculty, as well as between the students themselves. The teaching method relies on the ability to demonstrate techniques immediately and "in the moment", and these techniques are often done by hand even with the heavy use of computer software to assist with drafting, modeling and scripting. Over time, this has proven to be a very effective means of student learning, and has even been lauded as a means to teach courses in other, unrelated fields via "project-based learning". Furthermore, this direct and intense interaction develops what is commonly referred to as "studio culture", in which participants experience deep social bonds and camaraderie. Architecture faculty are reluctant to give this up if it means to replace it with a relatively disengaged, distant and virtual experience. Still, the pressure to offer studios online exists, and some online studios are already being offered. This paper reports on one such studio.

1.2 Course Description

The online studio course work analyzed in this paper is from a three credit sustainable design studio that met for 4 h per week for a total of 16 weeks. The typical weekly schedule was a 1 h group meeting during lunch on Fridays, along with 3 h of individual-professor meetings arranged by sign-up sheet. Otherwise, when required, group charrette times were scheduled, and students met outside of these scheduled times, online, to complete their work. By comparison, the on-campus version of this course met with all students for 4 h each week for 16 weeks, providing much more "face time" between the students and faculty. The reason for the difference in quality of meetings between the online and on-campus version of this course is predominantly the result of the type of student who enrolled in the on-line course.

1.3 Who the Students Are

The students in the online courses are from across the US, but with a cluster on the northern east coast. All had full time jobs, while some had dependent children. The students are highly and independently motivated, however they had limited time availability because of the conflict with work and family schedules. The online courses provided less of a time commitment that allowed them to pursue a graduate degree where the conventional on-campus courses would not permit it. As a result, it was difficult to find times during the week where all of them could meet simultaneously for group activities.

1.4 Course Workflow

The course met in a virtual classroom for group discussions, activities, workshops and topic introductions. Adobe Connect was the software used to accommodate this interaction. All group meetings were recorded and shared for those who were unable to attend. The course content and learning modules were held in Blackboard. Studio work submitted for the professor's review was done using Power Point. The students would submit a presentation via email that contained images of their design work. The professor then recorded a video of his feedback using the student's work and the built-in capabilities of Power Point to draw over their images and provide verbal feedback simultaneously, as shown in Fig. 1. Further discussion on this and other unconventional uses of software follows later in this paper. The recording was then posted to YouTube using privacy settings so only the intended student could view.

Workshops or other conventional in-class activities were held in the virtual classroom. Here the students and faculty could see and hear each other in real time using web cams and various audio capabilities similar to a web conference interaction. These were scheduled when most students could attend and recorded and shared for those who couldn't.

Finally, the design charrettes and juried feedback occurred twice during the semester and were conducted using Prezi, a web-based presentation software. The layout of the presentation was completed by both the faculty and the students, and

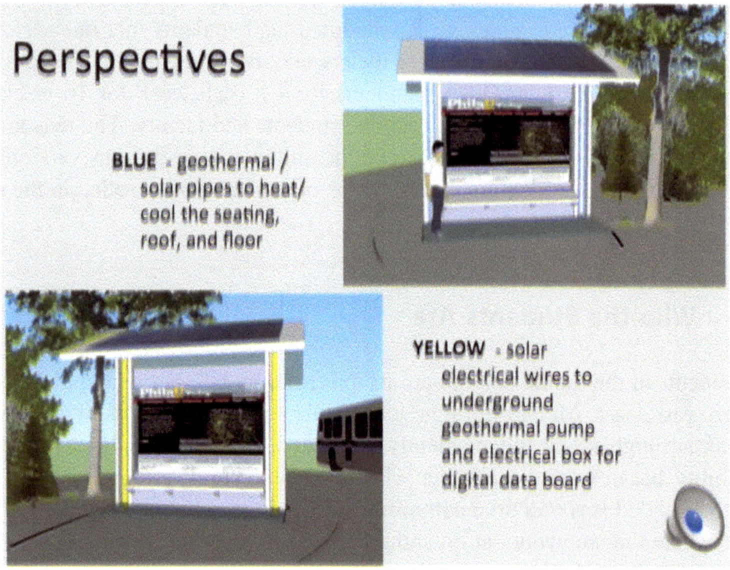

Fig. 1 Online student work showing a lack of site integration. *Source* Owens 2012

included jury bios, the student design work and distinct boundaries where the jury was to provide their feedback. The jury's feedback, as in all of the sustainable studios regardless of online or on-campus, were limited to a discussion of opportunities and challenges for the student to use to further develop their design work. This asynchronous review was convenient for the jury, who are also full time owners or employees of successful and busy practices.

2 Analysis of Outcomes

2.1 Studio Culture

Students in online studios spend much less time with each other and their faculty compared to the same course offered on-campus. The three credit online studio described in this paper was offered at the same time as the identical course was offered on-campus. The author was the faculty member teaching both courses. On average, the online students would spend between 1 and 4 h each week with either the faculty or their cohort. By comparison, the on-campus students were spending 4–8 h with either the faculty or their cohort. The time students spend together intensely working on design projects contributes to intellectual and emotional bonding, an essential element to what's commonly called "studio culture". Furthermore, students learn skills and methods directly from each other. When this time is decreased, so is the potential to bond and learn extracurricular skills and methods. However, some bonding was apparent, and those students who were located near each other did meet outside of class on several occasions. However, the connections made are neither as rich nor as deep as the connection between on-campus students. Because the types of students that enroll in these courses are students who otherwise do not have the time available for an on-campus experience because of work and/or family commitments, they still find value in the course without the development of a strong studio culture. So, while faculty are right to notice the difference in the culture and the impact it may have on the depth and richness of learning, it is important to also note that the students are learning more in online studios than they would by not taking any graduate level studios, which is likely their alternative.

2.2 Conceptual and Skill Based Learning Outcomes Versus Site Specific Design Development

Perhaps even more than in conventional architecture studio courses, sustainable design studios stress the importance of site integration. Climate, orientation, landscape, storm water, microclimate and thermal comfort are the foci of sustainable site design, and their integration is essential to successful design development. Repeated site visits are essential to the designer's ability to iteratively refine and propose designs that evolve the relationship between these components. Online students did not have the opportunity to visit the project site, and the results demonstrate a lack of integration and refinement. Figure 1 shows the work of an online student's bus stop proposal. Although it is technically and conceptually sophisticated because it includes radiant and conductive panels lined to a geoexchange system (not shown in figure), its formal development is not responsive to the unique characteristics of the site. This design could be placed almost anywhere and it would not loose any of its function. By comparison, Fig. 2 shows student work from the concurrent on-campus studio, where students visited the site repeatedly. The project site in both figures is identical. In this design proposal, there is a more sophisticated and developed relationship between the site, circulation, vegetation, orientation, climate, etc. Interestingly, Fig. 3 shows the work of another online student who coincidentally visited the site before it was assigned as the project site. This student was relatively familiar with the site and its activities, and the design reflects this knowledge.

2.3 Software and Work Flow

There is currently no standard software for the delivery and feedback of online studios. As a result, extensive and continuous experimentation was used in an attempt to discover methods that would replicate the benefits of face-to-face studio desk critiques while also working in a virtual environment. The market does provide solutions, but none of the products tested by the author so far work better than

Fig. 2 Thermal comfort used to locate vegetation and derive built form (student work). *Source* Dhayagude et al. 2012

Fig. 3 Online student work showing deeper levels of site integration after coincidently visiting the project site before starting the project. *Source* Pendleton 2012

the unconventional uses of existing software presented in this paper, such as Power Point mark-ups that are recorded and posted to YouTube or Prezi for charrette and jury feedback. The lack of a standard and viable software is confirmed via research of other US online architecture studios (Griffen 2014).

3 Insights into Digital Studios

3.1 Studio Culture

Although not all online students may be looking for a conventional studio culture experience, opportunities for social interaction were desired by the students and faculty. The few social events developed for this course were successful in bringing together most, but not all, of the students. These social events were timed to coincide with other events, like Greenbuild 2013 in Philadelphia. Other online programs at other universities have some residency component, if only for a few days or one week where social events are combined with final presentations and face-to-face juried reviews (Griffen 2014). If these are not already in place, it is worth considering them.

3.2 Conceptual and Skill Based Learning Outcomes Versus Site Specific Design Development

The concepts and skills associated with sustainable design seem to be achievable regardless of whether the course is offered online or on-campus. The students in this online course were able to comprehend, synthesize and create new designs with technical sophistication. However, site specific design development seemed to suffer. This appears to be the result of not visiting the project. It is worth seeing if there are alternatives to this. Perhaps it would be possible to arrange for projects that permit a site selection near each student. However, this could cause other complications, since it is then the faculty who will not have exposure to the site and its dynamics.

3.3 Software and Work Flow

Besides the lack of student culture and site specific design, the lack of adequate and appropriate software solutions specifically for studios is an obstacle to achieving learning outcomes. For the time being, there seems to be little alternative to faculty exploring, experimenting and testing new software and new uses for old software.

3.4 Implications for Sustainability and Sustainable Design Education

A major goal of sustainable design education is to impact the design and development of actual built projects. By increasing the number of sustainable projects, environmental, economic and social conditions improve, but also public awareness and demand increase, as the benefits of such projects become apparent. This, in turn, creates a reinforcing feedback loop that increases the demand for more architects, engineers, landscape architects, builders, etc. with the skill, knowledge and methods required for sustainable design. Online education is one means to support this dynamic and the goal of making more development sustainable.

Laws, regulations, green building standards, the flow of information and tax incentives have increased the number of sustainable projects. But within education, there appear to be three leverage points: increase the number of students in sustainable design programs, improve the learning outcomes and reduce any delays in the learning process between enrollment and graduation. Solutions to these leverage points can indirectly support the creation of new sustainable development projects or respond to the current demand. Online education is one potential solution to all three.

Online sustainable design programs can reach students that on-campus programs cannot. As an individual's time availability decreases and the cost of traditional education increases, fewer potential students are able to enroll in sustainable design programs. Online programs are able to accommodate these students, however. Meeting times are asynchronous and can be held at the student's convenience, while the tuition costs are often lower by comparison. This makes it more likely that potential students will enroll, perhaps even increasing the diversity of the students (and future practitioners) including those from different races and economic backgrounds. If more students enroll and graduate, this could introduce an additional beneficial feedback loop where these alumni will advocate for and seek out sustainable projects, thus increasing the total number of built sustainable projects.

Online education has the potential to improve sustainable design learning outcomes. This paper reports on one studio where the technical learning outcomes appear to be the same with the on-campus version of the course. However, the site-specific learning outcomes were not as strong. There is room for improvement here. Over time, it is reasonable to expect this outcome to improve as professors share best practices and software improves. But, as learning outcomes improve the built projects will also improve and so will the benefits.

Finally, online programs can assist in removing delays in the process of graduating students and thereby increase the number of advocates and skilled designers. As more professionals choose to design sustainably and are able to achieve results, the overall number of sustainable projects has the potential to increase through feedback loops. There are several circumstances that delay, distract or completely derail an enrolled student from graduating. Typical delays are most frequently caused by the demanding and competing commitments and circumstances of full-time employment and raising families, as well as financial issues encountered after enrollment. Online programs can be shorter while still achieving the required learning outcomes. This diminishes the exposure to financial shortages caused by unseen events while providing a shorter time period to juggle competing demands. As discussed, online programs can also be less expensive and more accommodating to hectic schedules.

4 Conclusion

The number of online studio courses is expected to increase in the future. To meet this need, communities of architecture faculty must discuss how to maintain the best qualities of on-campus sustainable studios and their learning outcomes, and seek ways to develop online software and means of interacting that maintain the high level of design standards and techniques common in on-campus sustainable studios. However, the faculty are, so far, reluctant to do so. Furthermore, online sustainable design programs can address three major leverage points: increase the number of enrolled students, improve learning outcomes and remove delays between enrollment and graduation. If solutions to these three opportunities are successful, it is likely that it will result in more sustainable development projects and the environmental, economic and social benefits associated with them. However, there appear to be at least two obstacles to online sustainable design education. The first is how to design sustainable sites that require site specific learning outcomes when the students in online programs are not able to visit project sites. The second obstacle is the lack of adequate software to provide the necessary interaction between students and the faculty during design studios. However, if architecture faculty respond to this need, it is reasonable that these issues can be solved.

References

Bidwell A (2013) College enrollment falls for second year in a row. Accessed from http://www.usnews.com/news/articles/2013/12/12/college-enrollment-falls-for-second-year-in-a-row

Griffen C (2014) Online studio methodology research. Philadelphia University, Philadelphia (Nexus Online Research Grant)

Dhayagude M, Herrero-Garcia V, Prakash A (2012) Project 1: the individual. Philadelphia University, Philadelphia

Owens S (2012) Project 1: the individual. Philadelphia University, Philadelphia

Pendleton S (2012) Project 1: the individual. Philadelphia University, Philadelphia

Supiano B (2013) Tuition increases slow down, but there's more to college affordability. Accessed from http://chronicle.com/article/Tuition-Increases-Slow-Down/142547/

Author Biography

Robert Fryer Over the past 20 years Professor Robert Fryer has practiced, taught, consulted and researched into sustainable design and architecture. Robert first practiced as an architect for 10 years before leaving to earn his graduate degree from the Architectural Association School of Architecture in the Environment and Energy Programme in London, where he completed his dissertation on deformable textile exterior wall assemblies that changed form according to climate stimuli. Since then he's held researcher and professor positions at Philadelphia University in Philadelphia, PA in the US. Most recently, Robert was head of developing an online version of the award winning on-campus MS Sustainable Design program, and was also a member of the group of faculty working across disciplines (Architecture, Engineering and Biology) to create a systems thinking course on biomimicry. He also consults with materials manufacturers on testing and creating new products and with communities and towns in using the integrated design charrette process to develop properties.

The Application of Systems Thinking to the Sustainable Design Process

Robert Fryer

Abstract

It is essential to evolve the sustainable design process in response to extraordinary socio-environmental issues that are associated with the design of the built environment. Issues such as the loss of biodiversity, ecosystem services, urban resilience, the consumption of large amounts of energy and material resources, among others, derive at least in part from design. Sustainable designers intend to address these issues, but the existing design model overlooks essential topics. The conceptual boundaries that separate what's explicitly considered from what's ignored need to be redrawn and expanded to include topics that lie outside the prevailing design process and fundamentally address these issues mentioned above. Systems thinking is a suitable means to model design processes because of its ability to simultaneously view diverse elements and the dynamic relationships between them. This paper applies systems thinking to reveal the prevailing design model's dynamics, externalities, feedback loops and leverage points. The leverage points unveil some opportunities for transformation, and a list of potential tools and skills are proposed. Among those explored are cross-scalar analysis between conceptual systems boundaries (individual/building/city-regional scales) and methods of including externalities (cross-disciplinary knowledge, thermal human comfort as design tool, and long term programming). Furthermore, these proposals are tested through student work and the methods, skills and tools are generalizable to other professors.

R. Fryer (✉)
MS Sustainable Design Program, College of Architecture and the Built Environment, Philadelphia University, 4201 Henry Avenue, Philadelphia, PA 19144, USA
e-mail: FryerR@philau.edu

W. Leal Filho (ed.), *Transformative Approaches to Sustainable Development at Universities*, World Sustainability Series, DOI 10.1007/978-3-319-08837-2_40

Keywords
Sustainable design · Design process · Systems thinking · Education

1 Introduction

Examining the prevailing green design approach from the alternative perspective of systems thinking reveals gaps between the desirable and actual outcomes of the green design process (see Fig. 1). This gap comes from the green design process, which most times is actually the conventional design approach overlaid with a green building standard, such as LEED.

Furthermore, the current green design process still misses essential elements of sustainable design, such as adaptive reuse, ecosystem services, social equity and occupant behavior.

Figure 1 also shows missing feedback loops (shown dashed). These feedback loops could provide the information required to adapt the design process and act to balance the undesirable outcomes if rigorously included in the process. However, there is little, if any, evidence that this is being done. By filling in these missing feedback loops and exploring new design methods, skills and tools, it is possible to advance the sustainable design process to include design issues that it currently misses.

2 Problem Explanation

Despite notable failures (Scofield 2009), green building standards, such as LEED, work within the limited boundaries and scales in which they are meant to be applied. However, their intended scales and boundaries are arbitrary and often are not inclusive enough to be truly sustainable and thus fail to address dynamic issues that affect the built environment.

For example, green buildings do not consider how to design for programmatic changes over time. Typically, a building is designed to fit the specific needs of the current client's programmatic requirements. However, buildings commonly outlast their originally intended function and architectural program. The option to reuse buildings for different programs, what is commonly referred to as "adaptive reuse", can be difficult and often impossible to exercise if a building is too tightly designed to one specific program. Designing for a "loose fit" is overlooked in the current green building design process. The LEED rating system does not address this, however LEED discusses flexibility in the design of hospitals to meet their future expansion needs (USGBC 2013). But there is no mention of cross-programmatic flexibility. Preservationists like to say, "The greenest building is the one already built" due to the cultural and historic significance of a building, but also due to the large energy and material investment. Designing for adaptive reuse and the multiple ways people care for and love buildings is not new (Brand 1995). Failing to design adaptive, well-liked buildings will result in buildings that either do not meet their function (and thus create

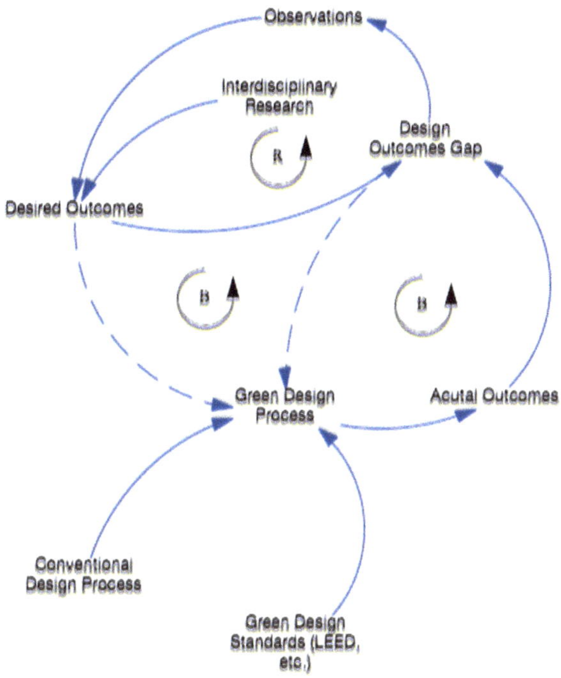

Fig. 1 Design outcomes gap. Causal loop diagram showing a gap between desired and actual design outcomes. This is a result of a *green* design process that is a product of a conventional design process and *green* design standards. There are missing feedback loops (shown *dashed*) that could be used to transform the design process

a lesser architecture) or they are demolished and replaced, which results in wasted embodied energy, increased landfill waste and site disturbance.

Green building design is often conflated with sustainable architecture, however they are different. The focus of green buildings is mostly high performance and efficiency. Limits to the availability of energy resources encourage the prudent use of petroleum, while limits to the availability of other resources, such as fresh water, encourage their efficient use, too. Because of this, it is often assumed that sustainability is synonymous with efficient use of resources. However, sustainability encompasses more than efficiency and optimization. Efficiency and optimization are necessary in reaching sustainable development, but are not themselves sufficient. Furthermore, efficiency and high performance in its most extreme forms can be counterproductive to sustainability. This is because both seek to minimize the number of elements within a system and to connect each element in a direct and simple relationship. In the pursuit of this goal, however, a myopic focus tends to neglect how each element is actually connected to other elements in often complex and indirect ways. Furthermore, efficiency and optimization eliminate redundancy and modularity which are fundamental characteristics of a dynamic system. Elimination of these produces a fragile system that is vulnerable to change and can suddenly and unexpectedly change behavior. As Walker and Salt write, "What this all adds up to is that there is no sustainable "optimal" state of an ecosystem, a social system, or the world. It is an illusion, a product of the way we look at and model the world. It is unattainable; in fact … it is counterproductive, and yet it is a widely pursued goal" (2006, p. 8).

When efficiency increases, redundancy decreases and buildings become exposed to extreme events and shocks. One example of this is the reliance on mechanical ventilation for fresh air supply in airtight buildings. In the case of power service interruptions, which can last weeks under uncomfortable thermal comfort conditions, a building is unable to supply fresh air to occupants without resort to opening doors and windows, further exacerbating an already thermally challenged event. Buildings are also exposed to electricity, water and natural gas service interruptions if they are solely reliant on a centralized distribution infrastructure, as is the convention.

These extreme events and shocks are predicted to increase in intensity, as occurred in 2012 on the East Coast of the US with Hurricane Sandy. The built environment is also threatened by climate change via exposure to increased temperatures and heat waves; sea level rise; and increased precipitation and flooding, as well as increased droughts. However, how to design for a dynamic climate is not addressed in current green design processes.

The leading cause of climate change is greenhouse gas (GHG) emissions. However, these emissions are not directly addressed in LEED. In fact, studies show, "It is possible to achieve LEED® Gold and Platinum status without significantly reducing the carbon emissions that result from operating energy. A building that is targeting Carbon Neutrality with respect to its operating Energy MUST achieve all of the credits associated with optimizing energy, and more" (emphasis in original) (Boake 2012). Without targeting the elimination of GHGs from building operations, green design contributes to the very conditions sustainability is attempting to resolve.

Another problem is that little, if any, consideration is given to the urban/suburban/rural continuum. Essential services flow across the conceptual boundaries of LEED project sites, and wastes flow out across them, too. Many ecosystem services lie across these boundaries and these must be considered if designing for sustainability. Some of these services are: storm buffers, food, clean air, water purification, recreation and materials.

Obviously, some of these issues may be beyond the scope of green building standards, but this is important because the standards inform the typical design approach used by the small portion of the architectural profession that engages in green and sustainable projects. Therefore, architectural designs, especially those in cities, continue to propagate the status quo that decreases that adaptability of the built environment, society and the biophysical systems at a time of unprecedented change. This results in diminished sustainability in its true sense.

3 Solutions Using Leverage Points

Educators have an essential role to play. Not only do most designers acquire design models while in school, but also academia can provide a rich and relatively risk-free laboratory to experiment with new design models. Access to expertise and research from other disciplines also helps. Empowering graduates with a new model as they enter or reenter practice, whether as employees in existing firms or by creating their own pioneering firms, can be a powerful system flow-changing leverage point to change the system. Furthermore, this could help establish a feedback loop between design outcomes and practice with the intent to provide useful information to change models (Meadows 2008).

3.1 Looking Across Scales

As Walker points out, understanding systems behavior depends on identifying and comprehending feedbacks. When feedbacks are misunderstood, ignored or even missing, systems behaviors can suddenly shift to new, undesirable paradigms. Resilience is a matter of feedback management to keep systems far enough away from thresholds to maintain their dynamic equilibrium. One of the most common mistakes and most challenging research topics is identifying and comprehending the feedbacks that work across different system scales (2006).

Through a series of projects, students are presented with design issues at different scales, such as individual, individual + building, and individual + building + city. They are tasked with considering the interactions across and within these scales. In one project, student teams select a city or region and begin research into system flows. The research typically reveals transportation, services (e.g. commercial, health, educational, food etc.), electricity, water, material resources and population systems as common areas to investigate. The teams diagram these systems and elements. Figure 2 shows one team's mapping of key town assets that are often flooded in the town they chose.

The teams uncover challenges to the systems through research. For example, one team found that Stockton, NJ was extremely vulnerable to power outages when flooded. The increasing storm frequency was leading to 100-year floods every 3 years. Furthermore, the service record of the power company supplying the town showed that it did not respond quickly to restore power when compared with other power providers in the region. Additionally, reports indicated the power company acknowledged it knew of this vulnerability, but still did not plan for it. It claimed that it would deplete profits, even though it was reporting strong profits at the same time.

The teams continue analysis to identify leverage points where design interventions would have the biggest impact. The team concluded, "At current, only initiatives at the city scale can solve the factors which threaten Stockton's resilience. Smaller and larger scale approaches have proven inadequate, short lived and characterized by inequity" (Pendleton and Zosky 2012).

The team synthesizes its research to propose design interventions. In this case, the team proposed several alternative renewable energy solutions. These proposals were evaluated using a standard SWOT analysis to arrive at a final design solution. The final proposal was a distributed photovoltaic array whose electricity would be stored in batteries located outside of the flood plain (see Fig. 3).

Fig. 2 Mapping of key assets that are located within the flood plain (student work). *Source* Pendleton and Zosky (2012)

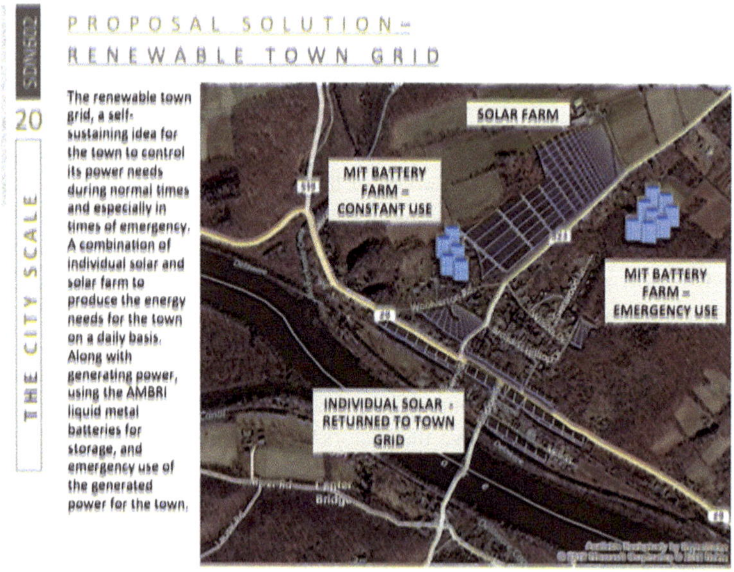

Fig. 3 Design proposal showing distributed solar energy PV arrays and batteries located outside of flood zones (student work). *Source* Pendleton and Zosky (2012)

At the end of the course, students are asked to reflect on their experience during the semester. One student wrote:

> The stressful relationship between resiliency and efficiency can create systems more vulnerable to shock when out of balance, as a narrow focus on one of the three [scales] comes at the expense of the other two.... This explains the need for designers to focus at the three scales simultaneously and with the intricacies of each factor in mind.... Rigidity in design not only exposes a system to decline but also misses the opportunities and resources which arise as [social, ecological] cycles shift.... When adaptive design is considered across the three scales [of individual, building and region], the issues are like an onion waiting to be peeled. Each issue is specific to its scale and may directly impact or be impacted from without. The practice of adaptive design strives to enhance the interrelationships between these scales and to design their capacity for change (Pendleton 2012).

3.2 Interdisciplinary Knowledge and Integrated Design Charrette

The institutionalization of disciplinary silos can be seen as a classic systems trap from the systems thinking perspective. Each discipline tends to see the project only through the lens of their own discipline, often resulting in multiple project goals, which sometimes conflict with each other. When a single system is pulled in different directions to achieve the stakeholders' different goals, the outcome is most often a result or behavior that none (or only one) of the stakeholders desires (Meadows 2008). This undermines project cohesion and the ultimate success of the project. To avoid or remove this trap requires the alignment of goals. One way to align goals is for the stakeholders to explicitly formulate them together, and then to work together to achieve them. This is a fundamental element of the integrated design charrette. Finally, clear communication is essential to the sharing of expertise and ideas, which requires honest presentation, listening and respect. Thus, the current practice design model does not result in the high levels of integration required to close the design outcomes gap. However, the studio seeks to avoid this systems trap by including multiple stakeholders and using the integrated design charrette process.

Students in the course are from different disciplinary backgrounds, such as Engineering, Architecture, Landscape Architecture, Interior Architecture and Construction Management. This approach tends to foster collaboration and communication as the students undertake design projects that require input from all of their backgrounds. Furthermore, experts from different disciplines teach studio and are present for the design charrettes. The project goals are formulated together at the beginning of the project and are pinned up on a wall for easy reference during the entire course. The interdisciplinary students and faculty continue to work as members of the same team to design and develop the project for the entire semester.

3.3 Thermal Comfort as Design Tool

The students use comfort prediction software before designing to understand the software and the project site. This is also the first step to immerse students into the topic of human comfort; as they alter the values of humidity, wind speed, temperature, etc. relationships and intuition are developed. The students identify the performance characteristics that change conditions enough to provide thermal comfort by "playing" with the values. Besides making their designs more realistic and effective, this step also enables creativity. For example, knowing that wind speed needs to be reduced to 5 m/s in winter opens new questions about how that can be achieved passively via different artificial or natural materials, textures, assemblies and form making (see Figs. 4 and 5). That is, designing with human comfort can create innovative aesthetics and spaces. Finally, students use this tool to evaluate and analyze their design.

3.4 Long Term Programming

Enabling architecture to adapt over time to suit different architectural programs is an essential, yet overlooked, component of sustainable design. However, it is difficult, if not impossible, to predict what changes a building, neighborhood or city may face in the future. Fortunately, there are exercises that can be applied to conventional programming that can begin to undermine myopic design processes that limit adaptability. For example, scenario planning, originally from military game simulations, is a tool students learn to help make choices as long-term time frames are considered.

The exercise starts by identifying key drivers that may affect the future of the organization, neighborhood or region. Key drivers are categorized among the social, technological, environmental, economic and political issues that may affect the project. After an exhaustive list of drivers is created, they are classified into important/not important and certain/uncertain categories. The drivers that are both important and uncertain are kept and used to create possible divergent future paths. In Fig. 6, these are lack of/increased business diversity and developers continue to revitalize/stop revitalization. These are used to create four different future worlds. The students then choose a strategy to deal with the four worlds. For example, if one world appears more likely than others to the team, it can choose to optimize a design for that world only. This is called the "Eggs in One Basket Strategy", and is the default position of all conventional programming exercises in practice. Other strategies include the "Blue Chip Strategy" (where all futures are designed for) and "The Hedge Strategy" (where two opposing worlds are designed for). Finally, the students create architectural programs based upon their strategy, and design to accommodate them accordingly.

Fig. 4 Thermal comfort used to locate vegetation and derive built form (student work). *Source* Dhayagude et al. (2012)

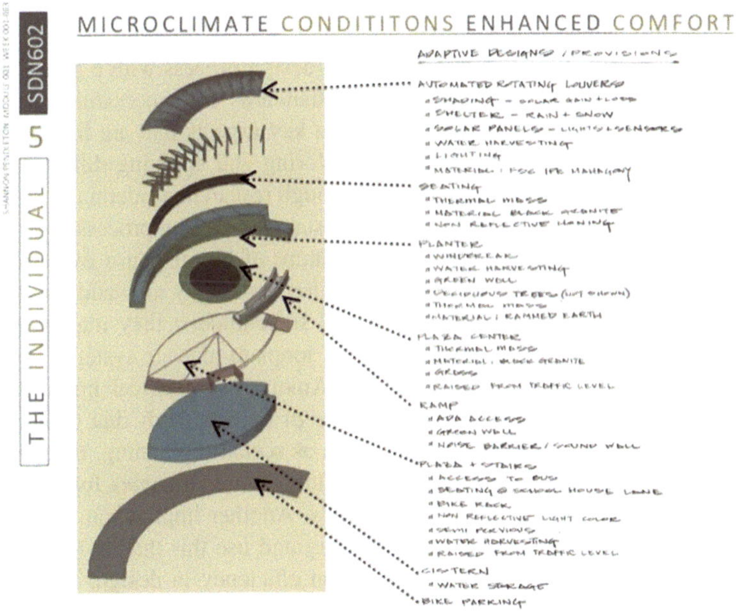

Fig. 5 The application of materials based upon the results of comfort simulations (student work). *Source* Pendleton (2012)

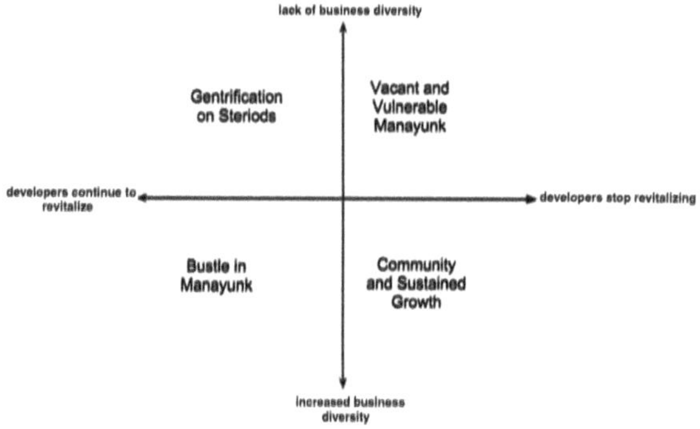

Fig. 6 The four worlds created through the scenario planning process for the town of Manayunk (student work). The *vertical axis* is labeled "Lack of business diversity" (*top*) and "Increased business diversity" (*bottom*). The *horizontal axis* is labeled "Developers continue to revitalize" (*left*) and "Developers stop revitalizing" (*right*). *Source* Pendleton and Zosky (2012)

4 Conclusion

For green buildings to become sustainable buildings their design process must be transformed to include new methods, tools and skills. Currently, the conventional green building process is the same as the old design process with a green building standard overlaid. Although green building standards are successful in what they intend to accomplish, they are also limited in key aspects that are intimately connected with sustainability. However, by exploring and applying different approaches, strategies and techniques uncovered through research, academia can provide a missing informational feedback loop to the sustainable design process. By so doing, design outcomes may reflect the dynamic nature of building use over time; they may creatively embrace thermal comfort requirements; they may address the needs and achieve the goals of a diverse group of stakeholders; they may look across scales to identify and apply missing feedback loops to manage systems of multiple scales to avoid unwanted paradigm shifts. Among the practical implications for sustainable design and development is the use of new methods that give insight to the built environment's dynamic nature, such as scenario planning, that provides a way to look into potential future uses of buildings and strategies for their design. This is easily incorporated into design studios. Another implication is the need to acknowledge the systemic behavior of the city and use this insight to avoid traps, such as disciplinary silos or over-focusing on efficiency in design. Seeking input from multiple, if not all, stakeholders in a project is essential to holistically addressing a design challenge. And finally, although there are others not mentioned here, the sustainable design studio may be impacted by the use of thermal comfort as part of an urban design agenda by understanding its impact on cross-scale issues, such as the heat island effect, to design for microclimatic interventions that could include introducing biodiversity and vegetation into a thermally massive urban tissue. These practical implications for sustainable design education point the way towards a design process that includes information from feedback loops that are currently missing.

References

Boake TM (2012) What is carbon neural design? LEED and carbon. http://www.tboake.com/carbon-aia/carbon_definition.html#LEED

Brand S (1995) How buildings learn: what happens after they're built. Penguin Books, New York

Dhayagude M, Herrero-Garcia V, Prakash A (2012) Project 1: the individual. Philadelphia University, Philadelphia

Meadows D (2008) Thinking in systems: a Primer. Chelsea Green Publishing, White River Junction

Pendleton S (2012) Project 1: the individual. Philadelphia University, Philadelphia

Pendleton S, Zosky M (2012) Project 3: city scale. Philadelphia University, Philadelphia

Scofield JH (2009) Do LEED-certified buildings save energy? Not really… energy and buildings. doi:10.1016/j.enbuild.2009.08.006

USGBC (2013) LEED reference guide for building design and construction. (LEED), v4 edn. United States Green Building Council, Washington, DC

Walker B, Salt D (2006) Resilience thinking: sustaining ecosystems and people in a changing world. Island Press, Washington

Author Biography

Robert Fryer Over the past 20 years Professor Robert Fryer has practiced, taught, consulted and researched into sustainable design and architecture. Robert first practiced as an architect for 10 years before leaving to earn his graduate degree from the Architectural Association School of Architecture in the Environment and Energy Programme in London, where he completed his dissertation on deformable textile exterior wall assemblies that changed form according to climate stimuli. Since then he's held researcher and professor positions at Philadelphia University in Philadelphia, PA in the US. Most recently, Robert was head of developing an online version of the award winning on-campus MS Sustainable Design program, and was also a member of the group of faculty working across disciplines (Architecture, Engineering and Biology) to create a systems thinking course on biomimicry. He also consults with materials manufacturers on testing and creating new products and with communities and towns in using the integrated design charrette process to develop properties.

MIX
Papier aus verantwortungsvollen Quellen
Paper from responsible sources
FSC® C105338

If you have any concerns about our products,
you can contact us on
ProductSafety@springernature.com

In case Publisher is established outside the EU,
the EU authorized representative is:
Springer Nature Customer Service Center GmbH
Europaplatz 3, 69115 Heidelberg, Germany

Printed by Libri Plureos GmbH
in Hamburg, Germany